*Probability and Statistics
for Engineers and Scientists*

RONALD E. WALPOLE
PROFESSOR OF MATHEMATICS AND STATISTICS, ROANOKE COLLEGE

RAYMOND H. MYERS
PROFESSOR OF STATISTICS, VIRGINIA POLYTECHNIC INSTITUTE AND STATE UNIVERSITY

Probability and Statistics for Engineers and Scientists

THIRD EDITION

MACMILLAN PUBLISHING COMPANY

NEW YORK

Collier Macmillan Publishers

LONDON

Macmillan Publishing Company
866 Third Avenue, New York, New York 10022

Collier Macmillan Canada, Inc.

Library of Congress Cataloging in Publication Data

Walpole, Ronald E.
   Probability and statistics for engineers and scientists.

   Bibliography: p.
   Includes index.
   1. Engineering—Statistical methods. 2. Probabilities.
I. Myers, Raymond H.   II. Title.
TA340.W35   1985   519'.02462   84–5738
ISBN 0–02–424170–9   (Hardcover Edition)
ISBN 0–02–946950–3   (International Edition)

Printing: 1 2 3 4 5 6 7 8   Year: 5 6 7 8 9 0 1 2 3

ISBN 0-02-424170-9

Preface

Like the previous editions, this third edition of *Probability and Statistics for Engineers and Scientists* has been written to serve as an introductory probability and statistics textbook for students majoring in engineering, mathematics, statistics, computer science, or one of the natural sciences. The objectives of the earlier editions have been maintained. That is, we have endeavored to achieve a balance between theory and applications based upon a prerequisite of a course in differential and integral calculus through partial differentiation and multiple integration.

The major changes in the third edition, which we feel will enhance the book's adaptability to the various scientific areas, are as follows:

1. Numerous real-life exercises and examples have been selected from actual statistical studies in a variety of scientific fields.
2. Unlike the previous editions in which exercise sets were placed only at the end of each chapter, they can now be found immediately following appropriate sections throughout the text.
3. Many sections have been extensively rewritten and a number of new figures have been added to help clarify some of the concepts.
4. New material on the use of stem and leaf plots for exploring the shape of a distribution of measurements supplements the material in Chapter 2 on empirical distributions.
5. Table A.1 of the Appendix has been expanded to give cumulative binomial probability sums for values of n from 1 through 20.
6. The material on hypothesis testing has been revised and now includes a discussion of P values.

7. The application of the chi-square distribution to tests of independence and homogeneity has been rewritten and clarified.
8. Two new sections on the study of residuals and the application of PRESS residuals to cross validation techniques have been added to the multiple linear regression material of Chapter 10.
9. Bartlett's test for the equality of several variances in an analysis-of-variance procedure is now based on the exact critical values in Table A.10 of the Appendix.
10. Latin square designs are now included in Chapter 11 to provide an expanded treatment of experimental designs.

The book contains sufficient material to allow for flexibility in the length of the course and the selection of topics. For those students who have time for only a one-semester course meeting three hours a week, the authors recommend the study of Chapters 1 through 5 and selected topics from Chapters 6, 7, and 8. Chapter 1 introduces the basic concepts of probability theory using elementary sample paces. Chapters 2 and 3 present an introduction to discrete and continuous random variables and their probability distributions, joint probability distributions, and mathematical expectations. Chapters 4 and 5 are devoted to a discussion of the particular discrete and continuous probability distributions that the engineer or scientist is most likely to apply in solving the various problems in his field of specialization. Perhaps unusual for a textbook at this level is the extensive use of transformation theory in our derivations of the sampling distributions in Chapter 6. However, the treatment of estimation procedures and hypothesis testing in Chapters 7 and 8 can only be appreciated and properly understood if one has gained an insight into the mathematical derivation of the test statistics involved.

For those students who wish to continue their training in statistics for an additional semester, the remainder of the book provides an excellent introduction to the study of regression theory, linear models, analysis-of-variance procedures, and the planning and analysis of various experimental designs. As a rule, students who take more than one semester of statistics also enroll in additional courses in mathematics and, perhaps, computer science. We have therefore included the use of matrices in our treatment of multiple and polynomial regression in Chapter 10 and assume the availability of at least a microcomputer. However, since matrix theory is limited primarily to Chapter 10, the professor could either omit this material completely or inject the basic concepts of matrix operations into the lecture sequence without requiring a formal course in matrix theory or linear algebra as a prerequisite.

Throughout the book we have demonstrated each new idea by an example. Only by solving a large number of exercises can the student be expected to develop an understanding of the basic concepts of probability theory and statistics. Therefore, we have included numerous exercises, both theoretical and applied, all of which are keyed to answers at the back of the book.

The authors wish to acknowledge their appreciation to all those who assisted in the preparation of this textbook. We are particularly grateful to

Debra Beard for typing and proofreading this revised third edition and the accompanying *Instructor's Solutions Supplement* and *Student's Solutions Manual*; to the science and engineering departments at the Virginia Polytechnic Institute and State University for providing the numerous data sets from actual research studies; to the Macmillan Publishing Company for their editorial assistance; and to the many teachers, students, and reviewers for their helpful suggestions and encouragement.

The authors are indebted to the literary executor of the late Sir Ronald A. Fisher, F.R.S., Cambridge, and to Oliver & Boyd Ltd., Edinburgh, for their permission to reprint a table from their book *Statistical Methods for Research Workers*; to Professor E. S. Pearson and the Biometrika trustees for permission to reprint in abridged form Tables 8 and 18 from *Biometrika Tables for Statisticians*, Vol. I; to Oliver & Boyd Ltd. for permission to reproduce tables from their book *Design and Analysis of Industrial Experiments* by O. L. Davies; to the McGraw-Hill Book Company for permission to reproduce Tables A-25d and A-25e from their book *Introduction to Statistical Analysis* by W. J. Dixon and F. J. Massey, Jr.; to C. Eisenhart, M. W. Hastay, and W. A. Wallis for permission to reproduce two tables from their book *Techniques of Statistical Analysis*. We wish also to express our appreciation for permission to reproduce tables from the *Annals of Mathematical Statistics*, from the *Bulletin of the Educational Research at Indiana University*, from a publication by the American Cyanamid Company, from *Biometrics*, from *Biometrika*, Vol. 38, and from the *Journal of the American Statistical Association*.

R. E. W.
R. H. M.

Contents

3. *Mathematical Expectation*   75

4. *Some Discrete Probability Distributions*   103

5. *Some Continuous Probability Distributions*   129

Omit

6. *Functions of Random Variables*   163

7. *Estimation Theory* 213

omit

8. *Tests of Hypotheses* 259

omit

9. *Linear Regression and Correlation* 315

10. *Multiple Linear Regression*   *353*

11. *Analysis of Variance*   *407*

12. *Factorial Experiments*   *471*

13. 2^k *Factorial Experiments*   *503*

14. *Nonparametric Statistics*   *529*

*Probability and Statistics
for Engineers and Scientists*

1

Probability

1.1 Sample Space

In the study of statistics we are basically concerned with the presentation and interpretation of **chance outcomes** that occur in a planned study or scientific investigation. For example, we may record the number of accidents that occur monthly at the intersection of Driftwood Lane and Royal Oak Drive, hoping to justify the installation of a traffic light; we might classify items coming off an assembly line as "defective" or "nondefective"; or we may be interested in the volume of gas released in a chemical reaction when the concentration of an acid is varied. Hence the statistician is usually dealing either with **numerical data**, representing **counts** or **measurements**, or perhaps with **categorical data** that can be classified according to some criterion.

We shall refer to any recording of information, whether it be numerical or categorical, as an **observation**. Thus the numbers 2, 0, 1, and 2, representing the number of accidents that occurred for each month from January through April during the past year at the intersection of Driftwood Lane and Royal Oak Drive, constitute a set of observations. Similarly, the categorical data N, D, N, N, and D, representing the items found to be defective or nondefective when five items are inspected, are recorded as observations.

Statisticians use the word *experiment* to describe any process that generates a set of data. A very simple example of a statistical experiment might be the tossing of a coin. In this experiment there are only two possible outcomes, heads or tails. Another experiment might be the launching of a missile and observing the velocity at specified times. The opinions of voters concerning a new sales tax can also be considered as observations of an experiment. We are particularly interested in the observations obtained by repeating the experiment several times. In most cases the outcomes will depend on chance and, therefore, cannot be predicted with certainty. If a chemist runs an analysis several times under the same conditions, he will obtain different measurements, indicating an element of chance in the experimental procedure. Even when a coin is tossed repeatedly, we cannot be certain that a given toss will result in a head. However, we do know the entire set of possibilities for each toss.

Definition 1.1   *The set of all possible outcomes of a statistical experiment is called the* **sample space** *and is represented by the symbol S.*

Each outcome in a sample space is called an **element** or a **member** of the sample space or simply a **sample point**. If the sample space has a finite number of elements, we may *list* the members separated by commas and enclosed in braces. Thus the sample space S, of possible outcomes when a coin is tossed, may be written

$$S = \{H, T\},$$

where H and T correspond to "heads" and "tails," respectively.

Example 1.1   Consider the experiment of tossing a die. If we are interested in the number that shows on the top face, the sample space would be

$$S_1 = \{1, 2, 3, 4, 5, 6\}.$$

If we are interested only in whether the number is even or odd, the sample space is simply

$$S_2 = \{\text{even, odd}\}.$$

Example 1.1 illustrates the fact that more than one sample space can be used to describe the outcomes of an experiment. In this case S_1 provides more information than S_2. If we know which element in S_1 occurs, we can tell which outcome in S_2 occurs; however, a knowledge of what happens in S_2 is of no help in determining which element in S_1 occurs. In general, it is desirable to use a sample space that gives the most information concerning the outcomes of the experiment.

In some experiments it will be helpful to list the elements of the sample space systematically by means of a **tree diagram**.

Example 1.2   An experiment consists of flipping a coin and then flipping it a second time if a head occurs. If a tail occurs on the first flip, then a die is tossed once. To list the elements of the sample space providing the most information, we construct the tree diagram of Figure 1.1. Now, the various paths along the branches of

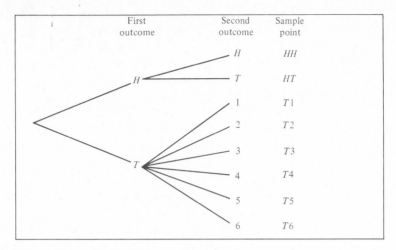

	First outcome	Second outcome	Sample point
		H	HH
	H	T	HT
		1	T1
		2	T2
	T	3	T3
		4	T4
		5	T5
		6	T6

Figure 1.1 Tree diagram for Example 1.2.

the tree give the distinct sample points. Starting with the top left branch and moving to the right along the first path, we get the sample point HH, indicating the possibility that heads occurs on two successive flips of the coin. Likewise, the sample point $T3$ indicates the possibility that the coin will show a tail followed by a 3 on the toss of the die. By proceeding along all paths, we see that the sample space is

$$S = \{HH, HT, T1, T2, T3, T4, T5, T6\}.$$

Example 1.3 Suppose that three items are selected at random from a manufacturing process. Each item is inspected and classified defective, D, or nondefective, N. To list the elements of the sample space providing the most information, we construct the tree diagram of Figure 1.2. Now, the various paths along the branches of the tree give the distinct sample points. Starting with the first path, we get the sample point DDD, indicating the possibility that all three items inspected are defective. As we proceed along the other paths, we see that the sample space is

$$S = \{DDD, DDN, DND, DNN, NDD, NDN, NND, NNN\}.$$

Sample spaces with a large or infinite number of sample points are best described by a *statement* or *rule*. For example, if the possible outcomes of an experiment are the set of cities in the world with a population over 1 million, our sample space is written

$$S = \{x \mid x \text{ is a city with a population over 1 million}\},$$

which reads "S is the set of all x such that x is a city with a population over 1 million." The vertical bar is read "such that." Similarly, if S is the set of all

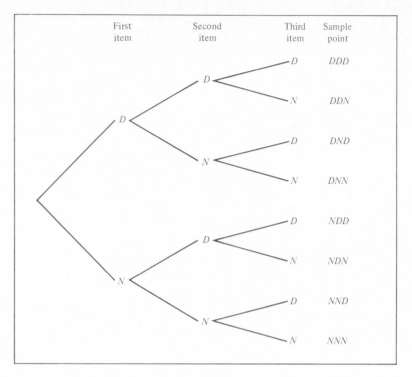

Figure 1.2   Tree diagram for Example 1.3.

points (x, y) on the boundary or the interior of a circle of radius 2 with center at the origin, we write

$$S = \{(x, y) \mid x^2 + y^2 \leq 4\}.$$

Whether we describe the sample space by the rule method or by listing the elements will depend on the specific problem at hand. The rule method has practical advantages, particularly in the many experiments where a listing becomes a very tedious chore.

1.2 Events

In any given experiment we may be interested in the occurrence of certain **events** rather than in the outcome of a specific element in the sample space. For instance, we might be interested in the event A that the outcome when a die is tossed is divisible by 3. This will occur if the outcome is an element of the subset $A = \{3, 6\}$ of the sample space S_1 in Example 1.1. As a further illustration, we might be interested in the event B that the number of defectives is greater than 1 in Example 1.3. This will occur if the outcome is an element of the subset $B = \{DDN, DND, NDD, DDD\}$ of the sample space S.

To each event we assign a collection of sample points, which constitute a subset of the sample space. This subset represents all the elements for which the event is true.

Definition 1.2   An **event** *is a subset of a sample space.*

Example 1.4   Given the sample space $S = \{t \mid t \geq 0\}$, where t is the life in years of a certain electronic component, then the event A that the component fails before the end of the fifth year is the subset $A = \{t \mid 0 \leq t < 5\}$.

It is conceivable that an event may be a subset that includes the entire sample space S, or a subset of S called the null set and denoted by the symbol $\varnothing$, which contains no elements at all. For instance, if we let A be the event of detecting a microscopic organism by the naked eye in a biological experiment, then $A = \varnothing$. Also, if $B = \{x \mid x$ is an even factor of $7\}$, then B must be the null set, since the only possible factors of 7 are the odd numbers 1 and 7.

Consider an experiment in which the smoking habits of the employees of some manufacturing firm are recorded. A possible sample space might classify an individual as a nonsmoker, a light smoker, a moderate smoker, or a heavy smoker. Let the subset of smokers be some event. Then all the nonsmokers correspond to a different event, also a subset of S, which is called the **complement** of the set of smokers.

Definition 1.3   *The* **complement** *of an event A with respect to S is the set of all elements of S that are not in A. We denote the complement of A by the symbol A'.*

Example 1.5   Let R be the event that a red card is selected from an ordinary deck of 52 playing cards, and let S be the entire deck. Then R' is the event that the card selected from the deck is not a red but a black card.

Example 1.6   Consider the sample space $S = \{$book, catalyst, cigarette, precipitate, engineer, rivet$\}$. Let $A = \{$catalyst, rivet, book, cigarette$\}$. Then $A' = \{$precipitate, engineer$\}$.

We now consider certain operations with events that will result in the formation of new events. These new events will be subsets of the same sample space as the given events. Suppose that A and B are two events associated with an experiment. In other words, A and B are subsets of the same sample space S. For example, in the tossing of a die we might let A be the event that an even number occurs and B the event that a number greater than 3 shows. Then the subsets $A = \{2, 4, 6\}$ and $B = \{4, 5, 6\}$, are subsets of the same sample space $S = \{1, 2, 3, 4, 5, 6\}$. Note that *both A and B* will occur on a given toss if the outcome is an element of the subset $\{4, 6\}$, which is just the intersection of A and B.

Definition 1.4   *The* **intersection** *of two events A and B, denoted by the symbol $A \cap B$, is the event containing all elements that are common to A and B.*

Example 1.7   Let P be the event that a person selected at random while dining at a popular cafeteria is a taxpayer, and let Q be the event that the person is over 65 years of age. Then the event $P \cap Q$ is the set of all taxpayers in the cafeteria who are over 65 years of age.

Example 1.8   Let $M = \{a, e, i, o, u\}$ and $N = \{r, s, t\}$; then it follows that $M \cap N = \varnothing$. That is, M and N have no elements in common and, therefore, cannot both occur simultaneously.

In certain statistical experiments it is by no means unusual to define two events A and B that cannot both occur simultaneously. The events A and B are then said to be **mutually exclusive**. Stated more formally, we have the following definition:

Definition 1.5   *Two events A and B are* **mutually exclusive** *if $A \cap B = \varnothing$, that is, if A and B have no elements in common.*

Example 1.9   A cable television company offers programs on eight different channels, three of which are affiliated with ABC, two with NBC, and one with CBS. The other two are an educational channel and the ESPN sports channel. Suppose that a person subscribing to this service turns on a television set without first selecting the channel. Let A be the event that the program belongs to the NBC network and B the event that it belongs to the CBS network. Since a television program cannot belong to more than one network, the events A and B have no programs in common. Therefore, the intersection $A \cap B$ contains no programs, and consequently the events A and B are mutually exclusive.

Often one is interested in the occurrence of at least one of two events associated with an experiment. Thus, in the die-tossing experiment, if $A = \{2, 4, 6\}$ and $B = \{4, 5, 6\}$, we might be interested in either A or B occurring, or both A and B occurring. Such an event, called the **union** of A and B, will occur if the outcome is an element of the subset $\{2, 4, 5, 6\}$.

Definition 1.6   *The* **union** *of the two events A and B, denoted by the symbol $A \cup B$, is the event containing all the elements that belong to A or B or to both.*

Example 1.10   Let $A = \{a, b, c\}$ and $B = \{b, c, d, e\}$; then $A \cup B = \{a, b, c, d, e\}$.

Example 1.11   Let P be the event that an employee selected at random from an oil drilling company smokes cigarettes. Let Q be the event that the employee selected drinks alcoholic beverages. Then the event $P \cup Q$ is the set of all employees who either drink or smoke, or who do both.

Example 1.12   If $M = \{x \mid 3 < x < 9\}$ and $N = \{y \mid 5 < y < 12\}$, then $M \cup N = \{z \mid 3 < z < 12\}$.

The relationship between events and the corresponding sample space can be illustrated graphically by means of **Venn diagrams**. In a Venn diagram we let the sample space be a rectangle and represent events by circles drawn inside the rectangle. Thus, in Figure 1.3, we see that

$$A \cap B = \text{regions 1 and 2,}$$
$$B \cap C = \text{regions 1 and 3,}$$
$$A \cup C = \text{regions 1, 2, 3, 4, 5, and 7,}$$
$$B' \cap A = \text{regions 4 and 7,}$$
$$A \cap B \cap C = \text{region 1,}$$
$$(A \cup B) \cap C' = \text{regions 2, 6, 7,}$$

and so forth. In Figure 1.4 we see that events A, B, and C are all subsets of the sample space S. It is also clear that event B is a subset of event A; event $B \cap C$ has no elements and hence B and C are mutually exclusive; event $A \cap C$ has at least one element; and event $A \cup B = A$. Figure 1.4 might, therefore, depict a situation in which we select a card at random from an ordinary deck of 52 playing cards and observe whether the following events occur:

    A: the card is red,

    B: the card is the jack, queen or king of diamonds,

    C: the card is an ace.

Clearly, the event $A \cap C$ consists only of the 2 red aces.

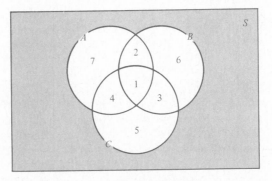

Figure 1.3   Events represented by various regions.

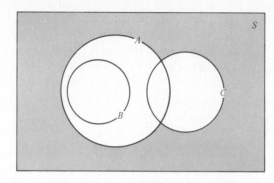

Figure 1.4   Events of the sample space *S*.

Several results that follow from the foregoing definitions, which may easily be verified by means of Venn diagrams, are as follows:

1. $A \cap \emptyset = \emptyset$.
2. $A \cup \emptyset = A$.
3. $A \cap A' = \emptyset$.
4. $A \cup A' = S$.
5. $S' = \emptyset$.
6. $\emptyset' = S$.
7. $(A')' = A$.
8. $(A \cap B)' = A' \cup B'$.
9. $(A \cup B)' = A' \cap B'$.

Exercises

1. List the elements of each of the following sample spaces:
   (a) The set of integers between 1 and 50 divisible by 8.
   (b) The set $S = \{x \mid x^2 + 4x - 5 = 0\}$.
   (c) The set of outcomes when a coin is tossed until a tail or three heads appear.
   (d) The set $S = \{x \mid x \text{ is a continent}\}$.
   (e) The set $S = \{x \mid 2x - 4 \geq 0 \text{ and } x < 1\}$.

2. Use the rule method to describe the sample space S consisting of all points in the first quadrant inside a circle of radius 3 with center at the origin.

3. Which of the following events are equal?
   (a) $A = \{1, 3\}$.
   (b) $B = \{x \mid x \text{ is a number on a die}\}$.
   (c) $C = \{x \mid x^2 - 4x + 3 = 0\}$.
   (d) $D = \{x \mid x \text{ is the number of heads when six coins are tossed}\}$.

4. An experiment involves tossing a pair of dice, 1 green and 1 red, and recording the numbers that come up. If x equals the outcome on the green die and y the outcome on the red die, describe the sample space S
   (a) by listing the elements (x, y);
   (b) by using the rule method.

5. An experiment consists of tossing a die and then flipping a coin once if the number on the die is even. If the number on the die is odd, the coin is flipped twice. Using the notation 4*H*, for example, to denote the event that the die comes up 4 and

then the coin comes up heads, and $3HT$ to denote the event that the die comes up 3 followed by a head and then a tail on the coin, construct a tree diagram to show the 18 elements of the sample space S.

6. Two jurors are selected from 4 alternates to serve at a murder trial. Using the notation $A_1 A_3$, for example, to denote the simple event that alternates 1 and 3 are selected, list the 6 elements of the sample space S.

7. Four students are selected at random from a chemistry class and classified as male or female. List the elements of the sample space S_1 using the letter M for "male" and F for "female." Define a second sample space S_2 where the elements represent the number of females selected.

8. For the sample space of Exercise 4,
    (a) list the elements corresponding to the event A that the sum is greater than 8;
    (b) list the elements corresponding to the event B that a 2 occurs on either die;
    (c) list the elements corresponding to the event C that a number greater than 4 comes up on the green die;
    (d) list the elements corresponding to the event $A \cap C$;
    (e) list the elements corresponding to the event $A \cap B$;
    (f) list the elements corresponding to the event $B \cap C$;
    (g) construct a Venn diagram to illustrate the intersections and unions of the events A, B, and C.

9. For the sample space of Exercise 5,
    (a) list the elements corresponding to the event A that a number less than 3 occurs on the die;
    (b) list the elements corresponding to the event B that 2 tails occur;
    (c) list the elements corresponding to the event A';
    (d) list the elements corresponding to the event $A' \cap B$;
    (e) list the elements corresponding to the event $A \cup B$.

10. An experiment consists of asking 3 women at random if they wash their dishes with brand X detergent.
    (a) List the elements of a sample space S using the letter Y for "yes" and N for "no."
    (b) List the elements of S corresponding to event E that at least 2 of the women use brand X.
    (c) Define an event that has as its elements the points $\{YYY, NYY, YYN, NYN\}$.

11. The résumés of 2 male applicants for a college teaching position in psychology are placed in the same file as the résumés of 2 female applicants. Two positions become available and the first, at the rank of assistant professor, is filled by selecting 1 of the 4 applicants at random. The second position, at the rank of instructor, is then filled by selecting at random one of the remaining 3 applicants. Using the notation $M_2 F_1$, for example, to denote the simple event that the first position is filled by the second male applicant and the second position is then filled by the first female applicant,
    (a) list the elements of the sample space S;
    (b) list the elements of S corresponding to event A that the position of assistant professor is filled by a male applicant;
    (c) list the elements of S corresponding to event B that exactly 1 of the 2 positions was filled by a male applicant;
    (d) list the elements of S corresponding to event C that neither position was filled by a male applicant;
    (e) list the elements of S corresponding to the event $A \cap B$;
    (f) list the elements of S corresponding to the event $A \cup C$;
    (g) construct a Venn diagram to illustrate the intersections and unions of the events A, B, and C.

12. A developer from Saudi Arabia has decided to invest large sums of money in real estate. Four states, Virginia, New York, Connecticut, and Massachusetts, are being considered for the construction of hotels, motels, and condominiums, all of which will be located either directly on the beach or at resorts in the mountains. Using the notation Cmb, for example, to denote the simple event that the developer selects Connecticut as the place to build a motel on a beach, construct a tree diagram to show the 24 elements of the sample space.

13. Construct a Venn diagram to illustrate the possible intersections and unions for the following events relative to the sample space S consisting of

all students at Roanoke College:

J: a student is a junior,

M: a student is a mathematics major,

W: a student is a woman.

14. If $S = \{0, 1, 2, 3, 4, 5, 6, 7, 8, 9\}$ and $A = \{0, 2, 4, 6, 8\}$, $B = \{1, 3, 5, 7, 9\}$, $C = \{2, 3, 4, 5\}$, and $D = \{1, 6, 7\}$, list the elements of the sets corresponding to the following events:
(a) $A \cup C$; (d) $(C' \cap D) \cup B$;
(b) $A \cap B$; (e) $(S \cap C)'$;
(c) C'; (f) $A \cap C \cap D'$.

15. Consider the sample space

$S = \{$copper, sodium, nitrogen, potassium, uranium, oxygen, zinc$\}$

and the events

$A = \{$copper, sodium, zinc$\}$,

$B = \{$sodium, nitrogen, potassium$\}$,

$C = \{$oxygen$\}$.

List the elements of the sets corresponding to the following events:
(a) A'; (d) $(B' \cap C')$;
(b) $A \cup C$; (e) $A \cap B \cap C$;
(c) $(A \cap B') \cup C'$; (f) $(A' \cup B') \cap (A' \cap C)$.

16. If $S = \{x \mid 0 < x < 12\}$, $M = \{x \mid 1 < x < 9\}$, and $N = \{x \mid 0 < x < 5\}$, find
(a) $M \cup N$;
(b) $M \cap N$;
(c) $M' \cap N'$.

17. Let A, B, and C be events relative to the sample space S. Using Venn diagrams, shade the areas representing the following events:
(a) $(A \cap B)'$;
(b) $(A \cup B)'$;
(c) $(A \cap C) \cup B$.

18. Which of the following pairs of events are mutually exclusive?
(a) A golfer scoring the lowest 18-hole round in a 72-hole tournament and losing the tournament.
(b) A poker player getting a flush (all cards in the same suit) and 3 of a kind on the same 5-card hand.
(c) A mother giving birth to a baby girl and a set of twin daughters on the same day.

(d) A chess player losing the last game and winning the match.

19. Suppose that a family is leaving on a summer vacation in their camper and that M is the event that they will experience mechanical problems, T is the event that they will receive a ticket for committing a traffic violation, and V is the event that they will arrive at a campsite with no vacancies. Referring to the Venn diagram of Figure 1.5, state in words the events represented by the following regions:
(a) Region 5.
(b) Region 3.
(c) Regions 1 and 2 together.
(d) Regions 4 and 7 together.
(e) Regions 3, 6, 7, and 8 together.

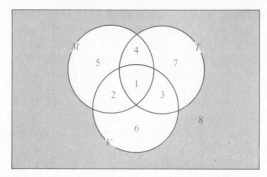

Figure 1.5   Venn diagram for Exercise 19.

20. Referring to Exercise 19 and the Venn diagram of Figure 1.5, list the numbers of the regions that represent the following events:
(a) The family will experience no mechanical problems and commit no traffic violation but will find a campsite with no vacancies.
(b) The family will experience both mechanical problems and trouble in locating a campsite with a vacancy but will not receive a ticket for a traffic violation.
(c) The family will either have mechanical trouble or find a campsite with no vacancies but will not receive a ticket for committing a traffic violation.
(d) The family will not arrive at a campsite with no vacancies.

21. By comparing appropriate regions of Venn diagrams, verify that
(a) $(A \cap B) \cup (A \cap B') = A$;
(b) $A' \cup (B' \cup C) = (A \cap B') \cup (A' \cup C)$.

1.3 Counting Sample Points

One of the problems that the statistician must consider and attempt to evaluate is the element of chance associated with the occurrence of certain events when an experiment is performed. These problems belong in the field of probability, a subject to be introduced in Section 1.4. In many cases we shall be able to solve a probability problem by counting the number of points in the sample space without actually listing each element. The fundamental principle of counting, often referred to as the **multiplication rule**, is stated as follows:

Theorem 1.1 *If an operation can be performed in n_1 ways, and if for each of these a second operation can be performed in n_2 ways, then the two operations can be performed together in $n_1 n_2$ ways.*

Example 1.13 How many sample points are in the sample space when a pair of dice are thrown once?

Solution. The first die can land in any one of $n_1 = 6$ ways. For each of these 6 ways the second die can also land in $n_2 = 6$ ways. Therefore, the pair of dice can land in

$$n_1 n_2 = (6)(6) = 36$$

possible ways.

Example 1.14 A developer of a new subdivision offers prospective home buyers a choice of Tudor, rustic, colonial, and traditional exterior styling in ranch, two-story, and split-level floor plans. In how many different ways can a buyer order one of these homes?

Solution. Since $n_1 = 4$ and $n_2 = 3$, a buyer must choose from

$$n_1 n_2 = (4)(3) = 12$$

possible homes.

The answers to the two preceding examples can be verified by constructing tree diagrams and counting the various paths along the branches. For instance, in Example 1.14 there will be $n_1 = 4$ branches corresponding to the different exterior styles, and then there will be $n_2 = 3$ branches extending from each of these 4 branches to represent the different floor plans. This tree diagram yields the $n_1 n_2 = 12$ choices of homes given by the paths along the branches as illustrated in Figure 1.6.

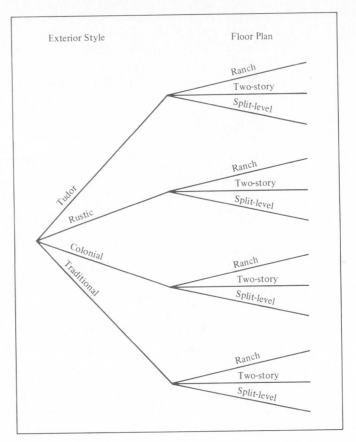

Exterior Style

Floor Plan

Ranch
Two-story
Split-level

Ranch
Two-story
Split-level

Ranch
Two-story
Split-level

Ranch
Two-story
Split-level

Tudor
Rustic
Colonial
Traditional

Figure 1.6   Tree diagram for Example 1.13.

The multiplication rule of Theorem 1.1 may be extended to cover any number of operations. Suppose, for instance, that a customer wishes to install a Trimline telephone and can choose from $n_1 = 10$ decorator colors, which we shall assume are available in any of $n_2 = 3$ optional cord lengths with $n_3 = 2$ types of dialing, namely, rotary or Touch-Tone. These three classifications result in

$$n_1 n_2 n_3 = (10)(3)(2) = 60$$

different ways for a customer to order one of these phones. The **generalized multiplication rule** covering k operations is stated in the following theorem.

Theorem 1.2   *If an operation can be performed in n_1 ways, and if for each of these a second operation can be performed in n_2 ways, and for each of the first two a third operation can be performed in n_3 ways, and so forth, then the sequence of k operations can be performed in $n_1 n_2, \ldots, n_k$ ways.*

Example 1.15 How many lunches consisting of a soup, sandwich, dessert, and a drink are possible if we can select from 4 soups, 3 kinds of sandwiches, 5 desserts, and 4 drinks?

Solution. Since $n_1 = 4$, $n_2 = 3$, $n_3 = 5$, and $n_4 = 4$, there are

$$n_1 \times n_2 \times n_3 \times n_4 = 4 \times 3 \times 5 \times 4 = 240$$

different ways to choose a lunch.

Example 1.16 How many even three-digit numbers can be formed from the digits 1, 2, 5, 6, and 9 if each digit can be used only once?

Solution. Since the number must be even, we have only $n_1 = 2$ choices for the units position. For each of these we have $n_2 = 4$ choices for the hundreds position and then $n_3 = 3$ choices for the tens position. Therefore, we can form a total of

$$n_1 n_2 n_3 = (2)(4)(3) = 24$$

even three-digit numbers.

Frequently, we are interested in a sample space that contains as elements all possible orders or arrangements of a group of objects. For example, we might want to know how many different arrangements are possible for sitting 6 people around a table, or we might ask how many different orders are possible for drawing 2 lottery tickets from a total of 20. The different arrangements are called **permutations**.

Definition 1.7 A **permutation** *is an arrangement of all or part of a set of objects.*

Consider the three letters a, b, and c. The possible permutations are *abc, acb, bac, bca, cab,* and *cba*. Thus we see that there are 6 distinct arrangements. Using Theorem 1.2, we could arrive at the answer 6 without actually listing the different orders. There are $n_1 = 3$ choices for the first position, then $n_2 = 2$ for the second, leaving only $n_3 = 1$ choice for the last position, giving a total of $n_1 n_2 n_3 = (3)(2)(1) = 6$ permutations. In general, n distinct objects can be arranged in $n(n-1)(n-2)\cdots(3)(2)(1)$ ways. We represent this product by the symbol $n!$, which is read "n factorial." Three objects can be arranged in $3! = (3)(2)(1) = 6$ ways. By definition $1! = 1$ and $0! = 1$.

Theorem 1.3 *The number of permutations of n distinct objects is n!.*

The number of permutations of the four letters a, b, c, and d will be $4! = 24$. Let us now consider the number of permutations that are possible by taking the four letters two at a time. These would be *ab, ac, ad, ba, ca, da, bc, cb, bd, db, cd,* and *dc*. Using Theorem 1.1 again, we have two positions to fill with

$n_1 = 4$ choices for the first and then $n_2 = 3$ choices for the second for a total of $n_1 n_2 = (4)(3) = 12$ permutations. In general, n distinct objects taken r at a time can be arranged in $n(n-1)(n-2)\cdots(n-r+1)$ ways. We represent this product by the symbol $_nP_r = n!/(n-r)!$.

Theorem 1.4   *The number of permutations of n distinct objects taken r at a time is*

$$_nP_r = \frac{n!}{(n-r)!}.$$

Example 1.17   Two lottery tickets are drawn from 20 for first and second prizes. Find the number of sample points in the space S.

Solution. The total number of sample points is

$$_{20}P_2 = \frac{20!}{18!} = (20)(19) = 380.$$

Example 1.18   How many ways can a local chapter of the American Chemical Society schedule 3 speakers for 3 different meetings if they are all available on any of 5 possible dates?

Solution. The total number of possible schedules is

$$_5P_3 = \frac{5!}{2!} = (5)(4)(3) = 60.$$

Permutations that occur by arranging objects in a circle are called **circular permutations**. Two circular permutations are not considered different unless corresponding objects in the two arrangements are preceded or followed by a different object as we proceed in a clockwise direction. For example, if 4 people are playing bridge, we do not have a new permutation if they all move one position in a clockwise direction. By considering one person in a fixed position and arranging the other three in 3! ways, we find that there are 6 distinct arrangements for the bridge game.

Theorem 1.5   *The number of permutations of n distinct objects arranged in a circle is $(n-1)!$.*

So far we have considered permutations of distinct objects. That is, all the objects were completely different or distinguishable. Obviously, if the letters b and c are both equal to x, then the six permutations of the letters a, b, c, become axx, axx, xax, xax, xxa, and xxa, of which only three are distinct. Therefore, with three letters, two being the same, we have $3!/2! = 3$ distinct permutations. With four different letters a, b, c, and d we have 24 distinct permutations. If we let $a = b = x$ and $c = d = y$, we can list only the following: *xxyy, xyxy, yxxy, yyxx, xyyx,* and *yxyx.* Thus we have $4!/2!2! = 6$ distinct permutations.

Theorem 1.6   *The number of distinct permutations of n things of which n_1 are of one kind, n_2 of a second kind, ..., n_k of a kth kind is*

$$\frac{n!}{n_1! \, n_2! \cdots n_k!}.$$

Example 1.19   How many different ways can 3 red, 4 yellow, and 2 blue bulbs be arranged in a string of Christmas tree lights with 9 sockets?

Solution. The total number of distinct arrangements is

$$\frac{9!}{3! \, 4! \, 2!} = 1260.$$

Often we are concerned with the number of ways of partitioning a set of n objects into r subsets called **cells**. A partition has been achieved if the intersection of every possible pair of the r subsets is the empty set $\varnothing$ and if the union of all subsets gives the original set. The order of the elements within a cell is of no importance. Consider the set $\{a, e, i, o, u\}$. The possible partitions into two cells in which the first cell contains 4 elements and the second cell 1 element are $\{(a, e, i, o), (u)\}, \{(a, i, o, u), (e)\}, \{(e, i, o, u), (a)\}, \{(a, e, o, u), (i)\},$ and $\{(a, e, i, u), (o)\}$. We see that there are 5 such ways to partition a set of 5 elements into two subsets or cells containing 4 elements in the first cell and 1 element in the second.

The number of partitions for this illustration is denoted by the symbol

$$\binom{5}{4, \ 1} = \frac{5!}{4! \, 1!} = 5,$$

where the top number represents the total number of elements and the bottom numbers represent the number of elements going into each cell. We state this more generally in the following theorem.

Theorem 1.7   *The number of ways of partitioning a set of n objects into r cells with n_1 elements in the first cell, n_2 elements in the second, and so forth, is*

$$\binom{n}{n_1, \ n_2, \ \ldots, \ n_r} = \frac{n!}{n_1! \, n_2! \cdots n_r!},$$

where $n_1 + n_2 + \cdots + n_r = n$.

Example 1.20   In how many ways can seven scientists be assigned to one triple and two double hotel rooms?

Solution. The total number of possible partitions would be

$$\binom{7}{3, 2, 2} = \frac{7!}{3!\, 2!\, 2!} = 210.$$

In many problems we are interested in the number of ways of *selecting r* objects from *n* without regard to order. These selections are called **combinations**. A combination is actually a partition with two cells, the one cell containing the *r* objects selected and the other cell containing the $(n - r)$ objects that are left.

The number of such combinations, denoted by $\binom{n}{r,\, n-r}$, is usually shortened to $\binom{n}{r}$, since the number of elements in the second cell must be $n - r$.

Theorem 1.8   *The number of combinations of n distinct objects taken r at a time is*

$$\binom{n}{r} = \frac{n!}{r!\,(n - r)!}.$$

Example 1.21   From 4 chemists and 3 physicists find the number of committees that can be formed consisting of 2 chemists and 1 physicist.

Solution. The number of ways of selecting 2 chemists from 4 is

$$\binom{4}{2} = \frac{4!}{2!\, 2!} = 6.$$

The number of ways of selecting 1 physicist from 3 is

$$\binom{3}{1} = \frac{3!}{1!\, 2!} = 3.$$

Using the multiplication rule of Theorem 1.1 with $n_1 = 6$ and $n_2 = 3$, we can form

$$n_1 n_2 = (6)(3) = 18$$

committees with 2 chemists and 1 physicist.

Exercises

1. Registrants at a large convention are offered 6 sightseeing tours on each of 3 days. In how many ways can a person arrange to go on a sightseeing tour planned by this convention?

2. In a medical study patients are classified in 8 ways according to whether they have blood type AB^+, AB^-, A^+, A^-, B^+, B^-, O^+, or O^-, and also according to whether their blood pressure is low, normal, or high. Find the number of ways in which a patient can be classified.

3. If an experiment consists of throwing a die and then drawing a letter at random from the English alphabet, how many points are in the sample space?

4. Students at a private liberal arts college are classified as being freshmen, sophomores, juniors, or seniors, and also according to whether they are male of female. Find the total number of possible classifications for the students of this college.

5. A certain shoe comes in 5 different styles with each style available in 4 distinct colors. If the store wishes to display pairs of these shoes showing all of its various styles and colors, how many different pairs would the store have on display?

6. A college freshman must take a science course, a humanities course, and a mathematics course. If she may select any of 6 science courses, any of 4 humanities, and any of 4 mathematics courses, in how many ways can she arrange her program?

7. A developer of a new subdivision offers a prospective home buyer a choice of 4 designs, 3 different heating systems, a garage or carport, and a patio or screened porch. How many different plans are available to this buyer?

8. A drug for the relief of asthma can be purchased from 5 different manufacturers in liquid, tablet, or capsule form, all of which come in regular and extra strength. In how many different ways can a doctor prescribe the drug for a patient suffering from asthma?

9. In a fuel economy study each of 3 race cars is tested using 5 different brands of gasoline at 7 test sites located in different regions of the country. If 2 drivers are used in the study, and test runs are made once under each distinct set of conditions, how many test runs are needed?

10. In how many different ways can a true–false test consisting of 9 questions be answered?

11. If a multiple choice test consists of 5 questions each with 4 possible answers of which only 1 is correct,
    (a) in how many different ways can a student check off one answer to each question?
    (b) in how many ways can a student check off one answer to each question and get all the answers wrong?

12. (a) How many distinct permutations can be made from the letters of the word *columns*?
    (b) How many of these permutations start with the letter *m*?

13. A witness to a hit-and-run accident told the police that the license number contained the letters RLH followed by three digits, the first of which was a five. If the witness cannot recall the last two digits, but is certain that all three digits are different, find the maximum number of automobile registrations that the police may have to check.

14. (a) In how many ways can 6 people be lined up to get on a bus?
    (b) If a certain 3 persons insist on following each other, how many ways are possible?
    (c) If a certain 2 persons refuse to follow each other, how many ways are possible?

15. A contractor wishes to build 9 houses, each different in design. In how many ways can he place these houses on a street if 6 lots are on one side of the street and 3 lots are on the opposite side?

16. (a) How many three-digit numbers can be formed from the digits 0, 1, 2, 3, 4, 5, and 6, if each digit can be used only once?
    (b) How many of these are odd numbers?
    (c) How many are greater than 330?

17. In how many ways can 4 boys and 5 girls sit in a row if the boys and girls must alternate?

18. Four married couples have bought 8 seats in a row for a concert. In how many different ways can they be seated
  (a) with no restrictions?
  (b) if each couple is to sit together?
  (c) if all the men sit together to the right of all the women?

19. In a regional spelling bee, the 8 finalists consist of 3 boys and 5 girls. Find the number of sample points in the space S for the number of possible orders at the conclusion of the contest for
  (a) all 8 finalists;
  (b) the first 3 positions.

20. In how many ways can the 5 starting positions on a basketball team be filled with 8 men who can play any of the positions?

21. Find the number of ways in which 6 teachers can be assigned to 4 sections of an introductory psychology course if no teacher is assigned to more than one section.

22. Three lottery tickets for first, second, and third prizes are drawn from a group of 40 tickets. Find the number of sample points in S for awarding the three prizes if each contestant holds only one ticket.

23. In how many ways can 5 different trees be planted in a circle?

24. In how many ways can a caravan of 8 covered wagons from Arizona be arranged in a circle?

25. How many distinct permutations can be made from the letters of the word *infinity*?

26. In how many ways can 3 oaks, 4 pines, and 2 maples be arranged along a property line if one does not distinguish between trees of the same kind?

27. A college plays 12 football games during a season. In how many ways can the team end the season with 7 wins, 3 losses, and 2 ties?

28. Nine people are going on a skiing trip in 3 cars that will hold 2, 4, and 5 passengers, respectively. In how many ways is it possible to transport the 9 people to the ski lodge using all cars?

29. How many ways are there to select 3 candidates from 8 equally qualified recent graduates for openings in an accounting firm?

30. In a California study, Dean Lester Breslow and Dr. James Enstrom of the University of California at Los Angeles' School of Public Health concluded that by following 7 simple health rules a man's life can be extended by 11 years on the average and a woman's life by seven years. These 7 rules are: no smoking, regular exercise, use alcohol moderately, get seven to eight hours of sleep, maintain proper weight, eat breakfast, and to not eat between meals. In how many ways can a person adopt 5 of these rules to follow
  (a) if the person presently violates all 7 rules?
  (b) if the person never drinks and always eats breakfast?

31. From a group of 4 men and 5 women, how many committees of size 3 are possible
  (a) with no restrictions?
  (b) with 1 man and 2 women?
  (c) with 2 men and 1 woman if a certain man must be on the committee?

32. How many bridge hands are possible containing 4 spades, 6 diamonds, 1 club, and 2 hearts?

33. From 4 red, 5 green, and 6 yellow apples, how many selections of 9 apples are possible if 3 of each color are to be selected?

34. A shipment of 12 television sets contains 3 defective sets. In how many ways can a hotel purchase 5 of these sets and receive at least 2 of the defective sets?

1.4 *Probability of an Event*

Perhaps it was man's unquenchable thirst for gambling that led to the early development of probability theory. In an effort to increase their winnings, gamblers called upon the mathematicians to provide optimum strategies for various games of chance. Some of the mathematicians providing these strategies were Pascal, Leibniz, Fermat, and James Bernoulli. As a result of this early development of probability theory, statistical inference, with all its predictions and generalizations, has branched out far beyond games of chance to encompass many other fields associated with chance occurrences such as politics, business, weather forecasting, and scientific research. For these predictions and generalizations to be reasonably accurate, an understanding of basic probability theory is essential.

What do we mean when we make the statements "John will probably win the tennis match," "I have a fifty-fifty chance of getting an even number when a die is tossed," "I am not likely to win at bingo tonight," or "Most of our graduating class will likely be married within 3 years"? In each case we are expressing an outcome of which we are not certain, but owing to past information or from an understanding of the structure of the experiment, we have some degree of confidence in the validity of the statement.

Throughout the remainder of this chapter we consider only those experiments for which the sample space contains a finite number of elements. The likelihood of the occurrence of an event resulting from such a statistical experiment is evaluated by means of a set of real numbers called **weights** or **probabilities** ranging from 0 to 1. To every point in the sample space we assign a probability such that the sum of all probabilities is 1. If we have reason to believe that a certain sample point is quite likely to occur when the experiment is conducted, the probability assigned should be close to 1. On the other hand, a probability closer to zero is assigned to a sample point that is not likely to occur. In many experiments, such as tossing a coin or a die, all the sample points have the same chance of occurring and are assigned equal probabilities. For points outside the sample space, that is, for simple events that cannot possibly occur, we assign a probability of zero.

To find the probability of an event A, we sum all the probabilities assigned to the sample points in A. This sum is called the **probability of A** and is denoted by $P(A)$.

Definition 1.8   *The **probability of an event** A is the sum of the weights of all sample points in A. Therefore,*

$$0 \le P(A) \le 1, \qquad P(\varnothing) = 0, \qquad and \qquad P(S) = 1.$$

Example 1.22   A coin is tossed twice. What is the probability that at least one head occurs?

Solution. The sample space for this experiment is

$$S = \{HH,\ HT,\ TH,\ TT\}.$$

If the coin is balanced, each of these outcomes would be equally likely to occur. Therefore, we assign a probability of w to each sample point. Then

$4w = 1$ or $w = 1/4$. If A represents the event of at least one head occurring, then

$$A = \{HH, HT, TH\}$$

and

$$P(A) = \tfrac{1}{4} + \tfrac{1}{4} + \tfrac{1}{4} = \tfrac{3}{4}.$$

Example 1.23 A die is loaded in such a way that an even number is twice as likely to occur as an odd number. If E is the event that a number less than 4 occurs on a single toss of the die, find $P(E)$.

Solution. The sample space is $S = \{1, 2, 3, 4, 5, 6\}$. We assign a probability of w to each odd number and a probability of $2w$ to each even number. Since the sum of the probabilities must be 1, we have $9w = 1$ or $w = 1/9$. Hence probabilities of $1/9$ and $2/9$ are assigned to each odd and even number, respectively. Therefore,

$$E = \{1, 2, 3\}$$

and

$$P(E) = \tfrac{1}{9} + \tfrac{2}{9} + \tfrac{1}{9} = \tfrac{4}{9}.$$

Example 1.24 In Example 1.23 let A be the event that an even number turns up and let B be the event that a number divisible by 3 occurs. Find $P(A \cup B)$ and $P(A \cap B)$.

Solution. For the events $A = \{2, 4, 6\}$ and $B = \{3, 6\}$ we have $A \cup B = \{2, 3, 4, 6\}$ and $A \cap B = \{6\}$. By assigning a probability of $1/9$ to each odd number and $2/9$ to each even number we have

$$P(A \cup B) = \tfrac{2}{9} + \tfrac{1}{9} + \tfrac{2}{9} + \tfrac{2}{9} = \tfrac{7}{9}$$

and

$$P(A \cap B) = \tfrac{2}{9}.$$

If the sample space for an experiment contains N elements, all of which are equally likely to occur, we assign a probability equal to $1/N$ to each of the N points. The probability of any event A containing n of these N sample points is then the ratio of the number of elements in A to the number of elements in S.

Theorem 1.9 *If an experiment can result in any one of N different equally likely outcomes, and if exactly n of these outcomes correspond to event A, then the probability of event A is*

$$P(A) = \frac{n}{N}.$$

Example 1.25 A mixture of candies contain 6 mints, 4 toffees, and 3 chocolates. If a person makes a random selection of one of these candies, find the probability of getting (a) a mint, or (b) a toffee or a chocolate.

Solution. Let M, T, and C represent the events that the person selects, respectively, a mint, toffee, or chocolate candy. The total number of candies is 13, all of which are equally likely to be selected.

(a) Since 6 of the 13 candies are mints, the probability of event M, selecting a mint at random, is

$$P(M) = \tfrac{6}{13}.$$

(b) Since 7 of the 13 candies are toffees or chocolates, it follows that

$$P(T \cup C) = \tfrac{7}{13}.$$

Example 1.26 In a poker hand consisting of 5 cards, find the probability of holding 2 aces and 3 jacks.

Solution. The number of ways of being dealt 2 aces from 4 is

$$\binom{4}{2} = \frac{4!}{2!\,2!} = 6$$

and the number of ways of being dealt 3 jacks from 4 is

$$\binom{4}{3} = \frac{4!}{3!\,1!} = 4.$$

By the multiplication rule of Theorem 1.1, there are $n = (6)(4) = 24$ hands with 2 aces and 3 jacks. The total number of 5-card poker hands, all of which are equally likely, is

$$N = \binom{52}{5} = \frac{52!}{5!\,47!} = 2{,}598{,}960.$$

Therefore, the probability of event C of getting 2 aces and 3 jacks in a 5-card poker hand is

$$P(C) = \frac{24}{2{,}598{,}960} = 0.9 \times 10^{-5}.$$

If the outcomes of an experiment are not equally likely to occur, the probabilities must be assigned on the basis of prior knowledge or experimental evidence. For example, if a coin is not balanced, we could estimate the probabilities of heads and tails by tossing the coin a large number of times and recording the outcomes. According to the **relative frequency** definition of probability, the true probabilities would be the fractions of heads and tails that occur in the long run.

To find a numerical value that represents adequately the probability of winning at tennis, we must depend on our past performance at the game as well as that of our opponent and to some extent in our belief in being able to win. Similarly, to find the probability that a horse will win a race, we must arrive at a probability based on the previous records of all the horses entered in the race as well as the records of the jockeys riding the horses. Intuition would undoubtedly also play a part in determining the size of the bet that we might be willing to wager. The use of intuition, personal beliefs, and other indirect information in arriving at probabilities is referred to as the **subjective** definition of probability.

1.5 Additive Rules

Often it is easier to calculate the probability of some event from known probabilities of other events. This may well be true if the event in question can be represented as the union of two other events or as the complement of some event. Several important laws that frequently simplify the computation of probabilities follow. The first, called the **additive rule**, applies to unions of events.

Theorem 1.10   *If A and B are any two events, then*

$$P(A \cup B) = P(A) + P(B) - P(A \cap B).$$

Proof. Consider the Venn diagram in Figure 1.7. The $P(A \cup B)$ is the sum of the probabilities of the sample points in $A \cup B$. Now $P(A) + P(B)$ is the sum of all the probabilities in A plus the sum of all the probabilities in B. Therefore, we have added the probabilities in $A \cap B$ twice. Since these probabilities add up to give $P(A \cap B)$, we must subtract this probability once to obtain the sum of the probabilities in $A \cup B$, which is $P(A \cup B)$.

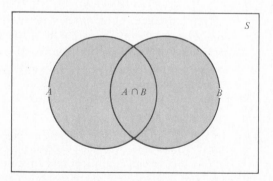

Figure 1.7   Additive rule of probability.

Corollary 1 *If A and B are mutually exclusive, then*

$$P(A \cup B) = P(A) + P(B).$$

Corollary 1 is an immediate result of Theorem 1.10, since if A and B are mutually exclusive, $A \cap B = \varnothing$ and then $P(A \cap B) = P(\varnothing) = 0$. In general, we write

Corollary 2 *If $A_1, A_2, A_3, \ldots, A_n$ are mutually exclusive, then*

$$P(A_1 \cup A_2 \cup \cdots \cup A_n) = P(A_1) + P(A_2) + \cdots + P(A_n).$$

Note that if $A_1, A_2, \ldots, A_n$ is a partition of a sample space S, then

$$P(A_1 \cup A_2 \cup \cdots \cup A_n) = P(A_1) + P(A_2) + \cdots + P(A_n)$$
$$= P(S)$$
$$= 1.$$

Example 1.27 The probability that Paula passes mathematics is 2/3, and the probability that she passes English is 4/9. If the probability of passing both courses is 1/4, what is the probability that Paula will pass at least one of these courses?

Solution. If M is the event "passing mathematics" and E the event "passing English," then by the additive rule we have

$$P(M \cup E) = P(M) + P(E) - P(M \cap E)$$
$$= \tfrac{2}{3} + \tfrac{4}{9} - \tfrac{1}{4}$$
$$= \tfrac{31}{36}.$$

Example 1.28 What is the probability of getting a total of 7 or 11 when a pair of dice are tossed?

Solution. Let A be the event that 7 occurs and B the event that 11 comes up. Now a total of 7 occurs for 6 of the 36 sample points and a total of 11 occurs for only 2 of the sample points. Since all sample points are equally likely, we have $P(A) = 1/6$ and $P(B) = 1/18$. The events A and B are mutually exclusive, since a total of 7 and 11 cannot both occur on the same toss. Therefore,

$$P(A \cup B) = P(A) + P(B)$$
$$= \tfrac{1}{6} + \tfrac{1}{18}$$
$$= \tfrac{2}{9}.$$

This result could also have been obtained by counting the total number of points for the event $A \cup B$, namely 8, and writing

$$P(A \cup B) = \frac{n}{N} = \frac{8}{36} = \frac{2}{9}.$$

Example 1.29   If the probabilities are, respectively, 0.09, 0.15, 0.21, and 0.23 that a person purchasing a new automobile will choose the color green, white, red, or blue, what is the probability that a given buyer will purchase a new automobile that comes in one of those colors?

Solution. Let G, W, R, and B be the events that a buyer selects, respectively, a green, white, red, or blue automobile. Since these four events are mutually exclusive, the probability is

$$P(G \cup W \cup R \cup B) = P(G) + P(W) + P(R) + P(B)$$
$$= 0.09 + 0.15 + 0.21 + 0.23$$
$$= 0.68.$$

Often it is more difficult to calculate the probability that an event occurs than it is to calculate the probability that the event does not occur. Should this be the case for some event A, we simply find $P(A')$ first and then using Theorem 1.11, find $P(A)$ by subtraction.

Theorem 1.11   *If A and A' are complementary events, then*

$$P(A) + P(A') = 1.$$

Proof. Since $A \cup A' = S$ and the sets A and A' are disjoint, then

$$1 = P(S)$$
$$= P(A \cup A')$$
$$= P(A) + P(A').$$

Example 1.30   If the probabilities that an automobile mechanic will service 3, 4, 5, 6, 7, or 8 or more cars on any given work day are, respectively, 0.12, 0.19, 0.28, 0.24, 0.10, and 0.07, what is the probability that he will service at least 5 cars on his next day at work?

Solution. Let E be the event that at least 5 cars are serviced. Now, $P(E) = 1 - P(E')$, where E' is the event that fewer than 5 cars are serviced. Since $P(E') = 0.12 + 0.19 = 0.31$, it follows from Theorem 1.11 that

$$P(E) = 1 - 0.31 = 0.69.$$

Exercises

---

1. Find the errors in each of the following statements:
   (a) The probabilities that an automobile sales-person will sell 0, 1, 2, or 3 cars on any given day in February are, respectively, 0.19, 0.38, 0.29, and 0.15.

(b) The probability that it will rain tomorrow is 0.40 and the probability that it will not rain tomorrow is 0.52.

(c) The probabilities that a printer will make 0, 1, 2, 3, or 4 or more mistakes in printing a document are, respectively, 0.19, 0.34, −0.25, 0.43, and 0.29.

(d) On a single draw from a deck of playing cards the probability of selecting a heart is 1/4, the probability of selecting a black card is 1/2, and the probability of selecting both a heart and a black card is 1/8.

2. Assuming that all elements of S in Exercise 8 on page 9 are equally likely to occur, find
   (a) the probability of event A;
   (b) the probability of event C;
   (c) the probability of event $A \cap C$.

3. A box contains 500 envelopes of which 75 contain $100 in cash, 150 contain $25, and 275 contain $10. An envelope may be purchased for $25. What is the sample space for the different amounts of money? Assign probabilities to the sample points and then find the probability that the first envelope purchased contains less than $100.

4. A die is constructed so that a 1 or 2 occurs twice as often as a 5, which occurs three times as often as a 3, 4, or 6. If the die is tossed once, find the probability that
   (a) the number is even;
   (b) the number is a perfect square;
   (c) the number is greater than 4.

5. If A and B are mutually exclusive events and $P(A) = 0.3$ and $P(B) = 0.5$, find
   (a) $P(A \cup B)$;
   (b) $P(A')$;
   (c) $P(A' \cap B)$.
   HINT: Construct Venn diagrams and fill in the probabilities associated with the various regions.

6. If A, B, and C are mutually exclusive events and $P(A) = 0.2$, $P(B) = 0.3$, and $P(C) = 0.2$, find
   (a) $P(A \cup B \cup C)$;
   (b) $P[A' \cap (B \cup C)]$;
   (c) $P(B \cup C')'$.
   HINT: Construct Venn diagrams as in Exercise 5.

7. If a letter is chosen at random from the English alphabet, find the probability that the letter
   (a) is a vowel exclusive of y;
   (b) is listed somewhere ahead of the letter j;
   (c) is listed somewhere after the letter g.

8. If a permutation of the word *white* is selected at random, find the probability that the permutation
   (a) begins with a consonant;
   (b) ends with a vowel;
   (c) has the consonant and vowels alternating.

9. If each coded item in a catalog begins with 3 distinct letters followed by 4 distinct nonzero digits, find the probability of randomly selecting one of these coded items with the first letter a vowel and the last digit even.

10. A pair of dice is tossed. Find the probability of getting
    (a) a total of 8;
    (b) at most a total of 5.

11. Two cards are drawn in succession from a deck without replacement. What is the probability that both cards are greater than 2 and less than 8?

12. If 3 books are picked at random from a shelf containing 5 novels, 3 books of poems, and a dictionary, what is the probability that
    (a) the dictionary is selected?
    (b) 2 novels and 1 book of poems are selected?

13. In a poker hand consisting of 5 cards, find the probability of holding
    (a) 3 aces;
    (b) 4 hearts and 1 club.

14. In a game of *Yahtzee*, where 5 dice are tossed simultaneously, find the probability of getting
    (a) 2 pairs;
    (b) 4 of a kind.

15. In a high school graduating class of 100 students, 54 studied mathematics, 69 studied history, and 35 studied both mathematics and history. If one of these students is selected at random, find the probability that
    (a) the student takes mathematics or history;
    (b) the student does not take either of these subjects;
    (c) the student takes history but not mathematics.

16. Referring to the important health practices advocated by the California study in Exercise 30 on page 18, suppose that in a senior college class of 500 students it is found that 210 smoke, 258 drink alcoholic beverages, 216 eat between meals, 122 smoke and drink alcoholic beverages, 83 eat between meals and drink alcoholic beverages, 97 smoke and eat between meals, and 52 engage in all three of these bad health practices. If a member of

this senior class is selected at random, find the probability that the student

(a) smokes but does not drink alcoholic beverages;
(b) eats between meals and drinks alcoholic beverages but does not smoke;
(c) neither smokes nor eats between meals.

17. The probability that an American industry will locate in Munich is 0.7, the probability that it will locate in Brussels is 0.4, and the probability that it will locate in either Munich or Brussels or both is 0.8. What is the probability that the industry will locate

(a) in both cities?
(b) in neither city?

18. From past experiences a stockbroker believes that under present economic conditions a customer will invest in tax-free bonds with a probability of 0.6, will invest in mutual funds with a probability of 0.3, and will invest in both tax-free bonds and

mutual funds with a probability of 0.15. At this time, find the probability that a customer will invest

(a) in either tax-free bonds or mutual funds;
(b) in neither tax-free bonds nor mutual funds.

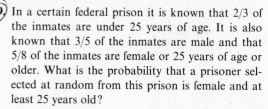

19. In a certain federal prison it is known that 2/3 of the inmates are under 25 years of age. It is also known that 3/5 of the inmates are male and that 5/8 of the inmates are female or 25 years of age or older. What is the probability that a prisoner selected at random from this prison is female and at least 25 years old?

20. The probabilities that a service station will pump gas into 0, 1, 2, 3, 4, or 5 or more cars during a certain 30-minute period are 0.03, 0.18, 0.24, 0.28, 0.10, and 0.17. Find the probability that in this 30-minute period

(a) more than 2 cars receive gas;
(b) at most 4 cars receive gas;
(c) 4 or more cars receive gas.

1.6 Conditional Probability

The probability of an event B occurring when it is known that some event A has occurred is called a **conditional probability** and is denoted by $P(B\,|\,A)$. The symbol $P(B\,|\,A)$ is usually read "the probability that B occurs given that A occurs" or simply "the probability of B, given A."

Consider the event B of getting a perfect square when a die is tossed. The die is constructed so that the even numbers are twice as likely to occur as the odd numbers. Based on the sample space $S = \{1, 2, 3, 4, 5, 6\}$, with probabilities of 1/9 and 2/9 assigned, respectively, to the odd and even numbers, the probability of B occurring is 1/3. Now suppose that it is known that the toss of the die resulted in a number greater than 3. We are now dealing with a reduced sample space $A = \{4, 5, 6\}$, which is a subset of S. To find the probability that B occurs, relative to the space A, we must first assign new probabilities to the elements of A proportional to their original probabilities such that their sum is 1. Assigning a probability of w to the odd number in A and a probability of $2w$ to the two even numbers, we have $5w = 1$ or $w = 1/5$. Relative to the space A, we find that B contains the single element 4. Denoting this event by the symbol $B\,|\,A$, we write $B\,|\,A = \{4\}$, and hence

$$P(B\,|\,A) = \tfrac{2}{5}.$$

This example illustrates that events may have different probabilities when considered relative to different sample spaces.

We can also write

$$P(B \mid A) = \frac{2}{5} = \frac{2/9}{5/9} = \frac{P(A \cap B)}{P(A)},$$

where $P(A \cap B)$ and $P(A)$ are found from the original sample space S. In other words, a conditional probability relative to a subspace A of S may be calculated directly from the probabilities assigned to the elements of the original sample space S.

Definition 1.9 *The* **conditional probability** *of B, given A, denoted by* $P(B \mid A)$, *is defined by*

$$P(B \mid A) = \frac{P(A \cap B)}{P(A)} \qquad if \quad P(A) > 0.$$

As an additional illustration, suppose that our sample space S is the population of adults in a small town who have completed the requirements for a college degree. We shall categorize them according to sex and employment status:

	Employed	Unemployed	Total
Male	460	40	500
Female	140	260	400
Total	600	300	900

One of these individuals is to be selected at random for a tour throughout the country to publicize the advantages of establishing new industries in the town. We shall be concerned with the following events:

M: a man is chosen,

E: the one chosen is employed.

Using the reduced sample space E, we find that

$$P(M \mid E) = \tfrac{460}{600} = \tfrac{23}{30}.$$

Let $n(A)$ denote the number of elements in any set A. Using this notation, we can write

$$P(M \mid E) = \frac{n(E \cap M)}{n(E)} = \frac{n(E \cap M)/n(S)}{n(E)/n(S)} = \frac{P(E \cap M)}{P(E)},$$

where $P(E \cap M)$ and $P(E)$ are found from the original sample space S. To verify this result, note that

$$P(E) = \frac{600}{900} = \tfrac{2}{3},$$

and

$$P(E \cap M) = \frac{460}{900} = \tfrac{23}{45}.$$

Hence

$$P(M \mid E) = \frac{23/45}{2/3} = \frac{23}{30},$$

as before.

Example 1.31   The probability that a regularly scheduled flight departs on time is $P(D) = 0.83$; the probability that it arrives on time is $P(A) = 0.92$; and the probability that it departs and arrives on time is $P(D \cap A) = 0.78$. Find the probability that a plane (a) arrives on time given that it departed on time, and (b) departed on time given that it has arrived on time.

Solution

(a) The probability that a plane arrives on time given that it departed on time is

$$P(A \mid D) = \frac{P(D \cap A)}{P(D)}$$

$$= \frac{0.78}{0.83} = 0.94.$$

(b) The probability that a plane departed on time given that it has arrived on time is

$$P(D \mid A) = \frac{P(D \cap A)}{P(A)}$$

$$= \frac{0.78}{0.92} = 0.85.$$

In the die-tossing experiment discussed on page 26, we noted that $P(B \mid A) = 2/5$ while $P(B) = 1/3$. That is, $P(B \mid A) \neq P(B)$, indicating that B *depends* on A. Now consider an experiment in which 2 cards are drawn in succession from an ordinary deck, with replacement. The events are defined as

A: the first card is an ace,

B: the second card is a spade.

Since the first card is replaced, our sample space for both the first and second draws consists of 52 cards, containing 4 aces and 13 spades. Hence

$$P(B\,|\,A) = \tfrac{13}{52} = \tfrac{1}{4}$$

and

$$P(B) = \tfrac{13}{52} = \tfrac{1}{4}.$$

That is, $P(B\,|\,A) = P(B)$. When this is true, the events A and B are said to be **independent**.

Definition 1.10   *Two events A and B are **independent** if and only if*

$$P(B\,|\,A) = P(B)$$

and

$$P(A\,|\,B) = P(A).$$

*Otherwise A and B are **dependent**.*

The condition $P(B\,|\,A) = P(B)$ implies that $P(A\,|\,B) = P(A)$, and conversely. For the card-drawing experiment, where we showed that $P(B\,|\,A) = P(B) = 1/4$, we also can see that $P(A\,|\,B) = P(A) = 1/13$.

1.7 Multiplicative Rules

Multiplying the formula in Definition 1.9 by $P(A)$, we obtain the following important **multiplicative rule**, which enables us to calculate the probability that two events will both occur.

Theorem 1.12   *If in an experiment the events A and B can both occur, then*

$$P(A \cap B) = P(A)P(B\,|\,A).$$

Thus the probability that both A and B occur is equal to the probability that A occurs multiplied by the probability that B occurs, given that A occurs. Since the events $A \cap B$ and $B \cap A$ are equivalent, it follows from Theorem 1.12 that we can also write

$$P(A \cap B) = P(B \cap A) = P(B)P(A\,|\,B).$$

In other words, it does not matter which event is referred to as A and which event is referred to as B.

Example 1.32 Suppose that we have a fuse box containing 20 fuses, of which 5 are defective. If 2 fuses are selected at random and removed from the box in succession without replacing the first, what is the probability that both fuses are defective?

Solution. We shall let A be the event that the first fuse is defective and B the event that the second fuse is defective; then we interpret $A \cap B$ as the event that A occurs, and then B occurs after A has occurred. The probability of first removing a defective fuse is 1/4; then the probability of removing a second defective fuse from the remaining 4 is 4/19. Hence

$$P(A \cap B) = (\tfrac{1}{4})(\tfrac{4}{19}) = \tfrac{1}{19}.$$

Example 1.33 One bag contains 4 white balls and 3 black balls, and a second bag contains 3 white balls and 5 black balls. One ball is drawn from the first bag and placed unseen in the second bag. What is the probability that a ball now drawn from the second bag is black?

Solution. Let B_1, B_2, and W_1 represent, respectively, the drawing of a black ball from bag 1, a black ball from bag 2, and a white ball from bag 1. We are interested in the union of the mutually exclusive events $B_1 \cap B_2$ and $W_1 \cap B_2$. The various possibilities and their probabilities are illustrated in Figure 1.8.

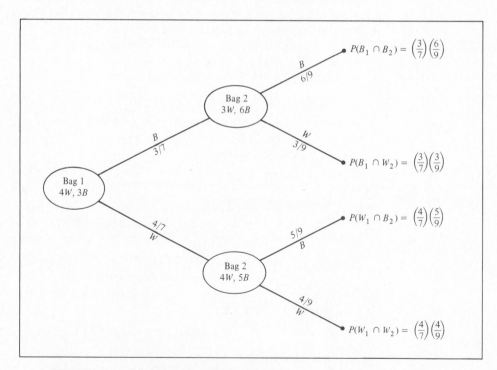

Figure 1.8 Tree diagram for Example 1.33.

Now,

$$P[(B_1 \cap B_2) \text{ or } (W_1 \cap B_2)] = P(B_1 \cap B_2) + P(W_1 \cap B_2)$$
$$= P(B_1)P(B_2 \mid B_1) + P(W_1)P(B_2 \mid W_1)$$
$$= (\tfrac{3}{7})(\tfrac{6}{9}) + (\tfrac{4}{7})(\tfrac{5}{9}) = \tfrac{38}{63}.$$

If, in Example 1.32, the first fuse is replaced and the fuses thoroughly rearranged before the second is removed, then the probability of a defective fuse on the second selection is still 1/4, that is, $P(B \mid A) = P(B)$ and the events A and B are independent. When this is true, we can substitute $P(B)$ for $P(B \mid A)$ in Theorem 1.12 to obtain the following **special multiplicative rule**.

Theorem 1.13   *Two events A and B are independent if and only if*

$$P(A \cap B) = P(A)P(B).$$

Therefore, to obtain the probability that two independent events will both occur, we simply find the product of their individual probabilities.

Example 1.34   A small town has one fire engine and one ambulance available for emergencies. The probability that the fire engine is available when needed is 0.98, and the probability that the ambulance is available when called is 0.92. In the event of an injury resulting from a burning building, find the probability that both the ambulance and the fire engine will be available.

Solution. Let A and B represent the respective events that the fire engine and the ambulance are available. Then

$$P(A \cap B) = P(A)P(B)$$
$$= (0.98)(0.92)$$
$$= 0.9016.$$

Example 1.35   A pair of dice are thrown twice. What is the probability of getting totals of 7 and 11?

Solution. Let A_1, A_2, B_1, and B_2 be the respective independent events that a 7 occurs on the first throw, a 7 occurs on the second throw, an 11 occurs on the first throw, and an 11 occurs on the second throw. We are interested in the probability of the union of the mutually exclusive events $A_1 \cap B_2$ and $B_1 \cap A_2$. Therefore,

$$P[(A_1 \cap B_2) \cup (B_1 \cap A_2)] = P(A_1 \cap B_2) + P(B_1 \cap A_2)$$
$$= P(A_1)P(B_2) + P(B_1)P(A_2)$$
$$= (\tfrac{1}{6})(\tfrac{1}{18}) + (\tfrac{1}{18})(\tfrac{1}{6})$$
$$= \tfrac{1}{54}.$$

Theorems 1.12 and 1.13 may be generalized to cover any number of events, as stated in the following theorem.

Theorem 1.14   *If, in an experiment, the events $A_1, A_2, A_3, \ldots, A_k$ can occur, then*

$$P(A_1 \cap A_2 \cap A_3 \cap \cdots \cap A_k = P(A_1)P(A_2 \mid A_1)P(A_3 \mid A_1 \cap A_2) \cdots$$

$$P(A_k \mid A_1 \cap A_2 \cap \cdots \cap A_{k-1}).$$

If the events $A_1, A_2, A_3, \ldots, A_k$ are independent, then

$$P(A_1 \cap A_2 \cap A_3 \cap \cdots \cap A_k = P(A_1)P(A_2)P(A_3) \cdots P(A_k).$$

Example 1.36   Three cards are drawn in succession, without replacement, from an ordinary deck of playing cards. Find the probability that the event $A_1 \cap A_2 \cap A_3$ occurs, where A_1 is the event that the first card is a red ace, A_2 is the event that the second card is a 10 or a jack, and A_3 is the event that the third card is greater than 3 but less than 7.

Solution.   First we define the events

A_1:   the first card is a red ace,

A_2:   the second card is a 10 or jack,

A_3:   the third card is greater than 3 but less than 7.

Now

$$P(A_1) = \tfrac{2}{52},$$
$$P(A_2 \mid A_1) = \tfrac{8}{51},$$
$$P(A_3 \mid A_1 \cap A_2) = \tfrac{12}{50},$$

and hence by Theorem 1.14,

$$P(A_1 \cap A_2 \cap A_3) = P(A_1)P(A_2 \mid A_1)P(A_3 \mid A_1 \cap A_2)$$
$$= (\tfrac{2}{52})(\tfrac{8}{51})(\tfrac{12}{50})$$
$$= \tfrac{8}{5525}.$$

Example 1.37   A coin is biased so that a head is twice as likely to occur as a tail. If the coin is tossed 3 times, what is the probability of getting 2 tails and 1 head?

Solution.   The sample space for the experiment consists of the 8 elements

$$S = \{HHH, HHT, HTH, THH, HTT, THT, TTH, TTT\}.$$

However, with an unbalanced coin it is no longer possible to assign equal probabilities to each sample point. To find the probabilities, first consider the sample space $S_1 = \{H, T\}$, which represents the outcomes when the coin is tossed once. Assigning probabilities of w and $2w$ for getting a tail and a head, respectively, we have $3w = 1$ or $w = 1/3$. Hence $P(H) = 2/3$ and $P(T) = 1/3$. Now let A be the event of getting 2 tails and 1 head in the 3 tosses of the coin. Then

$$A = \{TTH, THT, HTT\},$$

and since the outcomes on each of the 3 tosses are independent, it follows from Theorem 1.14 that

$$P(THH) = P(T \cap T \cap H) = P(T)P(T)P(H)$$
$$= (\tfrac{1}{3})(\tfrac{1}{3})(\tfrac{2}{3}) = \tfrac{2}{27}.$$

Similarly,

$$P(THT) = P(HTT) = \tfrac{2}{27}$$

and hence

$$P(A) = \tfrac{2}{27} + \tfrac{2}{27} + \tfrac{2}{27} = \tfrac{2}{9}.$$

Exercises

1. If R is the event that a convict committed armed robbery and D is the event that the convict pushed dope, state in words what probabilities are expressed by

   (a) $P(R \mid D)$;
   (b) $P(D' \mid R)$;
   (c) $P(R' \mid D')$.

2. A class in advanced physics is comprised of 10 juniors, 30 seniors, and 10 graduate students. The final grades showed that 3 of the juniors, 10 of the seniors, and 5 of the graduate students received an A for the course. If a student is chosen at random from this class and is found to have earned an A, what is the probability that he or she is a senior?

3. A random sample of 200 adults are classified below according to sex and the level of education attained.

Education	Male	Female
Elementary	38	45
Secondary	28	50
College	22	17

If a person is picked at random from this group, find the probability that

(a) the person is a male, given that the person has a secondary education;
(b) the person does not have a college degree, given that the person is a female.

4. In an experiment to study the dependence of hypertension on smoking habits, the following data were collected on 180 individuals:

	Non-smokers	Moderate Smokers	Heavy Smokers
Hypertension	21	36	30
No hyper-tension	48	26	19

If one of these individuals is selected at random, find the probability that the person is

(a) experiencing hypertension, given that the person is a heavy smoker;
(b) a nonsmoker, given that the person is experiencing no hypertension.

5. In the senior year of a high school graduating class of 100 students, 42 studied mathematics, 68

studied psychology, 54 studied history, 22 studied both mathematics and history, 25 studied both mathematics and psychology, 7 studied history but neither mathematics nor psychology, 10 studied all three subjects, and 8 did not take any of the three. If a student is selected at random, find the probability that

(a) a person enrolled in psychology takes all three subjects;

(b) a person not taking psychology is taking both history and mathematics.

6. A pair of dice is thrown. If it is known that one die shows a 4, what is the probability that
   (a) the other die shows a 5?
   (b) the total of both dice is greater than 7?

7. A card is drawn from an ordinary deck and we are told that it is red. What is the probability that the card is greater than 2 but less than 9?

8. The probability that an automobile being filled with gasoline will also need an oil change is 0.25; the probability that it needs a new oil filter is 0.40; and the probability that both the oil and filter need changing is 0.14.
   (a) If the oil had to be changed, what is the probability that a new oil filter is needed?
   (b) If a new oil filter is needed, what is the probability that the oil has to be changed?

9. The probability that a married man watches a certain television show is 0.4 and the probability that a married woman watches the show is 0.5. The probability that a man watches the show, given that his wife does, is 0.7. Find the probability that
   (a) a married couple watches the show;
   (b) a wife watches the show given that her husband does;
   (c) at least 1 person of a married couple will watch the show.

10. For married couples living in a certain city suburb the probability that the husband will vote on a bond referendum is 0.21, the probability that his wife will vote in the referendum is 0.28, and the probability that both the husband and wife will vote is 0.15. What is the probability that
    (a) at least one member of a married couple will vote?
    (b) a wife will vote, given that her husband will vote?

(c) a husband will vote, given that his wife does not vote?

11. The probability that a vehicle entering the Luray Caverns has Canadian license plates is 0.12; the probability that it is a camper is 0.28; and the probability that it is a camper with Canadian license plates is 0.09. What is the probability that
    (a) a camper entering the Luray Caverns has Canadian license plates?
    (b) a vehicle with Canadian license plates entering the Luray Caverns is a camper?
    (c) a vehicle entering the Luray Caverns does not have Canadian plates or is not a camper?

12. The probability that the lady of the house is home when the Avon representative calls is 0.6. Given that the lady of the house is home, the probability that she makes a purchase is 0.4. Find the probability that the lady of the house is home and makes a purchase when the Avon representative calls.

13. The probability that a doctor correctly diagnoses a particular illness is 0.7. Given that the doctor makes an incorrect diagnosis, the probability that the patient enters a law suit is 0.9. What is the probability that the doctor makes an incorrect diagnosis and the patient sues?

14. One bag contains 4 white balls and 3 black balls, and a second bag contains 3 white balls and 5 black balls. One ball is drawn at random from the second bag and is placed unseen in the first bag. What is the probability that a ball now drawn from the first bag is white?

15. A real estate agent has 8 master keys to open several new homes. Only 1 master key will open any given house. If 40% of these homes are usually left unlocked, what is the probability that the real estate agent can get into a specific home if the agent selects 3 master keys at random before leaving the office?

16. Two cards are drawn in succession from a deck without replacement. What is the probability that
    (a) both cards are red?
    (b) both cards are greater than 3 but less than 8?

17. A town has 2 fire engines operating independently. The probability that a specific engine is available when needed is 0.96.
    (a) What is the probability that neither is available when needed?

(b) What is the probability that a fire engine is available when needed?

18. The probability that Tom will be alive in 20 years is 0.7, and the probability that Nancy will be alive in 20 years is 0.9. If we assume independence for both, what is the probability that neither will be alive in 20 years?

19. One overnight case contains 2 bottles of aspirin and 3 bottles of thyroid tablets. A second tote bag contains 3 bottles of aspirin, 2 bottles of thyroid, and 1 bottle of laxative tablets. If 1 bottle of tablets is taken at random from each piece of luggage, find the probability that
(a) both bottles contain thyroid tablets;
(b) neither bottle contains thyroid tablets;
(c) the 2 bottles contain different tablets.

20. The probability that a person visiting his dentist will have an X-ray is 0.6; the probability that a person who has an X-ray will also have a cavity filled is 0.3; and the probability that a person who has had an X-ray and a cavity filled will also have a tooth extracted is 0.1. What is the probability that a person visiting his dentist will have an X-ray, a cavity filled, and a tooth extracted?

21. Find the probability of randomly selecting 4 good quarts of milk in succession from a cooler containing 20 quarts of which 5 have spoiled, by using
(a) the first formula of Theorem 1.14 on page 32;
(b) the formulas of Theorems 1.8 and 1.9 on pages 16 and 20, respectively.

22. From a box containing 6 black balls and 4 green balls, 3 balls are drawn in succession, each ball being replaced in the box before the next draw is made. What is the probability that
(a) all 3 are the same color?
(b) each color is represented?

23. The probability that a patient recovers from a delicate heart operation is 0.8. What is the probability that
(a) exactly 2 of the next 3 patients who have this operation survive?
(b) all of the next 3 patients who have this operation survive?

24. An allergist claims that 50% of the patients she tests are allergic to some type of weed. What is the probability that
(a) exactly 3 of her next 4 patients are allergic to weeds?
(b) none of her next 4 patients is allergic to weeds?

25. If the probability is 0.1 that a person will make a mistake on his or her state income tax return, find the probability that
(a) four totally unrelated persons each make a mistake;
(b) Mr. Jones and Ms. Clark both make a mistake, and Mr. Roberts and Ms. Williams do not make a mistake.

1.8 Bayes' Rule

Let us now return to the illustration of Section 1.6, where an individual is being selected at random from the adults of a small town to tour the country and publicize the advantages of establishing new industries in the town. Suppose that we are now given the additional information that 36 of those employed and 12 of those unemployed are members of the Rotary Club. We wish to find the probability of the event A that the individual selected is a member of the Rotary Club. Referring to Figure 1.9, we can write A as the union of the two mutually exclusive events $E \cap A$ and $E' \cap A$. Hence

$$A = (E \cap A) \cup (E' \cap A),$$

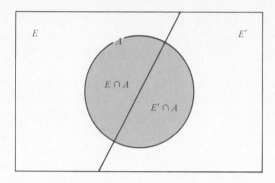

Figure 1.9   Venn diagram for the events A, E, and E'.

and by Corollary 1 of Theorem 1.10, and then Theorem 1.12, we can write

$$P(A) = P[(E \cap A) + (E' \cap A)]$$
$$= P(E \cap A) + P(E' \cap A)$$
$$= P(E)P(A \mid E) + P(E')P(A \mid E').$$

The data of Section 1.6, together with the additional data given above for the set A, enables us to compute

$$P(E) = \tfrac{600}{900} = \tfrac{2}{3}, \qquad P(A \mid E) = \tfrac{36}{600} = \tfrac{3}{50},$$

and

$$P(E') = \tfrac{1}{3}, \qquad P(A \mid E') = \tfrac{12}{300} = \tfrac{1}{25}.$$

If we display these probabilities by means of the tree diagram of Figure 1.10, in which the first branch yields the probability $P(E)P(A \mid E)$ and the second branch yields the probability $P(E')P(A \mid E')$, it follows that

$$P(A) = (\tfrac{2}{3})(\tfrac{3}{50}) + (\tfrac{1}{3})(\tfrac{1}{25})$$
$$= \tfrac{4}{75}.$$

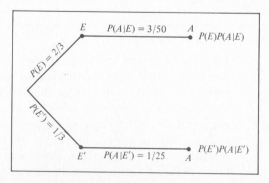

Figure 1.10   Tree diagram for the data on page 27.

A generalization of the foregoing illustration to the case where the sample space is partitioned into k subsets is covered by the following theorem, sometimes called the **theorem of total probability** or the **rule of elimination**.

Theorem 1.15   *If the events B_1, B_2, ..., B_k constitute a partition of the sample space S such that $P(B_i) \neq 0$ for $i = 1, 2, ..., k$, then for any event A of S*

$$P(A) = \sum_{i=1}^{k} P(B_i \cap A) = \sum_{i=1}^{k} P(B_i)P(A \mid B_i).$$

Proof. Consider the Venn diagram of Figure 1.11. The event A is seen to be the union of the mutually exclusive events $B_1 \cap A$, $B_2 \cap A$, ..., $B_k \cap A$; that is,

$$A = (B_1 \cap A) \cup (B_2 \cap A) \cup \cdots \cup (B_k \cap A).$$

Using Corollary 2 of Theorem 1.10, and then Theorem 1.12, we have

$$P(A) = P[(B_1 \cap A) \cup (B_2 \cap A) \cup \cdots \cup (B_k \cap A)]$$

$$= P(B_1 \cap A) + P(B_2 \cap A) + \cdots + P(B_k \cap A)$$

$$= \sum_{i=1}^{k} P(B_i \cap A) = \sum_{i=1}^{k} P(B_i)P(A \mid B_i).$$

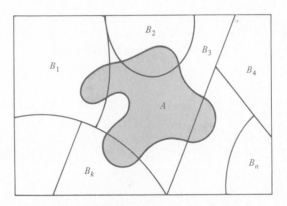

Figure 1.11   Partitioning the sample space S.

Example 1.38   Three members of a private country club have been nominated for the office of president. The probability that Mr. Adams will be elected is 0.3; the probability that Mr. Brown will be elected is 0.5; and the probability that Ms. Cooper will be elected is 0.2. Should Mr. Adams be elected, the probability for an increase in membership fees is 0.8. Should Mr. Brown or Ms. Cooper be elected, the corresponding probabilities for an increase in fees are 0.1 and 0.4. What is the probability that there will be an increase in membership fees?

Solution. Consider the following events:

A: membership fees are increased,

B_1: Mr. Adams is elected,

B_2: Mr. Brown is elected,

B_3: Ms. Cooper is elected.

Applying the rule of elimination, we can write

$$P(A) = P(B_1)P(A \mid B_1) + P(B_2)P(A \mid B_2) + P(B_3)P(A \mid B_3).$$

Referring to the tree diagram of Figure 1.12, we find that the three branches give the probabilities

$$P(B_1)P(A \mid B_1) = (0.3)(0.8) = 0.24,$$
$$P(B_2)P(A \mid B_2) = (0.5)(0.1) = 0.05,$$
$$P(B_3)P(A \mid B_3) = (0.2)(0.4) = 0.08,$$

and hence

$$P(A) = 0.24 + 0.05 + 0.08$$
$$= 0.37.$$

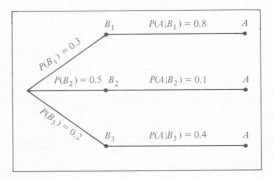

Figure 1.12   Tree diagram for Example 1.38.

Instead of asking for $P(A)$ by the rule of elimination, suppose that we now consider the problem of finding the conditional probability $P(B_3 \mid A)$ in Example 1.38. In other words, if it is known that membership fees have increased, what is the probability that Ms. Cooper was elected president of the club? Questions of this type can be answered by using the following theorem, called **Bayes' rule**:

Theorem 1.16 (Bayes' Rule)   *If the events $B_1, B_2, \ldots, B_k$ constitute a partition of the sample space S, where $P(B_i) \neq 0$ for $i = 1, 2, \ldots, k$, then for any event A in S such that $P(A) \neq 0$,*

$$P(B_r \mid A) = \frac{P(B_r \cap A)}{\sum\limits_{i=1}^{k} P(B_i \cap A)} = \frac{P(B_r)P(A \mid B_r)}{\sum\limits_{i=1}^{k} P(B_i)P(A \mid B_i)}$$

for $r = 1, 2, \ldots, k$.

Proof. By the definition of conditional probability,

$$P(B_r \mid A) = \frac{P(B_r \cap A)}{P(A)},$$

and then using Theorem 1.15 in the denominator, we have

$$P(B_r \mid A) = \frac{P(B_r \cap A)}{\sum\limits_{i=1}^{k} P(B_i \cap A)}.$$

Applying Theorem 1.12 to both numerator and denominator, we obtain the alternative form,

$$P(B_r \mid A) = \frac{P(B_r)P(A \mid B_r)}{\sum\limits_{i=1}^{k} P(B_i)P(A \mid B_i)},$$

which completes the proof.

Example 1.39   With reference to Example 1.38, if someone is considering joining the club but delays his or her decision for several weeks only to find out that the fees have been increased, what is the probability that Ms. Cooper was elected president of the club?

Solution. Using Bayes' rule to write

$$P(B_3 \mid A) = \frac{P(B_3)P(A \mid B_3)}{P(B_1)P(A \mid B_1) + P(B_2)P(A \mid B_2) + P(B_3)P(A \mid B_3)},$$

and then substituting the probabilities calculated in Example 1.38, we have

$$P(B_3 \mid A) = \frac{0.08}{0.24 + 0.05 + 0.08} = \frac{8}{37}.$$

In view of the fact that fees have increased, this result suggests that Ms. Cooper is probably not the president of the club.

Exercises

1. In a certain region of the country it is known from past experience that the probability of selecting an adult over 40 years of age with cancer is 0.02. If the probability of a doctor correctly diagnosing a person with cancer as having the disease is 0.78 and the probability of incorrectly diagnosing a person without cancer as having the disease is 0.06, what is the probability that a person is diagnosed as having cancer?

2. Police plan to enforce speed limits by using radar traps at 4 different locations within the city limits. The radar traps at each of the locations L_1, L_2, L_3, and L_4 are operated 40%, 30%, 20%, and 30% of the time, and if a person who is speeding on his way to work has probabilities of 0.2, 0.1, 0.5, and 0.2, respectively, of passing through these locations, what is the probability that he will receive a speeding ticket?

3. Referring to Exercise 1, what is the probability that a person diagnosed as having cancer actually has the disease?

4. If in Exercise 2 the person received a speeding ticket on his way to work, what is the probability that he passed through the radar trap located at L_2?

5. Suppose that colored balls are distributed in three indistinguishable boxes as follows:

	Box 1	Box 2	Box 3
Red	2	4	3
White	3	1	4
Blue	5	3	3

A box is selected at random from which a ball is selected at random and it is observed to be red. What is the probability that box 3 was selected?

6. A commuter owns two cars, one a compact and one a standard model. About three-fourths of the time he uses the compact to travel to work and about one-fourth of the time the larger car is used. When he uses the compact car he usually gets home by 5:30 P.M. about 75% of the time; if he uses the standard-sized car he gets home by 5:30 P.M. about 60% of the time (but he enjoys the air conditioner in the larger car). If he gets home after 5:30 P.M., what is the probability that he used the compact car?

7. A truth serum given to a suspect is known to be 90% reliable when the person is guilty and 99% reliable when the person is innocent. In other words, 10% of the guilty are judged innocent by the serum and 1% of the innocent are judged guilty. If the suspect was selected from a group of suspects of which only 5% have ever committed a crime, and the serum indicates that he is guilty, what is the probability that he is innocent?

8. A large industrial firm uses 3 local motels to provide overnight accommodations for its clients. From past experience it is known that 20% of the clients are assigned rooms at the Ramada Inn, 50% at the Sheraton, and 30% at the Lakeview Motor Lodge. If the plumbing is faulty in 5% of the rooms at the Ramada Inn, in 4% of the rooms at the Sheraton, and in 8% of the rooms at the Lakeview Motor Lodge, what is the probability that
   (a) a client will be assigned a room with faulty plumbing?
   (b) a person with a room having faulty plumbing was assigned accommodations at the Lakeview Motor Lodge?

2

Random Variables

2.1 Concept of a Random Variable

The term *statistical experiment* has been used to describe any process by which several chance observations are generated. Often we are not interested in the details associated with each sample point but only in some numerical description of the outcome. For example, the sample space giving a detailed description of each possible outcome when one tosses a coin 3 times may be written

$$S = \{HHH, HHT, HTH, THH, HTT, THT, TTH, TTT\}.$$

If one is concerned only with the number of heads that fall, then a numerical value of 0, 1, 2, or 3 will be assigned to each sample point.

The numbers 0, 1, 2, and 3 are random quantities determined by the outcome of the experiment. They may be thought of as values assumed by some **random variable** X, which in this case represents the number of heads when a coin is tossed 3 times.

Definition 2.1 *A **random variable** is a function that associates a real number with each element in the sample space.*

We shall use a capital letter, say X, to denote a random variable and its corresponding small letter, x in this case, for one of its values. In the coin-tossing illustration above, we notice that the random variable X assumes the value 2 for all elements in the subset

$$E = \{HHT, HTH, THH\}$$

of the sample space S. That is, each possible value of X represents an event that is a subset of the sample space for the given experiment.

Example 2.1   Two balls are drawn in succession without replacement from an urn containing 4 red balls and 3 black balls. The possible outcomes and the values y of the random variable Y, where Y is the number of red balls, are

Sample Space	y
RR	2
RB	1
BR	1
BB	0

Example 2.2   A stockroom clerk returns three safety helmets at random to three steel mill employees, who had previously checked them. If Smith, Jones, and Brown, in that order, receive one of the three hats, list the sample points for the possible orders of returning the helmets and find the values m of the random variable M that represents the number of correct matches.

Solution.   If S, J, and B stand for Smith's, Jones', and Brown's helmets, respectively, then the possible arrangements in which the helmets may be returned and the number of correct matches are

Sample Space	m
SJB	3
SBJ	1
JSB	1
JBS	0
BSJ	0
BJS	1

In each of the two preceding examples the sample space contains a finite number of elements. On the other hand, when a die is thrown until a 5 occurs, we obtain a sample space with an unending sequence of elements,

$$S = \{F, NF, NNF, NNNF, \ldots\},$$

where F and N represent, respectively, the occurrence and nonoccurrence of a 5. But even in this experiment, the number of elements can be equated to the number of whole numbers so that there is a first element, a second element, a third element, and so on, and in this sense can be counted.

Definition 2.2 *If a sample space contains a finite number of possibilities or an unending sequence with as many elements as there are whole numbers, it is called a* **discrete sample space**.

The outcomes of some statistical experiments may be neither finite nor countable. Such is the case, for example, when one conducts an investigation measuring the distances that a certain make of automobile will travel over a prescribed test course on 5 liters of gasoline. Assuming distance to be a variable measured to any degree of accuracy, then clearly we have an infinite number of possible distances in the sample space that cannot be equated to the number of whole numbers. Also, if one were to record the length of time for a chemical reaction to take place, once again the possible time intervals making up our sample space are infinite in number and uncountable. We see now that all sample spaces need not be discrete.

Definition 2.3 *If a sample space contains an infinite number of possibilities equal to the number of points on a line segment, it is called a* **continuous sample space**.

A random variable is called a **discrete random variable** if its set of possible outcomes is countable. Since the possible values of Y in Example 2.1 are 0, 1, and 2, and the possible values of M in Example 2.2 are 0, 1, and 3, it follows that Y and M are discrete random variables. When a random variable can take on values on a continuous scale, it is called a **continuous random variable**. Often the possible values of a continuous random variable are precisely the same values that are contained in the continuous sample space. Such is the case when the random variable represents the measured distance that a certain make of automobile will travel over a test course on 5 liters of gasoline.

In most practical problems, continuous random variables represent *measured* data, such as all possible heights, weights, temperatures, distances, or life periods, whereas discrete random variables represent *count* data, such as the number of defectives in a sample of k items or the number of highway fatalities per year in a given state. Note that the random variables Y and M of Examples 2.1 and 2.2 both represent count data, Y the number of red balls and M the number of correct hat matches.

2.2 Discrete Probability Distributions

A discrete random variable assumes each of its values with a certain probability. In the case of tossing a coin three times, the variable X, representing the number of heads, assumes the value 2 with probability 3/8, since 3 of the 8 equally likely sample points result in two heads and one tail. If one assumes equal weights for the simple events in Example 2.2, the probability that no employee gets back his right helmet, that is, the probability that M assumes the value zero, is 1/3. The possible values m of M and their probabilities are given by

m	0	1	3
$P(M = m)$	$\frac{1}{3}$	$\frac{1}{2}$	$\frac{1}{6}$

Note that the values of m exhaust all possible cases and hence the probabilities add to 1.

Frequently, it is convenient to represent all the probabilities of a random variable X by a formula. Such a formula would necessarily be a function of the numerical values x that we shall denote by $f(x)$, $g(x)$, $r(x)$, and so forth. Therefore, we write $f(x) = P(X = x)$; that is, $f(3) = P(X = 3)$. The set of ordered pairs $(x, f(x))$ is called the **probability function** or **probability distribution** of the discrete random variable X.

Definition 2.4 *The set of ordered pairs $(x, f(x))$ is a **probability function** or a **probability distribution** of the discrete random variable X if, for each possible outcome x,*

1. $f(x) \geq 0$.

2. $\sum\limits_{x} f(x) = 1$.

3. $P(X = x) = f(x)$.

Example 2.3 A shipment of 8 similar microcomputers to a retail outlet contains 3 that are defective. If a school makes a random purchase of 2 of these computers, find the probability distribution for the number of defectives.

Solution. Let X be a random variable whose values x are the possible numbers of defective computers purchased by the school. Then x can be any of the numbers 0, 1, and 2, Now,

$$f(0) = P(X = 0) = \frac{\binom{3}{0}\binom{5}{2}}{\binom{8}{2}} = \frac{10}{28},$$

$$f(1) = P(X = 1) = \frac{\binom{3}{1}\binom{5}{1}}{\binom{8}{2}} = \frac{15}{28},$$

$$f(2) = P(X = 2) = \frac{\binom{3}{2}\binom{5}{0}}{\binom{8}{2}} = \frac{3}{28}.$$

Thus the probability distribution of X is

x	0	1	2
$f(x)$	$\frac{10}{28}$	$\frac{15}{28}$	$\frac{3}{28}$

Example 2.4   If 50% of the automobiles sold by an agency for a certain foreign car are equipped with diesel engines, find a formula for the probability distribution of the number of diesel models among the next 4 cars sold by this agency.

Solution. Since the probability of selling a diesel model or a gasoline model is 0.5, the $2^4 = 16$ points in the sample space are equally likely to occur. Therefore, the denominator for all probabilities, and also for our function, will be 16. To obtain the number of ways of selling 3 diesel models, we need to consider the number of ways of partitioning 4 outcomes into two cells with 3 diesel models assigned to one cell and a gasoline model assigned to the other. This can be done in $\binom{4}{3} = 4$ ways. In general, the event of selling x diesel models and $4 - x$ gasoline models can occur in $\binom{4}{x}$ ways, where x can be 0, 1, 2, 3, or 4. Thus the probability distribution $f(x) = P(X = x)$ is

$$f(x) = \frac{\binom{4}{x}}{16} \quad \text{for } x = 0, 1, 2, 3, 4.$$

There are many problems in which we wish to compute the probability that the observed value of a random variable X will be less than or equal to some real number x. Writing $F(x) = P(X \leq x)$ for every real number x, we define $F(x)$ to be the **cumulative distribution** of the random variable X.

Definition 2.5   *The **cumulative distribution** $F(x)$ of a discrete random variable X with probability distribution $f(x)$ is given by*

$$F(x) = P(X \leq x) = \sum_{t \leq x} f(t) \quad \text{for } -\infty < x < \infty.$$

For the random variable M, the number of correct matches in Example 2.2, we have

$$F(2.4) = P(M \leq 2.4) = f(0) + f(1) = (\tfrac{1}{3}) + (\tfrac{1}{2}) = \tfrac{5}{6}.$$

The cumulative distribution of M is given by

$$F(m) = \begin{cases} 0 & \text{for } m < 0 \\ \frac{1}{3} & \text{for } 0 \le m < 1 \\ \frac{5}{6} & \text{for } 1 \le m < 3 \\ 1 & \text{for } m \ge 3. \end{cases}$$

One should pay particular notice to the fact that the cumulative distribution is defined not only for the values assumed by the given random variable but for all real numbers.

Example 2.5   Find the cumulative distribution of the random variable X in Example 2.4. Using $F(x)$, verify that $f(2) = 3/8$.

Solution. Direct calculations of the probability distribution of Example 2.4 give $f(0) = 1/16, f(1) = 1/4, f(2) = 3/8, f(3) = 1/4,$ and $f(4) = 1/16$. Therefore,

$$F(0) = f(0) = \tfrac{1}{16}$$
$$F(1) = f(0) + f(1) = \tfrac{5}{16}$$
$$F(2) = f(0) + f(1) + f(2) = \tfrac{11}{16}$$
$$F(3) = f(0) + f(1) + f(2) + f(3) = \tfrac{15}{16}$$
$$F(4) = f(0) + f(1) + f(2) + f(3) + f(4) = 1.$$

Hence

$$F(x) = \begin{cases} 0 & \text{for } x < 0 \\ \frac{1}{16} & \text{for } 0 \le x < 1 \\ \frac{5}{16} & \text{for } 1 \le x < 2 \\ \frac{11}{16} & \text{for } 2 \le x < 3 \\ \frac{15}{16} & \text{for } 3 \le x < 4 \\ 1 & \text{for } x \ge 4. \end{cases}$$

Now

$$f(2) = F(2) - F(1) = \tfrac{11}{16} - \tfrac{5}{16} = \tfrac{3}{8}.$$

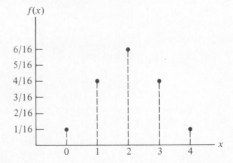

Figure 2.1   Bar chart.

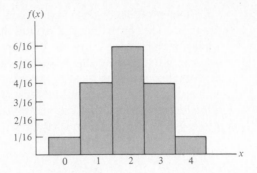

Figure 2.2   Probability histogram.

It is often helpful to look at a probability distribution in graphic form. One might plot the points $(x, f(x))$ of Example 2.4 to obtain Figure 2.1. By joining the points to the x axis either with a dashed or solid line, we obtain what is commonly called a **bar chart**. Figure 2.1 makes it very easy to see what values of X are most likely to occur, and it also indicates a perfectly symmetric situation in this case.

Instead of plotting the points $(x, f(x))$, we more frequently construct rectangles, as in Figure 2.2. Here the rectangles are constructed so that their bases of equal width are centered at each value x and their heights are equal to the corresponding probabilities given by $f(x)$. The bases are constructed so as to leave no space between the rectangles. Figure 2.2 is called a **probability histogram**.

Since each base in Figure 2.2 has unit width, the $P(X = x)$ is equal to the area of the rectangle centered at x. Even if the bases were not of unit width, we could adjust the heights of the rectangles to give areas that would still equal the probabilities of X assuming any of its values x. This concept of using areas to represent probabilities is necessary for our consideration of the probability distribution of a continuous random variable.

The graph of the cumulative distribution of Example 2.5, which appears as a step function in Figure 2.3, is obtained by plotting the points $(x, F(x))$.

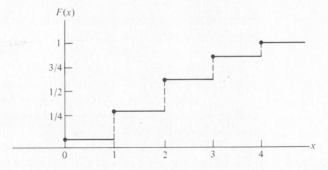

Figure 2.3   Discrete cumulative distribution.

Certain probability distributions are applicable to more than one physical situation. The probability distribution of Example 2.4, for example, also applies to the random variable Y, where Y is the number of heads when a coin is tossed 4 times, or to the random variable W, where W is the number of red cards that occur when 4 cards are drawn at random from a deck in succession with each card replaced and the deck shuffled before the next drawing. Special discrete distributions that can be applied to many different experimental situations will be considered in Chapter 4.

2.3 *Continuous Probability Distributions*

A continuous random variable has a probability of zero of assuming *exactly* any of its values. Consequently, its probability distribution cannot be given in tabular form. At first this may seem startling, but it becomes more plausible when we consider a particular example. Let us discuss a random variable whose values are the heights of all people over 21 years of age. Between any two values, say 163.5 and 164.5 centimeters, or even 163.99 and 164.01 centimeters, there are in infinite number of heights, one of which is 164 centimeters. The probability of selecting a person at random who is exactly 164 centimeters tall and not one of the infinitely large set of heights so close to 164 centimeters that you cannot humanly measure the difference is remote, and thus we assign a probability of zero to the event. This is not the case, however, if we talk about the probability of selecting a person who is at least 163 centimeters but not more than 165 centimeters tall. Now we are dealing with an interval rather than a point value of our random variable.

We shall concern ourselves with computing probabilities for various intervals of continuous random variables such as $P(a < X < b)$, $P(W > c)$, and so forth. Note that when X is continuous

$$P(a < X \le b) = P(a < X < b) + P(X = b)$$
$$= P(a < X < b).$$

That is, it does not matter whether we include an end point of the interval or not. This is not true, though, when X is discrete.

Although the probability distribution of a continuous random variable cannot be presented in tabular form, it can have a formula. Such a formula would necessarily be a function of the numerical values of the continuous variable X and as such will be represented by the functional notation $f(x)$. In dealing with continuous variables, $f(x)$ is usually called the **probability density function**, or simply the **density function** of X. Since X is defined over a continuous sample space, it is possible for $f(x)$ to have a finite number of discontinuities. However, most density functions that have practical applications in the analysis of statistical data are continuous and their graphs may take any of

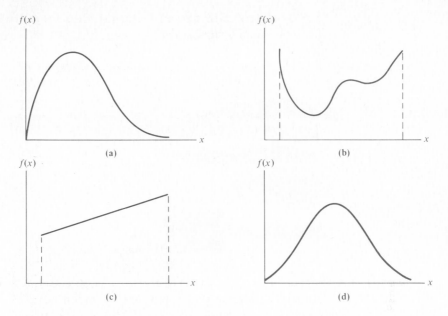

Figure 2.4 Typical density functions.

several forms, some of which are shown in Figure 2.4. Because areas will be used to represent probabilities and probabilities are positive numerical values, the density function must lie entirely above the x axis.

A probability density function is constructed so that the area under its curve bounded by the x axis is equal to 1 when computed over the range of X for which $f(x)$ is defined. Should this range of X be a finite interval, it is always possible to extend the interval to include the entire set of real numbers by defining $f(x)$ to be zero at all points in the extended portions of the interval. In Figure 2.5, the probability that X assumes a value between a and b is equal to

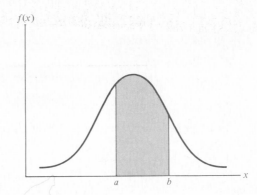

Figure 2.5 $P(a < X < b)$.

the shaded area under the density function between the ordinates at $x = a$ and $x = b$, and from integral calculus is given by

$$P(a < X < b) = \int_a^b f(x)\, dx.$$

Definition 2.6   *The function $f(x)$ is a **probability density function** for the continuous random variable X, defined over the set of real numbers R, if*

1. $f(x) \geq 0$ *for all $x \in R$.*

2. $\int_{-\infty}^{\infty} f(x)\, dx = 1$.

3. $P(a < X < b) = \int_a^b f(x)\, dx.$

Example 2.6   Suppose that the error in the reaction temperature, in °C, for a controlled laboratory experiment is a continuous random variable X having the probability density function

$$f(x) = \begin{cases} \dfrac{x^2}{3}, & -1 < x < 2 \\ 0, & \text{elsewhere.} \end{cases}$$

(a) Verify condition 2 of Definition 2.6.
(b) Find $P(0 < X \leq 1)$.

Solution

(a) $\displaystyle \int_{-\infty}^{\infty} f(x)\, dx = \int_{-1}^{2} \frac{x^2}{3}\, dx = \frac{x^3}{9}\Big|_{-1}^{2} = \frac{8}{9} + \frac{1}{9} = 1.$

(b) $\displaystyle P(0 < X \leq 1) = \int_0^1 \frac{x^2}{3}\, dx = \frac{x^3}{9}\Big|_0^1 = \frac{1}{9}.$

Definition 2.7   *The **cumulative distribution** $F(x)$ of a continuous random variable X with density function $f(x)$ is given by*

$$F(x) = P(X \leq x) = \int_{-\infty}^{x} f(t)\, dt \text{ for } -\infty < x < \infty.$$

As an immediate consequence of Definition 2.7 one can write the two results

$$P(a < X < b) = F(b) - F(a)$$

and

$$f(x) = \frac{dF(x)}{dx}$$

if the derivative exists.

Example 2.7 For the density function of Example 2.6 find $F(x)$ and use it to evaluate $P(0 < X \le 1)$.

Solution. For $-1 < x < 2$,

$$F(x) = \int_{-\infty}^{x} f(t) \, dt = \int_{-1}^{x} \frac{t^2}{3} \, dt = \frac{t^3}{9}\bigg|_{-1}^{x} = \frac{x^3 + 1}{9}.$$

Therefore,

$$F(x) = \begin{cases} 0, & x \le -1 \\ \dfrac{x^3 + 1}{9}, & -1 \le x < 2 \\ 1, & x \ge 2. \end{cases}$$

The cumulative distribution $F(x)$ is expressed graphically in Figure 2.6. Now,

$$P(0 < X \le 1) = F(1) - F(0) = \tfrac{2}{9} - \tfrac{1}{9} = \tfrac{1}{9},$$

which agrees with the result obtained by using the density function in Example 2.6.

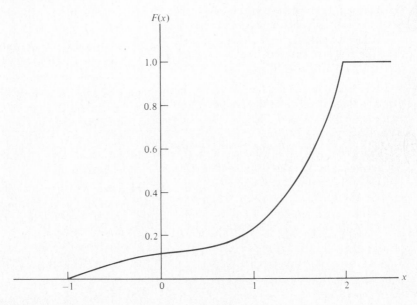

Figure 2.6 Continuous cumulative distribution.

Exercises

1. Classify the following random variables as discrete or continuous.
   - X: the number of automobile accidents per year in Virginia.
   - Y: the length of time to play 18 holes of golf.
   - M: the amount of milk produced yearly by a particular cow.
   - N: the number of eggs laid each month by a hen.
   - P: the number of building permits issued each month in a certain city.
   - Q: the weight of grain produced per acre.

2. An overseas shipment of 5 foreign automobiles contains 2 that have slight paint blemishes. If an agency receives 3 of these automobiles at random, list the elements of the sample space S using the letters B and N for "blemished" and "nonblemished," respectively; then to each sample point assign a value x of the random variable X representing the number of automobiles purchased by the agency with paint blemishes.

3. Let W be a random variable giving the number of heads minus the number of tails in three tosses of a coin. List the elements of the sample space S for the three tosses of the coin and to each sample point assign a value w of W.

4. A coin is flipped until 3 heads in succession occur. List only those elements of the sample space that require 6 or less tosses. Is this a discrete sample space? Explain.

5. Determine the value c so that each of the following functions can serve as a probability distribution of the discrete random variable X:
   - (a) $f(x) = c(x^2 + 4)$ for $x = 0, 1, 2, 3$;
   - (b) $f(x) = c\binom{2}{x}\binom{3}{3-x}$ for $x = 0, 1, 2$.

6. From a box containing 4 dimes and 2 nickels, 3 coins are selected at random without replacement. Find the probability distribution for the total T of the 3 coins. Express the probability distribution graphically as a probability histogram.

7. From a box containing 4 black balls and 2 green balls, 3 balls are drawn in succession, each ball being replaced in the box before the next draw is made. Find the probability distribution for the number of green balls.

8. Find the probability distribution of the random variable W in Exercise 3, assuming that the coin is biased so that a head is twice as likely to occur as a tail.

9. Find the probability distribution for the number of jazz records when 4 records are selected at random from a collection consisting of 5 jazz records, 2 classical records, and 3 polka records. Express your results by means of a formula.

10. Find a formula for the probability distribution of the random variable X representing the outcome when a single die is rolled once.

11. A shipment of 7 television sets contains 2 defective sets. A hotel makes a random purchase of 3 of the sets. If X is the number of defective sets purchased by the hotel, find the probability distribution of X. Express the results graphically as a probability histogram.

12. Three cards are drawn in succession from a deck without replacement. Find the probability distribution for the number of spades.

13. Find the cumulative distribution of the random variable W in Exercise 8. Using $F(w)$, find
   - (a) $P(W > 0)$;
   - (b) $P(-1 \le W < 3)$.

14. Construct a graph of the cumulative distribution of Exercise 13.

15. Find the cumulative distribution of the random variable X representing the number of defectives in Exercise 11. Using $F(x)$, find
   - (a) $P(X = 1)$;
   - (b) $P(0 < X \le 2)$.

16. Construct a graph of the cumulative distribution of Exercise 15.

17. The probability distribution of X, the number of imperfections per 10 meters of a synthetic fabric in

continuous rolls of uniform width, is given by

x	0	1	2	3	4
$f(x)$	0.41	0.37	0.16	0.05	0.01

Construct the cumulative distribution of X.

18. An investment firm offers its customers municipal bonds that mature after different numbers of years. Given that the cumulative distribution of T, the number of years to maturity for a randomly selected bond, is

$$F(t) = \begin{cases} 0, & t < 1 \\ \frac{1}{4}, & 1 \le t < 3 \\ \frac{1}{2}, & 3 \le t < 5 \\ \frac{3}{4}, & 5 \le t < 7 \\ 1, & t \ge 7, \end{cases}$$

find
(a) $P(T = 5)$;
(b) $P(T > 3)$;
(c) $P(1.4 < T < 6)$.

19. A continuous random variable X that can assume values between $x = 1$ and $x = 3$ has a density function given by $f(x) = 1/2$.
(a) Show that the area under the curve is equal to 1.
(b) Find $P(2 < X < 2.5)$.
(c) Find $P(X \le 1.6)$.

20. A continuous random variable X that can assume values between $x = 2$ and $x = 5$ has a density function given by $f(x) = 2(1 + x)/27$. Find
(a) $P(X < 4)$;
(b) $P(3 < X < 4)$.

21. The proportion of people who respond to a certain mail-order solicitation is a continuous random variable X that has the density function

$$f(x) = \begin{cases} \dfrac{2(x + 2)}{5}, & 0 < x < 1 \\ 0, & \text{elsewhere.} \end{cases}$$

(a) Show that $P(0 < X < 1) = 1$.
(b) Find the probability that more than 1/4 but fewer than 1/2 of the people contacted will respond to this type of solicitation.

22. The total number of hours, measured in units of 100 hours, that a family runs a vacuum cleaner over a period of one year is a continuous random variable X that has the density function

$$f(x) = \begin{cases} x, & 0 < x < 1 \\ 2 - x, & 1 \le x < 2 \\ 0, & \text{elsewhere.} \end{cases}$$

Find the probability that over a period of one year, a family runs their vacuum cleaner
(a) less than 120 hours;
(b) between 50 hours and 100 hours.

23. For the density function of Exercise 19, find $F(x)$ and use it to evaluate $P(2 < X < 2.5)$.

24. For the density function of Exercise 20, find $F(x)$ and use it to evaluate $P(3 \le X < 4)$.

25. Consider the density function

$$f(x) = \begin{cases} k\sqrt{x}, & 0 < x < 1 \\ 0, & \text{elsewhere.} \end{cases}$$

(a) Evaluate k.
(b) Find $F(x)$ and use it to evaluate

$$P(0.3 < X < 0.6).$$

26. The shelf life, in days, for bottles of a certain prescribed medicine is a random variable having the density function

$$f(x) = \begin{cases} \dfrac{20{,}000}{(x + 100)^3}, & x > 0 \\ 0, & \text{elsewhere.} \end{cases}$$

Find the probability that a bottle of this medicine will have a shelf life of
(a) at least 200 days;
(b) anywhere from 80 to 120 days.

27. The waiting time, in hours, between successive speeders spotted by a radar unit is a continuous random variable with cumulative distribution

$$F(x) = \begin{cases} 0, & x \le 0 \\ 1 - e^{-8x}, & x > 0. \end{cases}$$

Find the probability of waiting less than 12 minutes between successive speeders
(a) using the cumulative distribution of X;
(b) using the probability density function of X.

2.4 *Empirical Distributions*

Usually, in an experiment involving a continuous random variable the density function $f(x)$ is unknown and its form is assumed. For the choice of $f(x)$ to be reasonably valid, good judgment based on all available information is needed in its selection. Statistical data, generated in large masses, can be very useful in studying the behavior of the distribution if presented in a combined tabular and graphic display called a **stem and leaf plot**.

Table 2.1 Car Battery Lives

2.2	4.1	3.5	4.5	3.2	3.7	3.0	2.6
3.4	1.6	3.1	3.3	3.8	3.1	4.7	3.7
2.5	4.3	3.4	3.6	2.9	3.3	3.9	3.1
3.3	3.1	3.7	4.4	3.2	4.1	1.9	3.4
4.7	3.8	3.2	2.6	3.9	3.0	4.2	3.5

To illustrate the construction of a stem and leaf plot, consider the data of Table 2.1, which represents the lives of 40 similar car batteries recorded to the nearest tenth of a year. The batteries were guaranteed to last 3 years. First, split each observation into two parts consisting of a stem and a leaf such that the stem represents the digit preceding the decimal and the leaf corresponds to the decimal part of the number. In other words, for the number 3.7 the digit 3 is designated the stem and the digit 7 is the leaf. The four stems 1, 2, 3, and 4 for our data are listed consecutively on the left side of a vertical line in Table 2.2; the leaves are recorded on the right side of the line opposite the appropriate stem value. Thus the leaf 6 of the number 1.6 is recorded opposite the stem 1; the leaf 5 of the number 2.5 is recorded opposite the stem 2; and so forth. The number of leaves recorded opposite each stem is summarized under the frequency column.

Table 2.2 Stem and Leaf Plot of Battery Lives

Stems	Leaves	Frequency
1	69	2
2	25696	5
3	4318514723628297130097145	25
4	71354172	8

Table 2.3 Double-Stem and Leaf Plot of
Battery Lives

Stems	Leaves	Frequency
1·	69	2
2*	2	1
2·	5696	4
3*	431142322130014	15
3·	8576897975	10
4*	13412	5
4·	757	3

The stem and leaf plot of Table 2.2 contains only four stems and consequently does not provide an adequate picture of the distribution. To remedy this problem, we need to increase the number of stems in our plot. One simple way to accomplish this is to write each stem value twice on the left side of the vertical line and then record the leaves 0, 1, 2, 3, and 4 opposite the appropriate stem value where it appears for the first time; and the leaves 5, 6, 7, 8, and 9 opposite this same stem value where it appears for the second time. This modified double-stem and leaf plot is illustrated in Table 2.3, where the stems corresponding to leaves 0 through 4 have been coded by the symbol * and the stem corresponding to leaves 5 through 9 by the symbol ·.

A further increase in the number of stems may be achieved by writing each stem value five times on the left side of a vertical line, where we might now code the stems a for leaves 0 and 1, b for leaves 2 and 3, c for leaves 4 and 5, d for leaves 6 and 7, and e for leaves 8 and 9. For the data of Table 2.1 we would then use the stems $1d$, $1e$, $2a$, $2b$, $2c$, $2d$, and $2e$ to construct a five-stem and leaf plot.

In any given problem, we must decide on the appropriate stem values. This decision is made somewhat arbitrarily although we are guided by the size of our sample. Usually, we choose between 5 and 20 stems. The smaller the number of data available, the smaller is our choice for the number of stems. For example, if the data consist of numbers from 1 to 21 representing the number of people in a cafeteria line on 40 randomly selected workdays and we choose a double-stem and leaf plot, the stems would be 0*, 0·, 1*, 1·, and 2* so that the smallest observation 1 has stem 0* and leaf 1, the number 18 has stem 1· and leaf 8, and the largest observation 21 has stem 2* and leaf 1. On the other hand, if the data consist of numbers from $8800 to $9600 representing the best possible deals on 100 new automobiles from a certain dealership and we choose a single-stem and leaf plot, the stems would be 88, 89, 90, …, and 96 and the leaves would now each contain two digits. A car that sold for $9385 would have a stem value of 93 and the two-digit leaf 85. Multiple-digit leaves belonging to the same stem are usually separated by commas in the stem and leaf plot. Decimal points in the data are generally ignored when all the digits

Table 2.4 Relative Frequency Distribution of
Battery Lives

Class Interval	Class Midpoint	Frequency f	Relative Frequency
1.5–1.9	1.7	2	0.050
2.0–2.4	2.2	1	0.025
2.5–2.9	2.7	4	0.100
3.0–3.4	3.2	15	0.375
3.5–3.9	3.7	10	0.250
4.0–4.4	4.2	5	0.125
4.5–4.9	4.7	3	0.075

to the right of the decimal represent the leaf. Such was the case in Tables 2.2 and 2.3. However, if the data consist of numbers ranging from 21.8 to 74.9, we might choose the digits 2, 3, 4, 5, 6, and 7 as our stems so that a number such as 48.3 would have a stem value of 4 and a leaf of 8.3.

A **frequency distribution** in which the data are grouped into different classes or intervals can easily be constructed by counting the leaves belonging to each stem and noting that each stem defines a class interval. In Table 2.2 the stem 1 with 2 leaves defines the interval 1.0–1.9 containing 2 observations; the stem 2 with 5 leaves defines the interval 2.0–2.9 containing 5 observations; the stem 3 with 25 leaves defines the interval 3.0–3.9 with 25 observations; and the stem 4 with 8 leaves defines the interval 4.0–4.9 containing 8 observations. For the double-stem and leaf plot of Table 2.3, the stems define the seven class intervals 1.5–1.9, 2.0–2.4, 2.5–2.9, 3.0–3.4, 3.5–3.9, 4.0–4.4, and 4.5–4.9 with frequencies 2, 1, 4, 15, 10, 5, and 3, respectively. Dividing each class frequency by the total number of observations, we obtain the proportion of the set of observations in each of the classes. A table listing relative frequencies is called a **relative frequency distribution**. The relative frequency distribution for the data of Table 2.1, showing the midpoints of each class interval, is given in Table 2.4.

The information provided by a relative frequency distribution in tabular form is easier to grasp if presented graphically. Using the midpoints of each interval and the corresponding relative frequencies, we construct a **relative frequency histogram** (Figure 2.7) in exactly the same manner that we constructed the probability histogram of Section 2.2.

In Section 2.2 we suggested that the heights of the rectangles be adjusted so that the areas would represent probabilities. Once this is done, the vertical axis may be omitted. If we wish to estimate the probability distribution $f(x)$ of a continuous random variable X by a smooth curve as in Figure 2.8, it is important that the rectangles of the relative frequency histogram be adjusted so that the total area is equal to 1.

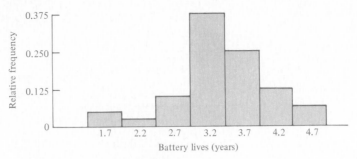

Figure 2.7   Relative frequency histogram.

The probability that a battery lasts between 3.45 and 4.45 years when selected at random from the infinite line of production of such batteries is given by the shaded area under the curve. Our estimated probability based on the recorded lives of the 40 batteries would be the sum of the areas contained in the rectangles between 3.45 and 4.45.

Although we have drawn an estimate of the shape of $f(x)$ in Figure 2.8, we still have no knowledge of its formula or equation and therefore cannot find the area that has been shaded. To help understand the method of estimating the formula for $f(x)$, let us recall some elementary analytic geometry. Parabolas, hyperbolas, circles, ellipses, and so forth, all have well-known forms of equations, and in each case we would recognize their graphs. Thinking in reverse, if we had only their graphs but recognized their form, then it is not difficult to estimate the unknown constants or parameters and arrive at the exact equation. For example, if the curve appeared to have the form of a parabola, then we know that it has an equation of the form $f(x) = ax^2 + bx + c$, where a, b, and c are parameters that can be determined by various estimation procedures.

Many continuous distributions can be represented graphically by the characteristic bell-shaped curve of Figure 2.8. The equation of the probability density function $f(x)$ in this case is as well known as that of a parabola or

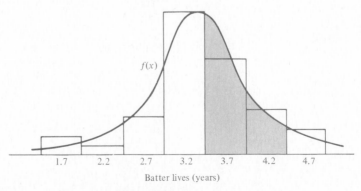

Figure 2.8   Estimating the probability density function.

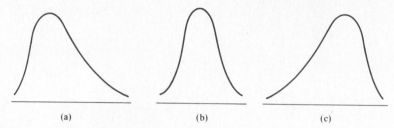

(a)   (b)   (c)

Figure 2.9   Skewness of data.

circle and depends only on the determination of two parameters. Once these parameters are estimated from the data, we can write the estimated equation, and then, using appropriate tables, find any probabilities we choose.

A distribution is said to be **symmetric** if it can be folded along a vertical axis so that the two sides coincide. A distribution that lacks symmetry with respect to a vertical axis is said to be **skewed**. The distribution illustrated in Figure 2.9(a) is said to be skewed to the right, since it has a long right tail and a much shorter left tail. In Figure 2.9(b) we see that the distribution is symmetric, while in Figure 2.9(c) it is skewed to the left.

By rotating a stem and leaf plot counterclockwise through an angle of 90°, we observe that the resulting columns of leaves form a picture that is similar to a histogram. Consequently, if our primary purpose in looking at the data is to determine the general shape or form of the density function, it will seldom be necessary to construct a relative frequency histogram. In Chapter 5 we shall consider most of the important density functions that are used in engineering and scientific investigations.

The cumulative distribution of X, where X represents the life of the car battery, can be estimated geometrically using the data of Table 2.4. To construct such a graph, we first arrange our data as in Table 2.5, a **relative**

Table 2.5 Relative Cumulative
Frequency Distribution of Battery Lives

Class Boundaries	Relative Cumulative Frequency
Less than 1.45	0.000
Less than 1.95	0.050
Less than 2.45	0.075
Less than 2.95	0.175
Less than 3.45	0.550
Less than 3.95	0.800
Less than 4.45	0.925
Less than 4.95	1.000

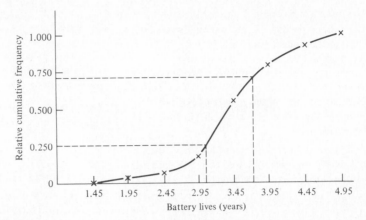

Figure 2.10 Continuous cumulative distribution.

cumulative frequency distribution, and then plot the relative cumulative frequency less than each class boundary against the corresponding class boundary as in Figure 2.10. We estimate $F(x)$ by drawing a smooth curve through the points.

Percentile, decile, and quartile points may be read quickly from the cumulative distribution. In Figure 2.10 the dashed lines indicate that the twenty-fifth percentile or first quartile and the seventh decile are approximately 3.05 and 3.70 years, respectively. This means that 25% or one-fourth of all the batteries of this type are expected to last less than 3.05 years, while 70% of such batteries can be expected to last less than 3.70 years.

Exercises

1. The following scores represent the final examination grade for an elementary statistics course:

23	60	79	32	57	74	52	70	82	36
80	77	81	95	41	65	92	85	55	76
52	10	64	75	78	25	80	98	81	67
41	71	83	54	64	72	88	62	74	43
60	78	89	76	84	48	84	90	15	79
34	67	17	82	69	74	63	80	85	61

(a) Construct a stem and leaf plot for the examination grades in which the stems are 1, 2, 3, ..., 9.
(b) Set up a relative frequency distribution.
(c) Construct a relative frequency histogram, draw an estimate of the graph of $f(x)$, and discuss the skewness of the distribution.
(d) Construct a relative cumulative frequency distribution.
(e) Draw an estimate of the graph of $F(x)$.
(f) Estimate the first quartile and the seventh decile.

2. The following data represent the length of life in years, measured to the nearest tenth, of 30 similar fuel pumps:

2.0	3.0	0.3	3.3	1.3	0.4
0.2	6.0	5.5	6.5	0.2	2.3
1.5	4.0	5.9	1.8	4.7	0.7
4.5	0.3	1.5	0.5	2.5	5.0
1.0	6.0	5.6	6.0	1.2	0.2

(a) Construct a stem and leaf plot for the fuel pump lives using the digit to the left of the decimal point as the stem for each observation.

(b) Set up a relative frequency distribution.

(c) Construct a relative cumulative frequency distribution.

(d) Draw an estimate of the graph of $F(x)$.

(e) Estimate the value below which two-thirds of the values fall.

3. The following data represent the length of life, in seconds, of 50 fruit flies subjected to a new spray in a controlled laboratory experiment:

17	20	10	9	23	13	12	19	18	24
12	14	6	9	13	6	7	10	13	7
16	18	8	13	3	32	9	7	10	11
13	7	18	7	10	4	27	19	16	8
7	10	5	14	15	10	9	6	7	15

(a) Construct a double-stem and leaf plot for the fruit fly lives using the stems 0∗, 0·, 1∗, 1·, 2∗, 2·, and 3∗ such that stems coded by the symbols ∗ and · are associated, respectively, with leaves 0 through 4 and 5 through 9.

(b) Set up a relative frequency distribution.

(c) Construct a relative frequency histogram, draw an estimate of the graph of $f(x)$, and discuss the skewness of the distribution.

(d) Construct a relative cumulative frequency distribution.

(e) Draw an estimate of the graph of $F(x)$.

(f) Estimate the seventy-fifth percentile.

4. Construct a stem and leaf plot for the data of Table 2.1 by writing each stem five times and then coding the stems as described on page 54.

5. The nicotine contents, in milligrams, for 40 cigarettes of a certain brand were recorded as follows:

1.09	1.92	2.31	1.79	2.28
1.74	1.47	1.97	0.85	1.24
1.58	2.03	1.70	2.17	2.55
2.11	1.86	1.90	1.68	1.51
1.64	0.72	1.69	1.85	1.82
1.79	2.46	1.88	2.08	1.67
1.37	1.93	1.40	1.64	2.09
1.75	1.63	2.37	1.75	1.69

(a) Construct a stem and leaf plot for the data in which the stems are the digits to the left of the decimal point, each repeated five times such that the double-digit leaves 00 through 19 are associated with stems coded by the letter a; leaves 20 through 39 are associated with stems coded by the letter b; and so forth. Thus a number such as 1.29 has a stem value of $1b$ and a leaf equal to 29.

(b) Set up a relative frequency distribution.

2.5 *Joint Probability Distributions*

Our study of random variables and their probability distributions in the preceding sections was restricted to one-dimensional sample spaces, in that we recorded outcomes of an experiment as values assumed by a single random variable. There will be situations, however, where we may find it desirable to record the simultaneous outcomes of several random variables. For example, we might measure the amount of precipitate P and volume V of gas released from a controlled chemical experiment giving rise to a two-dimensional sample space consisting of the outcomes (p, v), or one might be interested in the hardness H and tensile strength T of cold-drawn copper resulting in the outcomes (h, t). In a study to determine the likelihood of success in college, based on high school data, one might use a three-dimensional sample space and record for each individual his or her aptitude test score, high school rank in class, and grade-point average at the end of the freshman year in college.

If X are Y are two discrete random variables, the probability distribution for their simultaneous occurrence can be represented by a function with values $f(x, y)$ for any pair of values (x, y) within the range of the random variables X and Y. It is customary to refer to this function as the **joint probability distribution** of X and Y. Hence, in the discrete case,

$$f(x, y) = P(X = x, Y = y);$$

that is, the values $f(x, y)$ give the probability that outcomes x and y occur at the same time. For example, if a television set is to be serviced and X represents the age to the nearest year of the set and Y represents the number of defective tubes in the set, then $f(5, 3)$ is the probability that the television set is 5 years old and needs 3 new tubes.

Definition 2.8   *The function $f(x, y)$ is a* **joint probability distribution** *of the discrete random variables X and Y if*

1. $f(x, y) \geq 0$ *for all* (x, y).

2. $\sum_x \sum_y f(x, y) = 1$.

3. $P(X = x, Y = y) = f(x, y)$.

For any region A in the xy plane, $P[(X, Y) \in A] = \sum_A \sum f(x, y)$.

Example 2.8   Two refills for a ballpoint pen are selected at random from a box that contains 3 blue refills, 2 red refills, and 3 green refills. If X is the number of blue refills and Y is the number of red refills selected, find (a) the joint probability function $f(x, y)$, and (b) $P[(X, Y) \in A]$, where A is the region $\{(x, y) \mid x + y \leq 1\}$.

Solution

(a) The possible pairs of values (x, y) are (0, 0), (0, 1), (1, 0), (1, 1), (0, 2), and (2, 0). Now, $f(0, 1)$, for example, represents the probability that a red and a green refill are selected. The total number of equally likely ways of selecting any 2 refills from the 8 is $\binom{8}{2} = 28$. The number of ways of selecting 1 red from 3 red refills and 1 green from 3 green refills is $\binom{2}{1}\binom{3}{1} = 6$. Hence $f(0, 1) = \frac{6}{28} = \frac{3}{14}$. Similar calculations yield the probabilities for the other cases, which are presented in Table 2.6. Note that the probabilities sum

Table 2.6 Joint Probability
Distribution for Example 2.8

$f(x, y)$	x 0	1	2	Row Totals
0	$\frac{3}{28}$	$\frac{9}{28}$	$\frac{3}{28}$	$\frac{15}{28}$
y 1	$\frac{3}{14}$	$\frac{3}{14}$		$\frac{3}{7}$
2	$\frac{1}{28}$			$\frac{1}{28}$
Column Totals	$\frac{5}{14}$	$\frac{15}{28}$	$\frac{3}{28}$	1

to 1. In Chapter 3 it will become clear that the joint probability distribution of Table 2.6 can be represented by the formula

$$f(x, y) = \frac{\binom{3}{x}\binom{2}{y}\binom{3}{2 - x - y}}{\binom{8}{2}},$$

for $x = 0, 1, 2$; $y = 0, 1, 2$; $0 \le x + y \le 2$.
(b) $P[(X, Y) \in A] = P(X + Y \le 1)$
$\qquad\qquad\qquad = f(0, 0) + f(0, 1) + f(1, 0)$
$\qquad\qquad\qquad = \frac{3}{28} + \frac{3}{14} + \frac{9}{28}$
$\qquad\qquad\qquad = \frac{9}{14}.$

When X and Y are continuous random variables, the **joint density function**
$f(x, y)$ is a surface lying above the xy plane and $P[(X, Y) \in A]$, where A is any
region in the xy plane, is equal to the volume of the right cylinder bounded by
the base A and the surface.

Definition 2.9   *The function $f(x, y)$ is a* **joint density function** *of the continuous random variables X
and Y if*

1. $f(x, y) \ge 0$ *for all (x, y).*

2. $\displaystyle\int_{-\infty}^{\infty} \int_{-\infty}^{\infty} f(x, y)\, dx\, dy = 1.$

3. $\displaystyle P[(X, Y) \in A] = \iint_A f(x, y)\, dx\, dy$

for any region A in the xy plane.

Example 2.9 A candy company distributes boxes of chocolates with a mixture of creams, toffees, and nuts coated in both light and dark chocolate. For a randomly selected box, let X and Y, respectively, be the proportions of the light and dark chocolates that are creams and suppose that the joint density function is given by

$$f(x, y) = \begin{cases} \frac{2}{5}(2x + 3y), & 0 \le x \le 1, 0 \le y \le 1 \\ 0, & \text{elsewhere.} \end{cases}$$

(a) Verify condition 2 of Definition 2.9.
(b) Find $P[(X, Y) \in A]$, where A is the region $\{(x, y) \mid 0 < x < \frac{1}{2}, \frac{1}{4} < y < \frac{1}{2}\}$.

Solution

(a) $\displaystyle \int_{-\infty}^{\infty} \int_{-\infty}^{\infty} f(x, y) \, dx \, dy = \int_{0}^{1} \int_{0}^{1} \frac{2}{5}(2x + 3y) \, dx \, dy$

$$= \int_{0}^{1} \frac{2x^2}{5} + \frac{6xy}{5} \Big|_{x=0}^{x=1} dy$$

$$= \int_{0}^{1} \left(\frac{2}{5} + \frac{6y}{5} \right) dy = \frac{2y}{5} + \frac{3y^2}{5} \Big|_{0}^{1}$$

$$= \frac{2}{5} + \frac{3}{5} = 1.$$

(b) $P[(X, Y) \in A] = P(0 < X < \frac{1}{2}, \frac{1}{4} < Y < \frac{1}{2})$

$$= \int_{1/4}^{1/2} \int_{0}^{1/2} \frac{2}{5}(2x + 3y) \, dx \, dy$$

$$= \int_{1/4}^{1/2} \frac{2x^2}{5} + \frac{6xy}{5} \Big|_{x=0}^{x=1/2} dy$$

$$= \int_{1/4}^{1/2} \left(\frac{1}{10} + \frac{3y}{5} \right) dy = \frac{y}{10} + \frac{3y^2}{10} \Big|_{1/4}^{1/2}$$

$$= \frac{1}{10} \left[\left(\frac{1}{5} + \frac{3}{4} \right) - \left(\frac{1}{4} + \frac{3}{16} \right) \right] = \frac{13}{160}.$$

Given the joint probability distribution $f(x, y)$ of the discrete random variables X and Y, the probability distribution $g(x)$ of X alone is obtained by summing $f(x, y)$ over the values of Y. Similarly, the probability distribution $h(y)$ of Y alone is obtained by summing $f(x, y)$ over the values of X. We define $g(x)$ and $h(y)$ to be the **marginal distributions** of X and Y, respectively. When X and Y are continuous random variables, summations are replaced by integrals. We can now make the following general definition.

Definition 2.10   *The* **marginal distributions** *of X alone and of Y alone are given by*

$$g(x) = \sum_y f(x, y) \qquad and \qquad h(y) = \sum_x f(x, y)$$

for the discrete case and by

$$g(x) = \int_{-\infty}^{\infty} f(x, y)\, dy \qquad and \qquad h(y) = \int_{-\infty}^{\infty} f(x, y)\, dx$$

for the continuous case.

The term *marginal* is used here because, in the discrete case, the values of $g(x)$ and $h(y)$ are just the marginal totals of the respective columns and rows when the values of $f(x, y)$ are displayed in a rectangular table.

Example 2.10   Show that the column and row totals of Table 2.6 give the marginal distribution of X alone and of Y alone.

Solution. For the random variable X, we see that

$$P(X = 0) = g(0) = \sum_{y=0}^{2} f(0, y) = f(0, 0) + f(0, 1) + f(0, 2)$$

$$= \tfrac{3}{28} + \tfrac{3}{14} + \tfrac{1}{28} = \tfrac{5}{14},$$

$$P(X = 1) = g(1) = \sum_{y=0}^{2} f(1, y)$$

$$= f(1, 0) + f(1, 1) + f(1, 2)$$

$$= \tfrac{9}{28} + \tfrac{3}{14} + 0 = \tfrac{15}{28},$$

and

$$P(X = 2) = g(x) = \sum_{y=0}^{2} f(2, y) = f(2, 0) + f(2, 1) + f(2, 2)$$

$$= \tfrac{3}{28} + 0 + 0 = \tfrac{3}{28},$$

which are just the column totals of Table 2.6. In a similar manner we could show that the values of $h(y)$ are given by the row totals. In tabular form, these marginal distributions may be written as follows.

x	0	1	2
$g(x)$	$\frac{5}{14}$	$\frac{15}{28}$	$\frac{3}{28}$

y	0	1	2
$h(y)$	$\frac{15}{28}$	$\frac{3}{7}$	$\frac{1}{28}$

Example 2.11   Find $g(x)$ and $h(y)$ for the joint density function of Example 2.9.

Solution. By definition,

$$g(x) = \int_{-\infty}^{\infty} f(x, y) \, dy = \int_{0}^{1} \frac{2}{5} (2x + 3y) \, dy$$

$$= \frac{4xy}{5} + \frac{6y^2}{10} \Big|_{y=0}^{y=1} = \frac{4x + 3}{5}$$

for $0 \le x \le 1$, and $g(x) = 0$ elsewhere. Similarly,

$$h(y) = \int_{-\infty}^{\infty} f(x, y) \, dx = \int_{0}^{1} \frac{2}{5} (2x + 3y) \, dx$$

$$= \frac{2(1 + 3y)}{5}$$

for $0 \le y \le 1$, and $h(y) = 0$ elsewhere.

The fact that the marginal distributions $g(x)$ and $h(y)$ are indeed the probability distributions of the individual variables X and Y alone can easily be verified by showing that the conditions of Definition 2.4 or Definition 2.6 are satisfied. For example, in the continuous case

$$\int_{-\infty}^{\infty} g(x) \, dx = \int_{-\infty}^{\infty} \int_{-\infty}^{\infty} f(x, y) \, dy \, dx = 1$$

and

$$P(a < X < b) = P(a < X < b, -\infty < Y < \infty)$$

$$= \int_{a}^{b} \int_{-\infty}^{\infty} f(x, y) \, dy \, dx$$

$$= \int_{a}^{b} g(x) \, dx.$$

In Section 2.1 we stated that the value x of the random variable X represents an event that is a subset of the sample space. If we use the definition of conditional probability as given in Chapter 1,

$$P(B \mid A) = \frac{P(A \cap B)}{P(A)}, \qquad P(A) > 0,$$

where A and B are now the events defined by $X = x$ and $Y = y$, respectively, then

$$P(Y = y \mid X = x) = \frac{P(X = x, Y = y)}{P(X = x)}$$

$$= \frac{f(x, y)}{g(x)}, \qquad g(x) > 0,$$

when X and Y are discrete random variables.

It is not difficult to show that the function $f(x, y)/g(x)$, which is strictly a function of y with x fixed, satisfies all the conditions of a probability distribution. This is also true when $f(z, y)$ and $g(x)$ are the joint density and marginal distribution of continuous random variables. Expressing such a probability distribution by the symbol $f(y \mid x)$, we have the following definition.

Definition 2.11   *Let X and Y be two random variables, discrete or continuous. The* **conditional distribution** *of the random variable Y, given that $X = x$, is given by*

$$f(y \mid x) = \frac{f(x, y)}{g(x)}, \qquad g(x) > 0.$$

Similarly, the **conditional distribution** *of the random variable X, given that $Y = y$, is given by*

$$f(x \mid y) = \frac{f(x, y)}{h(y)}, \qquad h(y) > 0.$$

If one wished to find the probability that the discrete random variable X falls between a and b when it is known that the discrete variable $Y = y$, we evaluate

$$P(a < X < b \mid Y = y) = \sum_{x} f(x \mid y),$$

where the summation extends over all values of X between a and b. When X and Y are continuous, we evaluate

$$P(a < X < b \mid Y = y) = \int_{a}^{b} f(x \mid y) \, dx.$$

Example 2.12   Referring to Example 2.8, find the conditional distribution of X, given that $Y = 1$, and use it to determine $P(X = 0 \mid Y = 1)$.

Solution. We need to find $f(x \mid y)$, where $y = 1$. First we find that

$$h(1) = \sum_{x=0}^{2} f(x, 1) = \tfrac{3}{14} + \tfrac{3}{14} + 0 = \tfrac{3}{7}.$$

Now

$$f(x \mid 1) = \frac{f(x, 1)}{h(1)} = \frac{7}{3} f(x, 1), \qquad x = 0, 1, 2.$$

Therefore,

$$f(0 \mid 1) = \tfrac{7}{3} f(0, 1) = (\tfrac{7}{3})(\tfrac{3}{14}) = \tfrac{1}{2}$$
$$f(1 \mid 1) = \tfrac{7}{3} f(1, 1) = (\tfrac{7}{3})(\tfrac{3}{14}) = \tfrac{1}{2}$$
$$f(2 \mid 1) = \tfrac{7}{3} f(2, 1) = (\tfrac{7}{3})(0) = 0$$

and the conditional distribution of X, given that $Y = 1$, is

x	0	1	2
$f(x\,|\,1)$	$\frac{1}{2}$	$\frac{1}{2}$	0

Finally,

$$P(X = 0\,|\,Y = 1) = f(0\,|\,1) = \tfrac{1}{2}.$$

Therefore, if it is known that 1 of the 2 pen refills selected is red, we have a probability equal to 1/2 that the other refill is not blue.

Example 2.13 Suppose that the fraction X of male runners and the fraction Y of female runners who complete marathon races can be described by the joint density function

$$f(x, y) = \begin{cases} 8xy, & 0 \le x \le 1,\, 0 \le y \le x \\ 0, & \text{elsewhere.} \end{cases}$$

Find $g(x)$, $h(y)$, $f(y\,|\,x)$, and determine the probability that fewer than 1/8 of the women entered in a particular marathon actually finished if it is known that exactly 1/2 of the male runners completed the race.

Solution. By definition,

$$g(x) = \int_{-\infty}^{\infty} f(x, y)\, dy = \int_{0}^{x} 8xy\, dy$$

$$= 4xy^2 \Big|_{y=0}^{y=x} = 4x^3, \qquad 0 < x < 1,$$

and

$$h(y) = \int_{-\infty}^{\infty} f(x, y)\, dx = \int_{y}^{1} 8xy\, dx$$

$$= 4x^2 y \Big|_{x=y}^{x=1} = 4y(1 - y^2), \qquad 0 < y < 1.$$

Now,

$$f(y\,|\,x) = \frac{f(x, y)}{g(x)} = \frac{8xy}{4x^3} = \frac{2y}{x^2}, \qquad 0 < y < x,$$

and

$$P\!\left(Y < \frac{1}{8}\,\bigg|\, X = \frac{1}{2}\right) = \int_{0}^{1/8} 8y\, dy = \frac{1}{16}.$$

Example 2.14   Given the joint density function

$$f(x, y) = \begin{cases} \dfrac{x(1 + 3y^2)}{4}, & 0 < x < 2, 0 < y < 1 \\ 0, & \text{elsewhere,} \end{cases}$$

find $g(x)$, $h(y)$, $f(x\,|\,y)$, and evaluate $P(\tfrac{1}{4} < X < \tfrac{1}{2}\,|\,Y = \tfrac{1}{3})$.

Solution. By definition,

$$g(x) = \int_{-\infty}^{\infty} f(x, y)\, dy = \int_{0}^{1} \frac{x(1 + 3y^2)}{4}\, dy$$

$$= \frac{xy}{4} + \frac{xy^3}{4}\Big|_{y=0}^{y=1} = \frac{x}{2}, \qquad 0 < x < 2,$$

and

$$h(y) = \int_{-\infty}^{\infty} f(x, y)\, dx = \int_{0}^{2} \frac{x(1 + 3y^2)}{4}\, dx$$

$$= \frac{x^2}{8} + \frac{3x^2 y^2}{8}\Big|_{x=0}^{x=2} = \frac{1 + 3y^2}{2}, \qquad 0 < y < 1.$$

Therefore,

$$f(x\,|\,y) = \frac{f(x, y)}{h(y)} = \frac{x(1 + 3y^2)/4}{(1 + 3y^2)/2} = \frac{x}{2}, \qquad 0 < x < 2,$$

and

$$P\left(\frac{1}{4} < X < \frac{1}{2}\,\middle|\, Y = \frac{1}{3}\right) = \int_{1/4}^{1/2} \frac{x}{2}\, dx = \frac{3}{64}.$$

If $f(x\,|\,y)$ does not depend on y, as was the case in Example 2.14, then $f(x\,|\,y) = g(x)$ and $f(x, y) = g(x)h(y)$. The proof follows by substituting

$$f(x, y) = f(x\,|\,y)h(y)$$

into the marginal distribution of X. That is,

$$g(x) = \int_{-\infty}^{\infty} f(x, y)\, dy = \int_{-\infty}^{\infty} f(x\,|\,y)h(y)\, dy.$$

If $f(x\,|\,y)$ does not depend on y, we may write

$$g(x) = f(x\,|\,y) \int_{-\infty}^{\infty} h(y)\, dy.$$

Now

$$\int_{-\infty}^{\infty} h(y)\, dy = 1,$$

since $h(y)$ is the probability density function of Y. Therefore,

$$g(x) = f(x \mid y)$$

and then

$$f(x, y) = g(x)h(y),$$

which leads to the following definition.

Definition 2.12   *Let X and Y be two random variables, discrete or continuous, with joint probability distribution $f(x, y)$ and marginal distributions $g(x)$ and $h(y)$, respectively. The random variables X and Y are said to be **statistically independent** if and only if*

$$f(x, y) = g(x)h(y)$$

for all (x, y) within their range.

The continuous random variables of Example 2.14 are statistically independent, since the product of the two marginal distributions gives the joint density function. This is obviously not the case, however, for the continuous variables of Example 2.13. Checking for statistical independence of discrete random variables requires a more thorough investigation, since it is possible to have the product of the marginal distributions equal to the joint probability distribution for some but not all combinations of (x, y). If you can find any point (x, y) for which $f(x, y)$ is defined such that $f(x, y) \neq g(x)h(y)$, the discrete variables X and Y are not statistically independent.

Example 2.15   Show that the random variables of Example 2.8 are not statistically independent.

Solution. Let us consider the point $(0, 1)$. From Table 2.6 we find the three probabilities $f(0, 1)$, $g(0)$, and $h(1)$ to be

$$f(0, 1) = \tfrac{3}{14}$$

$$g(0) = \sum_{y=0}^{2} f(0, y) = \tfrac{3}{28} + \tfrac{3}{14} + \tfrac{1}{28} = \tfrac{5}{14}$$

$$h(1) = \sum_{x=0}^{2} f(x, 1) = \tfrac{3}{14} + \tfrac{3}{14} + 0 = \tfrac{3}{7}.$$

Clearly,

$$f(0, 1) \neq g(0)h(1),$$

and therefore X and Y are not statistically independent.

All the preceding definitions concerning two random variables can be generalized to the case of n random variables. Let $f(x_1, x_2, \ldots, x_n)$ be the joint probability function of the random variables $X_1, X_2, \ldots, X_n$. The marginal distribution of X_1, for example, is given by

$$g(x_1) = \sum_{x_2} \cdots \sum_{x_n} f(x_1, x_2, \ldots, x_n)$$

for the discrete case and by

$$g(x_1) = \int_{-\infty}^{\infty} \cdots \int_{-\infty}^{\infty} f(x_1, x_2, \ldots, x_n) \, dx_2 \, dx_3 \cdots dx_n$$

for the continuous case. We can now obtain **joint marginal distributions** such as $\phi(x_1, x_2)$, where

$$\phi(x_1, x_2) = \begin{cases} \displaystyle\sum_{x_3} \cdots \sum_{x_n} f(x_1, x_2, \ldots, x_n) & \text{(discrete case)} \\[2ex] \displaystyle\int_{-\infty}^{\infty} \cdots \int_{-\infty}^{\infty} f(x_1, x_2, \ldots, x_n) \, dx_3 \, dx_4 \cdots dx_n & \text{(continuous case).} \end{cases}$$

One could consider numerous conditional distributions. For example, the **joint conditional distribution** of X_1, X_2, and X_3, given that $X_4 = x_4$, $X_5 = x_5$, ..., $X_n = x_n$, is written

$$f(x_1, x_2, x_3 \mid x_4, x_5, \ldots, x_n) = \frac{f(x_1, x_2, \ldots, x_n)}{g(x_4, x_5, \ldots, x_n)},$$

where $g(x_4, x_5, \ldots, x_n)$ is the joint marginal distribution of the random variables $X_4, X_5, \ldots, X_n$.

A generalization of Definition 2.12 leads to the following definition for the mutually statistical independence of the variables $X_1, X_2, \ldots, X_n$.

Definition 2.13   *Let $X_1, X_2, \ldots, X_n$ be n random variables, discrete or continuous, with joint probability distribution $f(x_1, x_2, \ldots, x_n)$ and marginal distributions $f_1(x_1), f_2(x_2)$, ..., $f_n(x_n)$, respectively. The random variables $X_1, X_2, \ldots, X_n$ are said to be mutually **statistically independent** if and only if*

$$f(x_1, x_2, \ldots, x_n) = f_1(x_1) f_2(x_2) \cdots f_n(x_n)$$

for all $(x_1, x_2, \ldots, x_n)$ within their range.

Example 2.16   Suppose that the shelf life, in years, of a certain perishable food product packaged in cardboard containers is a random variable whose probability density function is given by

$$f(x) = \begin{cases} e^{-x}, & x > 0 \\ 0, & \text{elsewhere.} \end{cases}$$

Let X_1, X_2, and X_3 represent the shelf lives for three of these containers selected independently and find $P(X_1 < 2, 1 < X_2 < 3, X_3 > 2)$.

Solution. Since the containers were selected independently, we can assume that the random variables X_1, X_2, and X_3 are statistically independent having the joint probability density

$$f(x_1, x_2, x_3) = f(x_1)f(x_2)f(x_3)$$
$$= e^{-x_1}e^{-x_2}e^{-x_3}$$
$$= e^{-x_1-x_2-x_3}$$

for $x_1 > 0$, $x_2 > 0$, $x_3 > 0$, and $f(x_1, x_2, x_3) = 0$ elsewhere. Hence

$$P(X_1 < 2, 1 < X_2 < 3, X_3 > 2) = \int_2^\infty \int_1^3 \int_0^2 e^{-x_1-x_2-x_3} \, dx_1 \, dx_2 \, dx_3$$
$$= (1 - e^{-2})(e^{-1} - e^{-3})e^{-2}$$
$$= 0.0376.$$

Exercises

1. Determine the value of c so that the following functions represent joint probability distributions of the random variables X and Y:
   (a) $f(x, y) = cxy$, for $x = 1, 2, 3$; $y = 1, 2, 3$.
   (b) $f(x, y) = c|x - y|$, for $x = -2, 0, 2$; $y = -2, 3$.

2. If the joint probability distribution of X and Y is given by

   $$f(x, y) = \frac{(x + y)}{30}, \quad \text{for } x = 0, 1, 2, 3; \; y = 0, 1, 2,$$

   find
   (a) $P(X \le 2, Y = 1)$;   (c) $P(X > Y)$;
   (b) $P(X > 2, Y \le 1)$;   (d) $P(X + Y = 4)$.

3. From a sack of fruit containing 3 oranges, 2 apples, and 3 bananas a random sample of 4 pieces of fruit is selected. If X is the number of oranges and Y is the number of apples in the sample, find
   (a) the joint probability distribution of X and Y;
   (b) $P[(X, Y) \in A]$, where A is the region given by $\{(x, y) | x + y \le 2\}$.

4. Consider an experiment that consists of 2 rolls of a balanced die. If X is the number of 4's and Y is the number of 5's obtained in the 2 rolls of the die, find
   (a) the joint probability distribution of X and Y;
   (b) $P[(X, Y) \in A]$, where A is the region given by $\{(x, y) | 2x + y < 3\}$.

5. Let X denote the number of heads and Y the number of heads minus the number of tails when 3 coins are tossed. Find the joint probability distribution of X and Y.

6. Three cards are drawn without replacement from the 12 face cards (jacks, queens, and kings) of an ordinary deck of 52 playing cards. Let X be the number of kings selected and Y the number of jacks. Find
   (a) the joint probability distribution of X and Y;
   (b) $P[(X, Y) \in A]$, where A is the region given by $\{(x, y) | x + y \ge 2\}$.

7. Two random variables have the joint density given by

   $$f(x, y) = \begin{cases} 4xy, & 0 < x < 1, 0 < y < 1 \\ 0, & \text{elsewhere.} \end{cases}$$

   Find
   (a) $P(0 \le X \le 3/4, 1/8 \le Y \le 1/2)$;
   (b) $P(Y > X)$.

8. Two random variables have the joint density given by

$$f(x, y) = \begin{cases} k(x^2 + y^2), & 0 < x < 2, 1 < y < 4 \\ 0, & \text{elsewhere.} \end{cases}$$

(a) Find k.
(b) Find $P(1 < X < 2, 2 < Y \le 3)$.
(c) Find $P(1 \le X \le 2)$.
(d) Find $P(X + Y > 4)$.

9. Let X and Y have the joint density function

$$f(x, y) = \begin{cases} \dfrac{1}{y}, & 0 < x < y, 0 < y < 1 \\ 0, & \text{elsewhere.} \end{cases}$$

Find $P(X + Y > 1/2)$.

10. Referring to Exercise 2, find
(a) the marginal distribution of X;
(b) the marginal distribution of Y.

11. A coin is tossed twice. Let Z denote the number of heads on the first toss and W the total number of heads on the 2 tosses. If the coin is unbalanced and a head has a 40% chance of occurring, find
(a) the joint probability distribution of W and Z;
(b) the marginal distribution of W;
(c) the marginal distribution of Z;
(d) the probability that at least 1 head occurs.

12. Referring to Exercise 3, find
(a) $f(y \mid 2)$ for all values of y;
(b) $P(Y = 0 \mid X = 2)$.

13. Suppose that X and Y have the following joint probability distribution:

$f(x, y)$	x 1	2	3
y 1	0	$\frac{1}{6}$	$\frac{1}{12}$
2	$\frac{1}{5}$	$\frac{1}{9}$	0
3	$\frac{2}{15}$	$\frac{1}{4}$	$\frac{1}{18}$

(a) Evaluate the marginal distribution of X.
(b) Evaluate the marginal distribution of Y.
(c) Find $P(Y = 3 \mid X = 2)$.

14. Suppose that X and Y have the following joint probability distribution:

$f(x, y)$	x 2	4
y 1	0.10	0.15
3	0.20	0.30
5	0.10	0.15

(a) Find the marginal distribution of X.
(b) Find the marginal distribution of Y.

15. A privately owned liquor store operates both a drive-up facility as well as a walk-in facility. On a randomly selected day, let X and Y, respectively, be the proportions of the time that the drive-up and walk-in facilities are in use and suppose that the joint density function of these random variables is given by

$$f(x, y) = \begin{cases} \frac{2}{3}(x + 2y), & 0 \le x \le 1, 0 \le y \le 1 \\ 0, & \text{elsewhere.} \end{cases}$$

(a) Find the marginal density of X.
(b) Find the marginal density of Y.
(c) Find the probability that the drive-in facility is busy less than one-half of the time.

16. A candy company distributes boxes of chocolates with a mixture of creams, toffees, and cordials. Suppose that the weight of each box is 1 kilogram, but the individual weights of the creams, toffees, and cordials vary from box to box. For a randomly selected box, let X and Y represent the weights of the creams and the toffees, respectively, and suppose that the joint density function of these variables is given by

$$f(x, y) = \begin{cases} 24xy, & 0 \le x \le 1, 0 \le y \le 1, x + y \le 1 \\ 0, & \text{elsewhere.} \end{cases}$$

(a) Find the probability that in a given box the cordials account for more than 1/2 of the weight.
(b) Find the marginal density for the weight of the creams.
(c) Find the probability that the weight of the toffees in a box is less than 1/8 of a kilogram if it is known that creams constitute 3/4 of the weight.

17. Given the joint density function

$$f(x, y) = \begin{cases} \dfrac{6 - x - y}{8}, & 0 < x < 2, \, 2 < y < 4 \\ 0, & \text{elsewhere,} \end{cases}$$

find $P(1 < Y < 3 \,|\, X = 2)$.

18. Let X and Y denote the lengths of life, in years, of two components in an electronic system. If the joint density function of these variables is given by

$$f(x, y) = \begin{cases} e^{-(x+y)}, & x > 0, \, y > 0 \\ 0, & \text{elsewhere,} \end{cases}$$

find $P(0 < X < 1 \,|\, Y = 2)$.

19. Determine whether the two random variables of Exercise 13 are dependent or independent.

20. Determine whether the two random variables of Exercise 14 are dependent or independent.

21. The amount of kerosene, in thousands of liters, in a tank at the beginning of any day is a random amount Y from which a random amount X is sold during that day. Suppose that the tank is not resupplied during the day so that $x \le y$, and assume that the joint density function of these variables is given by

$$f(x, y) = \begin{cases} 2, & 0 < x < y, \, 0 < y < 1 \\ 0, & \text{elsewhere.} \end{cases}$$

(a) Determine if X and Y are independent.
(b) Find $P(1/4 < X < 1/2 \,|\, Y = 3/4)$.

22. The joint density function of the random variables X and Y is given by

$$f(x, y) = \begin{cases} 6x, & 0 < x < 1, \, 0 < y < 1 - x \\ 0, & \text{elsewhere.} \end{cases}$$

(a) Show that X and Y are not independent.
(b) Find $P(X > 0.3 \,|\, Y = 0.5)$.

23. Let X, Y, and Z have the joint probability density function

$$f(x, y, z) = \begin{cases} kxy^2z, & 0 < x < 1, \, 0 < y < 1, \\ & 0 < z < 2 \\ 0. & \text{elsewhere.} \end{cases}$$

(a) Find k.
(b) Find $P(X < 1/4, \, Y > 1/2, \, 1 < Z < 2)$.

24. Determine whether the two random variables of Exercise 7 are dependent or independent.

25. Determine whether the two random variables of Exercise 8 are dependent or independent.

26. The joint probability density function of the random variables X, Y, and Z is given by

$$f(x, y, z) = \begin{cases} \dfrac{4xyz^2}{9}, & 0 < x < 1, \, 0 < y < 1, \\ & 0 < z < 3 \\ 0, & \text{elsewhere.} \end{cases}$$

Find
(a) the joint marginal density function of Y and Z;
(b) the marginal density of Y;
(c) $P(1/4 < X < 1/2, \, Y > 1/3, \, 1 < Z < 2)$;
(d) $P(0 < X < 1/2 \,|\, Y = 1/4, \, Z = 2)$.

3

Mathematical Expectation

3.1 Mean of a Random Variable

If two coins are tossed 16 times and X is the number of heads that occur per toss, then the values of X can be 0, 1, and 2. Suppose that the experiment yields no heads, one head, and two heads a total of 4, 7, and 5 times, respectively. The average number of heads per toss of the two coins is then

$$\frac{(0)(4) + (1)(7) + (2)(5)}{16} = 1.06.$$

This is an average value and is not necessarily a possible outcome for the experiment. For instance, a salesman's average monthly income is not likely to be equal to any of his monthly paychecks.

Let us now restructure our computation for the average number of heads so as to have the following equivalent form:

$$(0)(\tfrac{4}{16}) + (1)(\tfrac{7}{16}) + (2)(\tfrac{5}{16}) = 1.06.$$

The numbers 4/16, 7/16, and 5/16 are the fractions of the total tosses resulting in 0, 1, and 2 heads, respectively. These fractions are also the relative frequencies for the different values of X in our experiment. In effect, then, we can calculate the mean or average of a set of data by knowing the distinct values that occur and their relative frequencies, without any knowledge of the total number of observations in our set of data. Therefore, if 4/16 or 1/4 of the tosses result in no heads, 7/16 of the tosses result in one head, and 5/16 of the tosses result in two heads, the mean number of heads per toss would be 1.06 no matter whether the total number of tosses was 16, 1000, or even 10,000.

Let us now use this method of relative frequencies to calculate the average number of heads per toss of two coins that we might expect in the long run. We shall refer to this average value as the **mean of the random variable X** or the **mean of the probability distribution of X** and write it as μ_X or simply as μ when it is clear to which random variable we refer. It is also common among statisticians to refer to this mean as the **mathematical expectation** or the **expected value** of the random variable X and denote it as $E(X)$.

Assuming that fair coins were tossed, we find that the sample space for our experiment is given by

$$S = \{HH, HT, TH, TT\}.$$

Since the 4 sample points are all equally likely, it follows that

$$P(X = 0) = P(TT) = \tfrac{1}{4},$$
$$P(X = 1) = P(TH) + P(HT) = \tfrac{1}{2},$$

and

$$P(X = 2) = P(HH) = \tfrac{1}{4},$$

where a typical element, say TH, indicates that the first toss resulted in a tail followed by a head on the second toss. Now, these probabilities are just the relative frequencies for the given events in the long run. Therefore,

$$\mu = E(X) = (0)(\tfrac{1}{4}) + (1)(\tfrac{1}{2}) + (2)(\tfrac{1}{4})$$
$$= 1.$$

This means that a person who tosses 2 coins over and over again will, on the average, get 1 head per toss.

The method described above for calculating the expected number of heads per toss of 2 coins suggests that the mean or expected value of any discrete random variable may be obtained by multiplying each of the values $x_1, x_2, \ldots,$ x_n of the random variable X by its corresponding probability $f(x_1), f(x_2), \ldots,$ $f(x_n)$ and summing the products. This is true, however, only if the random variable is discrete. In the case of continuous random variables, the definition of an expected value is essentially the same with summations replaced by integrations.

Definition 3.1   *Let X be a random variable with probability distribution $f(x)$. The* **mean** *or* **expected value** *of X is*

$$\mu = E(X) = \sum_x x f(x)$$

if X is discrete, and

$$\mu = E(X) = \int_{-\infty}^{\infty} x f(x)\, dx$$

if X is continuous.

Example 3.1 Find the expected number of chemists on a committee of size 3 selected at random from 4 chemists and 3 biologists.

Solution. Let X represent the number of chemists on the committee. The probability distribution of X is given by

$$f(x) = \frac{\binom{4}{x}\binom{3}{3-x}}{\binom{7}{3}}, \qquad x = 0, 1, 2, 3.$$

A few simple calculations yield $f(0) = 1/35$, $f(1) = 12/35$, $f(2) = 18/35$, and $f(3) = 4/35$. Therefore,

$$\mu = E(X) = (0)(\tfrac{1}{35}) + (1)(\tfrac{12}{35}) + (2)(\tfrac{18}{35}) + (3)(\tfrac{4}{35})$$
$$= \tfrac{12}{7} = 1.7.$$

Thus, if a committee of size 3 is selected at random over and over again from 4 chemists and 3 biologists, it would contain, on the average, 1.7 chemists.

Example 3.2 In a gambling game a man is paid \$5 if he gets all heads or all tails when three coins are tossed and he pays out \$3 if either one or two heads show. What is his expected gain?

Solution. The sample space for the possible outcomes when three coins are tossed simultaneously, or equivalently if 1 coin is tossed three times, is

$$S = \{HHH, HHT, HTH, THH, HTT, THT, TTH, TTT\}.$$

One can argue that each of these possibilities is equally likely and occurs with probability equal to 1/8. An alternative approach would be to apply the multiplicative rule of probability for independent events to each element of S. For example,

$$P(HHT) = P(H)P(H)P(T)$$
$$= (\tfrac{1}{2})(\tfrac{1}{2})(\tfrac{1}{2}) = \tfrac{1}{8}.$$

The random variable of interest is Y, the amount the gambler can win; and the possible values of Y are \$5 if event $E_1 = \{HHH, TTT\}$ occurs and $-\$3$ if event $E_2 = \{HHT, HTH, THH, HTT, THT, TTH\}$ occurs. Since E_1 and E_2 occur with probabilities 1/4 and 3/4, respectively, it follows that

$$\mu = E(Y) = (5)(\tfrac{1}{4}) + (-3)(\tfrac{3}{4}) = -1.$$

In this game the gambler will, on the average, lose \$1 per toss of the three coins.

A game is considered "fair" if the gambler will, on the average, come out even. Therefore, an expected gain of zero defines a fair game.

Example 3.3   Let X be the random variable that denotes the life in hours of a certain type of tube. The probability density function is given by

$$f(x) = \begin{cases} \dfrac{20,000}{x^3}, & x > 100 \\ \\ 0, & \text{elsewhere.} \end{cases}$$

Find the expected life of this type of tube.

Solution. Using Definition 3.1, we have

$$\mu = E(X) = \int_{100}^{\infty} x\,\frac{20,000}{x^3}\,dx$$

$$= \int_{100}^{\infty} \frac{20,000}{x^2}\,dx$$

$$= 200.$$

Therefore, we can expect this type of tube to last, on the average, 200 hours.

Now let us consider a new random variable $g(X)$, which depends on X; that is, each value of $g(X)$ is determined by knowing the values of X. For instance, $g(X)$ might be X^2 or $3X - 1$, so that whenever X assumes the value 2, $g(X)$ assumes the value $g(2)$. In particular, if X is a discrete random variable with probability distribution $f(x)$, $x = -1, 0, 1, 2$, and $g(X) = X^2$, then

$$P[g(X) = 0] = P(X = 0) = f(0)$$
$$P[g(X) = 1] = P(X = -1) + P(X = 1) = f(-1) + f(1)$$
$$P[g(X) = 4] = P(X = 2) = f(2)$$

so that the probability distribution of $g(X)$ may be written

$g(x)$	0	1	4
$P[g(X) = g(x)]$	$f(0)$	$f(-1) + f(1)$	$f(2)$

By the definition of an expected value of a random variable, we obtain

$$\begin{aligned} \mu_{g(X)} &= E[g(X)] \\ &= 0f(0) + 1[f(-1) + f(1)] + 4f(2) \\ &= (-1)^2 f(-1) + (0)^2 f(0) + (1)^2 f(1) + (2)^2 f(2) \\ &= \sum_x g(x) f(x). \\ &= \sum_x g(x) f(x). \end{aligned}$$

This result is generalized in Theorem 3.1 for both discrete and continuous random variables.

Theorem 3.1	Let X be a random variable with probability distribution $f(x)$. The mean or expected value of the random variable $g(X)$ is

$$\mu_{g(X)} = E[g(X)] = \sum_x g(x)f(x)$$

if X is discrete, and

$$\mu_{g(X)} = E[g(X)] = \int_{-\infty}^{\infty} g(x)f(x)\,dx$$

if X is continuous.

Example 3.4 Suppose that the number of cars, X, that pass through a car wash between 4:00 P.M. and 5:00 P.M. on any sunny Friday has the following probability distribution:

x	4	5	6	7	8	9
$P(X = x)$	$\frac{1}{12}$	$\frac{1}{12}$	$\frac{1}{4}$	$\frac{1}{4}$	$\frac{1}{6}$	$\frac{1}{6}$

Let $g(X) = 2X - 1$ represent the amount of money in dollars, paid to the attendant by the manager. Find the attendant's expected earnings for this particular time period.

Solution. By Theorem 3.1, the attendant can expect to receive

$$E[g(X)] = E(2X - 1) = \sum_{x=4}^{9} (2x - 1)f(x)$$

$$= (7)(\tfrac{1}{12}) + (9)(\tfrac{1}{12}) + (11)(\tfrac{1}{4}) + (13)(\tfrac{1}{4})$$

$$+ (15)(\tfrac{1}{6}) + (17)(\tfrac{1}{6})$$

$$= \$12.67.$$

Example 3.5 Let X be a random variable with density function

$$f(x) = \begin{cases} \dfrac{x^2}{3}, & -1 < x < 2 \\ 0, & \text{elsewhere.} \end{cases}$$

Find the expected value of $g(X) = 4X + 3$.

Solution. By Theorem 3.1, we have

$$E(4X + 3) = \int_{-1}^{2} \frac{(4x + 3)x^2}{3}\,dx$$

$$= \frac{1}{3} \int_{-1}^{2} (4x^3 + 3x^2)\,dx$$

$$= 8.$$

We shall now extend our concept of mathematical expectation to the case of two random variables X and Y with joint probability distribution $f(x, y)$.

Definition 3.2   *Let X and Y be random variables with joint probability distribution $f(x, y)$. The mean or expected value of the random variable $g(X, Y)$ is*

$$\mu_{g(X, Y)} = E[g(X, Y)] = \sum_x \sum_y g(x, y) f(x, y)$$

if X and Y are discrete, and

$$\mu_{g(X, Y)} = E[g(X, Y)] = \int_{-\infty}^{\infty} \int_{-\infty}^{\infty} g(x, y) f(x, y) \, dx \, dy$$

if X and Y are continuous.

Generalization of Definition 3.2 for the calculation of mathematical expectations of functions of several random variables is straightforward.

Example 3.6   Let X and Y be the random variables with joint probability distribution given by Table 2.6 on page 62. Find the expected value of $g(X, Y) = XY$.

Solution. By Definition 3.2, we write

$$E(XY) = \sum_{x=0}^{2} \sum_{y=0}^{2} xy f(x, y)$$

$$= (0)(0) f(0, 0) + (0)(1) f(0, 1) + (0)(2) f(0, 2)$$
$$+ (1)(0) f(1, 0) + (1)(1) f(1, 1)$$
$$+ (2)(0) f(2, 0)$$
$$= f(1, 1) = \tfrac{3}{14}.$$

Example 3.7   Find $E(Y/X)$ for the density function

$$f(x, y) = \begin{cases} \dfrac{x(1 + 3y^2)}{4}, & 0 < x < 2,\, 0 < y < 1 \\ 0, & \text{elsewhere.} \end{cases}$$

Solution. We have

$$E\left(\frac{Y}{X}\right) = \int_0^1 \int_0^2 \frac{y(1 + 3y^2)}{4} \, dx \, dy$$

$$= \int_0^1 \frac{y + 3y^3}{2} \, dy$$

$$= \frac{5}{8}.$$

Note that if $g(X, Y) = X$ in Definition 3.2, we have

$$E(X) = \begin{cases} \sum_x \sum_y xf(x, y) = \sum_x xg(x) & \text{(discrete case)} \\ \int_{-\infty}^{\infty} \int_{-\infty}^{\infty} xf(x, y) \, dx \, dy = \int_{-\infty}^{\infty} xg(x) \, dx & \text{(continuous case),} \end{cases}$$

where $g(x)$ is the marginal distribution of X. Therefore, in calculating $E(X)$ over a two-dimensional space, one may use either the joint probability distribution of X and Y or the marginal distribution of X.

Similarly, we define

$$E(Y) = \begin{cases} \sum_x \sum_y yf(x, y) = \sum_y yh(y) & \text{(discrete case)} \\ \int_{-\infty}^{\infty} \int_{-\infty}^{\infty} yf(x, y) \, dx \, dy = \int_{-\infty}^{\infty} yh(y) \, dy & \text{(continuous case),} \end{cases}$$

where $h(y)$ is the marginal distribution of the random variable Y.

Exercises

1. A shipment of 7 television sets contains 2 defectives. A hotel makes a random purchase of 3 of the sets. If X is the number of defective sets purchased by the hotel, find the mean of X.

2. The probability distribution of the discrete random variable X is

$$f(x) = \binom{3}{x}\left(\frac{1}{4}\right)^x\left(\frac{3}{4}\right)^{3-x}, \qquad x = 0, 1, 2, 3.$$

Find the mean of X.

3. Find the mean of the random variable T representing the total of the three coins in Exercise 6 on page 52.

4. A coin is biased so that a head is three times as likely to occur as a tail. Find the expected number of tails when this coin is tossed twice.

5. The probability distribution of X, the number of imperfections per 10 meters of a synthetic fabric in continuous rolls of uniform width, was given in Exercise 17 on page 52 as

x	0	1	2	3	4
$f(x)$	0.41	0.37	0.16	0.05	0.01

Find the average number of imperfections per 10 meters of this fabric.

6. An attendant at a car wash is paid according to the number of cars that pass through. Suppose the probabilities are 1/12, 1/12, 1/4, 1/4, 1/6, and 1/6, respectively, that the attendant receives $7, $9, $11, $13, $15, or $17 between 4:00 P.M. and 5:00 P.M. on any sunny Friday. Find the attendant's expected earnings for this particular period.

7. By investing in a particular stock, a person can make a profit in one year of $4000 with probability 0.3 or take a loss of $1000 with probability 0.7. What is this person's expected gain?

8. Suppose that an antique jewelry dealer is interested in purchasing a gold necklace for which the probabilities are 0.22, 0.36, 0.28, and 0.14, respectively, that she will be able to sell it for a profit of $250, sell it for a profit of $100, break even, or sell it for a loss of $150. What is her expected profit?

9. In a gambling game a woman is paid $3 if she draws a jack or a queen and $5 if she draws a king or an ace from an ordinary deck of 52 playing cards. If she draws any other card, she loses. How much should she pay to play if the game is fair?

10. A bowl contains 5 tags that cannot be distinguished. Three of the tags are marked $2 and the remaining 2 are marked $4 each. A player draws 2 tags at random from the bowl without replacement, and he is paid an amount equal to the sum of the values on the two tags that he draws. If it cost $5.60 to play, is this a fair game?

11. A private pilot wishes to insure his airplane for $50,000. The insurance company estimates that a total loss may occur with probability 0.002, a 50% loss with probability 0.01, and a 25% loss with probability 0.1. Ignoring all other partial losses, what premium should the insurance company charge each year to realize an average profit of $500?

12. If a dealer's profit, in units of $1000, on a new automobile can be looked upon as a random variable X having the density function

$$f(x) = \begin{cases} 2(1 - x), & 0 < x < 1 \\ 0, & \text{elsewhere,} \end{cases}$$

find the average profit per automobile.

13. The density function of coded measurements of pitch diameter of threads of a fitting is given by

$$f(x) = \begin{cases} \dfrac{4}{\pi(1 + x^2)}, & 0 < x < 1 \\ 0, & \text{elsewhere.} \end{cases}$$

Find the expected value of X.

14. What proportion of the people can be expected to respond to a certain mail-order solicitation if the proportion X has the density function

$$f(x) = \begin{cases} \dfrac{2(x + 2)}{5}, & 0 < x < 1 \\ 0, & \text{elsewhere?} \end{cases}$$

15. The density function of the continuous random variable X, the total number of hours, in units of 100 hours, that a family runs a vacuum cleaner over a period of one year, was given in Exercise 22 on page 53 as

$$f(x) = \begin{cases} x, & 0 < x < 1 \\ 2 - x, & 1 \le x < 2 \\ 0, & \text{elsewhere.} \end{cases}$$

Find the average number of hours per year that families run their vacuum cleaners.

16. Let X represent the outcome when a balanced die is tossed. Find $\mu_{g(X)}$, where $g(X) = 3X^2 + 4$.

17. Let X be a random variable with the following probability distribution:

x	-3	6	9
$f(x)$	$\frac{1}{6}$	$\frac{1}{2}$	$\frac{1}{3}$

Find $\mu_{g(X)}$, where $g(X) = (2X + 1)^2$.

18. Find the expected value of the random variable $g(X) = X^2$, where X has the probability distribution of Exercise 2.

19. A large industrial firm purchases several new typewriters at the end of each year, the exact number depending on the frequency of repairs in the previous year. Suppose that the number of typewriters, X, that are purchased each year has the following probability distribution:

x	0	1	2	3
$f(x)$	$\frac{1}{10}$	$\frac{3}{10}$	$\frac{2}{5}$	$\frac{1}{5}$

If the cost of the desired model will remain fixed at $1200 throughout this year and a discount of $50X^2$ dollars is credited toward any purchase, how much can this firm expect to spend on new typewriters at the end of this year?

20. A continuous random variable X has the density function

$$f(x) = \begin{cases} e^{-x}, & x > 0 \\ 0, & \text{elsewhere.} \end{cases}$$

Find the expected value of $g(X) = e^{2X/3}$.

21. What is the dealer's average profit per automobile if the profit on each automobile is given by $g(X) = X^2$, where X is a random variable having the density function of Exercise 12?

22. The hospital period, in days, for patients following treatment for a certain type of kidney disorder is a random variable $Y = X + 4$, where X has the density function

$$f(x) = \begin{cases} \dfrac{32}{(x + 4)^3}, & x > 0 \\ 0, & \text{elswhere.} \end{cases}$$

Find the average number of days that a person is hospitalized following treatment for this disorder.

23. Suppose that X and Y have the following joint probability function:

$f(x, y)$	x	
	2	4
y 1	0.10	0.15
3	0.20	0.30
5	0.10	0.15

(a) Find the expected value of $g(X, Y) = XY^2$.
(b) Find μ_X and μ_Y.

24. Referring to the random variables whose joint probability distribution is given in Exercise 3 on page 71.
(a) find $E(X^2 Y - 2XY)$;
(b) find μ_X and μ_Y.

25. Referring to the random variables whose joint probability distribution is given in Exercise 6 on page 71, find the mean for the total number of jacks and kings when 3 cards are drawn without replacement from the 12 face cards of an ordinary deck of 52 playing cards.

26. Let X and Y be random variables with joint density function

$$f(x, y) = \begin{cases} 4xy, & 0 < x < 1, 0 < y < 1 \\ 0, & \text{elsewhere.} \end{cases}$$

Find the expected value of $Z = \sqrt{X^2 + Y^2}$.

27. Suppose that X and Y are independent variables with probability densities

$$g(x) = \begin{cases} 8/x^3, & x > 2 \\ 0, & \text{elsewhere} \end{cases}$$

and

$$h(y) = \begin{cases} 2y, & 0 < y < 1 \\ 0, & \text{elsewhere.} \end{cases}$$

Find the expected value of $Z = XY$.

28. Referring to the random variables whose joint density function is given in Exercise 15 on page 72.
(a) find μ_X and μ_Y;
(b) find $E[(X + Y)/2]$.

29. Referring to the random variables whose joint probability density function is given in Exercise 16 on page 72, find the expected weight for the sum of the creams and toffees if one purchased a box of these chocolates.

30. Referring to the random variables whose joint probability density function is given in Exercise 21 on page 73, find the average amount of kerosene left in the tank at the end of the day.

3.2 Variance and Covariance

The mean or expected value of a random variable X is of special importance in statistics because it describes where the probability distribution is centered. By itself, however, the mean does not give an adequate description of the shape of the distribution. We need to know how the observations spread out from the mean. In Figure 3.1 we have the histograms of two discrete probability distributions with the same mean $\mu = 2$ that differ considerably in the variability or dispersion of their observations about the mean.

The most important measure of variability of a random variable X is obtained by letting $g(X) = (X - \mu)^2$ in Theorem 3.1. Because of its importance in statistics, it is referred to as the **variance of the random variable X** or the **variance of the probability distribution of X** and is denoted by Var(X) or the symbol σ_X^2, or simply by σ^2 when it is clear to which random variable we refer.

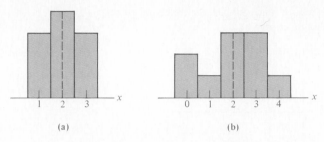

Figure 3.1 Distributions with equal means and different dispersions.

Definition 3.3 *Let X be a random variable with probability distribution $f(x)$ and mean μ. The* **variance** *of X is*

$$\sigma^2 = E[(X - \mu)^2] = \sum_x (x - \mu)^2 f(x)$$

if X is discrete, and

$$\sigma^2 = E[(X - \mu)^2] = \int_{-\infty}^{\infty} (x - \mu)^2 f(x) \, dx$$

if X is continuous. The positive square root of the variance, σ, is called the **standard deviation** *of X.*

The quantity $x - \mu$ in Definition 3.3 is called the **deviation of an observation from its mean**. Since these deviations are being squared and then averaged, σ^2 will be much smaller for a set of x values that are close to μ than it would be for a set of values that vary considerably from μ.

Example 3.8 Let the random variable X represent the number of automobiles that are used for official business purposes on any given workday. The probability distribution for Company A [Figure 3.1(a)] is given by

x	1	2	3
$f(x)$	0.3	0.4	0.3

and for Company B [Figure 3.1(b)] by

x	0	1	2	3	4
$f(x)$	0.2	0.1	0.3	0.3	0.1

Show that the variance of the probability distribution for Company B is greater than that of Company A.

Solution. For Company A, we find that

$$\mu = E(X) = (1)(0.3) + (2)(0.4) + (3)(0.3) = 2.0$$

and then

$$\sigma^2 = \sum_{x=1}^{3} (x-2)^2 f(x)$$

$$= (1-2)^2(0.3) + (2-2)^2(0.4) + (3-2)^2(0.3)$$

$$= 0.6.$$

For Company B, we have

$$\mu = E(X) = (0)(0.2) + (1)(0.1) + (2)(0.3) + (3)(0.3) + (4)(0.1)$$

$$= 2.0$$

and then

$$\sigma^2 = \sum_{x=0}^{4} (x-2)^2 f(x)$$

$$= (0-2)^2(0.2) + (1-2)^2(0.1) + (2-2)^2(0.3) + (3-2)^2(0.3) + (4-2)^2(0.1)$$

$$= 1.6.$$

Clearly, the variance of the number of automobiles that are used for official business purposes is greater for Company B than for Company A.

An alternative and preferred formula for finding σ^2, which often simplifies the calculations, is given in the following theorem.

Theorem 3.2 *The variance of a random variable X is given by*

$$\sigma^2 = E(X^2) - \mu^2.$$

Proof. For the discrete case we can write

$$\sigma^2 = \sum_x (x-\mu)^2 f(x) = \sum_x (x^2 - 2\mu x + \mu^2) f(x)$$

$$= \sum_x x^2 f(x) - 2\mu \sum_x x f(x) + \mu^2 \sum_x f(x).$$

Since $\mu = \sum_x x f(x)$ by definition, and $\sum_x f(x) = 1$ for any discrete probability distribution, it follows that

$$\sigma^2 = \sum_x x^2 f(x) - \mu^2$$

$$= E(X^2) - \mu^2.$$

For the continuous case the proof is step by step the same with summations replaced by integrations.

Example 3.9   The random variable X, representing the number of defective missiles when 3 missiles are fired, has the following probability distribution:

x	0	1	2	3
$f(x)$	0.51	0.38	0.10	0.01

Using Theorem 3.2, calculate σ^2.

Solution. First, we compute

$$\mu = (0)(0.51) + (1)(0.38) + (2)(0.10) + (3)(0.01)$$
$$= 0.61.$$

Now,

$$E(X^2) = (0)(0.51) + (1)(0.38) + (4)(0.10) + (9)(0.01)$$
$$= 0.87.$$

Therefore,

$$\sigma^2 = 0.87 - (0.61)^2 = 0.4979.$$

Example 3.10   The weekly demand for Pepsi, in thousands of liters, from a local chain of efficiency stores, is a continuous random variable X having the probability density

$$f(x) = \begin{cases} 2(x - 1), & 1 < x < 2 \\ 0, & \text{elsewhere.} \end{cases}$$

Find the mean and variance of X.

Solution

$$\mu = E(X) = 2 \int_1^2 x(x - 1)\, dx = \tfrac{5}{3}$$

and

$$E(X^2) = 2 \int_1^2 x^2(x - 1)\, dx = \tfrac{17}{6}.$$

Therefore,

$$\sigma^2 = \tfrac{17}{6} - (\tfrac{5}{3})^2 = \tfrac{1}{18}.$$

At this point the variance or standard deviation only has meaning when we compare two or more distributions that have the same units of measurement. Therefore, we could compare the variances of the distributions of contents, measured in liters, for two companies bottling orange juice, and the larger value would indicate the company whose product is more variable or less uniform. It would not be meaningful to compare the variance of a distribution of heights to the variance of a distribution of aptitude scores. In Section 3.4 we show how the standard deviation can be used to describe a single distribution of observations.

We shall now extend our concept of the variance of a random variable X to also include random variables related to X. For the random variable $g(X)$, the variance will be denoted by $\sigma_{g(X)}^2$ and is calculated by means of the following theorem.

Theorem 3.3 Let X be a random variable with probability distribution $f(x)$. The variance of the random variable $g(X)$ is

$$\sigma_{g(X)}^2 = E\{[g(X) - \mu_{g(X)}]^2\} = \sum_x [g(x) - \mu_{g(X)}]^2 f(x)$$

if X is discrete, and

$$\sigma_{g(X)}^2 = E\{[g(X) - \mu_{g(X)}]^2\} = \int_{-\infty}^{\infty} [g(x) - \mu_{g(X)}]^2 f(x)\, dx$$

if X is continuous.

Proof. Since $g(X)$ is itself a random variable with mean $\mu_{g(X)}$ as defined in Theorem 3.1, it follows from Definition 3.3 that

$$\sigma_{g(X)}^2 = E\{[g(X) - \mu_{g(X)}]^2\}.$$

Now, applying Theorem 3.1 again to the random variable $[g(X) - \mu_{g(X)}]^2$, the proof is complete.

Example 3.11 Calculate the variance of $g(X) = 2X + 3$, where X is a random variable with probability distribution

x	0	1	2	3
$f(x)$	$\frac{1}{4}$	$\frac{1}{8}$	$\frac{1}{2}$	$\frac{1}{8}$

Solution. First let us find the mean of the random variable $2X + 3$. According to Theorem 3.1,

$$\mu_{2X+3} = E(2X + 3) = \sum_{x=0}^{3} (2x + 3) f(x) = 6.$$

Now, using Theorem 3.3, we have

$$
\begin{aligned}
\sigma_{2X+3}^2 &= E\{[(2X + 3) - \mu_{2X+3}]^2\} \\
&= E\{[(2X + 3) - 6]^2\} \\
&= E(4X^2 - 12X + 9) \\
&= \sum_{x=0}^{3} (4x^2 - 12x + 9) f(x) \\
&= 4.
\end{aligned}
$$

Example 3.12   Let X be a random variable having the density function given in Example 3.5 on page 79. Find the variance of the random variable $g(X) = 4X + 3$.

Solution. In Example 3.5 we found $\mu_{4X+3} = 8$. Now, using Theorem 3.3,

$$\sigma^2_{4X+3} = E\{[(4X + 3) - 8]^2\}$$

$$= E[(4X - 5)^2]$$

$$= \int_{-1}^{2} (4x - 5)^2 \frac{x^2}{3}\, dx$$

$$= \frac{1}{3} \int_{-1}^{2} (16x^4 - 40x^3 + 25x^2)\, dx$$

$$= \frac{51}{5}.$$

If $g(X, Y) = (X - \mu_X)(Y - \mu_Y)$, where $\mu_X = E(X)$ and $\mu_Y = E(Y)$, Definition 3.2 yields an expected value called the **covariance** of X and Y, which we denote by σ_{XY} or cov(X, Y).

Definition 3.4   *Let X and Y be random variables with joint probability distribution $f(x, y)$. The* **covariance** *of X and Y is*

$$\sigma_{XY} = E[(X - \mu_X)(Y - \mu_Y)] = \sum_x \sum_y (x - \mu_X)(y - \mu_Y) f(x, y)$$

if X and Y are discrete, and

$$\sigma_{XY} = E[(X - \mu_X)(Y - \mu_Y)] = \int_{-\infty}^{\infty} \int_{-\infty}^{\infty} (x - \mu_X)(y - \mu_Y) f(x, y)\, dx\, dy$$

if X and Y are continuous.

The covariance will be positive when high values of X are associated with high values of Y and low values of X are associated with low values of Y. If low values of X are associated with high values of Y, and vice versa, then the covariance will be negative. When X and Y are statistically independent, it can be shown that the covariance is zero (see Theorem 3.10, Corollary 1). The converse, however, is not generally true. Two variables may have zero covariance and still not be statistically independent.

The alternative and preferred formula for σ_{XY} is given in the following theorem.

Theorem 3.4   *The covariance of two random variables X and Y with means μ_X and μ_Y, respectively, is given by*

$$\sigma_{XY} = E(XY) - \mu_X \mu_Y.$$

Proof. For the discrete case we can write

$$\sigma_{XY} = \sum_x \sum_y (x - \mu_X)(y - \mu_Y) f(x, y)$$

$$= \sum_x \sum_y (xy - \mu_X y - \mu_Y x + \mu_X \mu_Y) f(x, y)$$

$$= \sum_x \sum_y xy f(x, y) - \mu_X \sum_x \sum_y y f(x, y) - \mu_Y \sum_x \sum_y x f(x, y) + \mu_X \mu_Y \sum_x \sum_y f(x, y).$$

Since $\mu_X = \sum_x \sum_y x f(x, y)$ and $\mu_Y = \sum_x \sum_y y f(x, y)$ by definition, and in addition $\sum_x \sum_y f(x, y) = 1$ for any joint discrete probability distribution, it follows that

$$\sigma_{XY} = E(XY) - \mu_X \mu_Y - \mu_Y \mu_X + \mu_X \mu_Y$$
$$= E(XY) - \mu_X \mu_Y.$$

For the continuous case the proof is identical with summations replaced by integrals.

Example 3.13   The number of blue refills X and the number of red refills Y, when 2 refills for a ballpoint pen are selected at random from a certain box, was described in Example 2.8 on page 61 by the following joint probability distribution:

$f(x, y)$		0	1	2	$h(y)$
			x		
y	0	$\frac{3}{28}$	$\frac{9}{28}$	$\frac{3}{28}$	$\frac{15}{28}$
	1	$\frac{3}{14}$	$\frac{3}{14}$		$\frac{3}{7}$
	2	$\frac{1}{28}$			$\frac{1}{28}$
$g(x)$		$\frac{5}{14}$	$\frac{15}{28}$	$\frac{3}{28}$	

Find the covariance of X and Y.

Solution. From Example 3.6, we see that $E(XY) = 3/14$. Now,

$$\mu_X = E(X) = \sum_{x=0}^{2} \sum_{y=0}^{2} x f(x, y) = \sum_{x=0}^{2} x g(x)$$

$$= (0)(\tfrac{5}{14}) + (1)(\tfrac{15}{28}) + (2)(\tfrac{3}{28})$$

$$= \tfrac{3}{4}$$

and

$$\mu_Y = E(Y) = \sum_{x=0}^{2} \sum_{y=0}^{2} y f(x, y) = \sum_{y=0}^{2} y h(y)$$

$$= (0)(\tfrac{15}{28}) + (1)(\tfrac{3}{7}) + (2)(\tfrac{1}{28})$$

$$= \tfrac{1}{2}.$$

Therefore,

$$\sigma_{XY} = E(XY) - \mu_X \mu_Y$$
$$= \tfrac{3}{14} - (\tfrac{3}{4})(\tfrac{1}{2})$$
$$= -\tfrac{9}{56}.$$

Example 3.14   The fraction X of male runners and the fraction Y of female runners who complete marathon races was described in Example 2.13 on page 67 by the joint density function

$$f(x, y) = \begin{cases} 8xy, & 0 \le x \le 1, 0 \le y \le x \\ 0, & \text{elsewhere.} \end{cases}$$

Find the covariance of X and Y.

Solution. Using the marginal distributions

$$g(x) = \begin{cases} 4x^3, & 0 \le x \le 1 \\ 0, & \text{elsewhere} \end{cases}$$

and

$$h(y) = \begin{cases} 4y(1 - y^2), & 0 \le y \le 1 \\ 0, & \text{elsewhere} \end{cases}$$

from Example 2.13 and the joint density given above, we compute

$$\mu_X = E(X) = \int_0^1 \int_0^x 2x \, dx \, dy = \tfrac{1}{3}$$

$$\mu_Y = E(Y) = \int_0^1 \int_0^x 2y \, dx \, dy = \tfrac{2}{3}$$

and

$$E(XY) = \int_0^1 \int_0^x 2xy \, dx \, dy = \tfrac{1}{4}.$$

Then

$$\sigma_{XY} = E(XY) - \mu_X \mu_Y$$
$$= \tfrac{1}{4} - (\tfrac{1}{3})(\tfrac{2}{3})$$
$$= \tfrac{1}{36}.$$

Exercises

1. Use Definition 3.3 on page 84 to find the variance of the random variable X of Exercise 1 on page 81.

2. Let X be a random variable with the following probability distribution:

x	-2	3	5
$f(x)$	0.3	0.2	0.5

Find the standard deviation of X.

3. The random variable X, representing the number of chocolate chips in a cookie, has the following probability distribution:

x	2	3	4	5	6
$f(x)$	0.01	0.25	0.4	0.3	0.04

Using Theorem 3.2, find the variance of X.

4. Suppose that the probabilities are 0.4, 0.3, 0.2, and 0.1, respectively, that 0, 1, 2, or 3 power failures will hit a certain subdivision in any given year. Find the mean and variance of the random variable X representing the number of power failures hitting this subdivision.

5. The dealer's profit, in units of $1000, on a new automobile is a random variable X having the density function given in Exercise 12 on page 82. Find the variance of X.

6. The proportion of people who respond to a certain mail-order solicitation is a random variable X having the density function given in Exercise 14 on page 82. Find the variance of X.

7. The total number of hours, in units of 100 hours, that a family runs a vacuum cleaner over a period of one year is a random variable X having the density function given in Exercise 15 on page 82. Find the variance of X.

8. Referring to Exercise 16 on page 82, find $\sigma^2_{g(X)}$ for the function $g(X) = 3X^2 + 4$.

9. Find the standard deviation of the random variable $g(X) = (2X + 1)^2$ in Exercise 17 on page 82.

10. Using the results of Exercise 21 on page 82, find the variance of $g(X) = X^2$, where X is a random variable having the density function given in Exercise 12 on page 82.

11. The length of time, in minutes, for an airplane to wait for clearance to take off at a certain airport is a random variable $Y = 3X - 2$, where X has the density function

$$f(x) = \begin{cases} \frac{1}{4}e^{-x/4}, & x > 0 \\ 0, & \text{elsewhere.} \end{cases}$$

Find the mean and variance of the random variable Y.

12. Find the covariance of the random variables X and Y of Exercise 3 on page 71.

13. Find the covariance of the random variables X and Y of Exercise 13 on page 72.

14. Find the covariance of the random variables X and Y of Exercise 8 on page 72.

15. Referring to the random variables whose joint density function is given in Exercise 15 on page 72, and using your answers from Exercise 28 on page 83, find the covariance of X and Y.

16. Referring to the random variables whose joint density function is given in Exercise 16 on page 72, find the covariance between the weight of the creams and the weight of the toffees in these boxes of chocolates.

17. Find the covariance of the random variables X and Y having the joint probability density function

$$f(x, y) = \begin{cases} x + y, & 0 < x < 1, 0 < y < 1 \\ 0, & \text{elsewhere.} \end{cases}$$

18. Show that $\text{cov}(aX, bY) = ab\,\text{cov}(X, Y)$.

3.3 Properties of the Mean and Variance

We shall now develop some useful properties that will simplify the calculations of means and variances of random variables that appear in later chapters. These properties will permit us to calculate expectations in terms of other expectations that are either known or are easily computed. All the results that we present here are valid for both discrete and continuous random variables. Proofs are given only for the continuous case.

Theorem 3.5 *If a and b are constant, then*

$$E(aX + b) = aE(X) + b.$$

Proof. By the definition of an expected value,

$$E(aX + b) = \int_{-\infty}^{\infty} (ax + b)f(x)\, dx$$

$$= a \int_{-\infty}^{\infty} xf(x)\, dx + b \int_{-\infty}^{\infty} f(x)\, dx.$$

The first integral on the right is $E(X)$ and the second integral equals 1. Therefore, we have

$$E(aX + b) = aE(X) + b.$$

Corollary 1 *Setting a = 0, we see that E(b) = b.*

Corollary 2 *Setting b = 0, we see that E(aX) = aE(X).*

Example 3.15 Applying Theorem 3.5 to the discrete random variable $g(X) = 2X - 1$, rework Example 3.4 on page 79.

Solution. According to Theorem 3.5, we can write

$$E(2X - 1) = 2E(X) - 1.$$

Now,

$$\mu = E(X) = \sum_{x=4}^{9} xf(x)$$

$$= (4)(\tfrac{1}{12}) + (5)(\tfrac{1}{12}) + (6)(\tfrac{1}{4}) + (7)(\tfrac{1}{4}) + (8)(\tfrac{1}{6}) + (9)(\tfrac{1}{6})$$

$$= \tfrac{41}{6}.$$

Therefore,

$$\mu_{2X-1} = (2)(\tfrac{41}{6}) - 1 = \$12.67,$$

as before.

Example 3.16 Applying Theorem 3.5 to the continuous random variable $g(X) = 4X + 3$, rework Example 3.5 on page 79.

Solution. In Example 3.5 we may use Theorem 3.5 to write

$$E(4X + 3) = 4E(X) + 3.$$

Now

$$E(X) = \int_{-1}^{2} x\left(\frac{x^2}{3}\right) dx = \int_{-1}^{2} \frac{x^3}{3} \, dx = \frac{5}{4}.$$

Therefore,

$$E(4X + 3) = (4)\left(\frac{5}{4}\right) + 3 = 8,$$

as before.

Theorem 3.6 *The expected value of the sum or difference of two or more functions of a random variable X is the sum or difference of the expected values of the functions. That is,*

$$E[g(X) \pm h(Y)] = E[g(X)] \pm E[h(Y)].$$

Proof. By definition,

$$E[g(X) \pm h(X)] = \int_{-\infty}^{\infty} [g(x) \pm h(x)] f(x) \, dx$$

$$= \int_{-\infty}^{\infty} g(x) f(x) \, dx \pm \int_{-\infty}^{\infty} h(x) f(x) \, dx$$

$$= E[g(X)] \pm E[h(X)].$$

Example 3.17 Let X be a random variable with probability distribution as follows:

x	0	1	2	3
$f(x)$	$\frac{1}{3}$	$\frac{1}{2}$	0	$\frac{1}{6}$

Find the expected value of $Y = (X - 1)^2$.

Solution. Applying Theorem 3.6 to the function $Y = (X - 1)^2$, we can write

$$E[(X - 1)^2] = E(X^2 - 2X + 1) = E(X^2) - 2E(X) + E(1).$$

From Corollary 1 of Theorem 3.5, $E(1) = 1$, and by direct computation

$$E(X) = (0)(\tfrac{1}{3}) + (1)(\tfrac{1}{2}) + (2)(0) + (3)(\tfrac{1}{6}) = 1$$

and

$$E(X^2) = (0)(\tfrac{1}{3}) + (1)(\tfrac{1}{2}) + (4)(0) + (9)(\tfrac{1}{6}) = 2.$$

Hence

$$E\{(X - 1)^2\} = 2 - (2)(1) + 1 = 1.$$

Example 3.18   The weekly demand for ginger ale, in thousands of liters, from a local chain of efficiency stores is a continuous random variable $g(X) = X^2 + X - 2$, where X has the density function

$$f(x) = \begin{cases} 2(x - 1), & 1 < x < 2 \\ 0, & \text{elsewhere.} \end{cases}$$

Find the expected value of $g(X) = X^2 + X - 2$.

Solution.   By Theorem 3.6, we write

$$E(X^2 + X - 2) = E(X^2) + E(X) - E(2).$$

From Corollary 1 of Theorem 3.5, $E(2) = 2$, and by direct integration

$$E(X) = \int_1^2 2x(x - 1)\, dx = 2\int_1^2 (x^2 - x)\, dx = \tfrac{5}{3}$$

and

$$E(X^2) = \int_1^2 2x^2(x - 1)\, dx = 2\int_1^2 (x^3 - x^2)\, dx = \tfrac{17}{6}.$$

Now,

$$E(X^2 + X - 2) = \tfrac{17}{6} + \tfrac{5}{3} - 2 = \tfrac{5}{2},$$

so that the average weekly demand for ginger ale from this chain of efficiency stores is 2500 liters.

Suppose that we have two random variables X and Y with joint probability distribution $f(x, y)$. Two additional properties that will be very useful in succeeding chapters involve the expected values of the sum, difference, and product of these two random variables. First, however, let us prove a theorem on the expected value of the sum or difference of functions of the given variables. This, of course, is merely an extension of Theorem 3.6.

Theorem 3.7   *The expected value of the sum or difference of two or more functions of the random variables X and Y is the sum or difference of the expected values of the functions. That is,*

$$E[g(X,\, Y) \pm h(X,\, Y)] = E[g(X,\, Y)] \pm E[h(X,\, Y)].$$

Proof.   By Definition 3.2,

$$E[g(X,\, Y) \pm h(X,\, Y)] = \int_{-\infty}^{\infty} \int_{-\infty}^{\infty} [g(x, y) \pm h(x, y)]\, f(x, y)\, dx\, dy$$

$$= \int_{-\infty}^{\infty} \int_{-\infty}^{\infty} g(x, y)\, f(x, y)\, dx\, dy$$

$$\pm \int_{-\infty}^{\infty} \int_{-\infty}^{\infty} h(x, y)\, f(x, y)\, dx\, dy$$

$$= E[g(X,\, Y)] \pm E[h(X,\, Y)].$$

Corollary 1   *Setting $g(X, Y) = g(X)$ and $h(X, Y) = h(Y)$, we see that*

$$E[g(X) \pm h(X)] = E[g(X)] \pm E[h(X)].$$

Corollary 2   *Setting $g(X, Y) = X$ and $h(X, Y) = Y$, we see that*

$$E(X \pm Y) = E(X) \pm E(Y).$$

If X represents the daily production of some item from machine A and Y the daily production of the same kind of item from machine B, then $X + Y$ represents the total number of items produced daily from both machines. The second corollary of Theorem 3.7 states that the average daily production for both machines is equal to the sum of the average daily production of each machine.

Theorem 3.8   *Let X and Y be two independent random variables. Then*

$$E(XY) = E(X)E(Y).$$

Proof. By Definition 3.2,

$$E(XY) = \int_{-\infty}^{\infty} \int_{-\infty}^{\infty} xyf(x, y) \, dx \, dy.$$

Since X and Y are independent, we may write

$$f(x, y) = g(x)h(y),$$

where $g(x)$ and $h(y)$ are the marginal distributions of X and Y, respectively. Hence

$$E(XY) = \int_{-\infty}^{\infty} \int_{-\infty}^{\infty} xyg(x)h(y) \, dx \, dy$$

$$= \int_{-\infty}^{\infty} xg(x) \, dx \int_{-\infty}^{\infty} yh(y) \, dy$$

$$= E(X)E(Y).$$

Theorem 3.8 can be illustrated for discrete variables by tossing a green die and a red die. Let the random variable X represent the outcome on the green die and the random variable Y represent the outcome on the red die. Then XY represents the product of the numbers that occur on the pair of dice. In the long run, the average of the products of the numbers is equal to the product of the average number that occurs on the green die and the average number that occurs on the red die.

Example 3.19   Let X and Y be independent random variables with joint probability distribution

$$f(x, y) = \begin{cases} \dfrac{x(1 + 3y^2)}{4}, & 0 < x < 2, \, 0 < y < 1 \\[2mm] 0, & \text{elsewhere.} \end{cases}$$

Verify Theorem 3.8.

Solution.   Now

$$E(XY) = \int_0^1 \int_0^2 \frac{x^2 y(1 + 3y^2)}{4} \, dx \, dy$$

$$= \int_0^1 \frac{x^3 y(1 + 3y^2)}{12} \bigg|_{x=0}^{x=2} dy$$

$$= \int_0^1 \frac{2y(1 + 3y^2)}{3} \, dy$$

$$= \frac{5}{6}$$

$$E(X) = \int_0^1 \int_0^2 \frac{x^2(1 + 3y^2)}{4} \, dx \, dy$$

$$= \int_0^1 \frac{x^3(1 + 3y^2)}{12} \bigg|_{x=0}^{x=2} dy$$

$$= \int_0^1 \frac{2(1 + 3y^2)}{3} \, dy$$

$$= \frac{4}{3}$$

$$E(Y) = \int_0^1 \int_0^2 \frac{xy(1 + 3y^2)}{4} \, dx \, dy$$

$$= \int_0^1 \frac{x^2 y(1 + 3y^2)}{8} \bigg|_{x=0}^{x=2} dy$$

$$= \int_0^1 \frac{y(1 + 3y^2)}{2} \, dy$$

$$= \frac{5}{8}.$$

Hence

$$E(X)E(Y) = \left(\frac{4}{3}\right)\left(\frac{5}{8}\right) = \frac{5}{6} = E(XY).$$

We conclude this section by proving two theorems that are useful in calculating variances or standard deviations.

Theorem 3.9 *If a and b are constants, then*

$$\sigma^2_{aX+b} = a^2\sigma^2_X = a^2\sigma^2.$$

Proof. By definition

$$\sigma^2_{aX+b} = E\{[(aX+b) - \mu_{aX+b}]^2\}.$$

Now

$$\mu_{aX+b} = E(aX+b) = a\mu + b$$

by Theorem 3.5. Therefore,

$$\sigma^2_{aX+b} = E[(aX+b-a\mu-b)^2]$$
$$= a^2E[(X-\mu)^2]$$
$$= a^2\sigma^2.$$

Corollary 1 *Setting a = 1, we see that*

$$\sigma^2_{X+b} = \sigma^2_X = \sigma^2.$$

Corollary 2 *Setting b = 0, we see that*

$$\sigma^2_{aX} = a^2\sigma^2_X = a^2\sigma^2.$$

Corollary 1 states that the variance is unchanged if a constant is added to or subtracted from a random variable. The addition or subtraction of a constant simply shifts the values of X to the right or to the left but does not change their variability. However, if a random variable is multiplied or divided by a constant, then Corollary 2 states that the variance is multiplied or divided by the square of the constant.

Theorem 3.10 *If X and Y are random variables with joint probability distribution f(x, y), then*

$$\sigma^2_{aX+bY} = a^2\sigma^2_X + b^2\sigma^2_Y + 2ab\sigma_{XY}.$$

Proof. By definition

$$\sigma^2_{aX+bY} = E\{[(aX+bY) - \mu_{aX+bY}]^2\}.$$

Now

$$\mu_{aX+bY} = E(aX+bY) = aE(X) + bE(Y) = a\mu_X + b\mu_Y,$$

by using Corollary 2 of Theorem 3.7 followed by Corollary 2 of Theorem 3.5. Therefore,

$$\begin{aligned}
\sigma_{aX+bY}^2 &= E\{[(aX+bY)-(a\mu_X+b\mu_Y)]^2\} \\
&= E\{[a(X-\mu_X)+b(Y-\mu_Y)]^2\} \\
&= a^2E[(X-\mu_X)^2] + b^2E[(Y-\mu_Y)^2] + 2abE[(X-\mu_X)(Y-\mu_Y)] \\
&= a^2\sigma_X^2 + b^2\sigma_Y^2 + 2ab\sigma_{XY}.
\end{aligned}$$

Corollary 1   *If X and Y are independent random variables, then*

$$\sigma_{aX+bY}^2 = a^2\sigma_X^2 + b^2\sigma_Y^2.$$

The result given in Corollary 1 is obtained from Theorem 3.10 by proving the covariance of the independent variables X and Y to be zero. Hence, from Theorem 3.4,

$$\begin{aligned}
\sigma_{XY} &= E(XY) - \mu_X\mu_Y \\
&= 0,
\end{aligned}$$

since $E(XY) = E(X)E(Y)$ for independent variables.

Corollary 2   *If X and Y are independent random variables, then*

$$\sigma_{aX-bY}^2 = a^2\sigma_X^2 + b^2\sigma_Y^2.$$

Corollary 2 follows by replacing b by $-b$ in Corollary 1. Generalizing to a linear combination of n independent random variables, we write

---

Corollary 3   *If $X_1, X_2, \ldots, X_n$ are independent random variables, then*

$$\sigma_{a_1X_1+a_2X_2+\cdots+a_nX_n}^2 = a_1^2\sigma_{X_1}^2 + a_2^2\sigma_{X_2}^2 + \cdots + a_n^2\sigma_{X_n}^2.$$

Example 3.20   If X and Y are random variables with variances $\sigma_X^2 = 2$, $\sigma_Y^2 = 4$, and covariance $\sigma_{XY} = -2$, find the variance of the random variable $Z = 3X - 4Y + 8$.

Solution

$$\begin{aligned}
\sigma_Z^2 &= \sigma_{3X-4Y+8}^2 \\
&= \sigma_{3X-4Y}^2 &&\text{(by Theorem 3.9, Corollary 1)} \\
&= 9\sigma_X^2 + 16\sigma_Y^2 - 24\sigma_{XY} &&\text{(by Theorem 3.10)} \\
&= (9)(2) + (16)(4) - (24)(-2) \\
&= 130.
\end{aligned}$$

Example 3.21   If X and Y are independent random variables with variances $\sigma_X^2 = 1$ and $\sigma_Y^2 = 2$, find the variance of the random variable $Z = 3X - 2Y + 5$.

Solution

$$\sigma_Z^2 = \sigma_{3X-2Y+5}^2$$
$$= \sigma_{3X-2Y}^2 \qquad \text{(by Theorem 3.9, Corollary 1)}$$
$$= 9\sigma_X^2 + 4\sigma_Y^2 \qquad \text{(by Theorem 3.10, Corollary 2)}$$
$$= (9)(1) + (4)(2)$$
$$= 17.$$

3.4 Chebyshev's Theorem

In Section 3.2 we stated that the variance of a random variable tells us something about the variability of the observations about the mean. If a random variable has a small variance or standard deviation, we would expect most of the values to be grouped around the mean. Therefore, the probability that a random variable assumes a value within a certain interval about the mean is greater than for a similar random variable with a larger standard deviation. If we think of probability in terms of area, we would expect a continuous distribution with a small standard deviation to have most of its area close to μ, as in Figure 3.2(a). However, a large value of σ indicates a greater variability, and therefore we should expect the area to be more spread out, as in Figure 3.2(b).

We can argue the same way for a discrete distribution. The area in the probability histogram in Figure 3.3(b) is spread out much more than that of Figure 3.3(a) indicating a more variable distribution of measurements or outcomes.

The Russian mathematician P. L. Chebyshev (1821–1894) discovered that the fraction of the area between any two values symmetric about the mean is related to the standard deviation. Since the area under a probability distribution curve or in a probability histogram adds to 1, the area between any two numbers is the probability of the random variable assuming a value between these numbers.

The following theorem, due to Chebyshev, gives a conservative estimate of the probability that a random variable assumes a value within k standard

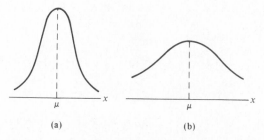

(a)   (b)

Figure 3.2 Variability of continuous observations about the mean.

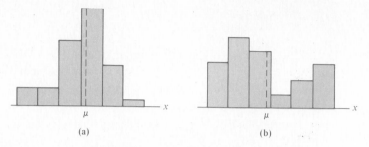

Figure 3.3   Variability of discrete observations about the mean.

deviations of its mean for any real number k. We shall give the proof only for the continuous case, leaving the discrete case as an exercise.

Theorem 3.11   **(Chebyshev's Theorem)**   *The probability that any random variable X will assume a value within k standard deviation of the mean is **at least** $1 - 1/k^2$. That is,*

$$P(\mu - k\sigma < X < \mu + k\sigma) \geq 1 - \frac{1}{k^2}.$$

Proof. By our previous definition of the variance of X we can write

$$\sigma^2 = E[(X - \mu)^2]$$

$$= \int_{-\infty}^{\infty} (x - \mu)^2 f(x)\, dx$$

$$= \int_{-\infty}^{\mu - k\sigma} (x - \mu)^2 f(x)\, dx + \int_{\mu - k\sigma}^{\mu + k\sigma} (x - \mu)^2 f(x)\, dx + \int_{\mu + k\sigma}^{\infty} (x - \mu)^2 f(x)\, dx$$

$$\geq \int_{-\infty}^{\mu - k\sigma} (x - \mu)^2 f(x)\, dx + \int_{\mu + k\sigma}^{\infty} (x - \mu)^2 f(x)\, dx,$$

since the second of the three integrals is nonnegative. Now, since $|x - \mu| \geq k\sigma$ wherever $x \geq \mu + k\sigma$ or $x \leq \mu - k\sigma$, we have $(x - \mu)^2 \geq k^2\sigma^2$ in both remaining integrals. It follows that

$$\sigma^2 \geq \int_{-\infty}^{\mu - k\sigma} k^2\sigma^2 f(x)\, dx + \int_{\mu + k\sigma}^{\infty} k^2\sigma^2 f(x)\, dx$$

and that

$$\int_{-\infty}^{\mu - k\sigma} f(x)\, dx + \int_{\mu + k\sigma}^{\infty} f(x)\, dx \leq \frac{1}{k^2}.$$

Hence

$$P(\mu - k\sigma < X < \mu + k\sigma) = \int_{\mu - k\sigma}^{\mu + k\sigma} f(x)\, dx \geq 1 - \frac{1}{k^2}$$

and the theorem is established.

For $k = 2$ the theorem states that the random variable X has a probability of at least $1 - 1/2^2 = 3/4$ of falling within two standard deviations of the mean. That is, three-fourths or more of the observations of any distribution lie in the interval $\mu \pm 2\sigma$. Similarly, the theorem says that at least eight-ninths of the observations of any distribution fall in the interval $\mu \pm 3\sigma$.

Example 3.22 A random variable X has a mean $\mu = 8$, a variance $\sigma^2 = 9$, and an unknown probability distribution. Find (a) $P(-4 < X < 20)$, and (b) $P(|X - 8| \geq 6)$.

Solution

(a)
$$P(-4 < X < 20) = P[8 - (4)(3) < X < 8 + (4)(3)]$$
$$\geq \tfrac{15}{16}.$$

(b)
$$P(|X - 8| \geq 6) = 1 - P(|X - 8| < 6)$$
$$= 1 - P(-6 < X - 8 < 6)$$
$$= 1 - P[8 - (2)(3) < X < 8 + (2)(3)]$$
$$\leq \tfrac{1}{4}.$$

Chebyshev's theorem holds for any distribution of observations and, for this reason, the results are usually weak. The value given by the theorem is a lower bound only. That is, we know that the probability of a random variable falling within two standard deviations of the mean can be *no less* than 3/4, but we never know how much more it might actually be. Only when the probability distribution is known can we determine exact probabilities.

Exercises

1. Referring to Exercise 3 on page 91, find the mean and variance of the discrete random variable $g(X) = 3X - 2$, where X represents the number of chocolate chips in a cookie.

2. Using Theorems 3.5 and 3.9, find the mean and variance of the random variable $g(X) = 5X + 3$, where X has the probability distribution of Exercise 4 on page 91.

3. Suppose that a grocery store purchases 5 cartons of skim milk at the wholesale price of $1.20 per carton and retails the milk at $1.65 per carton. After the expiration date, the unsold milk is removed from the shelf and the grocer receives a credit from the distributor equal to three-fourths of the wholesale price. If the probability distribution of the random variable X, the number of cartons that are sold from this lot, is given by

x	0	1	2	3	4	5
$f(x)$	$\tfrac{1}{15}$	$\tfrac{2}{15}$	$\tfrac{2}{15}$	$\tfrac{3}{15}$	$\tfrac{4}{15}$	$\tfrac{3}{15}$

find the expected profit.

4. Repeat Exercise 11 on page 91 by applying Theorems 3.5 and 3.9.

5. Let X be a random variable with the following probability distribution:

x	-3	6	9
$f(x)$	$\tfrac{1}{6}$	$\tfrac{1}{2}$	$\tfrac{1}{3}$

Find $E(X)$ and $E(X^2)$ and then, using these values, evaluate $E[(2X + 1)^2]$.

6. The total time, measured in units of 100 hours, that a teenager runs her stereo set over a period of one year is a continuous random variable X that has the density function

$$f(x) = \begin{cases} x, & 0 < x < 1 \\ 2 - x, & 1 \le x < 2 \\ 0, & \text{elsewhere.} \end{cases}$$

Use Theorem 3.6 to evaluate the mean of the random variable $Y = 60X^2 + 39X$, where Y is equal to the number of kilowatt hours expended annually.

7. If a random variable X is defined such that $E[(X-1)^2] = 10$, $E[(X-2)^2] = 6$, find μ and σ^2.

8. Suppose that X and Y are independent random variables having the joint probability distribution

		x
$f(x, y)$	2	4
y 1	0.10	0.15
3	0.20	0.30
5	0.10	0.15

Find
(a) $E(2X - 3Y)$;
(b) $E(XY)$.

9. Use Theorem 3.7 to evaluate $E(2XY^2 - X^2Y)$ for the joint probability distribution given in Table 2.6 on page 62.

10. Let X represent the number that occurs when a red die is tossed and Y the number that occurs when a green die is tossed. Find
(a) $E(X + Y)$;
(b) $E(X - Y)$;
(c) $E(XY)$.

11. Repeat Exercise 27 on page 83 by applying Theorem 3.8.

12. If the joint density function of X and Y is given by

$$f(x, y) = \begin{cases} \frac{2}{7}(x + 2y), & 0 < x < 1, 1 < y < 2 \\ 0, & \text{elsewhere,} \end{cases}$$

find the expected value of $g(X, Y) = (X/Y^3) + X^2Y$.

13. Let X represent the number that occurs when a green die is tossed and Y the number that occurs when a red die is tossed. Find the variance of the random variable
(a) $2X - Y$;
(b) $X + 3Y - 5$.

14. If X and Y are independent random variables with variances $\sigma_X^2 = 5$ and $\sigma_Y^2 = 3$, find the variance of the random variable $Z = -2X + 4Y - 3$.

15. Repeat Exercise 14 if X and Y are not independent and $\sigma_{XY} = 1$.

16. A random variable X has a mean $\mu = 12$, a variance $\sigma^2 = 9$, and an unknown probability distribution. Using Chebyshev's theorem, find
(a) $P(6 < X < 18)$;
(b) $P(3 < X < 21)$.

17. A random variable X has a mean $\mu = 10$ and a variance $\sigma^2 = 4$. Using Chebyshev's theorem, find
(a) $P(|X - 10| \ge 3)$;
(b) $P(|X - 10| < 3)$;
(c) $P(5 < X < 15)$;
(d) the value of the constant c such that

$$P(|X - 10| \ge c) \le 0.04.$$

18. Compute the $P(\mu - 2\sigma < X < \mu + 2\sigma)$, where X has the density function

$$f(x) = \begin{cases} 6x(1 - x), & 0 < x < 1 \\ 0, & \text{elsewhere} \end{cases}$$

and compare with the result given by Chebyshev's theorem.

19. Prove Chebyshev's theorem when X is a discrete random variable.

4

Some Discrete Probability Distributions

4.1 Introduction

No matter whether a discrete probability distribution is represented graphically by a histogram, in tabular form, or by means of a formula, the behavior of a random variable is described. Often, the observations generated by different statistical experiments have the same general type of behavior. Consequently, discrete random variables associated with these experiments can be described by essentially the same probability distribution and therefore can be represented by a single formula. In fact, one needs only a handful of important probability distributions to describe most discrete random variables encountered in practice.

4.2 Discrete Uniform Distribution

The simplest of all discrete probability distributions is one in which the random variable assumes each of its values with an equal probability. Such a probability distribution is called the **discrete uniform distribution**.

Discrete Uniform Distribution   *If the random variable X assumes the values* $x_1, x_2, \ldots, x_k$, *with equal probabilities, then the discrete uniform distribution is given by*

$$f(x; k) = \frac{1}{k}, \qquad x = x_1, x_2, \ldots, x_k.$$

We have used the notation $f(x; k)$ instead of $f(x)$ to indicate that the uniform distribution depends on the *parameter k*.

Example 4.1   When a light bulb is selected at random from a box that contains a 40-watt bulb, a 60-watt bulb, a 75-watt bulb, and a 100-watt bulb, each element of the sample space $S = \{40, 60, 75, 100\}$ occurs with probability 1/4. Therefore, we have a uniform distribution, with

$$f(x; 4) = \tfrac{1}{4}, \qquad x = 40, 60, 75, 100.$$

Example 4.2   When a die is tossed, each element of the sample space $S = \{1, 2, 3, 4, 5, 6\}$ occurs with probability 1/6. Therefore, we have a uniform distribution, with

$$f(x; 6) = \tfrac{1}{6}, \qquad x = 1, 2, 3, 4, 5, 6.$$

The graphic representation of the uniform distribution by means of a histogram always turns out to be a set of rectangles with equal heights. The histogram for Example 4.2 is shown in Figure 4.1.

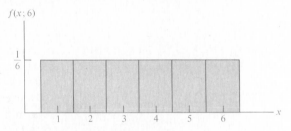

Figure 4.1   Histogram for the tossing of a die.

Theorem 4.1   *The mean and variance of the discrete uniform distribution* $f(x; k)$ *are*

$$\mu = \frac{\sum_{i=1}^{k} x_i}{k} \qquad and \qquad \sigma^2 = \frac{\sum_{i=1}^{k} (x_i - \mu)^2}{k}.$$

Proof. By definition

$$\mu = E(X) = \sum_{i=1}^{k} x_i f(x_i; k) = \sum_{i=1}^{k} \frac{x_i}{k}$$

$$= \frac{\sum_{i=1}^{k} x_i}{k}.$$

Also, by definition,

$$\sigma^2 = E[(X - \mu)^2] = \sum_{i=1}^{k} (x_i - \mu)^2 f(x_i; k)$$

$$= \sum_{i=1}^{k} \frac{(x_i - \mu)^2}{k} = \frac{\sum_{i=1}^{k} (x_i - \mu)^2}{k}.$$

Example 4.3   Referring to Example 4.2, we find that

$$\mu = \frac{1 + 2 + 3 + 4 + 5 + 6}{6} = 3.5$$

and

$$\sigma^2 = \frac{(1 - 3.5)^2 + (2 - 3.5)^2 + \cdots + (6 - 3.5)^2}{6} = \frac{35}{12}.$$

4.3 Binomial and Multinomial Distributions

An experiment often consists of repeated trials, each with two possible outcomes that may be labeled **success** and **failure**. This is true in testing items as they come off an assembly line, where each test or trial may indicate a defective or a nondefective item. We may choose to define either outcome as a success. It is also true if cards are drawn in succession from an ordinary deck and each trial is labeled a success or a failure, depending on whether the card is a heart or not a heart. If each card is replaced and the deck shuffled before the next drawing, the two experiments just described have similar properties, in that the repeated trials are independent and the probability of a success remains constant from trial to trial. Experiments of this type are known as **binomial experiments**. Observe in the card-drawing example that the probabilities of a success for the repeated trials change if the cards are not replaced. This is, the probability of selecting a heart on the first draw is 1/4, but on the second draw it is a conditional probability having a value of 13/51 or 12/51, depending on whether or not a heart occurred on the first draw. This, then, would no longer be considered a binomial experiment.

A binomial experiment is one that possesses the following properties:

1. The experiment consists of *n* repeated trials.
2. Each trial results in an outcome that may be classified as a success or a failure.

3. The probability of success, denoted by p, remains constant from trial to trial.

4. The repeated trials are independent.

Consider the binomial experiment where three items are selected at random from a manufacturing process, inspected, and classified defective or nondefective. A defective item is designated a success. The number of successes is a random variable X assuming integral values from zero through 3. The eight possible outcomes and the corresponding values of X are

Outcome	x
NNN	0
NDN	1
NND	1
DNN	1
NDD	2
DND	2
DDN	2
DDD	3

Since the items are selected independently from a process that we shall assume produces 25% defectives,

$$P(NDN) = P(N)P(D)P(N) = (\tfrac{3}{4})(\tfrac{1}{4})(\tfrac{3}{4}) = \tfrac{9}{64}.$$

Similar calculations yield the probabilities for the other possible outcomes. The probability distribution of X is therefore given by

x	0	1	2	3
$f(x)$	$\frac{27}{64}$	$\frac{27}{64}$	$\frac{9}{64}$	$\frac{1}{64}$

The number X of successes in n trials of a binomial experiment is called a **binomial random variable**. The probability distribution of this discrete random variable is called the **binomial distribution** and its values will be denoted by $b(x; n, p)$, since they depend on the number of trials and the probability of a success on a given trial. Thus, for the probability distribution of X, the number of defectives,

$$P(X = 2) = f(2) = b(2; 3, \tfrac{1}{4}) = \tfrac{9}{64}.$$

Let us now generalize the above illustration to yield a formula for $b(x; n, p)$. That is, we wish to find a formula that gives the probability of x successes in n trials for a binomial experiment. First, consider the probability of x successes and $n - x$ failures in a specified order. Since the trials are independent, we can multiply all the probabilities corresponding to the different outcomes. Each success occurs with probability p and each failure with probability $q = 1 - p$. Therefore, the probability for the specified order is $p^x q^{n-x}$. We must now determine the total number of sample points in the experiment that have x successes and $n - x$ failures. This number is equal to the number of partitions of n outcomes into two groups with x in one group and $n - x$ in the other and is given by $\binom{n}{x}$. Because these partitions are mutually exclusive, we add the probabilities of all the different partitions to obtain the general formula, or simply multiply $p^x q^{n-x}$ by $\binom{n}{x}$.

Binomial Distribution   *If a binomial trial can result in a success with probability p and a failure with probability $q = 1 - p$, then the probability distribution of the binomial random variable X, the number of successes in n independent trials, is*

$$b(x; n, p) = \binom{n}{x} p^x q^{n-x}, \qquad x = 0, 1, 2, \ldots, n.$$

Note that when $n = 3$ and $p = 1/4$, the probability distribution of X, the number of defectives, may be written as

$$b\left(x; 3, \frac{1}{4}\right) = \binom{3}{x}\left(\frac{1}{4}\right)^x\left(\frac{3}{4}\right)^{3-x}, \qquad x = 0, 1, 2, 3,$$

rather than in the tabular form above.

Example 4.4   The probability that a certain kind of component will survive a given shock test is 3/4. Find the probability that exactly 2 of the next 4 components tested survive.

Solution. Assuming that the tests are independent and $p = 3/4$ for each of the 4 tests, we obtain

$$b\left(2; 4, \frac{3}{4}\right) = \binom{4}{2}\left(\frac{3}{4}\right)^2\left(\frac{1}{4}\right)^2$$

$$= \frac{4!}{2!\,2!} \cdot \frac{3^2}{4^4}$$

$$= \frac{27}{128}.$$

The binomial distribution derives its name from the fact that the $n + 1$ terms in the binomial expansion of $(q + p)^n$ correspond to the various values of

$b(x; n, p)$ for $x = 0, 1, 2, \ldots, n$. That is,

$$(q + p)^n = \binom{n}{0}q^n + \binom{n}{1}pq^{n-1} + \binom{n}{2}p^2q^{n-2} + \cdots + \binom{n}{n}p^n$$

$$= b(0; n, p) + b(1; n, p) + b(2; n, p) + \cdots + b(n; n, p).$$

Since $p + q = 1$, we see that $\sum_{x=0}^{n} b(x; n, p) = 1$, a condition that must hold for any probability distribution.

Frequently, we are interested in problems where it is necessary to find $P(X < r)$ or $P(a \leq X \leq b)$. Fortunately, binomial sums $B(r; n, p) = \sum_{x=0}^{r} b(x; n, p)$ are available and are given in Table A.1 of the Appendix for $n = 1, 2, \ldots, 20$, and selected values of p from 0.1 to 0.9. We illustrate the use of Table A.1 with the following example.

Example 4.5 The probability that a patient recovers from a rare blood disease is 0.4. If 15 people are known to have contracted this disease, what is the probability that (a) at least 10 survive, (b) from 3 to 8 survive, and (c) exactly 5 survive?

Solution. Let X be the number of people that survive.

(a) $P(X \geq 10) = 1 - P(X < 10)$

$$= 1 - \sum_{x=0}^{9} b(x; 15, 0.4)$$

$$= 1 - 0.9662$$

$$= 0.0338.$$

(b) $P(3 \leq X \leq 8) = \sum_{x=3}^{8} b(x; 15, 0.4)$

$$= \sum_{x=0}^{8} b(x; 15, 0.4) - \sum_{x=0}^{2} b(x; 15, 0.4)$$

$$= 0.9050 - 0.0271$$

$$= 0.8779.$$

(c) $P(X = 5) = b(5; 15, 0.4)$

$$= \sum_{x=0}^{5} b(x; 15, 0.4) - \sum_{x=0}^{4} b(x; 15, 0.4)$$

$$= 0.4032 - 0.2173$$

$$= 0.1859.$$

Since the probability distribution of any binomial random variable depends only on the values assumed by the parameters n, p, and q, it would seem reasonable to assume that the mean and variance of a binomial random variable also depend on the values assumed by these parameters. Indeed this is true, and in Theorem 4.2 we derive general formulas as functions of n, p, and q that can be used to compute the mean and variance of any binomial random variable.

Theorem 4.2   *The mean and variance of the binomial distribution $b(x; n, p)$ are*

$$\mu = np \quad and \quad \sigma^2 = npq.$$

Proof. Let the outcome on the jth trial be represented by the random variable I_j, which assumes the values 0 and 1 with probabilities q and p, respectively. This is called a **Bernoulli variable** or perhaps more appropriately, an **indicator variable**, since $I_j = 0$ indicates a failure and $I_j = 1$ indicates a success.

Therefore, in a binomial experiment the number of successes can be written as the sum of the n independent indicator variables. Hence

$$X = I_1 + I_2 + \cdots + I_n.$$

The mean of any I_j is $E(I_j) = 0 \cdot q + 1 \cdot p = p$. Therefore, using Corollary 2 of Theorem 3.7 on page 95, the mean of the binomial distribution is

$$\mu = E(X) = E(I_1) + E(I_2) + \cdots + E(I_n)$$
$$= \underbrace{p + p + \cdots + p}_{n \text{ terms}}$$
$$= np.$$

The variance of any I_j is given by

$$\sigma^2_{I_j} = E[(I_j - p)^2] = E(I_j^2) - p^2$$
$$= (0)^2 q + (1)^2 p - p^2$$
$$= p(1 - p) = pq.$$

By extending Corollary 2 of Theorem 3.10 on page 98 to the case of n independent variables, the variance of the binomial distribution is

$$\sigma^2_X = \sigma^2_{I_1} + \sigma^2_{I_2} + \cdots + \sigma^2_{I_n}$$
$$= \underbrace{pq + pq + \cdots + pq}_{n \text{ terms}}$$
$$= npq.$$

Example 4.6   Find the mean and variance of the binomial random variable of Exercise 4.5, and then use Chebyshev's theorem to interpret the interval $\mu \pm 2\sigma$.

Solution. Since Example 4.5 was a binomial experiment with $n = 15$ and $p = 0.4$, by Theorem 4.2, we have

$$\mu = (15)(0.4) = 6 \qquad \text{and} \qquad \sigma^2 = (15)(0.4)(0.6) = 3.6.$$

Taking the square root of 3.6, we find that $\sigma = 1.897$. Hence the required interval is $6 \pm (2)(1.897)$, or from 2.206 to 9.794. Chebyshev's theorem states that the number of recoveries among 15 patients subjected to the given disease has a probability of at least 3/4 of falling between 2.206 and 9.794, or, because the data is discrete, between 3 and 9 inclusive.

The binomial experiment becomes a **multinomial experiment** if we let each trial have more than 2 possible outcomes. Hence the classification of a manufactured product as being light, heavy, or acceptable and the recording of accidents at a certain intersection according to the day of the week constitute multinomial experiments. The drawing of a card from a deck *with replacement* is also a multinomial experiment if the 4 suits are the outcomes of interest.

In general, if a given trial can result in any one of k possible outcomes E_1, E_2, ..., E_k with probabilities p_1, p_2, ..., p_k, then the **multinomial distribution** will give the probability that E_1 occurs x_1 times; E_2 occurs x_2 times; ...; and E_k occurs x_k times in n independent trials, where

$$x_1 + x_2 + \cdots + x_k = n.$$

We shall denote this joint probability distribution by $f(x_1, x_2, \ldots, x_k; p_1, p_2, \ldots, p_k, n)$. Clearly, $p_1 + p_2 + \cdots + p_k = 1$, since the result of each trial must be one of the k possible outcomes.

To derive the general formula, we proceed as in the binomial case. Since the trials are independent, any specified order yielding x_1 outcomes for E_1, x_2 for E_2, ..., x_k for E_k will occur with probability $p_1^{x_1} p_2^{x_2} \cdots p_k^{x_k}$. The total number of orders yielding similar outcomes for the n trials is equal to the number of partitions of n items into k groups with x_1 in the first group; x_2 in the second group; ...; and x_k in the kth group. This can be done in

$$\binom{n}{x_1, x_2, \ldots, x_k} = \frac{n!}{x_1! x_2! \cdots x_k!}$$

ways. Since all the partitions are mutually exclusive and occur with equal probability, we obtain the multinomial distribution by multiplying the probability for a specified order by the total number of partitions.

Multinomial Distribution   *If a given trial can result in the k outcomes E_1, E_2, ..., E_k with probabilities p_1, p_2, ..., p_k, then the probability distribution of the random variables X_1, X_2, ..., X_k, representing the number of occurrences for E_1, E_2, ..., E_k in n independent trials is*

$$f(x_1, x_2, \ldots, x_k; p_1, p_2, \ldots, p_k, n) = \binom{n}{x_1, x_2, \ldots, x_k} p_1^{x_1} p_2^{x_2} \cdots p_k^{x_k}$$

with

$$\sum_{i=1}^{k} x_i = n \qquad and \qquad \sum_{i=1}^{k} p_i = 1.$$

The multinomial distribution derives its name from the fact that the terms of the multinomial expansion of $(p_1 + p_2 + \cdots + p_k)^n$ correspond to all the possible values of $f(x_1, x_2, \ldots, x_k; p_1, p_2, \ldots, p_k, n)$.

Example 4.7   If a pair of dice is tossed 6 times, what is the probability of obtaining a total of 7 or 11 twice, a matching pair once, and any other combination 3 times?

Solution. We list the following possible events,

E_1:   a total of 7 or 11 occurs,

E_2:   a matching pair occurs,

E_3:   neither a pair nor a total of 7 or 11 occurs.

The corresponding probabilities for a given trial are $p_1 = 2/9$, $p_2 = 1/6$, and $p_3 = 11/18$. These values remain constant for all 6 trials. Using the multinomial distribution with $x_1 = 2$, $x_2 = 1$, and $x_3 = 3$, we find that the required probability is

$$f\left(2, 1, 3; \frac{2}{9}, \frac{1}{6}, \frac{11}{18}, 6\right) = \binom{6}{2, 1, 3}\left(\frac{2}{9}\right)^2\left(\frac{1}{6}\right)^1\left(\frac{11}{18}\right)^3$$

$$= \frac{6!}{2!\,1!\,3!} \cdot \frac{2^2}{9^2} \cdot \frac{1}{6} \cdot \frac{11^3}{18^3}$$

$$= 0.1127.$$

Exercises

1. An employee is selected from a staff of 10 to supervise a certain project by selecting a tag at random from a box containing 10 tags numbered from 1 to 10. Find the formula for the probability distribution of X representing the number on the tag that is drawn. What is the probability that the number drawn is less than 4?

2. A roulette wheel is divided into 25 sectors of equal area numbered from 1 to 25. Find a formula for the probability distribution of X, the number that occurs when the wheel is spun.

3. Find the mean and variance of the random variable X of Exercise 1.

4. In a certain city district the need for money to buy drugs is given as the reason for 75% of all thefts. Find the probability that among the next 5 theft cases reported in this district
   (a) exactly 2 resulted from the need for money to buy drugs;
   (b) at most 3 resulted from the need for money to buy drugs.

5. A fruit grower claims that 2/3 of his peach crop has been contaminated by the medfly infestation. Find the probability that among 4 peaches inspected by this grower
   (a) all 4 have been contaminated by the medfly;
   (b) anywhere from 1 to 3 have been contaminated.

6. According to a survey by the Administrative Management Society, 1/3 of U.S. companies give employees four weeks of vacation after they have been with the company for 15 years. Find the probability that among 6 companies surveyed at random, the number that give employees 4 weeks of vacation after 15 years of employment is
   (a) anywhere from 2 to 5;
   (b) fewer than 3.

7. If we define the random variable X to be the number of heads that occur when a balanced coin is flipped once, find the probability distribution of X. Is this probability distribution the discrete uniform distribution, the binomial distribution, or both?

8. According to a study published by a group of University of Massachusetts sociologists approximately 60% of the Valium users in the state of Massachusetts first took Valium for psychological problems. Find the probability that among the next 8 users interviewed from this state
   (a) exactly 3 began taking Valium for psychological problems;
   (b) at least 5 began taking Valium for problems that were not psychological.

9. In testing a certain kind of truck tire over a rugged terrain, it is found that 25% of the trucks fail to complete the test run without a blowout. Of the next 15 trucks tested, find the probability that
   (a) from 3 to 6 have blowouts;
   (b) fewer than 4 have blowouts;
   (c) more than 5 have blowouts.

10. A nationwide survey of seniors by the University of Michigan reveals that almost 70% disapprove of daily pot smoking according to a report in *Parade*, September 14, 1980. If 12 seniors are selected at random and asked their opinion, find the probability that the number who disapprove of smoking pot daily is
    (a) anywhere from 7 to 9;
    (b) at most 5;
    (c) not less than 8.

11. The probability that a patient recovers from a delicate heart operation is 0.9. What is the probability that exactly 5 of the next 7 patients having this operation survive?

12. A traffic control engineer reports that 75% of the vehicles passing through a check point are from within the state. What is the probability that fewer than 4 of the next 9 vehicles are from out of the state?

13. A study conducted at George Washington University and the National Institute of Health examined national attitudes about tranquilizers. The study revealed that approximately 70% believe "tranquilizers don't really cure anything, they just cover up the real trouble." According to this study, what is the probability that at least 3 of the next 5 people selected at random will be of this opinion?

14. A survey of the residents in a United States city showed that 20% preferred a white telephone over any other color available. What is the probability that more than half of the next 20 telephones installed in this city will be white?

15. It is known that 50% of mice inoculated with a serum are protected from a certain disease. If 5 mice are inoculated, find the probability that
    (a) none contracts the disease;
    (b) fewer than 2 contract the disease;
    (c) more than 3 contract the disease.

16. Suppose that airplane engines operate independently and fail with probability equal to 0.4. Assuming that a plane makes a safe flight if at least one-half of its engines run, determine whether a 4-engine plane or a 2-engine plane has the higher probability for a successful flight.

17. Repeat Exercise 16 when the probability of engine failure is 0.2.

18. Find the mean and variance of the binomial random variable in Exercise 11.

19. Find the mean and variance of the binomial random variable in Exercise 14.

20. (a) Find the mean and variance of the binomial random variable in Exercise 10.
    (b) According to Chebyshev's theorem, what fraction of the time will the number of seniors who disapprove of daily pot smoking, among groups of 12, be anywhere from 6 to 11?

21. If X represents the number of people in Exercise 13 who believe that tranquilizers do not cure but only cover up the real problem, find the mean and variance of X when 5 people are selected at random and then use Chebyshev's theorem to interpret the interval $\mu \pm 2\sigma$.

22. (a) In Exercise 9 how many of the 15 trucks would you expect to have blowouts?
    (b) According to Chebyshev's theorem, there is a probability of at least 3/4 that the number of trucks among the next 15 that have blowouts will fall in what interval?

23. (a) How many heads can we expect if a coin is tossed 64 times?
    (b) According to Chebyshev's theorem, what is the probability that the number of heads will fall between 20 and 44 if we repeat this experiment over again?

24. A card is drawn from a well-shuffled deck of 52 playing cards, the result recorded, and the card replaced. If the experiment is repeated 5 times, what is the probability of obtaining 2 spades and 1 heart?

25. The surface of a circular dart board has a small center circle called the bull's-eye and 20 pie-shaped regions numbered from 1 to 20. Each of the pie-shaped regions is further divided into three

parts such that a person throwing a dart that lands on a specified number scores the value of the number, double the number, or triple the number, depending on which of the three parts the dart falls. If a person hits the bull's-eye with probability 0.01, hits a double with probability 0.10, hits a triple with probability 0.05, and misses the dart board with probability 0.02, what is the probability that 7 throws will result in no bull's-eyes, no triples, a double twice, and a complete miss once?

26. According to the theory of genetics, a certain cross of guinea pigs will result in red, black, and white offspring in the ratio 8 : 4 : 4. Find the probability that among 8 offspring 5 will be red, 2 black, and 1 white.

27. The probabilities are 0.4, 0.2, 0.3, and 0.1, respectively, that a delegate to a certain convention arrived by air, bus, automobile, or train. What is the probability that among 9 delegates randomly selected at this convention, 3 arrived by air, 3 arrived by bus, 1 arrived by automobile, and 2 arrived by train?

4.4 Hypergeometric Distribution

If we wish to find the probability of observing 3 red cards in 5 draws from an ordinary deck of 52 playing cards, the binomial distribution of Section 4.3 does not apply unless each card is replaced and the deck reshuffled before the next drawing is made. To solve the problem of sampling without replacement, let us restate the problem. If 5 cards are drawn at random, we are interested in the probability of selecting 3 red cards from the 26 available and 2 black cards from the 26 black cards available in the deck. There are $\binom{26}{3}$ ways of selecting 3 red cards, and for each of these ways we can choose 2 black cards in $\binom{26}{2}$ ways. Therefore, the total number of ways to select 3 red and 2 black cards in 5 draws is the product $\binom{26}{3}\binom{26}{2}$. The total number of ways to select any 5 cards from the 52 that are available is $\binom{52}{5}$. Hence the probability of selecting

5 cards without replacement of which 3 are red and 2 are black is given by

$$\frac{\binom{26}{3}\binom{26}{2}}{\binom{52}{5}} = \frac{(26!/3!\,23!)(26!/2!\,24!)}{(52!/5!\,47!)} = 0.3251.$$

In general we are interested in the probability of selecting x successes from the k items labeled success and $n - x$ failures from the $N - k$ items labeled failures when a random sample of size n is selected from N items. This is known as a **hypergeometric experiment**.

A hypergeometric experiment is one that possesses the following two properties:

1. A random sample of size n is selected from N items.
2. k of the N items may be classified as successes and $N - k$ are classified as failures.

The number X of successes in a hypergeometric experiment is called a **hypergeometric random variable**. Accordingly, the probability distribution of the hypergeometric variable is called the **hypergeometric distribution** and its values will be denoted by $h(x; N, n, k)$, since they depend on the number of successes k in the set N from which we select n items.

Example 4.8   A committee of size 5 is to be selected at random from 3 chemists and 5 physicists. Find the probability distribution for the number of chemists on the committee.

Solution. Let the random variable X be the number of chemists on the committee. The two properties of a hypergeometric experiment are satisfied. Hence

$$P(X = 0) = h(0; 8, 5, 3) = \frac{\binom{3}{0}\binom{5}{5}}{\binom{8}{5}} = \frac{1}{56}$$

$$P(X = 1) = h(1; 8, 5, 3) = \frac{\binom{3}{1}\binom{5}{4}}{\binom{8}{5}} = \frac{15}{56}$$

$$P(X = 2) = h(2; 8, 5, 3) = \frac{\binom{3}{2}\binom{5}{3}}{\binom{8}{5}} = \frac{30}{56}$$

$$P(X = 3) = h(3; 8, 5, 3) = \frac{\binom{3}{3}\binom{5}{2}}{\binom{8}{5}} = \frac{10}{56}.$$

In tabular form the hypergeometric distribution of X is as follows:

x	0	1	2	3
$h(x; 8, 5, 3)$	$\frac{1}{56}$	$\frac{15}{56}$	$\frac{30}{56}$	$\frac{10}{56}$

It is not difficult to see that the probability distribution can be given by the formula

$$h(x; 8, 5, 3) = \frac{\binom{3}{x}\binom{5}{5-x}}{\binom{8}{5}}, \qquad x = 0, 1, 2, 3.$$

Let us now generalize Example 4.8 to find a formula for $h(x; N, n, k)$. The total number of samples of size n chosen from N items is $\binom{N}{n}$. These samples are assumed to be equally likely. There are $\binom{k}{x}$ ways of selecting x successes from the k that are available and for each of these ways we can choose the $n - x$ failures in $\binom{N-k}{n-x}$ ways. Thus the total number of favorable samples among the $\binom{N}{n}$ possible samples is given by $\binom{k}{x}\binom{N-k}{n-x}$. Hence we have the following definition.

Hypergeometric Distribution   *The probability distribution of the hypergeometric random variable X, the number of successes in a random sample of size n selected from N items of which k are labeled* **success** *and $N - k$ labeled* **failure**, *is*

$$h(x; N, n, k) = \frac{\binom{k}{x}\binom{N-k}{n-x}}{\binom{N}{n}}, \qquad x = 0, 1, 2, \ldots, n.$$

Example 4.9   Lots of 40 components each are called acceptable if they contain no more than 3 defectives. The procedure for sampling the lot is to select 5 components at random and to reject the lot if a defective is found. What is the probability that exactly 1 defective will be found in the sample if there are 3 defectives in the entire lot?

Solution. Using the hypergeometric distribution with $n = 5$, $N = 40$, $k = 3$, and $x = 1$, we find the probability of obtaining one defective to be

$$h(1; 40, 5, 3) = \frac{\binom{3}{1}\binom{37}{4}}{\binom{40}{5}} = 0.3011.$$

Theorem 4.3   *The mean and variance of the hypergeometric distribution $h(x; N, n, k)$ are*

$$\mu = \frac{nk}{N}$$

and

$$\sigma^2 = \frac{N-n}{N-1} \cdot n \cdot \frac{k}{N}\left(1 - \frac{k}{N}\right).$$

Proof. To find the mean of the hypergeometric distribution, we write

$$E(X) = \sum_{x=0}^{n} x \, \frac{\binom{k}{x}\binom{N-k}{n-x}}{\binom{N}{n}}$$

$$= k \sum_{x=1}^{n} \frac{(k-1)!}{(x-1)!(k-x)!} \cdot \frac{\binom{N-k}{n-x}}{\binom{N}{n}}$$

$$= k \sum_{x=1}^{n} \frac{\binom{k-1}{x-1}\binom{N-k}{n-x}}{\binom{N}{n}}.$$

Letting $y = x - 1$, we find that this becomes

$$E(X) = k \sum_{y=0}^{n-1} \frac{\binom{k-1}{y}\binom{N-k}{n-1-y}}{\binom{N}{n}}.$$

Writing

$$\binom{N-k}{n-1-y} = \binom{(N-1)-(k-1)}{n-1-y}$$

and

$$\binom{N}{n} = \frac{N!}{n!(N-n)!} = \frac{N}{n}\binom{N-1}{n-1},$$

we obtain

$$E(X) = \frac{nk}{N}\sum_{y=0}^{n-1}\frac{\binom{k-1}{y}\binom{(N-1)-(k-1)}{n-1-y}}{\binom{N-1}{n-1}}$$

$$= \frac{nk}{N},$$

since the summation represents the total of all probabilities in a hypergeometric experiment when $n-1$ items are selected at random from $N-1$, of which $k-1$ are labeled success.

To find the variance of the hypergeometric distribution, we proceed along the same steps as above to obtain

$$E[X(X-1)] = \frac{k(k-1)n(n-1)}{N(N-1)}.$$

Now, by Theorem 3.2,

$$\sigma_{\cdot}^2 = E(X^2) - \mu^2$$

$$= E[X(X-1)] + \mu - \mu^2$$

$$= \frac{k(k-1)n(n-1)}{N(N-1)} + \frac{nk}{N} - \frac{n^2k^2}{N^2}$$

$$= \frac{nk(N-k)(N-n)}{N^2(N-1)}$$

$$= \frac{N-n}{N-1} \cdot n \cdot \frac{k}{N}\left(1 - \frac{k}{N}\right).$$

Example 4.10   Find the mean and variance of the random variable of Example 4.9, and then use Chebyshev's theorem to interpret the interval $\mu \pm 2\sigma$.

Solution. Since Example 4.9 was a hypergeometric experiment with $N = 40$, $n = 5$, and $k = 3$, then by Theorem 4.3 we have

$$\mu = \frac{(5)(3)}{40} = \frac{3}{8} = 0.375$$

and

$$\sigma^2 = \left(\frac{40-5}{39}\right)(5)\left(\frac{3}{40}\right)\left(1-\frac{3}{40}\right)$$

$$= 0.3113.$$

Taking the square root of 0.3113, we find that $\sigma = 0.558$. Hence the required interval is $0.375 \pm (2)(0.558)$, or from -0.741 to 1.491. Chebyshev's theorem states that the number of defectives obtained when 5 components are selected at random from a lot of 40 components of which 3 are defective has a probability of at least 3/4 of falling between -0.741 and 1.491. That is, at least three-fourths of the time, the 5 components include less than 2 defectives.

If n is small relative to N, the probability for each drawing will change only slightly. Hence we essentially have a binomial experiment and can approximate the hypergeometric distribution by using the binomial distribution with $p = k/N$. The mean and variance can also be approximated by the formulas

$$\mu = np = \frac{nk}{N}$$

$$\sigma^2 = npq = n \cdot \frac{k}{N}\left(1-\frac{k}{N}\right).$$

Comparing these formulas with those of Theorem 4.3, we see that the mean is the same while the variance differs by a correction factor of $(N-n)/(N-1)$. This is negligible when n is small relative to N.

Example 4.11   A manufacturer of automobile tires reports that among a shipment of 5000 sent to a local distributor, 1000 are slightly blemished. If one purchases 10 of these tires at random from the distributor, what is the probability that exactly 3 will be blemished?

Solution. Since $N = 5000$ is large relative to the sample size $n = 10$, we shall approximate the desired probability by using the binomial distribution. The probability of obtaining a blemished tire is 0.2. Therefore, the probability of obtaining exactly 3 blemished tires is

$$h(3; 5000, 10, 1000) \simeq b(3; 10, 0.2)$$

$$= \sum_{x=0}^{3} b(x; 10, 0.2) - \sum_{x=0}^{2} b(x; 10, 0.2)$$

$$= 0.8791 - 0.6778$$

$$= 0.2013.$$

The hypergeometric distribution can be extended to treat the case where the N items can be partitioned into k cells $A_1, A_2, \ldots, A_k$ with a_1 elements in the first cell a_2 elements in the second cell, ..., a_k elements in the kth cell. We are now interested in the probability that a random sample of size n yields x_1 elements from A_1, x_2 elements from A_2, ..., and x_k elements from A_k. Let us represent this probability by

$$f(x_1, x_2, \ldots, x_k; a_1, a_2, \ldots, a_k, N, n).$$

To obtain a general formula, we note that the total number of samples that can be chosen of size n from N items is still $\binom{N}{n}$. There are $\binom{a_1}{x_1}$ ways of selecting x_1 items from the items in A_1, and for each of these we can choose x_2 items from the items in A_2 in $\binom{a_2}{x_2}$ ways. Therefore, we can select x_1 items from A_1 and x_2 items from A_2 in $\binom{a_1}{x_1}\binom{a_2}{x_2}$ ways. Continuing in this way, we can select all n items consisting of x_1 from A_1, x_2 from A_2, ..., and x_k from A_k in $\binom{a_1}{x_1}\binom{a_2}{x_2}\cdots\binom{a_k}{x_k}$ ways. The required probability distribution is now defined as follows.

Multivariate Hypergeometric Distribution *If N items can be partitioned into the k cells A_1, A_2, ..., A_k with a_1, a_2, ..., a_k elements, respectively, then the probability distribution of the random variables X_1, X_2, ..., X_k, representing the number of elements selected from $A_1, A_2, \ldots, A_k$ in a random sample of size n, is*

$$f(x_1, x_2, \ldots, x_k; a_1, a_2, \ldots, a_k, N, n) = \frac{\binom{a_1}{x_1}\binom{a_2}{x_2}\cdots\binom{a_k}{x_k}}{\binom{N}{n}}$$

with $\sum_{i=1}^{k} x_i = n$ and $\sum_{i=1}^{k} a_i = N$.

Example 4.12   A group of 10 individuals is being used in a biological case study. The group contains 3 people with blood type O, 4 with blood type A, and 3 with blood type B. What is the probability that a random sample of 5 will contain 1 person with blood type O, 2 people with blood type A, and 2 people with blood type B?

Solution. Using the extension of the hypergeometric distribution with $x_1 = 1$, $x_2 = 2$, $x_3 = 2$, $a_1 = 3$, $a_2 = 4$, $a_3 = 3$, $N = 10$, and $n = 5$, the desired probability is

$$f(1, 2, 2; 3, 4, 3, 10, 5) = \frac{\binom{3}{1}\binom{4}{2}\binom{3}{2}}{\binom{10}{5}} = \frac{3}{14}.$$

Exercises

1. If 7 cards are dealt from an ordinary deck of 52 playing cards, what is the probability that
   (a) exactly 2 of them will be face cards?
   (b) at least 1 of them will be a queen?

2. To avoid detection at customs, a traveler has placed 6 narcotic tablets in a bottle containing 9 vitamin pills that are similar in appearance. If the customs official selects 3 of the tablets at random for analysis, what is the probability that the traveler will be arrested for illegal possession of narcotics?

3. A homeowner plants 6 bulbs selected at random from a box containing 5 tulip bulbs and 4 daffodil bulbs. What is the probability that he planted 2 daffodil bulbs and 4 tulip bulbs?

4. From a lot of 10 missiles, 4 are selected at random and fired. If the lot contains 3 defective missiles that will not fire, what is the probability that
   (a) all 4 will fire?
   (b) at most 2 will not fire?

5. A random committee of size 3 is selected from 4 doctors and 2 nurses. Write a formula for the probability distribution of the random variable X representing the number of doctors on the committee. Find $P(2 \leq X \leq 3)$.

6. What is the probability that a waitress will refuse to serve alcoholic beverages to only 2 minors if she randomly checks the I.D.'s of 5 students from among 9 students of which 4 are not of legal age?

7. A company is interested in evaluating its current inspection procedure on shipments of 50 identical items. The procedure is to take a sample of 5 and pass the shipment if no more than 2 are found to be defective. What proportion of 20% defective shipments will be accepted?

8. A manufacturing company uses an acceptance scheme on production items before they are shipped. The plan is a two-stage one. Boxes of 25 are readied for shipment and a sample of 3 are tested for defectives. If any defectives are found, the entire box is sent back for 100% screening. If no defectives are found, the box is shipped.

(a) What is the probability that a box containing 3 defectives will be shipped?
(b) What is the probability that a box containing only 1 defective will be sent back for screening?

9. Suppose that the manufacturing company in Exercise 8 decided to change its acceptance scheme. Under the new scheme an inspector takes one at random, inspects it, and then replaces it in the box; a second inspector does likewise. Finally, a third inspector goes through the same procedure. The box is not shipped if any of the three find a defective. Answer parts (a) and (b) of Exercise 8 under this new plan.

10. In Exercise 4 how many defective missiles might we expect to be included among the 4 that are selected? Use Chebyshev's theorem to describe the variability of the number of defective missiles included when 4 are selected from several lots each of size 10 containing 3 defective missiles.

11. If a person is dealt 13 cards from an ordinary deck of 52 playing cards several times, how many hearts per hand can he expect? Between what two values would you expect the number of hearts to fall at least 75% of the time?

12. It is estimated that 4000 of the 10,000 voting residents of a town are against a new sales tax. If 15 eligible voters are selected at random and asked their opinion, what is the probability that at most 7 favor the new tax?

13. An annexation suit is being considered against a county subdivision of 1200 residences by a neighboring city. If tne occupants of half the residences object to being annexed, what is the probability that in a random sample of 10 at least 3 favor the annexation suit?

14. Among 150 IRS employees in a large city, only 30 are women. If 10 of the applicants are chosen at random to provide free tax assistance for the residents of this city, use the binomial approximation to the hypergeometric to find the probability that at least 3 women are selected.

15. A nationwide survey of 17,000 seniors by the University of Michigan reveals that almost 70% disapprove of daily pot smoking according to a report in *Parade*, September 14, 1980. If 18 of these seniors are selected at random and asked their opinions, what is the probability that more than 9 but less than 14 disapprove of smoking pot?

16. Find the probability of being dealt a bridge hand of 13 cards containing 5 spades, 2 hearts, 3 diamonds, and 3 clubs.

17. A foreign student club lists as its members 2 Canadians, 3 Japanese, 5 Italians, and 2 Germans. If a committee of 4 is selected at random, find the probability that

(a) all nationalities are represented;
(b) all nationalities except the Italians are represented.

18. An urn contains 3 green balls, 2 blue balls, and 4 red balls. In a random sample of 5 balls, find the probability that both blue balls and at least 1 red ball are selected.

19. A car rental agency at a local airport has available 5 Fords, 7 Chevrolets, 4 Dodges, 3 Datsuns, and 4 Toyotas. If the agency randomly selects 9 of these cars to chauffeur delegates from the airport to the downtown convention center, find the probability that 2 Fords, 3 Chevrolets, 1 Dodge, 1 Datsun, and 2 Toyotas are used.

4.5 Negative Binomial and Geometric Distributions

Let us consider an experiment in which the properties are the same as those listed for a binomial experiment, with the exception that the trials will be repeated until a *fixed* number of successes occur. Therefore, instead of finding the probability of x successes in n trials, where n is fixed, we are now interested in the probability that the kth success occurs on the xth trial. Experiments of this kind are called **negative binomial experiments**.

As an illustration, consider the use of a drug that is known to be effective in 60% of the cases in which it is used. The use of the drug will be considered a success if it is effective in bringing some degree of relief to the patient. We are interested in finding the probability that the fifth patient to experience relief is the seventh patient to receive the drug during a given week. Designating a success by S and a failure by F, a possible order of achieving the desired result is $SFSSSFS$, which occurs with probability $(0.6)(0.4)(0.6)(0.6)(0.6)(0.4)(0.6) = (0.6)^5(0.4)^2$. We could list all possible orders by rearranging the F's and S's except for the last outcome, which must be the fifth success. The total number of possible orders is equal to the number of partitions of the first six trials into two groups with 2 failures assigned to the one group and the 4 successes assigned to the other group. This can be done in $\binom{6}{4} = 15$ mutually exclusive ways. Hence, if X represents the outcome on which the fifth success occurs, then

$$P(X = 7) = \binom{6}{4}(0.6)^5(0.4)^2 = 0.1866.$$

The number X of trials to produce k successes in a negative binomial experiment is called a **negative binomial random variable** and its probability distribution is called the **negative binomial distribution**. Since its probabilities depend on the number of successes desired and the probability of a success on a given trial, we shall denote them by the symbol $b^*(x; k, p)$. To obtain the general formula for $b^*(x; k, p)$, consider the probability of a success on the xth trial preceded by $k - 1$ successes and $x - k$ failures in some specified order. Since the trials are independent, we can multiply all the probabilities corresponding to each desired outcome. Each success occurs with probability p and each failure with probability $q = 1 - p$. Therefore, the probability for the specified order, ending in a success, is $p^{k-1}q^{x-k}p = p^k q^{x-k}$. The total number of sample points in the experiment ending in a success, after the occurrence of $k - 1$ successes and $x - k$ failures in any order, is equal to the number of partitions of $x - 1$ trials into two groups with $k - 1$ successes corresponding to one group and $x - k$ failures corresponding to the other group. This number is given by the term $\binom{x-1}{k-1}$, each mutually exclusive and occurring with equal probability $p^k q^{x-k}$. We obtain the general formula by multiplying $p^k q^{x-k}$ by $\binom{x-1}{k-1}$.

Negative Binomial Distribution   *If repeated independent trials can result in a success with probability p and a failure with probability q = 1 − p, then the probability distribution of the random variable X, the number of the trial on which the kth success occurs, is given by*

$$b^*(x; k, p) = \binom{x-1}{k-1} p^k q^{x-k}, \qquad x = k, k+1, k+2, \ldots.$$

Example 4.13   Find the probability that a person tossing three coins will get either all heads or all tails for the second time on the fifth toss.

Solution. Using the negative binomial distribution with $x = 5$, $k = 2$, and $p = 1/4$, we have

$$b^*\left(5; 2, \frac{1}{4}\right) = \binom{4}{1}\left(\frac{1}{4}\right)^2\left(\frac{3}{4}\right)^3$$

$$= \frac{4!}{1!\,3!} \cdot \frac{3^3}{4^5}$$

$$= \frac{27}{256}.$$

The negative binomial distribution derives its name from the fact that each term in the expansion of $p^k(1 - q)^{-k}$ corresponds to the values of $b^*(x; k, p)$ for $x = k, k+1, k+2, \ldots$.

If we consider the special case of the negative binomial distribution where $k = 1$, we have a probability distribution for the number of trials required for a single success. An example would be the tossing of a coin until a head occurs. We might be interested in the probability that the first head occurs on the fourth toss. The negative binomial distribution reduces to the form $b^*(x; 1, p) = pq^{x-1}$, $x = 1, 2, 3, \ldots$. Since the successive terms constitute a geometric progression, it is customary to refer to this special case as the **geometric distribution** and denote its values by $g(x; p)$.

Geometric Distribution   *If repeated independent trials can result in a success with probability p and a failure with probability $q = 1 - p$, then the probability distribution of the random variable X, the number of the trial on which the first success occurs, is given by*

$$g(x; p) = pq^{x-1}, \qquad x = 1, 2, 3, \ldots.$$

Example 4.14   In a certain manufacturing process it is known that, on the average, 1 in every 100 items is defective. What is the probability that the fifth item inspected is the first defective item found?

Solution. Using the geometric distribution with $x = 5$ and $p = 0.01$, we have

$$g(5; 0.01) = (0.01)(0.99)^4$$
$$= 0.0096.$$

4.6 Poisson Distribution

Experiments yielding numerical values of a random variable X, the number of outcomes occurring during a given time interval or in a specified region, are often called **Poisson experiments**. The given time interval may be of any length, such as a minute, a day, a week, a month, or even a year. Hence a Poisson experiment might generate observations for the random variable X representing the number of telephone calls per hour received by an office, the number of days school is closed due to snow during the winter, or the number of postponed games due to rain during a baseball season. The specified region could be a line segment, an area, a volume, or perhaps a piece of material. In this case X might represent the number of field mice per acre, the number of bacteria in a given culture, or the number of typing errors per page.

A Poisson experiment is one that possesses the following properties:

1. The number of outcomes occurring in one time interval or specified region is independent of the number that occurs in any other disjoint time interval or region of space.

2. The probability that a single outcome will occur during a very short time interval or in a small region is proportional to the length of the time interval or the size of the region and does not depend on the number of outcomes occurring outside this time interval or region.

3. The probability that more than one outcome will occur in such a short time interval or fall in such a small region is negligible.

The number X of outcomes occurring in a Poisson experiment is called a **Poisson random variable** and its probability distribution is called the **Poisson distribution**. Since its probabilities depend only on μ, the average number of outcomes occurring in the given time interval or specified region, we shall denote them by the symbol $p(x; \mu)$. The derivation of the formula for $p(x; \mu)$, based on the properties for a Poisson experiment listed above, is beyond the scope of this text. We list the result in the following definition.

Poisson Distribution   *The probability distribution of the Poisson random variable X, representing the number of outcomes occurring in a given time interval or specified region, is*

$$p(x; \mu) = \frac{e^{-\mu}\mu^x}{x!}, \qquad x = 0, 1, 2, \ldots,$$

where μ is the average number of outcomes occurring in the given time interval or specified region and $e = 2.71828 \cdots$.

Table A.2 contains Poisson probability sums $P(r; \mu) = \sum\limits_{x=0}^{r} p(x; \mu)$ for a few selected values of μ ranging from 0.1 to 18. We illustrate the use of this table with the following two examples.

Example 4.15   The average number of radioactive particles passing through a counter during 1 millisecond in a laboratory experiment is 4. What is the probability that 6 particles enter the counter in a given millisecond?

Solution. Using the Poisson distribution with $x = 6$ and $\mu = 4$, we find from Table A.2 that

$$p(6; 4) = \frac{e^{-4}4^6}{6!} = \sum_{x=0}^{6} p(x; 4) - \sum_{x=0}^{5} p(x; 4) = 0.8893 - 0.7851$$

$$= 0.1042.$$

Example 4.16   The average number of oil tankers arriving each day at a certain port city is known to be 10. The facilities at the port can handle at most 15 tankers per day. What is the probability that on a given day tankers will have to be sent away?

Solution. Let X be the number of tankers arriving each day. Then, using Table A.2, we have

$$P(X > 15) = 1 - P(X \leq 15)$$

$$= 1 - \sum_{x=0}^{15} p(x; 10)$$

$$= 1 - 0.9513$$

$$= 0.0487.$$

Theorem 4.4   *The mean and variance of the Poisson distribution $p(x; \mu)$ both have the value μ.*

Proof. To verify that the mean is indeed μ, we write

$$E(X) = \sum_{x=0}^{\infty} x \cdot \frac{e^{-\mu}\mu^x}{x!} = \sum_{x=1}^{\infty} x \cdot \frac{e^{-\mu}\mu^x}{x!}$$

$$= \mu \sum_{x=1}^{\infty} \frac{e^{-\mu}\mu^{x-1}}{(x-1)!}.$$

Now, let $y = x - 1$ to give

$$E(X) = \mu \sum_{y=0}^{\infty} \frac{e^{-\mu}\mu^y}{y!} = \mu,$$

since

$$\sum_{y=0}^{\infty} \frac{e^{-\mu}\mu^y}{y!} = \sum_{y=0}^{\infty} p(y; \mu) = 1.$$

The variance of the Poisson distribution is obtained by first finding

$$E[X(X-1)] = \sum_{x=0}^{\infty} x(x-1) \frac{e^{-\mu}\mu^x}{x!} = \sum_{x=2}^{\infty} x(x-1) \frac{e^{-\mu}\mu^x}{x!}$$

$$= \mu^2 \sum_{x=2}^{\infty} \frac{e^{-\mu}\mu^{x-2}}{(x-2)!}.$$

Setting $y = x - 2$, we have

$$E[X(X-1)] = \mu^2 \sum_{y=0}^{\infty} \frac{e^{-\mu}\mu^y}{y!} = \mu^2.$$

Hence

$$\sigma^2 = E[X(X-1)] + \mu - \mu^2$$

$$= \mu^2 + \mu - \mu^2$$

$$= \mu.$$

In Example 4.15, where $\mu = 4$, we also have $\sigma^2 = 4$ and hence $\sigma = 2$. Using Chebyshev's theorem, we can state that our random variable has a probability of at least 3/4 of falling in the interval $\mu \pm 2\sigma = 4 \pm (2)(2)$, or from 0 to 8. Therefore, we conclude that at least three-fourths of the time the number of radioactive particles entering the counter will be anywhere from 0 to 8 during a given millisecond.

We shall now derive the Poisson distribution as a limiting form of the binomial distribution when $n \to \infty$, $p \to 0$, and np remains constant. Hence, if n is large and p is close to 0, the Poisson distribution can be used, with $\mu = np$, to approximate binomial probabilities. If p is close to 1, we can still use the Poisson distribution to approximate binomial probabilities by interchanging what we have defined to be a success and a failure, thereby changing p to a value close to 0.

Theorem 4.5   *Let X be a binomial random variable with probability distribution $b(x; n, p)$. When $n \to \infty$, $p \to 0$, and $\mu = np$ remains constant,*

$$b(x; n, p) \to p(x; \mu).$$

Proof. The binomial distribution can be written

$$b(x; n, p) = \binom{n}{x} p^x q^{n-x}$$

$$= \frac{n!}{x!\,(n-x)!}\, p^x (1-p)^{n-x}$$

$$= \frac{n(n-1)\,\cdots\,(n-x+1)}{x!}\, p^x (1-p)^{n-x}.$$

Substituting $p = \mu/n$, we have

$$b(x; n, p) = \frac{n(n-1)\,\cdots\,(n-x+1)}{x!} \left(\frac{\mu}{n}\right)^x \left(1 - \frac{\mu}{n}\right)^{n-x}$$

$$= 1 \left(1 - \frac{1}{n}\right) \cdots \left(1 - \frac{x-1}{n}\right) \frac{\mu^x}{x!} \left(1 - \frac{\mu}{n}\right)^n \left(1 - \frac{\mu}{n}\right)^{-x}.$$

As $n \to \infty$ while x and μ remain constant,

$$\lim_{n \to \infty} 1 \left(1 - \frac{1}{n}\right) \cdots \left(1 - \frac{x-1}{n}\right) = 1,$$

$$\lim_{n \to \infty} \left(1 - \frac{\mu}{n}\right)^{-x} = 1,$$

and from the definition of the number e,

$$\lim_{n \to \infty} \left(1 - \frac{\mu}{n}\right)^n = \lim_{n \to \infty} \left\{ \left[1 + \frac{1}{(-n)/\mu}\right]^{-n/\mu} \right\}^{-\mu} = e^{-\mu}.$$

Hence, under the given limiting conditions,

$$b(x; n, p) \to \frac{e^{-\mu}\mu^x}{x!}, \qquad x = 0, 1, 2, \ldots.$$

Example 4.17   In a manufacturing process in which glass items are being produced, defects or bubbles occur, occasionally rendering the piece undesirable for marketing. It is known that on the average 1 in every 1000 of these items produced has one or more bubbles. What is the probability that a random sample of 8000 will yield fewer than 7 items possessing bubbles?

Solution. This is essentially a binomial experiment with $n = 8000$ and $p = 0.001$. Since p is very close to zero and n is quite large, we shall approximate with the Poisson distribution using $\mu = (8000)(0.001) = 8$. Hence, if X represents the number of bubbles, we have

$$P(X < 7) = \sum_{x=0}^{6} b(x; 8000, 0.001)$$

$$\simeq \sum_{x=0}^{6} p(x; 8)$$

$$= 0.3134.$$

Exercises

1. The probability that a person living in a certain city owns a dog is estimated to be 0.3. Find the probability that the tenth person randomly interviewed in this city is the fifth one to own a dog.

2. A scientist inoculates several mice, one at a time, with a disease germ until he finds 2 that have contracted the disease. If the probability of contracting the disease is 1/6, what is the probability that 8 mice are required?

3. Suppose the probability is 0.8 that any given person will believe a tale about the transgressions of a famous actress. What is the probability that
   (a) the sixth person to hear this tale is the fourth one to believe it?
   (b) the third person to hear this tale is the first one to believe it?

4. Find the probability that a person flipping a coin gets
   (a) the third head on the seventh flip;
   (b) the first head on the fourth flip.

5. Three people toss a coin and the odd man pays for the coffee. If the coins all turn up the same, they are tossed again. Find the probability that fewer than 4 tosses are needed.

6. According to a study published by a group of University of Massachusetts sociologists, about two-thirds of the 20 million persons in this country who take Valium are women. Assuming this figure to be a valid estimate, find the probability that on a given day the fifth prescription written by a doctor for Valium is
   (a) the first prescribing Valium for a woman;
   (b) the third prescribing Valium for a woman.

7. The probability that a student pilot passes the written test for his private pilot's license is 0.7. Find the probability that a person passes the test
   (a) on the third try;
   (b) before the fourth try.

8. On the average a certain intersection results in 3 traffic accidents per month. What is the probabil-

ity that in any given month at this intersection

(a) exactly 5 accidents will occur?

(b) less than 3 accidents will occur?

(c) at least 2 accidents will occur?

9. A secretary makes 2 errors per page on the average. What is the probability that on the next page she makes
(a) 4 or more errors?
(b) no errors?

10. A certain area of the eastern United States is, on the average, hit by 6 hurricanes a year. Find the probability that in a given year this area will be hit by
(a) fewer than 4 hurricanes;
(b) anywhere from 6 to 8 hurricanes.

11. In an inventory study it was determined that, on the average, demands for a particular item at a warehouse were made 5 times per day. What is the probability that on a given day this item is requested
(a) more than 5 times?
(b) not at all?

12. The average number of field mice per acre in a 5-acre wheat field is estimated to be 12. Find the probability that fewer than 7 field mice are found
(a) on a given acre;
(b) on 2 of the next 3 acres inspected.

13. A restaurant prepares a tossed salad containing on the average 5 vegetables. Find the probability that the salad contains more than 5 vegetables
(a) on a given day;
(b) on 3 of the next 4 days;
(c) for the first time in April on April 5.

14. The probability that a person dies from a certain respiratory infection is 0.002. Find the probability that fewer than 5 of the next 2000 so infected will die.

15. Suppose that on the average 1 person in 1000 makes a numerical error in preparing his or her income tax return. If 10,000 forms are selected at random and examined, find the probability that 6, 7, or 8 of the forms will be in error.

16. The probability that a student fails the screening test for scoliosis (curvature of the spine) at a local high school is known to be 0.004. Of the next 1875 students who are screened for scoliosis, find the probability that
(a) fewer than 5 fail the test;
(b) 8, 9, or 10 fail the test.

17. (a) Find the mean and variance in Exercise 14 of the random variable X representing the number of persons among 2000 that die from the respiratory infection.

(b) According to Chebyshev's theorem, there is a probability of at least 3/4 that the number of persons to die among 2000 persons infected will fall within what interval?

18. (a) Find the mean and variance in Exercise 15 of the random variable X representing the number of persons among 10,000 who make an error in preparing their income tax returns.

(b) According to Chebyshev's theorem, there is a probability of at least 8/9 that the number of persons who make errors in preparing their income tax returns among 10,000 returns will be within what interval?

5

Some Continuous Probability Distributions

5.1 Normal Distribution

The most important continuous probability distribution in the entire field of statistics is the **normal distribution**. Its graph, called the **normal curve**, is the bell-shaped curve of Figure 5.1, which describes so many sets of data that occur in nature, industry, and research. In 1733 Abraham DeMoivre developed the mathematical equation of the normal curve. It provided a basis upon which much of the theory of inductive statistics is founded. The normal distribution is often referred to as the **Gaussian distribution** in honor of Karl

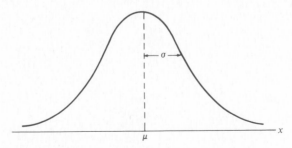

Figure 5.1   The normal curve.

Friedrich Gauss (1777–1855), who also derived its equation from a study of errors in repeated measurements of the same quantity.

A continuous random variable X having the bell-shaped distribution of Figure 5.1 is called a **normal random variable**. The mathematical equation for the probability distribution of the normal variable depends upon the two parameters μ and σ, its mean and standard deviation. Hence we denote the values of the density function of X by $n(x; \mu, \sigma)$.

Normal Distribution   *The density function of the normal random variable X, with mean μ and variance σ^2, is*

$$n(x; \mu, \sigma) = \frac{1}{\sqrt{2\pi}\,\sigma}\, e^{-(1/2)[(x-\mu)/\sigma]^2}, \qquad -\infty < x < \infty.$$

where $\pi = 3.14159\ldots$ and $e = 2.71828\ldots$.

Once μ and σ are specified, the normal curve is completely determined. For example, if $\mu = 50$ and $\sigma = 5$, then the ordinates $n(x; 50, 5)$ can easily be computed for various values of x and the curve drawn. In Figure 5.2 we have sketched two normal curves having the same standard deviation but different means. The two curves are identical in form but are centered at different positions along the horizontal axis.

In Figure 5.3 we have sketched two normal curves with the same mean but different standard deviations. This time we see that the two curves are centered at exactly the same position on the horizontal axis, but the curve with the larger standard deviation is lower and spreads out farther. Remember that the

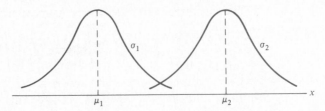

Figure 5.2   Normal curves with $\mu_1 < \mu_2$ and $\sigma_1 = \sigma_2$.

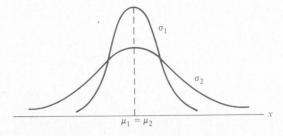

Figure 5.3   Normal curves with $\mu_1 = \mu_2$ and $\sigma_1 < \sigma_2$.

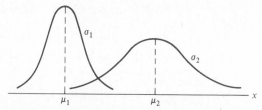

Figure 5.4   Normal curves with $\mu_1 < \mu_2$ and $\sigma_1 < \sigma_2$.

area under a probability curve must be equal to 1 and therefore the more variable the set of observations the lower and wider the corresponding curve will be.

Figure 5.4 shows the results of sketching two normal curves having different means and different standard deviations. Clearly, they are centered at different positions on the horizontal axis and their shapes reflect the two different values of σ.

From an inspection of Figures 5.1 through 5.4 and by examination of the first and second derivatives of $n(x; \mu, \sigma)$, we list the following properties of the normal curve:

1. The **mode**, which is the point on the horizontal axis where the curve is a maximum, occurs at $x = \mu$.
2. The curve is symmetric about a vertical axis through the mean μ.
3. The curve has its points of inflection at $x = \mu \pm \sigma$, is concave downward if $\mu - \sigma < X < \mu + \sigma$, and is concave upward otherwise.
4. The normal curve approaches the horizontal axis asymptotically as we proceed in either direction away from the mean.
5. The total area under the curve and above the horizontal axis is equal to 1.

We shall now show that the parameters μ and σ^2 are indeed the mean and the variance of the normal distribution. To evaluate the mean, we write

$$E(X) = \frac{1}{\sqrt{2\pi}\,\sigma} \int_{-\infty}^{\infty} x e^{-(1/2)[(x-\mu)/\sigma]^2} \, dx.$$

Setting $z = (x - \mu)/\sigma$ and $dx = \sigma \, dz$, we obtain

$$E(X) = \frac{1}{\sqrt{2\pi}} \int_{-\infty}^{\infty} (\mu + \sigma z)e^{-z^2/2} \, dz$$

$$= \mu \frac{1}{\sqrt{2\pi}} \int_{-\infty}^{\infty} e^{-z^2/2} \, dz + \frac{\sigma}{\sqrt{2\pi}} \int_{-\infty}^{\infty} z e^{-z^2/2} \, dz.$$

The first term on the right is μ times the area under a normal curve with mean zero and variance 1, and hence equal to μ. By straightforward integration, the second term is equal to 0. Hence

$$E(X) = \mu.$$

The variance of the normal distribution is given by

$$E[(X - \mu)^2] = \frac{1}{\sqrt{2\pi}\,\sigma} \int_{-\infty}^{\infty} (x - \mu)^2 e^{-(1/2)[(x-\mu)/\sigma]^2} \, dx.$$

Again setting $z = (x - \mu)/\sigma$ and $dx = \sigma \, dz$, we obtain

$$E[(X - \mu)^2] = \frac{\sigma^2}{\sqrt{2\pi}} \int_{-\infty}^{\infty} z^2 e^{-z^2/2} \, dz.$$

Integrating by parts with $u = z$ and $dv = ze^{-z^2/2} \, dz$ so that $du = dz$ and $v = -e^{-z^2/2}$, we find that

$$E[(X - \mu)^2] = \frac{\sigma^2}{\sqrt{2\pi}} \left(-ze^{-z^2/2} \Big|_{-\infty}^{\infty} + \int_{-\infty}^{\infty} e^{-z^2/2} \, dz \right)$$

$$= \sigma^2(0 + 1)$$

$$= \sigma^2.$$

Many random variables have probability distributions that can be described adequately by the normal curve once μ and σ^2 are specified. In this chapter we shall assume that these two parameters are known, perhaps from previous investigations. Later we shall make statistical inferences when μ and σ^2 are unknown and have been estimated from the available experimental data.

5.2 Areas Under the Normal Curve

The curve of any continuous probability distribution or density function is constructed so that the area under the curve bounded by the two ordinates $x = x_1$ and $x = x_2$ equals the probability that the random variable X assumes a value between $x = x_1$ and $x = x_2$. Thus, for the normal curve in Figure 5.5,

$$P(x_1 < X < x_2) = \int_{x_1}^{x_2} n(x; \mu, \sigma) \, dx$$

$$= \frac{1}{\sqrt{2\pi}\,\sigma} \int_{x_1}^{x_2} e^{-(1/2)[(x-\mu)/\sigma]^2} \, dx$$

is represented by the area of the shaded region.

In Figures 5.2, 5.3, and 5.4 we saw how the normal curve is dependent on the mean and the standard deviation of the distribution under investigation.

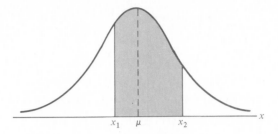

Figure 5.5 $P(x_1 < X < x_2)$ = area of the shaded region.

The area under the curve between any two ordinates must then also depend on the values μ and σ. This is evident in Figure 5.6, where we have shaded regions corresponding to $P(x_1 < X < x_2)$ for two curves with different means and variances. The $P(x_1 < X < x_2)$, where X is the random variable describing distribution I, is indicated by the darker shaded area. If X is the random variable describing distribution II, then $P(x_1 < X < x_2)$ is given by the entire shaded region. Obviously, the two shaded regions are different in size; therefore, the probability associated with each distribution will be different for the two given values of X.

The difficulty encountered in solving integrals of normal density functions necessitates the tabulation of normal curve areas for quick reference. However, it would be a hopeless task to attempt to set up separate tables for every conceivable value of μ and σ. Fortunately, we are able to transform all the observations of any normal random variable X to a new set of observations of a normal random variable Z with mean zero and variance 1. This can be done by means of the transformation

$$Z = \frac{X - \mu}{\sigma}.$$

Whenever X assumes a value x, the corresponding value of Z is given by $z = (x - \mu)/\sigma$. Therefore, if X falls between the values $x = x_1$ and $x = x_2$, the random variable Z will fall between the corresponding values $z_1 = (x_1 - \mu)/\sigma$

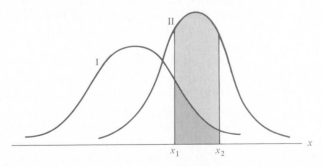

Figure 5.6 $P(x_1 < X < x_2)$ for different normal curves.

and $z_2 = (x_2 - \mu)/\sigma$. Consequently, we may write

$$P(x_1 < X < x_2) = \frac{1}{\sqrt{2\pi}\,\sigma} \int_{x_1}^{x_2} e^{-(1/2)[(x-\mu)/\sigma]^2}\, dx$$

$$= \frac{1}{\sqrt{2\pi}} \int_{z_1}^{z_2} e^{-z^2/2}\, dz$$

$$= \int_{z_1}^{z_2} n(z;\, 0,\, 1)\, dz = P(z_1 < Z < z_2),$$

where Z is seen to be a normal random variable with mean zero and variance 1.

Definition 5.1   *The distribution of a normal random variable with mean zero and variance 1 is called a* **standard normal distribution**.

The original and transformed distributions are illustrated in Figure 5.7. Since all the values of X falling between x_1 and x_2 have corresponding z values between z_1 and z_2, the area under the X curve between the ordinates $x = x_1$ and $x = x_2$ in Figure 5.7 equals the area under the Z curve between the transformed ordinates $z = z_1$ and $z = z_2$.

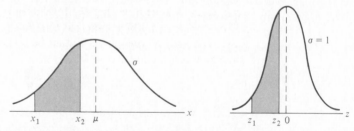

Figure 5.7   The original and transformed normal distributions.

We have now reduced the required number of tables of normal-curve areas to one, that of the standard normal distribution. Table A.3 gives the area under the standard normal curve corresponding to $P(Z < z)$ for values of z ranging from -3.49 to 3.49. To illustrate the use of this table, let us find the probability that Z is less than 1.74. First, we locate a value of z equal to 1.7 in the left column, then move across the row to the column under 0.04, where we read 0.9591. Therefore, $P(Z < 1.74) = 0.9591$. To find a z value corresponding to a given probability, the process is reversed. For example, the z value leaving an area of 0.2148 under the curve to the left of z is seen to be -0.79.

Example 5.1   Given a standard normal distribution, find the area under the curve that lies (a) to the right of $z = 1.84$, and (b) between $z = -1.97$ and $z = 0.86$.

Solution

(a) The area in Figure 5.8(a) to the right of $z = 1.84$ is equal to 1 minus the area in Table A.3 to the left of $z = 1.84$, namely, $1 - 0.9671 = 0.0329$.

(b) The area in Figure 5.8(b) between $z = -1.97$ and $z = 0.86$ is equal to the area to the left of $z = 0.86$ minus the area to the left of $z = -1.97$. From Table A.3 we find the desired area to be $0.8051 - 0.0244 = 0.7807$.

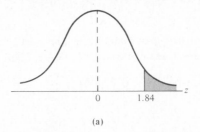
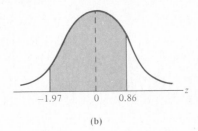

(a)   (b)

Figure 5.8   Areas for Example 5.1.

Example 5.2   Given a standard normal distribution, find the value of k such that (a) $P(Z > k) = 0.3015$ and (b) $P(k < Z < -0.18) = 0.4197$.

Solution

(a) In Figure 5.9(a) we see that the k value leaving an area of 0.3015 to the right must then leave an area of 0.6985 to the left. From Table A.3 it follows that $k = 0.52$.

(b) From Table A.3 we note that the total area to the left of -0.18 is equal to 0.4286. In Figure 5.9(b) we see that the area between k and -0.18 is 0.4197 so that the area to the left of k must be $0.4286 - 0.4197 = 0.0089$. Hence, from Table A.3, we have $k = -2.37$.

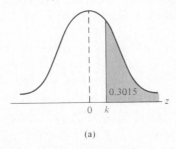
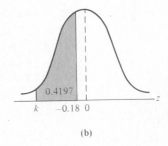

(a)   (b)

Figure 5.9   Areas for Example 5.2.

Example 5.3   Given a normal distribution with $\mu = 50$ and $\sigma = 10$, find the probability that X assumes a value between 45 and 62.

Solution.   The z values corresponding to $x_1 = 45$ and $x_2 = 62$ are

$$z_1 = \frac{45 - 50}{10} = -0.5$$

and

$$z_2 = \frac{62 - 50}{10} = 1.2.$$

Therefore,

$$P(45 < X < 62) = P(-0.5 < Z < 1.2).$$

The $P(-0.5 < Z < 1.2)$ is given by the area of the shaded region in Figure 5.10. This area may be found by subtracting the area to the left of the ordinate $z = -0.5$ from the entire area to the left of $z = 1.2$. Using Table A.3, we have

$$\begin{aligned} P(45 < X < 62) &= P(-0.5 < Z < 1.2) \\ &= P(Z < 1.2) - P(Z < -0.5) \\ &= 0.8849 - 0.3085 \\ &= 0.5764. \end{aligned}$$

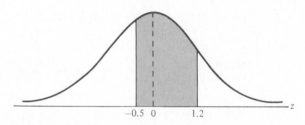

Figure 5.10   Area for Example 5.3.

Example 5.4   Given a normal distribution with $\mu = 300$ and $\sigma = 50$, find the probability that X assumes a value greater than 362.

Solution. The normal probability distribution showing the desired area is given in Figure 5.11. To find the $P(X > 362)$, we need to evaluate the area under the normal curve to the right of $x = 362$. This can be done by trans-

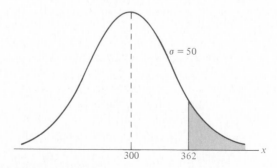

Figure 5.11   Area for Example 5.4.

forming $x = 362$ to the corresponding z value, obtaining the area to the left of z from Table A.3, and then subtracting this area from 1. We find that

$$z = \frac{362 - 300}{50} = 1.24.$$

Hence

$$
\begin{aligned}
P(X > 362) &= P(Z > 1.24) \\
&= 1 - P(Z < 1.24) \\
&= 1 - 0.8925 \\
&= 0.1075.
\end{aligned}
$$

According to Chebyshev's theorem, the probability that a random variable assumes a value within 2 standard deviations of the mean is at least 3/4. If the random variable has a normal distribution, the z values corresponding to $x_1 = \mu - 2\sigma$ and $x_2 = \mu + 2\sigma$ are easily computed to be

$$z_1 = \frac{(\mu - 2\sigma) - \mu}{\sigma} = -2$$

and

$$z_2 = \frac{(\mu + 2\sigma) - \mu}{\sigma} = 2.$$

Hence

$$
\begin{aligned}
P(\mu - 2\sigma < X < \mu + 2\sigma) &= P(-2 < Z < 2) \\
&= P(Z < 2) - P(Z < -2) \\
&= 0.9772 - 0.0228 \\
&= 0.9544,
\end{aligned}
$$

which is a much stronger statement than that given by Chebyshev's theorem.

Occasionally, we are required to find the value of z corresponding to a specified probability that falls between values listed in Table A.3 (see Example 5.5). For convenience, we shall always choose the z value corresponding to the tabular probability that comes closest to the specified probability. However, if the given probability falls midway between two tabular probabilities, we shall choose for z the value falling midway between the corresponding values of z. For instance, to find the z value corresponding to a probability of 0.7975, which falls between 0.7967 and 0.7995 in Table A.3, we choose $z = 0.83$, since 0.7975 is closer to 0.7967. On the other hand, for a probability of 0.7981, which falls midway between 0.7967 and 0.7995, we take $z = 0.835$.

The preceding two examples were solved by going first from a value of x to a z value and then computing the desired area. In Example 5.5 we reverse the process and begin with a known area or probability, find the z value, and then determine x by rearranging the formula

$$z = \frac{x - \mu}{\sigma} \qquad \text{to give} \qquad x = \sigma z + \mu.$$

Example 5.5   Given a normal distribution with $\mu = 40$ and $\sigma = 6$, find the value of x that has (a) 45% of the area to the left, and (b) 14% of the area to the right.

Solution

(a) An area of 0.45 to the left of the desired x value is shaded in Figure 5.12(a). We require a z value that leaves an area of 0.45 to the left. From Table A.3 we find $P(Z < -0.13) = 0.45$ so that the desired z value is -0.13. Hence

$$x = (6)(-0.13) + 40$$
$$= 39.22.$$

(b) In Figure 5.12(b) we shade an area equal to 0.14 to the right of the desired x value. This time we require a z value that leaves 0.14 of the area to the right and hence an area of 0.86 to the left. Again, from Table A.3, we find $P(Z > 1.08) = 0.86$ so that the desired z value is 1.08 and

$$x = (6)(1.08) + 40$$
$$= 46.48.$$

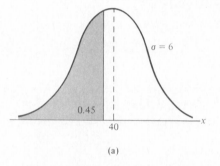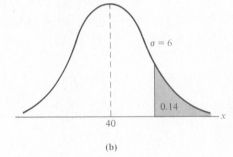

(a)   (b)

Figure 5.12   Areas for Example 5.5.

5.3 Applications of the Normal Distribution

Some of the many problems to which the normal distribution is applicable are treated in the following examples. The use of the normal curve to approximate binomial probabilities will be considered in Section 5.4.

Example 5.6   A certain type of storage battery lasts on the average 3.0 years, with a standard deviation of 0.5 year. Assuming that the battery lives are normally distributed, find the probability that a given battery will last less than 2.3 years.

Solution. First construct a diagram such as Figure 5.13, showing the given distribution of battery lives and the desired area. To find the $P(X < 2.3)$, we

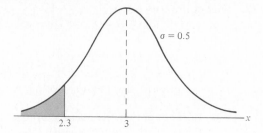

Figure 5.13 Area for Example 5.6.

need to evaluate the area under the normal curve to the left of 2.3. This is accomplished by finding the area to the left of the corresponding z value. Hence we find that

$$z = \frac{2.3 - 3}{0.5} = -1.4,$$

and then using Table A.3 we have

$$P(X < 2.3) = P(Z < -1.4)$$
$$= 0.0808.$$

Example 5.7 An electrical firm manufactures light bulbs that have a length of life that is normally distributed with mean equal to 800 hours and a standard deviation of 40 hours. Find the probability that a bulb burns between 778 and 834 hours.

Solution. The distribution of light bulbs is illustrated in Figure 5.14. The z values corresponding to $x_1 = 778$ and $x_2 = 834$ are

$$z_1 = \frac{778 - 800}{40} = -0.55$$

and

$$z_2 = \frac{834 - 800}{40} = 0.85.$$

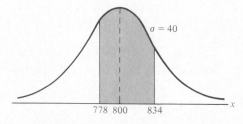

Figure 5.14 Area for Example 5.7.

Hence

$$P(778 < X < 834) = P(-0.55 < Z < 0.85)$$
$$= P(Z < 0.85) - P(Z < -0.55)$$
$$= 0.8023 - 0.2912$$
$$= 0.5111.$$

Example 5.8   On an examination the average grade was 74 and the standard deviation was 7. If 12% of the class are given A's, and the grades are curved to follow a normal distribution, what is the lowest possible A and the highest possible B?

Solution. In this example we begin with a known area or probability, find the z value, and then determine x from the formula $x = \sigma z + \mu$. An area of 0.12, corresponding to the fraction of students receiving A's, is shaded in Figure 5.15. We require a z value that leaves 0.12 of the area to the right and hence an area of 0.88 to the left. From Table A.3, $P(Z < 1.175) = 0.88$ so that the desired z value is 1.175. Hence

$$x = (7)(1.175) + 74$$
$$= 82.225.$$

Therefore, the lowest A is 83 and the highest B is 82.

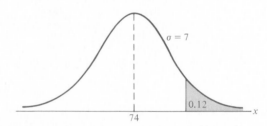

Figure 5.15   Area for Example 5.8.

Example 5.9   Refer to Example 5.8 and find the sixth decile.

Solution. The sixth decile, written D_6, is the x value that leaves 60% of the area to the left as shown in Figure 5.16. From Table A.3 we find

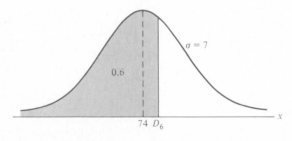

Figure 5.16   Area for Example 5.9.

$P(Z < 0.25) = 0.6$ so that the desired z value is 0.25. Now

$$x = (7)(0.25) + 74$$
$$= 75.75.$$

Hence $D_6 = 75.75$. That is, 60% of the grades are 75 or less.

Example 5.10 Gauges are used to reject all components in which a certain dimension is not within the specification $1.50 \pm d$. It is known that this measurement is normally distributed with mean 1.50 and standard deviation 0.2. Determine the value d such that the specifications "cover" 95% of the measurements.

Solution. From Table A.3 we know that

$$P(-1.96 < Z < 1.96) = 0.95.$$

Therefore,

$$1.96 = \frac{(1.50 + d) - 1.50}{0.2}$$

from which we obtain

$$d = (0.2)(1.96) = 0.392.$$

An illustration of the specifications is given in Figure 5.17.

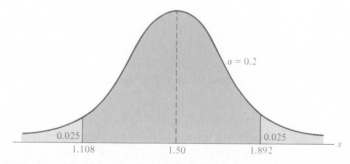

Figure 5.17 Specifications for Example 5.10.

Example 5.11 A certain machine makes electrical resistors having a mean resistance of 40 ohms and a standard deviation of 2 ohms. Assuming that the resistance follows a normal distribution and can be measured to any degree of accuracy, what percentage of resistors will have a resistance that exceeds 43 ohms?

Solution. A percentage is found by multiplying the relative frequency by 100%. Since the relative frequency for an interval is equal to the probability of falling in the interval, we must find the area to the right of $x = 43$ in Figure 5.18. This can be done by transforming $x = 43$ to the corresponding z value,

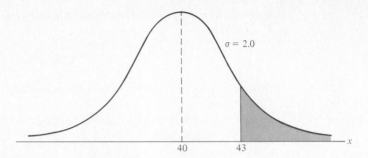

Figure 5.18   Area for Example 5.11.

obtaining the area to the left of z from Table A.3, and then subtracting this area from 1. We find

$$z = \frac{43 - 40}{2} = 1.5.$$

Hence

$$P(X > 43) = P(Z > 1.5)$$
$$= 1 - P(Z < 1.5)$$
$$= 1 - 0.9332$$
$$= 0.0668.$$

Therefore, 6.68% of the resistors will have a resistance exceeding 43 ohms.

Example 5.12   Find the percentage of resistances exceeding 43 ohms in Example 5.11 if the resistance is measured to the nearest ohm.

Solution. This problem differs from Example 5.11 in that we now assign a measurement of 43 ohms to all resistors whose resistances are greater than 42.5 and less than 43.5. We are actually approximating a discrete distribution

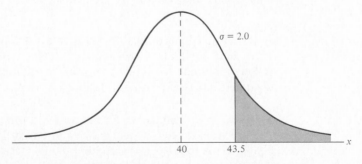

Figure 5.19   Area for Example 5.12.

by means of a continuous normal distribution. The required area is the region shaded to the right of 43.5 in Figure 5.19. We now find that

$$z = \frac{43.5 - 40}{2} = 1.75.$$

Hence

$$P(X > 43.5) = P(Z > 1.75)$$
$$= 1 - P(Z < 1.75)$$
$$= 1 - 0.9599$$
$$= 0.0401.$$

Therefore, 4.01% of the resistances exceed 43 ohms when measured to the nearest ohm. The difference 6.68% − 4.01% = 2.67% between this answer and that of Example 5.11 represents all those resistors having a resistance greater than 43 and less than 43.5 that are now being recorded as 43 ohms.

Example 5.13 The quality grade-point averages of 300 college freshmen follow approximately a normal distribution with a mean of 2.1 and a standard deviation of 0.6. How many of these freshmen would you expect to have a score between 2.5 and 3.5 inclusive if the grade-point averages are computed to the nearest tenth?

Solution. Since the scores are recorded to the nearest tenth, we require the area between $x_1 = 2.45$ and $x_2 = 3.55$, as indicated in Figure 5.20. The corresponding z values are

$$z_1 = \frac{2.45 - 2.1}{0.6} = 0.58$$

and

$$z_2 = \frac{3.55 - 2.1}{0.6} = 2.42.$$

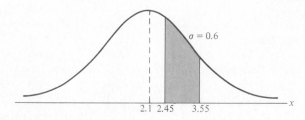

Figure 5.20 Area for Example 5.13.

Therefore,

$$P(2.45 < X < 3.55) = P(0.58 < Z < 2.42)$$
$$= P(Z < 2.42) - P(Z < 0.58)$$
$$= 0.9922 - 0.7190$$
$$= 0.2732.$$

Hence 27.32%, or approximately 82 of the 300 freshmen, should have a score between 2.5 and 3.5 inclusive.

Exercises

1. Given a standard normal distribution, find the area under the curve which lies
   (a) to the left of $z = 1.43$;
   (b) to the right of $z = -0.89$;
   (c) between $z = -2.16$ and $z = -0.65$;
   (d) to the left of $z = -1.39$;
   (e) to the right of $z = 1.96$;
   (f) between $z = -0.48$ and $z = 1.74$.

2. Find the value of z if the area under a standard normal curve
   (a) to the right of z is 0.3622;
   (b) to the left of z is 0.1131;
   (c) between 0 and z, with $z > 0$, is 0.4838;
   (d) between $-z$ and z, with $z > 0$, is 0.9500.

3. Given a standard normal distribution, find the value of k such that
   (a) $P(Z < k) = 0.0427$;
   (b) $P(Z > k) = 0.2946$;
   (c) $P(-0.93 < Z < k) = 0.7235$.

4. Given a normal distribution with $\mu = 30$ and $\sigma = 6$, find
   (a) the normal-curve area to the right of $x = 17$;
   (b) the normal-curve area to the left of $x = 22$;
   (c) the normal-curve area between $x = 32$ and $x = 41$;
   (d) the value of x that has 80% of the normal-curve area to the left;
   (e) the two values of x that contain the middle 75% of the normal-curve area.

5. Given the normally distributed variable X with mean 18 and standard deviation 2.5, find
   (a) $P(X < 15)$;
   (b) the value of k such that $P(X < k) = 0.2236$;

(c) the value of k such that $P(X > k) = 0.1814$;
(d) $P(17 < X < 21)$.

6. According to Chebyshev's theorem, the probability that any random variable assumes a value within 3 standard deviations of the mean is at least 8/9. If it is known that the probability distribution of a random variable X is normal with mean μ and variance σ^2, what is the exact value of $P(\mu - 3\sigma < X < \mu + 3\sigma)$?

7. A UCLA research scientist reports that mice will live an average of 40 months when their diets are sharply restricted and then enriched with vitamins and proteins. Assuming that the lifetimes of such mice are normally distributed with a standard deviation of 6.3 months, find the probability that a given mouse will live
   (a) more than 32 months;
   (b) less than 28 months;
   (c) between 37 and 49 months.

8. The loaves of rye bread distributed to local stores by a certain bakery have an average length of 30 centimeters and a standard deviation of 2 centimeters. Assuming that the lengths are normally distributed, what percentage of the loaves are
   (a) longer than 31.7 centimeters?
   (b) between 29.3 and 33.5 centimeters in length?
   (c) shorter than 25.5 centimeters?

9. A soft-drink machine is regulated so that it discharges an average of 200 milliliters per cup. If the amount of drink is normally distributed with a standard deviation equal to 15 milliliters,
   (a) what fraction of the cups will contain more than 224 milliliters?

(b) what is the probability that a cup contains between 191 and 209 milliliters?

(c) how many cups will probably overflow if 230-milliliter cups are used for the next 1000 drinks?

(d) below what value do we get the smallest 25% of the drinks?

10. The finished inside diameter of a piston ring is normally distributed with a mean of 10 centimeters and a standard deviation of 0.03 centimeter.

(a) What proportion of rings will have inside diameters exceeding 10.075 centimeters?

(b) What is the probability that a piston ring will have an inside diameter between 9.97 and 10.03 centimeters?

(c) Below what value of inside diameter will 15% of the piston rings fall?

11. A lawyer commutes daily from his suburban home to his midtown office. On the average the trip one way takes 24 minutes, with a standard deviation of 3.8 minutes. Assume the distribution of trip times to be normally distributed.

(a) What is the probability that a trip will take at least 1/2 hour?

(b) If the office opens at 9:00 A.M. and he leaves his house at 8:45 A.M. daily, what percentage of the time is he late for work?

(c) If he leaves the house at 8:35 A.M. and coffee is served at the office from 8:50 A.M. until 9:00 A.M., what is the probability that he misses coffee?

(d) Find the length of time above which we find the slowest 15% of the trips.

(e) Find the probability that 2 of the next 3 trips will take at least 1/2 hour.

12. If a set of grades on a statistics examination are approximately normally distributed with a mean of 74 and a standard deviation of 7.9, find

(a) the lowest passing grade if the lowest 10% of the students are given F's;

(b) the highest B if the top 5% of the students are given A's;

(c) the lowest B if the top 10% of the students are given A's and the next 25% are given B's.

13. In a mathematics examination the average grade was 82 and the standard deviation was 5. All students with grades from 88 to 94 received a grade of B. If the grades are approximately normally distributed and 8 students received a B grade, how many students took the examination?

14. The heights of 1000 students are normally distributed with a mean of 174.5 centimeters and a standard deviation of 6.9 centimeters. Assuming that the heights are recorded to the nearest half-centimeter, how many of these students would you expect to have heights

(a) less than 160.0 centimeters?

(b) between 171.5 and 182.0 centimeters inclusive?

(c) equal to 175.0 centimeters?

(d) greater than or equal to 188.0 centimeters?

15. A company pays its employees an average wage of $9.25 an hour with a standard deviation of 60 cents. If the wages are approximately normally distributed and paid to the nearest cent,

(a) what percentage of the workers receive wages between $8.75 and $9.69 an hour inclusive?

(b) the highest 5% of the employee hourly wages is greater than what amount?

16. The weights of a large number of miniature poodles are approximately normally distributed with a mean of 8 kilograms and a standard deviation of 0.9 kilogram. If measurements are recorded to the nearest tenth of a kilogram, find the fraction of these poodles with weights

(a) over 9.5 kilograms;

(b) at most 8.6 kilograms;

(c) between 7.3 and 9.1 kilograms inclusive.

17. The tensile strength of a certain metal component is normally distributed with a mean of 10,000 kilograms per square centimeter and a standard deviation of 100 kilograms per square centimeter. Measurements are recorded to the nearest 50 kilograms per square centimeter.

(a) What proportion of these components exceed 10,150 kilograms per square centimeter in tensile strength?

(b) If specifications require that all components have tensile strength between 9800 and 10,200 kilograms per square centimeter inclusive, what proportion of pieces would we expect to scrap?

18. If a set of observations are normally distributed, what percent of these differ from the mean by

(a) more than 1.3σ?

(b) less than 0.52σ?

19. The IQ's of 600 applicants to a certain college are approximately normally distributed with a mean

of 115 and a standard deviation of 12. If the college requires an IQ of at least 95, how many of these students will be rejected on this basis regardless of their other qualifications?

20. The average rainfall, recorded to the nearest hundredth of a centimeter, in Roanoke, Virginia, for the month of March is 9.22 centimeters. Assuming a normal distribution with a standard deviation of 2.83 centimeters, find the probability that next March Roanoke receives
   (a) less than 1.84 centimeters of rain;
   (b) more than 5 centimeters but not over 7 centimeters of rain;
   (c) more than 13.8 centimeters of rain.

21. The average life of a certain type of small motor is 10 years with a standard deviation of 2 years. The manufacturer replaces free all motors that fail while under guarantee. If he is willing to replace only 3% of the motors that fail, how long a guarantee should he offer? Assume that the lives of the motors follow a normal distribution.

5.4 Normal Approximation to the Binomial

Probabilities associated with binomial experiments are readily obtainable from the formula $b(x; n, p)$ of the binomial distribution or from Table A.1 when n is small. If n is large, we usually compute the binomial probabilities by approximation procedures. In Section 4.6 we illustrated how the Poisson distribution can be used to approximate bonomial probabilities when n is quite large and p is very close to 0 or 1. Both the binomial and Poisson distributions are discrete. The first application of a continuous probability distribution to approximate probabilities over a discrete sample space was demonstrated in Examples 5.12 and 5.13, where the normal curve was used. We now state a theorem that allows us to use areas under the normal curve to approximate binomial probabilities when n is sufficiently large.

Theorem 5.1   *If X is a binomial random variable with mean $\mu = np$ and variance $\sigma^2 = npq$, then the limiting form of the distribution of*

$$Z = \frac{X - np}{\sqrt{npq}},$$

as $n \to \infty$, is the standard normal distribution $n(z; 0, 1)$.

It turns out that the normal distribution with $\mu = np$ and $\sigma^2 = np(1 - p)$ not only provides a very accurate approximation to the binomial distribution when n is large and p is not extremely close to 0 or 1, but also provides a fairly good approximation even when n is small and p is reasonably close to 1/2.

To investigate the normal approximation to the binomial distribution, we first draw the histogram for $b(x; 15, 0.4)$ and then superimpose the particular normal curve having the same mean and variance as the binomial variable X.

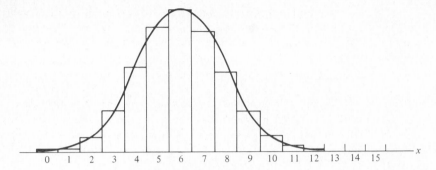

Figure 5.21   Normal approximation of $b(x; 15, 0.4)$.

Hence we draw a normal curve with

$$\mu = np = (15)(0.4) = 6$$

and

$$\sigma^2 = npq = (15)(0.4)(0.6) = 3.6.$$

The histogram of $b(x; 15, 0.4)$ and the corresponding superimposed normal curve, which is completely determined by its mean and variance, are illustrated in Figure 5.21.

The exact probability that the binomial random variable X assumes a given value x is equal to the area whose base is centered at x. For example, the exact probability that X assumes the value 4 is equal to the area of the rectangle with base centered at $x = 4$. Using Table A.1, we find this area to be

$$P(X = 4) = b(4; 15, 0.4) = 0.1268,$$

which is approximately equal to the area of the shaded region under the normal curve between the two ordinates $x_1 = 3.5$ and $x_2 = 4.5$ in Figure 5.22.

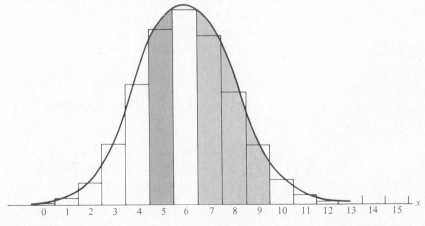

Figure 5.22   Normal approximation of $b(4; 15, p.4)$ and $\sum_{x=7}^{9} b(x; 15, 0.4)$.

Converting to z values, we have

$$z_1 = \frac{3.5 - 6}{1.897} = -1.32$$

and

$$z_2 = \frac{4.5 - 6}{1.897} = -0.79.$$

If X is a binomial random variable and Z a standard normal variable, then

$$
\begin{aligned}
P(X = 4) &= b(4; 15, 0.4) \\
&\simeq P(-1.32 < Z < -0.79) \\
&= P(Z < -0.79) - P(Z < -1.32) \\
&= 0.2148 - 0.0934 \\
&= 0.1214.
\end{aligned}
$$

This agrees very closely with the exact value of 0.1268.

The normal approximation is most useful in calculating binomial sums for large values of n, which, without tables of binomial sums, is an impossible task. Referring to Figure 5.22, we might be interested in the probability that X assumes a value from 7 to 9 inclusive. The exact probability is given by

$$
\begin{aligned}
P(7 \leq X \leq 9) &= \sum_{x=7}^{9} b(x; 15, 0.4) \\
&= \sum_{x=0}^{9} b(x; 15, 0.4) - \sum_{x=0}^{6} b(x; 15, 0.4) \\
&= 0.9662 - 0.6098 \\
&= 0.3564,
\end{aligned}
$$

which is equal to the sum of the areas of the rectangles with bases centered at $x = 7$, 8, and 9. For the normal approximation we find the area of the shaded region under the curve between the ordinates $x_1 = 6.5$ and $x_2 = 9.5$ in Figure 5.22. The corresponding z values are

$$z_1 = \frac{6.5 - 6}{1.897} = 0.26$$

and

$$z_2 = \frac{9.5 - 6}{1.897} = 1.85.$$

Now,

$$
\begin{aligned}
P(7 \leq X \leq 9) &\simeq P(0.26 < Z < 1.85) \\
&= P(Z < 1.85) - P(Z < 0.26) \\
&= 0.9678 - 0.6026 \\
&= 0.3652.
\end{aligned}
$$

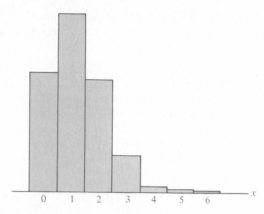

Figure 5.23 Histogram for $b(x; 6, 0.2)$.

Once again, the normal curve approximation provides a value that agrees very closely with the exact value of 0.3564. The degree of accuracy, which depends on how well the curve fits the histogram, will increase as n increases. This is particularly true when p is not very close to $1/2$ and the histogram is no longer symmetric. Figures 5.23 and 5.24 show the histograms for $b(x; 6, 0.2)$ and $b(x; 15, 0.2)$, respectively. It is evident that a normal curve would fit the histogram when $n = 15$ considerably better than when $n = 6$.

In summary, we use the normal approximation to evaluate binomial probabilities whenever p is not close to 0 or 1. The approximation is excellent when n is large and fairly good for small values of n if p is reasonably close to $1/2$. One possible guide to determine when the normal approximation may be used is provided by calculating np and nq. If both np and nq are greater than or equal to 5 the approximation will be good.

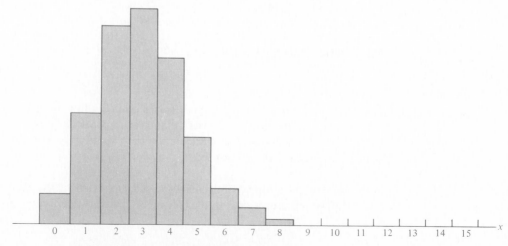

Figure 5.24 Histogram for $b(x; 15, 0.2)$.

Example 5.14 The probability that a patient recovers from a rare blood disease is 0.4. If 100 people are known to have contracted this disease, what is the probability that less than 30 survive?

Solution. Let the binomial variable X represent the number of patients that survive. Since $n = 100$, we should obtain fairly accurate results using the normal-curve approximation with

$$\mu = np = (100)(0.4) = 40$$

and

$$\sigma = \sqrt{npq} = \sqrt{(100)(0.4)(0.6)} = 4.899.$$

To obtain the desired probability, we have to find the area to the left of $x = 29.5$. The z value corresponding to 29.5 is

$$z = \frac{29.5 - 40}{4.899} = -2.14,$$

and the probability of fewer than 30 of the 100 patients surviving is given by the shaded region in Figure 5.25. Hence

$$P(X < 30) \simeq P(Z < -2.14) = 0.0162.$$

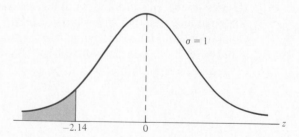

Figure 5.25   Area for Example 5.14.

Example 5.15 A multiple-choice quiz has 200 questions each with 4 possible answers of which only 1 is the correct answer. What is the probability that sheer guess-work yields from 25 to 30 correct answers for 80 of the 200 problems about which the student has no knowledge?

Solution. The probability of a correct answer for each of the 80 questions is $p = 1/4$. If X represents the number of correct answers due to guesswork, then

$$P(25 \le X \le 30) = \sum_{x=25}^{30} b(x; 80, \tfrac{1}{4}).$$

Using the normal curve approximation with

$$\mu = np = (80)(\tfrac{1}{4}) = 20$$

and

$$\sigma = \sqrt{npq} = \sqrt{(80)(\tfrac{1}{4})(\tfrac{3}{4})} = 3.873,$$

we need the area between $x_1 = 24.5$ and $x_2 = 30.5$. The corresponding z values are

$$z_1 = \frac{24.5 - 20}{3.873} = 1.16$$

and

$$z_2 = \frac{30.5 - 20}{3.873} = 2.71.$$

The probability of correctly guessing from 25 to 30 questions is given by the shaded region in Figure 5.26. From Table A.3 we find that

$$P(25 \le X \le 30) = \sum_{x=25}^{30} b(x;\ 80,\ \tfrac{1}{4})$$

$$\simeq P(1.16 < Z < 2.71)$$

$$= P(Z < 2.71) - P(Z < 1.16)$$

$$= 0.9966 - 0.8770$$

$$= 0.1196.$$

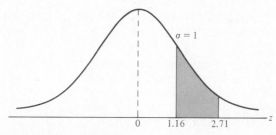

Figure 5.26 Area for Example 5.15.

Exercises

Any 3

1. Evaluate $P(1 \le X \le 4)$ for a binomial variable with $n = 15$ and $p = 0.2$ by using
   (a) Table A.1 in the Appendix;
   (b) the normal-curve approximation.

2. A coin is tossed 400 times. Use the normal curve approximation to find the probability of obtaining
   (a) between 185 and 210 heads inclusive;
   (b) exactly 205 heads;
   (c) less than 176 or more than 227 heads.

3. A pair of dice is rolled 180 times. What is the probability that a total of 7 occurs
   (a) at least 25 times?
   (b) between 33 and 41 times inclusive?
   (c) exactly 30 times?

4. A process yields 10% defective items. If 100 items are randomly selected from the process, what is the probability that the number of defectives
   (a) exceeds 13?
   (b) is less than 8?

5. The probability that a patient recovers from a delicate heart operation is 0.9. Of the next 100 patients having this operation, what is the probability that
   (a) between 84 and 95 inclusive survive?
   (b) fewer than 86 survive?

6. Researchers at George Washington University and the National Institute of Health claim that approximately 75% of the people believe "tranquilizers work very well to make a person more calm and relaxed." Of the next 80 people interviewed, what is the probability that
   (a) at least 50 are of this opinion?
   (b) at most 56 are of this opinion?

7. If 20% of the residents in a United States city prefer a white telephone over any other color available, what is the probability that among the next 1000 telephones installed in this city
   (a) between 170 and 185 inclusive will be white?
   (b) at least 210 but not more than 225 will be white?

8. A drug manufacturer claims that a certain drug cures a blood disease on the average 80% of the time. To check the claim, government testers used the drug on a sample of 100 individuals and decided to accept the claim if 75 or more are cured.
   (a) What is the probability that the claim will be rejected when the cure probability is, in fact, 0.8?
   (b) What is the probability that the claim will be accepted by the government when the cure probability is as low as 0.7?

9. One-sixth of the male freshmen entering a large state school are out-of-state students. If the students are assigned at random to the dormitories, 180 to a building, what is the probability that in a given dormitory at least one-fifth of the students are from out of state?

10. A pharmaceutical company knows that approximately 5% of its birth-control pills have an ingredient that is below the minimum strength, thus rendering the pill ineffective. What is the probability that fewer than 10 in a sample of 200 pills will be ineffective?

11. Statistics released by the National Highway Traffic Safety Administration and the National Safety Council show that on an average weekend night, 1 out of every 10 drivers on the road is drunk. If 400 drivers are randomly checked next Saturday night, what is the probability that the number of drunk drivers will be
    (a) less than 32?
    (b) more than 49?
    (c) at least 35 but less than 47?

12. According to a study published by a group of sociologists at the University of Massachusetts, approximately 49% of the Valium users in the state of Massachusetts are white-collar workers. What is the probability that between 482 and 510, inclusive, of the next 1000 randomly selected Valium users from this state will be white-collar workers?

13. According to the 1981 May/June issue of *Consumers Digest*, census figures show that in 1978 almost 53% of all households in the United States were composed of only one or two persons. Assuming that this percentage is still valid today, what is the probability that between 490 and 515 inclusive of the next 1000 randomly selected households in America will consist of either one or two persons?

5.5 *Gamma, Exponential, and Chi-Square Distributions*

Although the normal distribution can be used to solve many problems in engineering and science, there are still numerous situations that require different types of density functions. Three such density functions, the **gamma**, **exponential**, and **chi-square distributions**, will be discussed in this section. Additional densities are presented in Exercises 9 and 15 on pages 160 and 161 and in Sections 5.6, 6.8, and 6.9.

The gamma distribution derives its name from the well-known **gamma function**, studied in many areas of mathematics. Before we proceed to the gamma distribution, let us review this function and some of its important properties.

Definition 5.2   *The* **gamma function** *is defined by*

$$\Gamma(\alpha) = \int_0^\infty x^{\alpha-1} e^{-x}\, dx$$

for $\alpha > 0$.

Integrating by parts with $u = x^{\alpha-1}$ and $dv = e^{-x}\, dx$, we obtain

$$\Gamma(\alpha) = -e^{-x} x^{\alpha-1} \Big|_0^\infty + \int_0^\infty e^{-x}(\alpha-1)x^{\alpha-2}\, dx$$

$$= (\alpha-1) \int_0^\infty x^{\alpha-2} e^{-x}\, dx$$

for $\alpha > 1$, which yields the recursion formula

$$\Gamma(\alpha) = (\alpha-1)\Gamma(\alpha-1).$$

Repeated application of the recursion formula gives

$$\Gamma(\alpha) = (\alpha-1)(\alpha-2)\Gamma(\alpha-2)$$
$$= (\alpha-1)(\alpha-2)(\alpha-3)\Gamma(\alpha-3),$$

and so forth. Note that when $\alpha = n$, where n is a positive integer,

$$\Gamma(n) = (n-1)(n-2), \ldots, \Gamma(1).$$

However, by Definition 5.2,

$$\Gamma(1) = \int_0^\infty e^{-x}\, dx = 1$$

and hence

$$\Gamma(n) = (n-1)!.$$

One important property of the gamma function, left for the reader to verify (see Exercise 3 on page 160), is that $\Gamma(1/2) = \sqrt{\pi}$.

We shall now include the gamma function in our definition of the gamma distribution.

Gamma Distribution   *The continuous random variable X has a* **gamma distribution**, *with parameters α and β, if its density function is given by*

$$f(x) = \begin{cases} \dfrac{1}{\beta^{\alpha}\Gamma(\alpha)}\, x^{\alpha-1} e^{-x/\beta}, & x > 0 \\[2mm] 0, & elsewhere, \end{cases}$$

where $\alpha > 0$ and $\beta > 0$.

Graphs of several gamma distributions are shown in Figure 5.27 for certain specified values of the parameters α and β. The special gamma distribution for which $\alpha = 1$ is called the **exponential distribution**.

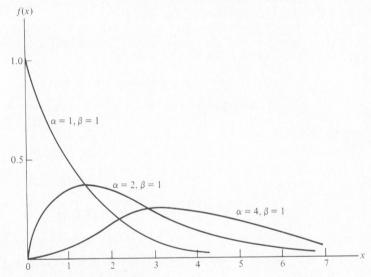

Figure 5.27   Gamma distributions.

Exponential Distribution   *The continuous random variable X has an* **exponential distribution**, *with parameter β, if its density function is given by*

$$f(x) = \begin{cases} \dfrac{1}{\beta}\, e^{-x/\beta}, & x > 0 \\[2mm] 0, & elsewhere, \end{cases}$$

where $\beta > 0$.

The exponential distribution has many applications in the field of statistics, particularly in the areas of **reliability theory** and **waiting times** or **queueing problems**. One interesting application of the exponential distribution to reliability theory is given in the following example. Additional applications are given in Exercises 16 and 17 on page 161.

Example 5.16 Suppose that a system contains a certain type of component whose time in years to failure is given by the random variable T, distributed exponentially with parameter $\beta = 5$. If 5 of these components are installed in different systems, what is the probability that at least 2 are still functioning at the end of 8 years?

Solution. The probability that a given component is still functioning after 8 years is given by

$$P(T > 8) = \frac{1}{5} \int_8^\infty e^{-t/5} \, dt$$

$$= e^{-8/5}$$

$$\simeq 0.2.$$

Let X represent the number of components functioning after 8 years. Then, using the bonomial distribution,

$$P(X \geq 2) = \sum_{x=2}^{5} b(x; 5, 0.2)$$

$$= 1 - \sum_{x=0}^{1} b(x; 5, 0.2)$$

$$= 1 - 0.7373$$

$$= 0.2627.$$

A second special case of the gamma distribution is obtained by letting $\alpha = v/2$ and $\beta = 2$, where v is a positive integer. The probability density so obtained is called the **chi-square distribution with v degrees of freedom**.

Chi-Square Distribution   *The continuous random variable X has a* **chi-square distribution**, *with v degrees of freedom, if its density function is given by*

$$f(x) = \frac{1}{2^{v/2}\Gamma(v/2)} x^{v/2-1} e^{-x/2}, \qquad x > 0$$

$$0, \qquad\qquad\qquad\qquad\qquad elsewhere,$$

vhere v is a positive integer.

The chi-square distribution is one of our main tools in the area of hypothesis testing, a subject to be studied in later chapters. We shall now derive the mean and variance of these three density functions.

Theorem 5.2   *The mean and variance of the gamma distribution are*

$$\mu = \alpha\beta \quad \text{and} \quad \sigma^2 = \alpha\beta^2.$$

Proof. To find the mean of the gamma distribution, we write

$$\mu = E(X) = \frac{1}{\beta^\alpha\Gamma(\alpha)} \int_0^\infty x^\alpha e^{-x/\beta}\, dx.$$

Now, let $y = x/\beta$, to give

$$\mu = \frac{\beta}{\Gamma(\alpha)} \int_0^\infty y^\alpha e^{-y}\, dy$$

$$= \frac{\beta\Gamma(\alpha + 1)}{\Gamma(\alpha)} = \alpha\beta.$$

To find the variance of the gamma distribution, we proceed as above to obtain

$$E(X^2) = \frac{\beta^2\Gamma(\alpha + 2)}{\Gamma(\alpha)} = (\alpha + 1)\alpha\beta^2,$$

and then

$$\sigma^2 = E(X^2) - \mu^2 = (\alpha + 1)\alpha\beta^2 - \alpha^2\beta^2$$

$$= \alpha\beta^2.$$

Corollary 1   *The mean and variance of the exponential distribution are*

$$\mu = \beta \quad \text{and} \quad \sigma^2 = \beta^2.$$

Corollary 2   *The mean and variance of the chi-square distribution are*

$$\mu = v \quad \text{and} \quad \sigma^2 = 2v.$$

5.6 *Weibull Distribution*

Modern technology has enabled us to design many complicated systems whose operation, or perhaps safety, depend on the reliability of the various components making up the systems. For example, a fuse may burn out, a steel column may buckle, or a heat-sensing device may fail. Identical components subjected to identical environmental conditions will fail at different and unpredictable times. One distribution that has been used extensively in recent years to deal with such problems is the **Weibull distribution** introduced by the Swedish physicist Waloddi Weibull in 1939.

Weibull Distribution   *The continuous random variable X has a* **Weibull distribution**, *with parameters α and β, if its density function is given by*

$$f(x) = \begin{cases} \alpha\beta x^{\beta-1}e^{-\alpha x^{\beta}} & x > 0 \\ 0, & elsewhere, \end{cases}$$

where α > 0 and β > 0.

The graphs of the Weibull distribution for $\alpha = 1$ and various values of the parameter β are illustrated in Figure 5.28. We see that the curves change in shape considerably for different values of the parameters, particularly the parameter β. If we let $\beta = 1$, the Weibull distribution reduces to the exponential distribution. For values of $\beta > 1$, the curves become somewhat bell-shaped and resemble the normal curves, but display some skewness.

The mean and variance of the Weibull distribution are stated in the following theorem. The reader is asked to provide the proof in Exercise 13 on page 160.

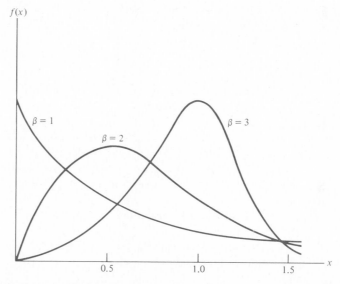

Figure 5.28   Weibull distributions ($\alpha = 1$).

Theorem 5.3   *The mean and variance of the Weibull distribution are*

$$\mu = \alpha^{-1/\beta}\Gamma\left(1 + \frac{1}{\beta}\right)$$

$$\sigma^2 = \alpha^{-2/\beta}\left\{\Gamma\left(1 + \frac{2}{\beta}\right) - \left[\Gamma\left(1 + \frac{1}{\beta}\right)\right]^2\right\}.$$

The Weibull distribution is often applied to reliability and life testing problems such as the **time to failure** or **life length** of a component, measured from some specified time until it fails. Let us represent this time to failure by the continuous random variable T with probability density function $f(t)$, where $f(t)$ is the Weibull distribution.

To apply the Weibull distribution to reliability theory, let us first define the reliability of a component or product as the *probability that it will function properly for at least a specified time under specified experimental conditions.* Therefore, if $R(t)$ is defined to be the reliability of the given component at time t, we may write

$$R(t) = P(T > t)$$

$$= \int_t^\infty f(t)\, dt$$

$$= 1 - F(t),$$

where $F(t)$ is the cumulative distribution of T. The conditional probability that a component will fail in the interval from $T = t$ to $T = t + \Delta t$, given that it survived to time t, is given by

$$\frac{F(t + \Delta t) - F(t)}{R(t)}.$$

Dividing this ratio by Δt and taking the limit as $\Delta t \to 0$, we get the **failure rate**, denoted by $Z(t)$. Hence

$$Z(t) = \lim_{\Delta t \to 0} \frac{F(t + \Delta t) - F(t)}{\Delta t} \frac{1}{R(t)}$$

$$= \frac{F'(t)}{R(t)} = \frac{f(t)}{R(t)} = \frac{f(t)}{1 - F(t)},$$

which expresses the failure rate in terms of the distribution of the time to failure.

From the fact that $R(t) = 1 - F(t)$ and then $R'(t) = -F'(t)$, we can write the differential equation

$$Z(t) = \frac{-R'(t)}{R(t)} = \frac{-d[\ln R(t)]}{dt}$$

and then solving,

$$\ln R(t) = -\int Z(t)\, dt + \ln c$$

or

$$R(t) = ce^{-\int Z(t)\,dt},$$

where c satisfies the initial assumption that $R(0) = 1$ or $F(0) = 1 - R(0) = 0$. Thus we see that a knowledge of either the density function $f(t)$ or the failure rate $Z(t)$ uniquely determines the other.

Example 5.17 Show that the failure-rate function is given by

$$Z(t) = \alpha\beta t^{\beta-1}, \qquad t > 0,$$

if and only if the time to failure distribution is the Weibull distribution

$$f(t) = \alpha\beta t^{\beta-1} e^{-\alpha t^\beta}, \qquad t > 0.$$

Solution. Assume that $Z(t) = \alpha\beta t^{\beta-1}$, $t > 0$. Then we can write

$$f(t) = Z(t)R(t),$$

where

$$R(t) = ce^{-\int Z(t)\,dt} = ce^{-\int \alpha\beta t^{\beta-1}\,dt} = ce^{-\alpha t^\beta}.$$

From the condition that $R(0) = 1$, we find that $c = 1$. Hence

$$R(t) = e^{-\alpha t^\beta}$$

and

$$f(t) = \alpha\beta t^{\beta-1} e^{-\alpha t^\beta}, \qquad t > 0.$$

Now, if we assume

$$f(t) = \alpha\beta t^{\beta-1} e^{-\alpha t^\beta}, \qquad t > 0,$$

then $Z(t)$ is determined by writing

$$Z(t) = \frac{f(t)}{R(t)},$$

where

$$R(t) = 1 - F(t) = 1 - \int_0^t \alpha\beta x^{\beta-1} e^{-\alpha x^\beta}\,dx$$

$$= 1 + \int_0^t de^{-\alpha x^\beta}$$

$$= e^{-\alpha t^\beta}.$$

Then

$$Z(t) = \frac{\alpha\beta t^{\beta-1} e^{-\alpha t^\beta}}{e^{-\alpha t^\beta}} = \alpha\beta t^{\beta-1}, \qquad t > 0.$$

In Example 5.17 the failure rate is seen to decrease with time if $\beta < 1$, increases with time if $\beta > 1$, and is constant if $\beta = 1$. In view of the fact that the Weibull distribution with $\beta = 1$ reduces to the exponential distribution, the assumption of constant failure rate is often referred to as the **exponential assumption**.

Exercises

1. If a random variable X has the gamma distribution with $\alpha = 2$ and $\beta = 1$, find $P(1.8 < X < 2.4)$.

2. In a certain city, the daily consumption of water (in millions of liters) follows approximately a gamma distribution with $\alpha = 2$ and $\beta = 3$. If the daily capacity of this city is 9 million liters of water, what is the probability that on any given day the water supply is inadequate?

3. Use the gamma function with $y = \sqrt{2x}$ to show that $\Gamma(1/2) = \sqrt{\pi}$.

4. Suppose that the time, in hours, taken to repair a heat pump is a random variable X having a gamma distribution with parameters $\alpha = 2$ and $\beta = 1/2$. What is the probability that the next service call will require
   (a) at most 1 hour to repair the heat pump?
   (b) at least 2 hours to repair the heat pump?

5. (a) Find the mean and variance of the daily water consumption in Exercise 2.
   (b) According to Chebyshev's theorem, there is a probability of at least 3/4 that the water consumption on any given day will fall within what interval?

6. In a certain city, the daily consumption of electric power, in millions of kilowatt-hours, is a random variable X having a gamma distribution with mean $\mu = 6$ and variance $\sigma^2 = 12$.
   (a) Find the values of α and β.
   (b) Find the probability that on any given day the daily power consumption will exceed 12 million kilowatt-hours.

7. The length of time for one individual to be served at a cafeteria is a random variable having an exponential distribution with a mean of 4 minutes. What is the probability that a person is served in less than 3 minutes on at least 4 of the next 6 days?

8. The life in years of a certain type of electrical switch has an exponential distribution with a failure rate of $\beta = 2$. If 100 of these switches are installed in different systems, what is the probability that at most 30 fail during the first year?

9. A random variable X has the **continuous uniform distribution** if its density function is given by

$$f(x) = \begin{cases} \dfrac{1}{\beta - \alpha}, & \alpha < x < \beta \\ 0, & \text{elsewhere.} \end{cases}$$

For a continuous uniform distribution with $\alpha = 2$ and $\beta = 7$, find
(a) $P(X \geq 4)$;
(b) $P(3 < X < 5.5)$.

10. The daily amount of coffee, in liters, dispensed by a machine located in an airport lobby is a random variable X having a continuous uniform distribution (see Exercise 9) with $\alpha = 7$ and $\beta = 10$. Find the probability that on a given day the amount of coffee dispensed by this machine will be
    (a) at most 8.8 liters;
    (b) more than 7.4 liters but less than 9.5 liters;
    (c) at least 8.5 liters.

11. Given a continuous uniform distribution, show that

(a) $\mu = \dfrac{\alpha + \beta}{2}$;

(b) $\sigma^2 = \dfrac{(\beta - \alpha)^2}{12}$.

12. Suppose that the service life, in years, of a hearing aid battery is a random variable having a Weibull distribution with $\alpha = 1/2$ and $\beta = 2$.
    (a) How long can such a battery be expected to last?
    (b) What is the probability that such a battery will still be operating after 2 years?

13. Derive the mean and variance of the Weibull distribution.

14. The lives of a certain automobile seal have the Weibull distribution with failure rate $Z(t) = 1/\sqrt{t}$. Find the probability that such a seal is still in use after 4 years.

15. The continuous random variable X has the **beta distribution** with parameters α and β if its density function is given by

$$f(x) = \begin{cases} \dfrac{\Gamma(\alpha + \beta)}{\Gamma(\alpha)\Gamma(\beta)} x^{\alpha-1}(1-x)^{\beta-1}, & 0 < x < 1 \\ 0, & \text{elsewhere}, \end{cases}$$

where $\alpha > 0$ and $\beta > 0$. If the proportion of a brand of television sets requiring service during the first year of operation is a random variable having a beta distribution with $\alpha = 3$ and $\beta = 2$, what is the probability that at least 80% of the new models sold this year of this brand will require service during their first year of operation?

16. The exponential distribution is frequently applied to the waiting times between successes in a Poisson process. If the number of calls received per hour by a telephone answering service is a Poisson random variable with parameter $\mu = 6$, it can be shown that the time, in hours, between successive calls has an exponential distribution with parameter $\beta = 1/6$. What is the probability of waiting more than 15 minutes between any two successive calls?

17. When α is a positive integer n, the gamma distribution is also known as the **Erlang distribution**. Setting $\alpha = n$ in the gamma distribution on page 154, the Erlang distribution is

$$f(x) = \begin{cases} \dfrac{x^{n-1}e^{-x/\beta}}{\beta^n(n-1)!}, & x > 0 \\ 0, & \text{elsewhere}. \end{cases}$$

It can be shown that if the times between successive events are independent, each having an exponential distribution with parameter β, then the total elapsed waiting time X until all n events occur has the Erlang distribution. Referring to Exercise 16, what is the probability that the next 3 calls will be received within the next 30 minutes?

6

Functions of Random Variables

6.1 Transformations of Variables

Frequently, in statistics, one encounters the need to derive the probability distribution of a function of one or more random variables. For example, suppose that X is a discrete random variable with probability distribution $f(x)$ and suppose further that $Y = u(X)$ defines a one-to-one transformation between the values of X and Y. We wish to find the probability distribution of Y. It is important to note that the one-to-one transformation implies that each value x is related to one, and only one, value $y = u(x)$ and that each value y is related to one, and only one, value $x = w(y)$, where $w(y)$ is obtained by solving $y = u(x)$ for x in terms of y.

From our discussion of discrete probability distributions in Chapter 2 it is clear that the random variable Y assumes the value y when X assumes the value $w(y)$. Consequently, the probability distribution of Y is given by

$$g(y) = P(Y = y) = P[X = w(y)] = f[w(y)].$$

Theorem 6.1 *Suppose that X is a **discrete** random variable with probability distribution $f(x)$. Let $Y = u(X)$ define a one-to-one transformation between the values of X and Y so* *that the equation $y = u(x)$ can be uniquely solved for x in terms of y, say $x = w(y)$. Then the probability distribution of Y is*

$$g(y) = f[w(y)].$$

Example 6.1   Let X be a geometric random variable with probability distribution $f(x) = \frac{3}{4}(\frac{1}{4})^{x-1}$, $x = 1, 2, 3, \dots$. Find the probability distribution of the random variable $Y = X^2$.

Solution. Since the values of X are all positive, the transformation defines a one-to-one correspondence between the x and y values, $y = x^2$ and $x = \sqrt{y}$. Hence

$$g(y) = \begin{cases} f(\sqrt{y}) = \frac{3}{4}(\frac{1}{4})^{\sqrt{y}-1}, & y = 1, 4, 9, \dots \\ 0, & \text{elsewhere} \end{cases}$$

Consider a problem where X_1 and X_2 are two discrete random variables with joint probability distribution $f(x_1, x_2)$ and we wish to find the joint probability distribution $g(y_1, y_2)$ of the two new random variables $Y_1 = u_1(X_1, X_2)$ and $Y_2 = u_2(X_1, X_2)$, which define a one-to-one transformation between the set of points (x_1, x_2) and (y_1, y_2). Solving the equations $y_1 = u_1(x_1, x_2)$ and $y_2 = u_2(x_1, x_2)$ simultaneously, we obtain the unique inverse solutions $x_1 = w_1(y_1, y_2)$ and $x_2 = w_2(y_1, y_2)$. Hence the random variables Y_1 and Y_2 assume the values y_1 and y_2, respectively, when X_1 assumes the value $w_1(y_1, y_2)$ and X_2 assumes the value $w_2(y_1, y_2)$. The joint probability distribution of Y_1 and Y_2 is then

$$\begin{aligned} g(y_1, y_2) &= P(Y_1 = y_1, Y_2 = y_2) \\ &= P[X_1 = w_1(y_1, y_2), X_2 = w_2(y_1, y_2)] \\ &= f[w_1(y_1, y_2), w_2(y_1, y_2)]. \end{aligned}$$

Theorem 6.2   *Suppose that X_1 and X_2 are **discrete** random variables with joint probability distribution $f(x_1, x_2)$. Let $Y_1 = u_1(X_1, X_2)$ and $Y_2 = u_2(X_1, X_2)$ define a one-to-one transformation between the points (x_1, x_2) and (y_1, y_2) so that the equations $y_1 = u_1(x_1, x_2)$ and $y_2 = u_2(x_1, x_2)$ may be uniquely solved for x_1 and x_2 in terms of y_1 and y_2, say $x_1 = w_1(y_1, y_2)$ and $x_2 = w_2(y_1, y_2)$. Then the joint probability distribution of Y_1 and Y_2 is*

$$g(y_1, y_2) = f[w_1(y_1, y_2), w_2(y_1, y_2)].$$

Theorem 6.2 is extremely useful in finding the distribution of some random variable $Y_1 = u_1(X_1, X_2)$, where X_1 and X_2 are discrete random variables with joint probability distribution $f(x_1, x_2)$. We simply define a second function, say $Y_2 = u_2(X_1, X_2)$, maintaining a one-to-one correspondence between the points (x_1, x_2) and (y_1, y_2), and obtain the joint probability distribution $g(y_1, y_2)$. The distribution of Y_1 is just the marginal distribution of $g(y_1, y_2)$, found by summing over the y_2 values. Denoting the distribution of Y_1 by $h(y_1)$, we can then write

$$h(y_1) = \sum_{y_2} g(y_1, y_2).$$

Example 6.2 Let X_1 and X_2 be two independent random variables having Poisson distributions with parameters μ_1 and μ_2, respectively. Find the distribution of the random variable $Y_1 = X_1 + X_2$.

Solution. Since X_1 and X_2 are independent, we can write

$$f(x_1, x_2) = f(x_1)f(x_2)$$

$$= \frac{e^{-\mu_1}\mu_1^{x_1}}{x_1!} \frac{e^{-\mu_2}\mu_2^{x_2}}{x_2!}$$

$$= \frac{e^{-(\mu_1 + \mu_2)}\mu_1^{x_1}\mu_2^{x_2}}{x_1!\,x_2!}.$$

where $x_1 = 0, 1, 2, \ldots$ and $x_2 = 0, 1, 2, \ldots$. Let us now define a second random variable, say $Y_2 = X_2$. The inverse functions are given by $x_1 = y_1 - y_2$ and $x_2 = y_2$. Using Theorem 6.2, we find the joint probability distribution of Y_1 and Y_2 to be

$$g(y_1, y_2) = \frac{e^{-(\mu_1 + \mu_2)}\mu_1^{y_1 - y_2}\mu_2^{y_2}}{(y_1 - y_2)!\,y_2!},$$

where $y_1 = 0, 1, 2, \ldots$ and $y_2 = 0, 1, 2, \ldots, y_1$. Note that since $x_1 > 0$, the transformation $x_1 = y_1 - x_2$ implies that x_2 and hence y_2 must always be less than or equal to y_1. Consequently, the marginal probability distribution of Y_1 is

$$h(y_1) = \sum_{y_2 = 0}^{y_1} g(y_1, y_2)$$

$$= e^{-(\mu_1 + \mu_2)} \sum_{y_2 = 0}^{y_1} \frac{\mu_1^{y_1 - y_2}\mu_2^{y_2}}{(y_1 - y_2)!\,y_2!}$$

$$= \frac{e^{-(\mu_1 + \mu_2)}}{y_1!} \sum_{y_2 = 0}^{y_1} \frac{y_1!}{y_2!(y_1 - y_2)!} \mu_1^{y_1 - y_2}\mu_2^{y_2}$$

$$= \frac{e^{-(\mu_1 + \mu_2)}}{y_1!} \sum_{y_2 = 0}^{y_1} \binom{y_1}{y_2} \mu_1^{y_1 - y_2}\mu_2^{y_2}.$$

Recognizing this sum as the binomial expansion of $(\mu_1 + \mu_2)^{y_1}$, we obtain

$$h(y_1) = \frac{e^{-(\mu_1 + \mu_2)}(\mu_1 + \mu_2)^{y_1}}{y_1!}, \qquad y_1 = 0, 1, 2, \ldots,$$

from which we conclude that the sum of the two independent random variables having Poisson distributions, with parameters μ_1 and μ_2, has a Poisson distribution with parameter $\mu_1 + \mu_2$.

To find the probability distribution of the random variable $Y = u(X)$ when X is a continuous random variable and the transformation is one to one, we shall need Theorem 6.3.

Theorem 6.3   *Suppose that X is a **continuous** random variable with probability distribution $f(x)$. Let $Y = u(X)$ define a one-to-one correspondence between the values of X and Y so that the equation $y = u(x)$ can be uniquely solved for x in terms of y, say $x = w(y)$. Then the probability distribution of Y is*

$$g(y) = f[w(y)] |J|,$$

*where $J = w'(y)$ and is called the **Jacobian** of the transformation.*

Proof. (a) Suppose that $y = u(x)$ is an increasing function as in Figure 6.1. Then we see that whenever Y falls between a and b, the random variable X must fall between $w(a)$ and $w(b)$. Hence

$$P(a < Y < b) = P[w(a) < X < w(b)]$$

$$= \int_{w(a)}^{w(b)} f(x) \, dx.$$

Changing the variable of integration from x to y by the relation $x = w(y)$, we obtain $dx = w'(y) \, dy$, and hence

$$P(a < Y < b) = \int_{a}^{b} f[w(y)] w'(y) \, dy.$$

Since the integral gives the desired probability for every $a < b$ within the permissible set of y values, then the probability distribution of Y is

$$g(y) = f[w(y)] w'(y) = f[w(y)] \, J.$$

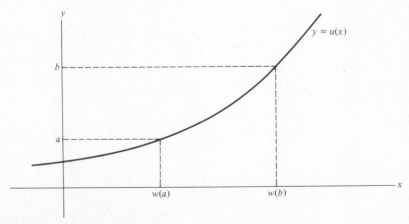

Figure 6.1   Increasing function.

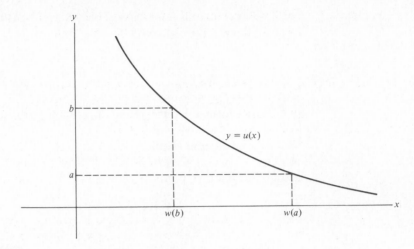

Figure 6.2 Decreasing function.

If we recognize $J = w'(y)$ as the reciprocal of the slope of the tangent line to the curve of the increasing function $y = u(x)$, it is then obvious that $J = |J|$. Hence

$$g(y) = f[w(y)]|J|.$$

(b) Suppose that $y = u(x)$ is a decreasing function as in Figure 6.2. Then we write

$$P(a < Y < b) = P[w(b) < X < w(a)]$$

$$= \int_{w(b)}^{w(a)} f(x)\ dx.$$

Again changing the variable of integration to y, we obtain

$$P(a < Y < b) = \int_b^a f[w(y)]w'(y)\ dy$$

$$= -\int_a^b f[w(y)]w'(y)\ dy,$$

from which we conclude that

$$g(y) = -f[w(y)]w'(y) = -f[w(y)]J.$$

In this case the slope of the curve is negative and $J = -|J|$. Hence

$$g(y) = f[w(y)]|J|,$$

as before.

Example 6.3   Let X be a continuous random variable with probability distribution

$$f(x) = \begin{cases} \dfrac{x}{12}, & 1 < x < 5 \\ 0, & \text{elsewhere.} \end{cases}$$

Find the probability distribution of the random variable $Y = 2X - 3$.

Solution. The inverse solution of $y = 2x - 3$ yields $x = (y + 3)/2$, from which we obtain $J = w'(y) = dx/dy = 1/2$. Therefore, using Theorem 6.3, we find the density function of Y to be

$$g(y) = \begin{cases} \dfrac{(y + 3)/2}{12} \left(\dfrac{1}{2} \right) = \dfrac{y + 3}{48}, & -1 < y < 7 \\ 0, & \text{elsewhere.} \end{cases}$$

To find the joint probability distribution of the random variables $Y_1 = u_1(X_1, X_2)$ and $Y_2 = u_2(X_1, X_2)$ when X_1 and X_2 are continuous and the transformation is one to one, we need an additional theorem, analogous to Theorem 6.2, which we state without proof.

Theorem 6.4   *Suppose that X_1 and X_2 are **continuous** random variables with joint probability distribution $f(x_1, x_2)$. Let $Y_1 = u_1(X_1, X_2)$ and $Y_2 = u_2(X_1, X_2)$ define a one-to-one transformation between the points (x_1, x_2) and (y_1, y_2) so that the equations $y_1 = u_1(x_1, x_2)$ and $y_2 = u_2(x_1, x_2)$ may be uniquely solved for x_1 and x_2 in terms of y_1 and y_2, say $x_1 = w_1(y_1, y_2)$ and $x_2 = w_2(y_1, y_2)$. Then the joint probability distribution of Y_1 and Y_2 is*

$$g(y_1, y_2) = f[w_1(y_1, y_2), w_2(y_1, y_2)] \, |J|,$$

where the Jacobian is the 2×2 determinant

$$J = \begin{vmatrix} \partial x_1/\partial y_1 & \partial x_1/\partial y_2 \\ \partial x_2/\partial y_1 & \partial x_2/\partial y_2 \end{vmatrix}$$

and $\partial x_1/\partial y_1$ is simply the derivative of $x_1 = w_1(y_1, y_2)$ with respect to y_1 with y_2 held constant, referred to in calculus as the partial derivative of x_1 with respect to y_1. The other partial derivatives are defined in a similar manner.

Example 6.4   Let X_1 and X_2 be two continuous random variables with joint probability distribution

$$f(x_1, x_2) = \begin{cases} 4x_1x_2, & 0 < x_1 < 1, \, 0 < x_2 < 1 \\ 0, & \text{elsewhere.} \end{cases}$$

Find the joint probability distribution of $Y_1 = X_1^2$ and $Y_2 = X_1X_2$.

Solution. The inverse solutions of $y_1 = x_1^2$ and $y_2 = x_1 x_2$ are $x_1 = \sqrt{y_1}$ and $x_2 = y_2/\sqrt{y_1}$, from which we obtain

$$J = \begin{vmatrix} 1/2\sqrt{y_1} & 0 \\ -y_2/2y_1^{3/2} & 1/\sqrt{y_1} \end{vmatrix} = \frac{1}{2y_1}.$$

To determine the set B of points in the y_1y_2-plane into which the set A of points in the x_1x_2-plane is mapped, we write

$$x_1 = \sqrt{y_1} \qquad \text{and} \qquad x_2 = y_2/\sqrt{y_1}$$

and then setting $x_1 = 0$, $x_2 = 0$, $x_1 = 1$, and $x_2 = 1$ the boundaries of set A are transformed to $y_1 = 0$, $y_2 = 0$, $y_1 = 1$, and $y_2 = \sqrt{y_1}$ or $y_2^2 = y_1$. The two regions are illustrated in Figure 6.3. Clearly, the transformation is one to one, mapping the set $A = \{(x_1, x_2) | 0 < x_1 < 1, 0 < x_2 < 1\}$ into the set $B = \{(y_1, y_2) | y_2^2 < y_1 < 1, 0 < y_2 < 1\}$. From Theorem 6.4 the joint probability distribution of Y_1 and Y_2 is

$$g(y_1, y_2) = 4(\sqrt{y_1}) \frac{y_2}{\sqrt{y_1}} \frac{1}{2y_1}$$

$$= \begin{cases} \dfrac{2y_2}{y_1}, & y_2^2 < y_1 < 1, 0 < y_2 < 1 \\ \\ 0, & \text{elsewhere.} \end{cases}$$

Problems frequently arise when we wish to find the probability distribution of the random variable $Y = u(X)$ when X is a continuous random variable and the transformation is not one to one. That is, to each value x there corresponds exactly one value y, but to each y value there corresponds more than one x value. For example, suppose that $f(x)$ is positive over the interval $-1 < x < 2$ and zero elsewhere. Consider the transformation $y = x^2$. In this case $x = \pm\sqrt{y}$ for $0 < y < 1$ and $x = \sqrt{y}$ for $1 < y < 4$. For the interval

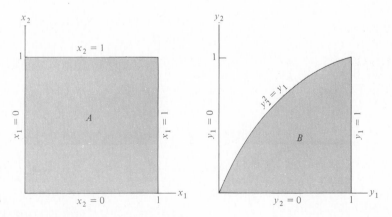

Figure 6.3   Mapping set A into set B.

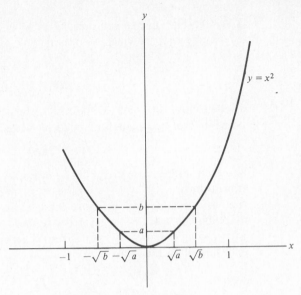

Figure 6.4   Decreasing and increasing function.

$1 < y < 4$, the probability distribution of Y is found as before, using Theorem 6.3. That is,

$$g(y) = f[w(y)]\,|J| = \frac{f(\sqrt{y})}{2\sqrt{y}}, \qquad 1 < y < 4.$$

However, when $0 < y < 1$, we may partition the interval $-1 < x < 1$ to obtain the two inverse functions

$$x = -\sqrt{y}, \qquad -1 < x < 0$$

and

$$x = \sqrt{y}, \qquad 0 < x < 1.$$

Then to every y value there corresponds a single x value for each partition. From Figure 6.4 we see that

$$P(a < Y < b) = P(-\sqrt{b} < X < -\sqrt{a}) + P(\sqrt{a} < X < \sqrt{b})$$

$$= \int_{-\sqrt{b}}^{-\sqrt{a}} f(x)\,dx + \int_{\sqrt{a}}^{\sqrt{b}} f(x)\,dx.$$

Changing the variable of integration from x to y, we obtain

$$P(a < Y < b) = \int_{b}^{a} f(-\sqrt{y})J_1\,dy + \int_{a}^{b} f(\sqrt{y})J_2\,dy$$

$$= -\int_{a}^{b} f(-\sqrt{y})J_1\,dy + \int_{a}^{b} f(\sqrt{y})J_2\,dy,$$

where

$$J_1 = \frac{d(-\sqrt{y})}{dy} = \frac{-1}{2\sqrt{y}} = -|J_1|$$

and

$$J_2 = \frac{d(\sqrt{y})}{dy} = \frac{1}{2\sqrt{y}} = |J_2|.$$

Hence we can write

$$P(a < Y < b) = \int_a^b [f(-\sqrt{y})|J_1| + f(\sqrt{y})|J_2|] \, dy,$$

and then

$$g(y) = f(-\sqrt{y})|J_1| + f(\sqrt{y})|J_2|$$
$$= \frac{[f(-\sqrt{y}) + f(\sqrt{y})]}{2\sqrt{y}}, \qquad 0 < y < 1.$$

The probability distribution of Y for $0 < y < 4$ may now be written

$$g(y) = \begin{cases} \dfrac{[f(-\sqrt{y}) + f(\sqrt{y})]}{2\sqrt{y}}, & 0 < y < 1 \\[2ex] \dfrac{f(\sqrt{y})}{2\sqrt{y}}, & 1 < y < 4 \\[2ex] 0, & \text{elsewhere.} \end{cases}$$

This procedure for finding $g(y)$ when $0 < y < 1$ is generalized in Theorem 6.5 for k inverse functions. For transformations not one to one of functions of several variables, the reader is referred to *Introduction to Mathematical Statistics* by Hogg and Craig (see the Bibliography).

Theorem 6.5 *Suppose that X is a **continuous** random variable with probability distribution $f(x)$. Let $Y = u(X)$ define a transformation between the values of X and Y that is not one to one. If the interval over which X is defined can be partitioned into k mutually disjoint sets such that each of the inverse functions $x_1 = w_1(y)$, $x_2 = w_2(y)$, ..., $x_k = w_k(y)$ of $y = u(x)$ defines a one-to-one correspondence, then the probability distribution of Y is*

$$g(y) = \sum_{i=1}^{k} f[w_i(y)] |J_i|,$$

where $J_i = w_i'(y)$, $i = 1, 2, \ldots, k$.

Example 6.5 Show that $Y = (X - \mu)^2/\sigma^2$ has a chi-square distribution with 1 degree of freedom when X has a normal distribution with mean μ and variance σ^2.

Solution. Let $Z = (X - \mu)/\sigma$, where the random variable Z has the standard normal distribution

$$f(z) = \frac{1}{\sqrt{2\pi}} e^{-z^2/2}, \qquad -\infty < z < \infty.$$

We shall now find the distribution of the random variable $Y = Z^2$. The inverse solutions of $y = z^2$ are $z = \pm\sqrt{y}$. If we designate $z_1 = -\sqrt{y}$ and $z_2 = \sqrt{y}$, then $J_1 = -1/2\sqrt{y}$ and $J_2 = 1/2\sqrt{y}$. Hence, by Theorem 6.5, we have

$$g(y) = \frac{1}{\sqrt{2\pi}} e^{-y/2} \left| \frac{-1}{2\sqrt{y}} \right| + \frac{1}{\sqrt{2\pi}} e^{-y/2} \left| \frac{1}{2\sqrt{y}} \right|$$

$$= \frac{1}{2^{1/2}\sqrt{\pi}} y^{1/2 - 1} e^{-y/2}, \qquad y > 0.$$

Since $g(y)$ is a density function, it follows that

$$1 = \frac{1}{2^{1/2}\sqrt{\pi}} \int_0^\infty y^{1/2 - 1} e^{-y/2} \, dy$$

$$= \frac{\Gamma(1/2)}{\sqrt{\pi}} \int_0^\infty \frac{1}{2^{1/2}\Gamma(1/2)} y^{1/2 - 1} e^{-y/2} \, dy$$

$$= \frac{\Gamma(1/2)}{\sqrt{\pi}},$$

the integral being the area under a gamma probability curve with parameters $\alpha = 1/2$ and $\beta = 2$. Therefore, $\sqrt{\pi} = \Gamma(1/2)$ and the probability distribution of Y is given by

$$g(y) = \begin{cases} \dfrac{1}{2^{1/2}\Gamma(1/2)} y^{1/2 - 1} e^{-y/2}, & y > 0 \\[2ex] 0, & \text{elsewhere,} \end{cases}$$

which is seen to be a chi-square distribution with 1 degree of freedom.

6.2 *Moments and Moment-Generating Functions*

If $g(X) = X^r$ for $r = 0, 1, 2, 3, \ldots$, Theorem 3.1 on page 79 yields an expected value called the rth **moment about the origin** of the random variable X, which we denote by μ_r'.

Definition 6.1 *The rth **moment about the origin** of the random variable X is given by*

$$\mu'_r = E(X^r) = \begin{cases} \sum_x x^r f(x) & \text{if } X \text{ is discrete} \\ \int_{-\infty}^{\infty} x^r f(x)\,dx & \text{if } X \text{ is continuous.} \end{cases}$$

Since the first and second moments about the origin are given by $\mu'_1 = E(X)$ and $\mu'_2 = E(X^2)$, we can write the mean and variance of a random variable as

$$\mu = \mu'_1 \qquad \text{and} \qquad \sigma^2 = \mu'_2 - \mu^2.$$

Although the moments of a random variable can be determined directly from Definition 6.1, an alternative procedure exists that may sometimes simplify the summations or integrals involved in the calculations. This procedure requires us to utilize a **moment-generating function**.

Definition 6.2 *The **moment-generating function** of the random variable X is given bt $E(e^{tX})$ and is denoted by $M_X(t)$. Hence*

$$M_X(t) = E(e^{tX}) = \begin{cases} \sum_x e^{tx} f(x) & \text{if } X \text{ is discrete} \\ \int_{-\infty}^{\infty} e^{tx} f(x)\,dx & \text{if } X \text{ is continuous.} \end{cases}$$

Moment-generating functions will exist only if the sum or integral of Definition 6.2 converges. If a moment-generating function of a random variable X does exist, it can be used to generate all the moments of that variable. The method is described in Theorem 6.6.

Theorem 6.6 *Let X be a random variable with moment-generating function $M_X(t)$. Then*

$$\left. \frac{d^r M_X(t)}{dt^r} \right|_{t=0} = \mu'_r.$$

Proof. Assuming that we can differentiate inside summation and integral signs, we obtain

$$\frac{d^r M_X(t)}{dt^r} = \begin{cases} \sum_x x^r e^{tx} f(x) & \text{if } X \text{ is discrete} \\ \int_{-\infty}^{\infty} x^r e^{tx} f(x)\,dx & \text{if } X \text{ is continuous.} \end{cases}$$

Setting $t = 0$, we see that both cases reduce to $E(X^r) = \mu'_r$.

Example 6.6 Find the moment-generating function of the binomial random variable X and then use it to verify that $\mu = np$ and $\sigma^2 = npq$.

Solution. From Definition 6.2 we have

$$M_X(t) = \sum_{x=0}^{n} e^{tx} \binom{n}{x} p^x q^{n-x}$$

$$= \sum_{x=0}^{n} \binom{n}{x} (pe^t)^x q^{n-x}.$$

Recognizing this last sum as the binomial expansion of $(pe^t + q)^n$, we obtain

$$M_X(t) = (pe^t + q)^n.$$

Now

$$\frac{dM_X(t)}{dt} = n(pe^t + q)^{n-1} pe^t$$

and

$$\frac{d^2 M_X(t)}{dt^2} = np[e^t(n-1)(pe^t + q)^{n-2} pe^t + (pe^t + q)^{n-1} e^t].$$

Setting $t = 0$, we get

$$\mu_1' = np \qquad \text{and} \qquad \mu_2' = np[(n-1)p + 1].$$

Therefore,

$$\mu = \mu_1' = np$$

and

$$\sigma^2 = \mu_2' - \mu^2 = np(1-p) = npq.$$

which agrees with the results obtained in Chapter 4.

Example 6.7 Show that the moment-generating function of the random variable X having a normal probability distribution with mean μ and variance σ^2 is given by $M_X(t) = e^{\mu t + \sigma^2 t^2/2}$.

Solution. From Definition 6.2 the moment-generating function of the normal random variable X is

$$M_X(t) = \int_{-\infty}^{\infty} e^{tx} \frac{1}{\sqrt{2\pi}\,\sigma} e^{-(1/2)[(x-\mu)/\sigma]^2}\, dx$$

$$= \int_{-\infty}^{\infty} \frac{1}{\sqrt{2\pi}\,\sigma} e^{-[x^2 - 2(\mu + t\sigma^2)x + \mu^2]/2\sigma^2}\, dx.$$

Completing the square in the exponent, we can write

$$x^2 - 2(\mu + t\sigma^2)x + \mu^2 = [x - (\mu + t\sigma^2)]^2 - 2\mu t\sigma^2 - t^2\sigma^4$$

and then

$$M_X(t) = \int_{-\infty}^{\infty} \frac{1}{\sqrt{2\pi}\,\sigma} e^{-\{[x-(\mu+t\sigma^2)]^2 - 2\mu t\sigma^2 - t^2\sigma^4\}/2\sigma^2}\, dx$$

$$= e^{\mu t + \sigma^2 t^2/2} \int_{-\infty}^{\infty} \frac{1}{\sqrt{2\pi}\,\sigma} e^{-(1/2)\{[x-(\mu+t\sigma^2)]/\sigma\}^2}\, dx.$$

Let $w = [x - (\mu + t\sigma^2)]/\sigma$; then $dx = \sigma\, dw$ and

$$M_X(t) = e^{\mu t + \sigma^2 t^2/2} \int_{-\infty}^{\infty} \frac{1}{\sqrt{2\pi}} e^{-w^2/2}\, dw$$

$$= e^{\mu t + \sigma^2 t^2/2},$$

since the last integral represents the area under a standard normal density curve and hence equals 1.

Example 6.8 Show that the moment-generating function of the random variable X having a chi-square distribution with v degrees of freedom is $M_X(t) = (1 - 2t)^{-v/2}$.

Solution. The chi-square distribution was obtained as a special case of the gamma distribution by setting $\alpha = v/2$ and $\beta = 2$. Substituting for $f(x)$ in Definition 6.2, we obtain

$$M_X(t) = \int_0^{\infty} e^{tx} \frac{1}{2^{v/2}\Gamma(v/2)} x^{v/2-1} e^{-x/2}\, dx$$

$$= \frac{1}{2^{v/2}\Gamma(v/2)} \int_0^{\infty} x^{v/2-1} e^{-x(1-2t)/2}\, dx.$$

Writing $y = x(1 - 2t)/2$ and $dx = [2/(1 - 2t)]\, dy$, we get

$$M_X(t) = \frac{1}{2^{v/2}\Gamma(v/2)} \int_0^{\infty} \left(\frac{2y}{1-2t}\right)^{v/2-1} e^{-y} \frac{2}{1-2t}\, dy$$

$$= \frac{1}{\Gamma(v/2)(1-2t)^{v/2}} \int_0^{\infty} y^{v/2-1} e^{-y}\, dy$$

$$= (1 - 2t)^{-v/2},$$

since the last integral equals $\Gamma(v/2)$.

Although the method of transforming variables provides an effective way of finding the distribution of a function of several variables, there is an alternative and often preferred procedure when the function in question is a linear combination of independent random variables. This procedure utilizes the properties of moment-generating functions discussed in the following four theorems. In keeping with the mathematical scope of this text, we state Theorem 6.7 without proof.

Theorem 6.7 (Uniqueness Theorem)   *Let X and Y be two random variables with moment-generating functions $M_X(t)$ and $M_Y(t)$, respectively. If $M_X(t) = M_Y(t)$ for all values of t, then X and Y have the same probability distribution.*

Theorem 6.8 $M_{X+a}(t) = e^{at} M_X(t)$.

Proof

$$M_{X+a}(t) = E[e^{t(X+a)}]$$
$$= e^{at} E(e^{tX}) = e^{at} M_X(t).$$

Theorem 6.9 $M_{aX}(t) = M_X(at)$.

Proof

$$M_{aX}(t) = E[e^{t(aX)}] = E[e^{(at)X}]$$
$$= M_X(at).$$

Theorem 6.10   *If $X_1, X_2, \ldots, X_n$ are independent random variables with moment-generating functions $M_{X_1}(t), M_{X_2}(t), \ldots, M_{X_n}(t)$, respectively, and $Y = X_1 + X_2 + \cdots + X_n$, then*

$$M_Y(t) = M_{X_1}(t) M_{X_2}(t) \cdots M_{X_n}(t).$$

Proof. For the continuous case

$$M_Y(t) = E(e^{tY}) = E[e^{t(X_1 + X_2 + \cdots + X_n)}]$$

$$= \int_{-\infty}^{\infty} \cdots \int_{-\infty}^{\infty} e^{t(X_1 + X_2 + \cdots + X_n)} f(x_1, x_2, \ldots, x_n) \, dx_1 \, dx_2 \cdots dx_n.$$

Since the variables are independent, we have

$$f(x_1, x_2, \ldots, x_n) = f_1(x_1) f_2(x_2) \cdots f_n(x_n)$$

and then

$$M_Y(t) = \int_{-\infty}^{\infty} e^{tx_1} f_1(x_1) \, dx_1 \int_{-\infty}^{\infty} e^{tx_2} f_2(x_2) \, dx_2 \cdots \int_{-\infty}^{\infty} e^{tx_n} f_n(x_n) \, dx_n$$

$$= M_{X_1}(t) M_{X_2}(t) \cdots M_{X_n}(t).$$

The proof for the discrete case is obtained in a similar manner by replacing integrations with summations.

One might use Theorems 6.7 and 6.10 along with Exercise 19 on page 180 as an alternative method to Example 6.2 in finding the distribution of the sum of two independent Poisson random variables. For example, if X_1 and X_2 are independent Poisson variables with moment-generating functions given by

$$M_{X_1}(t) = e^{\mu_1(e^t - 1)} \qquad \text{and} \qquad M_{X_2}(t) = e^{\mu_2(e^t - 1)},$$

respectively, then according to Theorem 6.10, the moment-generating function of the random variable $Y_1 = X_1 + X_2$ is

$$\begin{aligned} M_{Y_1}(t) &= M_{X_1}(t)M_{X_2}(t) \\ &= e^{\mu_1(e^t - 1)}e^{\mu_2(e^t - 1)} \\ &= e^{(\mu_1 + \mu_2)(e^t - 1)}, \end{aligned}$$

which we immediately identify as the moment-generating function of a random variable having a Poisson distribution with the parameter $\mu_1 + \mu_2$. Hence, according to Theorem 6.7, we again conclude that the sum of two independent random variables having Poisson distributions, with parameters μ_1 and μ_2, has a Poisson distribution with parameter $\mu_1 + \mu_2$.

In applied statistics one frequently needs to know the probability distribution of a linear combination of independent normal random variables. Let us obtain the distribution of the random variable $Y = a_1 X_1 + a_2 X_2$ when X_1 is a normal variable with mean μ_1 and variance σ_1^2 and X_2 is also a normal variable but independent of X_1, with mean μ_2 and variance σ_2^2. First, by Theorem 6.10, we find

$$M_Y(t) = M_{a_1 X_1}(t)M_{a_2 X_2}(t),$$

and then, using Theorem 6.9,

$$M_Y(t) = M_{X_1}(a_1 t)M_{X_2}(a_2 t).$$

Substituting $a_1 t$ for t in the moment-generating function of the normal distribution derived in Example 6.7 and then $a_2 t$ for t, we have

$$\begin{aligned} M_Y(t) &= e^{a_1\mu_1 t + a_1^2\sigma_1^2 t^2/2}\, e^{a_2\mu_2 t + a_2^2\sigma_2^2 t^2/2} \\ &= e^{(a_1\mu_1 + a_2\mu_2)t + (a_1^2\sigma_1^2 + a_2^2\sigma_2^2)t^2/2}, \end{aligned}$$

which we recognize as the moment-generating function of a distribution that is normal with mean $a_1\mu_1 + a_2\mu_2$ and variance $a_1^2\sigma_1^2 + a_2^2\sigma_2^2$.

Generalizing to the case of n independent normal variables, we state the following result.

Theorem 6.11 *If X_1, X_2, ..., X_n are independent random variables having normal distributions with means μ_1, μ_2, ..., μ_n and variances σ_1^2, σ_2^2, ..., σ_n^2, respectively, then the random variable*

$$Y = a_1 X_1 + a_2 X_2 + \cdots + a_n X_n$$

has a normal distribution with mean

$$\mu_Y = a_1\mu_1 + a_2\mu_2 + \cdots + a_n\mu_n$$

and variance

$$\sigma_Y^2 = a_1^2\sigma_1^2 + a_2^2\sigma_2^2 + \cdots + a_n^2\sigma_n^2.$$

It is now evident that the Poisson distribution and the normal distribution possess a reproductive property in that the sums of independent random variables having either of these distributions is a random variable that also has the same type of distribution. This reproductive property is also possessed by the chi-square distribution.

Theorem 6.12   *If X_1, X_2, ..., X_n are mutually independent random variables that have, respectively, chi-square distributions with v_1, v_2, ..., v_n degrees of freedom, then the random variable*

$$Y = X_1 + X_2 + \cdots + X_n$$

has a chi-square distribution with $v = v_1 + v_2 + \cdots + v_n$ degrees of freedom.

Proof. By Theorem 6.10,

$$M_Y(t) = M_{X_1}(t)M_{X_2}(t) \cdots M_{X_n}(t).$$

From Example 6.8,

$$M_{X_i}(t) = (1 - 2t)^{-v_i/2}, \qquad i = 1, 2, \ldots, n.$$

Therefore,

$$M_Y(t) = (1 - 2t)^{-v_1/2}(1 - 2t)^{-v_2/2} \cdots (1 - 2t)^{-v_n/2}$$
$$= (1 - 2t)^{-(v_1 + v_2 + \cdots + v_n)/2},$$

which we recognize as the moment-generating function of a chi-square distribution with $v = v_1 + v_2 + \cdots + v_n$ degrees of freedom.

Corollary   *If X_1, X_2, ..., X_n are independent random variables having identical normal distributions with mean μ and variance σ^2, then the random variable*

$$\overset{\cdot}{Y} = \sum_{i=1}^{n} \left(\frac{X_i - \mu}{\sigma} \right)^2$$

has a chi-square distribution with $v = n$ degrees of freedom.

This corollary is an immediate consequence of Example 6.5, which states that each of the n independent random variables $[(X_i - \mu)/\sigma]^2$, $i = 1, 2, \ldots, n$, has a chi-square distribution with 1 degree of freedom.

Exercises

---

1. Let X be a random variable with probability distribution

$$f(x) = \begin{cases} \frac{1}{3}, & x = 1, 2, 3 \\ 0, & \text{elsewhere.} \end{cases}$$

Find the probability distribution of the random variable $Y = 2X - 1$.

2. Let X be a binomial random variable with probability distribution

$$f(x) = \begin{cases} \dbinom{3}{x}\left(\dfrac{2}{5}\right)^x\left(\dfrac{3}{5}\right)^{3-x}, & x = 0, 1, 2, 3 \\ 0, & \text{elsewhere.} \end{cases}$$

Find the probability distribution of the random variable $Y = X^2$.

3. Let X_1 and X_2 be discrete random variables with the multinomial distribution

$$f(x_1, x_2)$$

$$= \binom{2}{x_1, x_2, 2 - x_1 - x_2} \left(\frac{1}{4}\right)^{x_1} \left(\frac{1}{3}\right)^{x_2} \left(\frac{5}{12}\right)^{2 - x_1 - x_2}$$

for $x_1 = 0, 1, 2$; $x_2 = 0, 1, 2$; $x_1 + x_2 \leq 2$; and zero elsewhere. Find the joint probability distribution of $Y_1 = X_1 + X_2$ and $Y_2 = X_1 - X_2$.

4. Let X_1 and X_2 be discrete random variables with joint probability distribution

$$f(x_1, x_2) = \begin{cases} \dfrac{x_1 x_2}{18}, & x_1 = 1, 2, x_2 = 1, 2, 3 \\ 0, & \text{elsewhere.} \end{cases}$$

Find the probability distribution of the random variable $Y = X_1 X_2$.

5. Let X have the probability distribution

$$f(x) = \begin{cases} 1, & 0 < x < 1 \\ 0, & \text{elsewhere.} \end{cases}$$

Show that the random variable $Y = -2 \ln X$ has a chi-square distribution with 2 degrees of freedom.

6. Given the random variable X with probability distribution

$$f(x) = \begin{cases} 2x, & 0 < x < 1 \\ 0, & \text{elsewhere,} \end{cases}$$

find the probability distribution of Y, where $Y = 8X^3$.

7. The speed of a molecule in a uniform gas at equilibrium is a random variable V whose probability distribution is given by

$$f(v) = \begin{cases} kv^2 e^{-bv^2}, & v > 0 \\ 0, & \text{elsewhere,} \end{cases}$$

where k is an appropriate constant and b depends on the absolute temperature and mass of the molecule. Find the probability distribution of the kinetic energy of the molecule W, where $W = mV^2/2$.

8. In Exercise 21 on page 82 a dealer's profit, in units of $1000, on a new automobile is given by

$Y = X^2$, where X is a random variable having the density function

$$f(x) = \begin{cases} 2(1 - x), & 0 < x < 1 \\ 0, & \text{elsewhere.} \end{cases}$$

(a) Find the probability density function of the random variable Y.

(b) Using the density function of Y, find the probability that the profit will be less than $250 on the next new automobile sold by this dealership.

9. In Exercise 22 on page 82 the hospital period, in days, for patients following treatment for a certain type of kidney disorder is a random variable $Y = X + 4$, where X has the density function

$$f(x) = \begin{cases} \dfrac{32}{(x + 4)^3}, & x > 0 \\ 0, & \text{elsewhere.} \end{cases}$$

(a) Find the probability density function of the random variable Y.

(b) Using the density function of Y, find the probability that the hospital period for a patient following this treatment will exceed 8 days.

10. In Exercise 16 on page 72 the random variables X and Y, representing the weights of creams and toffees in 1 kilogram boxes of chocolates containing a mixture of creams, toffees, and cordials, has the joint density function

$$f(x, y) = \begin{cases} 24xy, & 0 \leq x \leq 1, 0 \leq y \leq 1, \\ & x + y \leq 1 \\ 0, & \text{elsewhere.} \end{cases}$$

(a) Find the probability density function of the random variable $Z = X + Y$.

(b) Using the density function of Z, find the probability that in a given box the sum of the creams and toffees account for at least $1/2$ but less than $3/4$ of the total weight.

11. In Exercise 21 on page 73 the amount of kerosene, in thousands of liters, in a tank at the beginning of any day is a random amount Y from which a random amount X is sold during that day. Assume that the joint density function of these variables is given by

$$f(x, y) = \begin{cases} 2, & 0 < x < y, 0 < y < 1 \\ 0, & \text{elsewhere.} \end{cases}$$

Find the probability density function for the amount of kerosene left in the tank at the end of the day.

12. Let X_1 and X_2 be independent random variables each having the probability distribution

$$f(x) = \begin{cases} e^{-x}, & x > 0 \\ 0, & \text{elsewhere.} \end{cases}$$

Show that the random variables Y_1 and Y_2 are independent when $Y_1 = X_1 + X_2$ and $Y_2 = X_1/(X_1 + X_2)$.

13. A current of I amperes flowing through a resistance of R ohms varies according to the probability distribution

$$f(i) = \begin{cases} 6i(1 - i), & 0 < i < 1 \\ 0, & \text{elsewhere.} \end{cases}$$

If the resistance varies independently of the current according to the probability distribution

$$g(r) = \begin{cases} 2r, & 0 < r < 1 \\ 0, & \text{elsewhere,} \end{cases}$$

find the probability distribution for the power $W = I^2 R$ watts.

14. Let X be a random variable with probability distribution

$$f(x) = \begin{cases} \dfrac{1 + x}{2}, & -1 < x < 1 \\ 0, & \text{elsewhere.} \end{cases}$$

Find the probability distribution of the random variable $Y = X^2$.

15. Let X have the probability distribution

$$f(x) = \begin{cases} \dfrac{2(x + 1)}{9}, & -1 < x < 2 \\ 0, & \text{elsewhere.} \end{cases}$$

Find the probability distribution of the random variable $Y = X^2$.

16. Show that the rth moment about the origin of the gamma distribution defined on page 154 is given by

$$\mu_r' = \frac{\beta^r \Gamma(\alpha + r)}{\Gamma(\alpha)}.$$

HINT: Substitute $y = x/\beta$ in the integral defining μ_r' and then use the gamma function to evaluate the integral.

17. A random variable X has the discrete uniform distribution

$$f(x; k) = \begin{cases} \dfrac{1}{k}, & x = 1, 2, 3, \ldots, k \\ 0, & \text{elsewhere.} \end{cases}$$

Show that the moment-generating function of X is given by

$$M_X(t) = \frac{e^t(1 - e^{kt})}{k(1 - e^t)}.$$

18. A random variable X has the geometric distribution $g(x; p) = pq^{x-1}$ for $x = 1, 2, 3, \ldots$. Show that the moment-generating function of X is given by

$$M_X(t) = \frac{pe^t}{1 - qe^t}$$

and then use $M_X(t)$ to find the mean and variance of the geometric distribution.

19. A random variable X has the Poisson distribution $p(x; \mu) = e^{-\mu}\mu^x/x!$ for $x = 0, 1, 2, \ldots$. Show that the moment-generating function of X is given by

$$M_X(t) = e^{\mu(e^t - 1)}.$$

Using $M_X(t)$, find the mean and variance of the Poisson distribution.

20. The moment-generating function of a certain Poisson random variable X is given by

$$M_X(t) = e^{4(e^t - 1)}.$$

Find $P(\mu - 2\sigma < X < \mu + 2\sigma)$.

21. Using the moment-generating function of Example 6.8, show that the mean and variance of the chi-square distribution with v degrees of freedom are, respectively, v and $2v$.

22. By expanding e^{tx} in a Maclaurin series and integrating term by term, show that

$$M_X(t) = \int_{-\infty}^{\infty} e^{tx} f(x)\, dx$$

$$= 1 + \mu t + \mu_2' \frac{t^2}{2!} + \cdots + \mu_r' \frac{t^r}{r!} + \cdots.$$

23. If $X_1, X_2, \ldots, X_n$ are independent random variables having identical exponential distributions with parameter θ, show that the density function of the random variable $Y = X_1 + X_2 + \cdots + X_n$

is a gamma distribution with parameters $\alpha = n$ and $\beta = \theta$.

24. If the number of hurricanes that hit a certain area of the eastern United States per year is a random variable having a Poisson distribution with $\mu = 6$, find the probability that this area will be hit by
   (a) exactly 15 hurricanes in 2 years;
   (b) at most 9 hurricanes in 2 years.

25. If the number of minutes it takes a pharmacist to fill a prescription is a random variable having an exponential distribution with parameter $\theta = 5$, use the results of Exercise 23 to find the probability that the pharmacist will take
   (a) less than 8 minutes to fill 2 prescriptions;
   (b) at least 12 minutes to fill 3 prescriptions.

6.3 *Random Sampling*

The outcome of a statistical experiment may be recorded either as a numerical value or as a descriptive representation. When a pair of dice are tossed and the total is the outcome of interest, we record a numerical value. However, if the students in a certain school are given blood tests and the type of blood is of interest, then a descriptive representation might be the most useful. A person's blood can be classified in 8 ways. It must be AB, A, B, or O, with a plus or minus sign, depending on the presence or absence of the Rh antigen.

The statistician is primarily concerned with the analysis of numerical data. For the classification of blood types, it may be convenient to use numbers from 1 to 8 to represent each blood type and then record the appropriate number for each student. In any particular study, the number of possible observations may be small, large but finite, or infinite. For example, in the classification of blood types we can only have as many observations as there are students in the school. The project, therefore, results in a finite number of observations. On the other hand, if we could toss a pair of dice indefinitely and record the totals that occur, we would obtain an infinite set of values, each value representing the result of a single toss of a pair of dice.

The totality of observations with which we are concerned, whether their number be finite or infinite, constitutes what we call a **population**. In past years the word *population* referred to observations obtained from statistical studies involving people. Today, the statistician uses the term to refer to observations relevant to anything of interest, whether it be groups of people, animals, or objects.

Definition 6.3   A **population** *consists of the totality of the observations with which we are concerned.*

The number of observations in the population is defined to be the **size** of the population. If there are 600 students in the school that we classified according to blood type, we say that we have a population of size 600. The numbers on the cards in a deck, the heights of residents in a certain city, and the lengths of fish in a particular lake are examples of populations with finite size. In each case the total number of observations is a finite number. The die-tossing experiment generates a population whose size is infinite. Similarly, the observations obtained by measuring the atmospheric pressure every day from the past on into the future, or all measurements on the depth of a lake from any conceivable position, are examples of populations whose sizes are infinite. Some finite populations are so large that in theory we assume them to be infinite. This is true if you consider the population of lifetimes of a certain type of storage battery being manufactured for mass distribution throughout the country.

Each observation in a population is a value of a random variable X having some probability distribution $f(x)$. If one is inspecting items coming off an assembly line for defects, then each observation in the population might be a value 0 or 1 of the binomial random variable X with probability distribution

$$b(x; 1, p) = p^x q^{1-x}, \qquad x = 0, 1,$$

where 0 indicates a nondefective item and 1 indicates a defective item. Of course, it is assumed that p, the probability of any item being defective, remains constant from trial to trial. In the blood-type experiment the random variable X represents the type of blood by assuming a value from 1 to 8. Each student is given one of the values of the discrete random variable. The lives of the storage batteries are values assumed by a continuous random variable having perhaps a normal distribution. When we speak hereafter about a "binomial population," a "normal population," or, in general, the "population $f(x)$," we shall mean a population whose observations are values of a random variable having a binomial distribution, a normal distribution, or the probability distribution $f(x)$. Hence the mean and variance of a random variable or probability distribution are also referred to as the mean and variance of the corresponding population.

In the field of statistical inference the statistician is interested in arriving at conclusions concerning a population when it is impossible or impractical to observe the entire set of observations that make up the population. For example, in attempting to determine the average length of life of a certain brand of light bulb, it would be impossible to test all such bulbs if we are to have any left to sell. Exorbitant costs can also be a prohibitive factor in studying the entire population. Therefore, we must depend on a subset of observations from the population to help us make inferences concerning that same population. This takes us into the theory of sampling.

Definition 6.4   *A **sample** is a subset of a population.*

If our inferences from the sample to the population are to be valid, we must obtain samples that are representative of the population. All too often we are tempted to choose a sample by selecting the most convenient members of the population. Such a procedure may lead to erroneous inferences concerning the population. Any sampling procedure that produces inferences that consistently overestimate or consistently underestimate some characteristic of the population is said to be **biased**. To eliminate any possibility of bias in the sampling procedure, it is desirable to choose a **random sample** in the sense that the observations are made independently and at random.

In selecting a random sample of size n from a population $f(x)$, let us define the random variable X_i, $i = 1, 2, \ldots, n$, to represent the ith measurement or sample value that we observe. The random variables $X_1, X_2, \ldots, X_n$ will then constitute a random sample from the population $f(x)$ with numerical values $x_1, x_2, \ldots, x_n$ if the measurements are obtained by repeating the experiment n independent times under essentially the same conditions. Because of the identical conditions under which the elements of the sample are selected, it is reasonable to assume that the n random variables $X_1, X_2, \ldots, X_n$ are independent and that each has the same probability distribution $f(x)$. That is, the probability distributions of $X_1, X_2, \ldots, X_n$ are, respectively, $f(x_1), f(x_2), \ldots, f(x_n)$ and their joint probability distribution is

$$f(x_1, x_2, \ldots, x_n) = f(x_1)f(x_2) \cdots f(x_n).$$

The concept of a random sample is defined formally in the following definition.

Definition 6.5 *Let $X_1, X_2, \ldots, X_n$ be n independent random variables each having the same probability distribution $f(x)$. We then define $X_1, X_2, \ldots, X_n$ to be a **random sample** of size n from the population $f(x)$ and write its joint probability distribution as*

$$f(x_1, x_2, \ldots, x_n) = f(x_1)f(x_2) \cdots f(x_n).$$

If one makes a random selection of $n = 8$ storage batteries from a manufacturing process, which has maintained the same specifications, and records the length of life for each battery with the first measurement x_1 being a value of X_1, the second measurement x_2 a value of X_2, and so forth, then $x_1, x_2, \ldots, x_8$ are the values of the random sample $X_1, X_2, \ldots, X_8$. If we assume the population of battery lives to be normal, the possible values of any X_i, $i = 1, 2, \ldots, 8$, will be precisely the same as those in the original population, and hence X_i has the same identical normal distribution as X.

6.4 Some Important Statistics

Our main purpose in selecting random samples is to elicit information about the unknown population parameters. Suppose, for example, that we wish to arrive at a conclusion concerning the proportion of coffee-drinking people in the United States who prefer a certain brand of coffee. It would be impossible to question every coffee-drinking American in order to compute the value of the parameter p representing the population proportion. Instead, a large random sample is selected and the proportion $\hat{p}$ of people in this sample favoring the brand of coffee in question is calculated. The value $\hat{p}$ is now used to make an inference concerning the true proportion p.

Now, $\hat{p}$ is a function of the observed values in the random sample; since many random samples are possible from the same population, we would expect $\hat{p}$ to vary somewhat from sample to sample. That is, $\hat{p}$ is a value of a random variable that we represent by $\hat{P}$. Such a random variable is called a **statistic**.

Definition 6.6   *Any function of the random variables constituting a random sample is called a* **statistic**.

In Chapter 3 we introduced the two parameters μ and σ^2, which measure the center and the variability of a probability distribution. These are constant population parameters and are in no way affected or influenced by the observations of a random sample. We shall, however, define some important statistics that describe corresponding measures of a random sample. The most commonly used statistics for measuring the center of a set of data, arranged in order of magnitude, are the **mean**, **median**, and **mode**. The most important of these and the one we shall consider first is the mean.

Definition 6.7   *If X_1, X_2, ..., X_n represent a random sample of size n, then the* **sample mean** *is defined by the statistic*

$$\bar{X} = \frac{\sum\limits_{i=1}^{n} X_i}{n}.$$

Note that the statistic $\bar{X}$ assumes the value $\bar{x} = \sum\limits_{i=1}^{n} x_i/n$ when X_1 assumes the value x_1, X_2 assumes the value x_2, and so forth. In practice the value of a statistic is usually given the same name as the statistic. For instance, the term *sample mean* is applied to both the statistic $\bar{X}$ and also to its computed value $\bar{x}$.

Example 6.9   A food inspector examined a random sample of 7 cans of a certain brand of tuna to determine the percent of foreign impurities. The following data were recorded: 1.8, 2.1, 1.7, 1.6, 0.9, 2.7, and 1.8. Compute the sample mean.

Solution. The observed value $\bar{x}$ of the statistic $\bar{X}$ is

$$\bar{x} = \frac{1.8 + 2.1 + 1.7 + 1.6 + 0.9 + 2.7 + 1.8}{7} = 1.8\%.$$

The second most useful statistic for measuring the center of a set of data is the median. We shall designate the median by the symbol $\tilde{X}$.

Definition 6.8   *If $X_1, X_2, \ldots, X_n$ represent a random sample of size n, arranged in increasing order of magnitude, then the* **sample median** *is defined by the statistic*

$$\tilde{X} = \begin{cases} X_{(n+1)/2} & \text{if n is odd} \\ \dfrac{X_{n/2} + X_{(n/2)+1}}{2} & \text{if n is even.} \end{cases}$$

Example 6.10   The number of foreign ships arriving at an East coast port on 7 randomly selected days were 8, 3, 9, 5, 6, 8, and 5. Find the sample median.

> **Solution.** Arranging the observations in increasing order of magnitude, we get
>
> $$3 \quad 5 \quad 5 \quad 6 \quad 8 \quad 8 \quad 9$$
>
> and hence $\tilde{x} = 6$.

Example 6.11   The nicotine contents for a random sample of 6 cigarettes of a certain brand are found to be 2.3, 2.7, 2.5, 2.9, 3.1, and 1.9 milligrams. Find the median.

> **Solution.** If we arrange these nicotine contents in an increasing order of magnitude, we get
>
> $$1.9 \quad 2.3 \quad 2.5 \quad 2.7 \quad 2.9 \quad 3.1$$
>
> and the median is then the mean of 2.5 and 2.7. Therefore,
>
> $$\tilde{x} = \frac{2.5 + 2.7}{2} = 2.6 \text{ milligrams.}$$

The third and final statistic for measuring the center of a random sample that we shall discuss is the mode, designated by the statistic M.

Definition 6.9   *If $X_1, X_2, \ldots, X_n$, not necessarily all different, represent a random sample of size n, then the* **mode** *M is that value of the sample that occurs most often or with the greatest frequency. The mode may not exist, and when it does it is not necessarily unique.*

The mode does not always exist. This is certainly true when all observations occur with the same frequency. For some sets of data there may be several values occurring with the greatest frequency in which case we have more than one mode.

Example 6.12   If the donations of a random sample of residents of Fairway Forest toward the Virginia Lung Association are recorded as 9, 10, 5, 9, 9, 7, 8, 6, 10, and 11 dollars, then the mode is $m = \$9$, the value that occurs with the greatest frequency.

Example 6.13   The number of movies attended last month by a random sample of 12 high school students were recorded as follows: 2, 0, 3, 1, 2, 4, 2, 5, 4, 0, 1, and 4. In this case, there are two modes, 2 and 4, since both 2 and 4 occur with the greatest frequency. The distribution is said to be **bimodal**.

Example 6.14   No mode exists for the nicotine contents of Example 6.11, since each measurement occurs only once.

In summary, let us consider the relative merits of the mean, median, and mode. The mean is the most commonly used measure of central location in statistics. It is easy to calculate and it employs all available information. The distributions of means obtained in repeated sampling from a population are well known, and consequently the methods used in statistical inference for estimating μ are based on the sample mean. The only real disadvantage to the mean is that it may be affected adversely by extreme values. In Example 6.12 the mean contribution to the Virginia Lung Association was $8.40, which is fairly close to the mode or median, both of which are $9. However, if one of the contributions had been much larger, say $90 instead of $11, then the mean contribution is $16.30, a value considerably higher than the majority of gifts.

The median has the advantage of being easy to compute if the number of observations is relatively small. It is not influenced by extreme values and consequently in Example 6.12 gives a truer average, namely $9, if the highest contribution is $90 rather than $11. In dealing with samples selected from populations, the sample means usually will not vary as much from sample to sample as will the medians. Therefore, if we are attempting to estimate the center of a population based on a sample value, the mean is more stable than the median. Hence a sample mean is likely to be closer than the sample median to the population mean.

The mode is the least used measure of the three. For small sets of data its value is almost useless if, in fact, it exists at all. Only in the case of a large mass of data does it have a significant meaning. Its two main advantages are that (1) it requires no calculation, and (2) it can be used for qualitative as well as quantitative data. Thus, if jogging is the preferred form of exercise expressed by most people, we say that jogging is the **modal choice**.

The three measures of central location defined above do not by themselves give an adequate description of our data. We need to know how the observations spread out from the average. It is quite possible to have two sets of observations with the same mean or median that differ considerably in the variability of their measurements about the average.

Consider the following measurements, in liters, for two samples of orange juice bottled by companies A and B:

Sample A	0.97	1.00	0.94	1.03	1.11
Sample B	1.06	1.01	0.88	0.91	1.14

Both samples have the same mean, 1.00 liters. It is quite obvious that company *A* bottles orange juice with a more uniform content than company *B*. We say that the variability or the dispersion of the observations from the average is less for sample *A* than for sample *B*. Therefore, in buying orange juice, we would feel more confident that the bottle we select will be closer to the advertised average if we buy from company *A*.

The most important statistics for measuring the variability of a random sample are the **range** and the **variance**. The simpler of these to compute is the range.

Definition 6.10   *The **range** of a random sample X_1, X_2, ..., X_n is defined by the statistic $X_{(n)} - X_{(1)}$, where $X_{(n)}$ and $X_{(1)}$ are, respectively, the largest and smallest observations in the sample.*

Example 6.15   The IQ's of a random sample of five members of a sorority are 108, 112, 127, 118, and 113. Find the range.

Solution. The range of the five IQ's is $127 - 108 = 19$.

In the case of the companies bottling orange juice, the range for company A is 0.17 liters compared to a range of 0.26 liters for company B, indicating a greater spread in the values for company B.

The range can be a poor measure of variability, particularly if the size of the sample or population is large. It considers only the extreme values and tells us nothing about the distribution of values in between. Consider, for example, the following two sets of data, both with a range of 12:

	Observed Data								
Set *A*	3	4	5	6	8	9	10	12	15
Set *B*	3	7	7	7	8	8	8	9	15

In the first set the mean and median are both 8, but the numbers vary over the entire interval from 3 to 15. In the second set the mean and median are also both 8, but most of the values are closer to the center of the data. Although the range fails to measure this variability between the upper and lower observations, it does have some useful applications. In industry the range might be predetermined by specifying in advance that a particular measurement on items coming off an assembly line must fall within a certain interval. As long as all the measurements fall within the specified interval, the process is said to be in control.

To overcome the disadvantage of the range, we shall consider a measure of variability, namely, the **sample variance**, that considers the position of each observation relative to the sample mean.

Definition 6.11   *If $X_1, X_2, \ldots, X_n$ represent a random sample of size n, then the **sample variance** is defined by the statistic*

$$S^2 = \frac{\sum\limits_{i=1}^{n} (X_i - \bar{X})^2}{n-1}.$$

The computed value of S^2 for a given sample is denoted by s^2. Note that S^2 is essentially defined to be the average of the squares of the deviations of the observations from their mean. The reason for using $n - 1$ as a divisor rather than the more obvious choice n will become apparent in Chapter 7.

Example 6.16   A comparison of coffee prices at 4 randomly selected grocery stores in San Diego showed increases from the previous month of 12, 15, 17, and 20 cents for a 200-gram jar. Find the variance of this random sample of price increases.

Solution. Calculating the sample mean, we get

$$\bar{x} = \frac{12 + 15 + 17 + 20}{4} = 16 \text{ cents.}$$

Therefore,

$$s^2 = \frac{\sum\limits_{i=1}^{4} (x_i - 16)^2}{3}$$

$$= \frac{(12 - 16)^2 + (15 - 16)^2 + (17 - 16)^2 + (20 - 16)^2}{3}$$

$$= \frac{(-4)^2 + (-1)^2 + (1)^2 + (4)^2}{3}$$

$$= \frac{34}{3}.$$

If $\bar{x}$ is a decimal number that has been rounded off, we accumulate a large error using the sample variance formula of Definition 6.11. To avoid this, let us derive the more useful computational formula, as given in the following theorem.

Theorem 6.13   *If S^2 is the variance of a random sample of size n, we may write*

$$S^2 = \frac{n \sum\limits_{i=1}^{n} X_i^2 - \left(\sum\limits_{i=1}^{n} X_i \right)^2}{n(n-1)}.$$

Proof. By definition,

$$S^2 = \frac{\sum_{i=1}^{n} (X_i - \bar{X})^2}{n - 1}$$

$$= \frac{\sum_{i=1}^{n} (X_i^2 - 2\bar{X}X_i + \bar{X}^2)}{n - 1}$$

$$= \frac{\sum_{i=1}^{n} X_i^2 - 2\bar{X}\sum_{i=1}^{n} X_i + n\bar{X}^2}{n - 1}.$$

Replacing $\bar{X}$ by $\sum_{i=1}^{n} X_i/n$ and multiplying numerator and denominator by n, we obtain the more useful computational formula

$$S^2 = \frac{n\sum_{i=1}^{n} X_i^2 - \left(\sum_{i=1}^{n} X_i\right)^2}{n(n - 1)}.$$

Definition 6.12   *The* **sample standard deviation**, *denoted by S, is the positive square root of the sample variance.*

Example 6.17   Find the variance of the data 3, 4, 5, 6, 6, and 7, representing the number of trout caught by a random sample of 6 fishermen on June 19, 1984, at Lake Muskoka.

Solution. We find that $\sum_{i=1}^{6} x_i^2 = 171$, $\sum_{i=1}^{6} x_i = 31$, $n = 6$. Hence

$$s^2 = \frac{(6)(171) - (31)^2}{(6)(5)} = \frac{13}{6}.$$

Exercises

---

1. Define suitable populations from which the following samples are selected:
   (a) Persons in 200 homes are called by telephone in the city of Richmond and asked to name the candidate that they favor for election to the school board.
   (b) A coin is tossed 100 times and 34 tails are recorded.

   (c) Two hundred pairs of a new type of tennis shoe were tested on the professional tour and, on the average, lasted 4 months.
   (d) On five different occasions it took a lawyer 21, 26, 24, 22, and 21 minutes to drive from her suburban home to her midtown office.

2. The number of tickets issued for traffic violations by 8 state troopers during the Memorial Day

weekend are 5, 4, 7, 7, 6, 3, 8, and 6.

   (a) If these values represent the number of tickets issued by a random sample of 8 state troopers from Montgomery County in Virginia, define a suitable population.

   (b) If the values represent the number of tickets issued by a random sample of 8 state troopers from South Carolina, define a suitable population.

3. The numbers of incorrect answers on a true-false competency test for a random sample of 15 students were recorded as follows: 2, 1, 3, 0, 1, 3, 6, 0, 3, 3, 5, 2, 1, 4, and 2. Find
   (a) the mean;
   (b) the median;
   (c) the mode.

4. The lengths of time, in minutes, that 10 patients waited in a doctor's office before receiving treatment were recorded as follows: 5, 11, 9, 5, 10, 15, 6, 10, 5, and 10. Treating the data as a random sample, find
   (a) the mean;
   (b) the median;
   (c) the mode.

5. The reaction times for a random sample of 9 subjects to a stimulant were recorded as 2.5, 3.6, 3.1, 4.3, 2.9, 2.3, 2.6, 4.1, and 3.4 seconds. Calculate
   (a) the mean;
   (b) the median.

6. According to ecology writer Jacqueline Killeen, phosphates contained in household detergents pass right through our sewer systems causing lakes to turn into swamps that eventually dry up into deserts. The following data show the amount of phosphates per load of laundry, in grams, for a random sample of various types of detergents used according to the prescribed directions:

Laundry Detergent	Phosphates per Load (gm)
A & P Blue Sail	48
Dash	47
Concentrated All	42
Cold Water All	42
Breeze	41
Oxydol	34
Ajax	31

Laundry Detergent	Phosphates per Load (gm)
Sears	30
Fab	29
Cold Power	29
Bold	29
Rinso	26

For the given phosphate data, find
   (a) the mean;
   (b) the median;
   (c) the mode.

7. A random sample of employees from a local manufacturing plant pledged the following donations, in dollars, to the United Fund: 10, 40, 25, 5, 20, 10, 25, 50, 30, 10, 5, 15, 25, 50, 10, 30, 5, 25, 45, and 15. Calculate
   (a) the mean;
   (b) the mode.

8. Find the mean, median, and mode for the sample whose observations, 15, 7, 8, 95, 19, 12, 8, 22, and 14, represent the number of sick days claimed on 9 Federal income tax returns. Which value appears to be the best measure of the center of our data? Give reasons for your preference.

9. With reference to the lengths of time that 10 patients waited in a doctor's office before receiving treatment in Exercise 4, find
   (a) the range;
   (b) the standard deviation.

10. With reference to the sample of reaction times for the 9 subjects receiving the stimulant in Exercise 5, calculate
   (a) the range;
   (b) the variance using the formula of Definition 6.11 on page 188.

11. With reference to the random sample of incorrect answers on a true–false competency test for the 15 students in Exercise 3, calculate the variance using the formula
   (a) of Definition 6.11 on page 188;
   (b) of Theorem 6.13 on page 188.

12. The tar contents of 8 brands of cigarettes selected at random from the latest list released by the Federal Trade Commission are as follows: 7.3, 8.6, 10.4, 16.1, 12.2, 15.1, 14.5, and 9.3 milligrams. Calculate

(a) the mean;

(b) the variance.

13. The grade-point averages of 20 college seniors selected at random from a graduating class are as follows:

3.2	1.9	2.7	2.4
2.8	2.9	3.8	3.0
2.5	3.3	1.8	2.5
3.7	2.8	2.0	3.2
2.3	2.1	2.5	1.9

Calculate the standard deviation.

14. (a) Show that the sample variance is unchanged if a constant c is added to or subtracted from each value in the sample.

(b) Show that the sample variance becomes c^2 times its original value if each observation in the sample is multiplied by c.

15. Verify that the variance of the sample 4, 9, 3, 6, 4, and 7 is 5.1, and using this fact along with the results of Exercise 14 find

(a) the variance of the sample 12, 27, 9, 18, 12, and 21;

(b) the variance of the sample 9, 14, 8, 11, 9, and 12.

16. In testing for carbon monoxide in a certain brand of cigarette, the data, in milligrams per cigarette, were coded by subtracting 12 from each observation. Use the results of Exercise 14 to find the standard deviation for the carbon monoxide contents of a random sample of 15 cigarettes of this brand if the coded measurements are 3.8, -0.9, 5.4, 4.5, 5.2, 5.6, 2.7, -0.1, -0.3, -1.7, 5.7, 3.3, 4.4, -0.5, and 1.9.

17. A random sample of 5 bank presidents indicated annual salaries of \$63,000, \$48,000, \$52,000, \$35,000, and \$41,000. Use the results of Exercise 14 to find the variance of this set of data by first dividing each salary by 1000 and then subtracting 50.

18. A taxi company tested a random sample of 10 steel-belted radial tires of a certain brand and recorded the following tread wear: 48,000, 53,000, 45,000, 61,000, 59,000, 56,000, 63,000, 49,000, 53,000, and 54,000 kilometers. Use the results of Exercise 14 to find the standard deviation of this set of data by first dividing each observation by 1000 and then subtracting 55.

6.5 *Sampling Distributions*

The field of statistical inference is basically concerned with generalizations and predictions. For example, we might claim, based on the opinions of several people interviewed on the street, that in a forthcoming election 60% of the eligible voters in the city of Detroit favor a certain candidate. In this case we are dealing with a random sample of opinions from a very large finite population. As a second illustration we might state that the average cost to build a residence in Charleston, South Carolina, is between \$90,000 and \$95,000, based on the estimates of 3 contractors selected at random from the 30 now building in this city. The population being sampled here is again finite but very small. Finally, let us consider a soft-drink dispensing machine in which the average amount of drink dispensed is being held to 240 milliliters. A company official computes the mean of 40 drinks to obtain $\bar{x} = 236$ milliliters, and on the basis of this value decides that the machine is still dispensing drinks with an average content of $\mu = 240$ milliliters. The 40 drinks represent a sample from the infinite population of possible drinks that will be dispensed by this machine.

In each of the examples above we have computed a statistic from a sample selected from the population, and from these statistics we made various statements concerning the values of population parameters that may or may not be true. The company official made the decision that the soft-drink machine dispenses drinks with an average content of 240 milliliters, even though the sample mean was 236 milliliters, because he knows from sampling theory that such a sample value is likely to occur. In fact, if he ran similar tests, say every hour, he would expect the values of $\bar{x}$ to fluctuate above and below $\mu = 240$ milliliters. Only when the value of $\bar{x}$ is substantially different from 240 milliliters will the company official initiate action to adjust the machine.

Since a statistic is a random variable that depends only on the observed sample, it must have a probability distribution.

Definition 6.13   *The probability distribution of a statistic is called a* **sampling distribution**.

The probability distribution of $\bar{X}$ is called the **sampling distribution of the mean**. It is customary to refer to the standard deviation of the sampling distribution as the **standard error** of the statistic. Therefore, the standard error of the mean is just the standard deviation of the sampling distribution of $\bar{X}$. Also, the standard error of the sample standard deviation for all possible samples of size n selected from a specified population is the standard deviation of the statistic S.

The sampling distribution of a statistic will depend on the size of the population, the size of the samples, and the method of choosing the samples. For the remainder of this chapter we study several of the more important sampling distributions of frequently used statistics. Applications of these sampling distributions to problems of statistical inference are considered throughout most of the remaining chapters.

6.6 *Sampling Distributions of Means*

The first important sampling distribution to be considered is that of the mean $\bar{X}$. Suppose that a random sample of n observations is taken from a normal population with mean μ and variance σ^2. Each observation X_i, $i = 1, 2, \ldots, n$, of the random sample will then have the same normal distribution as the population being sampled. Hence, by the reproductive property of the normal distribution established in Theorem 6.11, we conclude that

$$\bar{X} = \frac{X_1 + X_2 + \cdots + X_n}{n}$$

has a normal distribution with mean

$$\mu_{\bar{X}} = \frac{\mu + \mu + \cdots + \mu}{n} = \mu$$

and variance

$$\sigma_{\bar{X}}^2 = \frac{\sigma^2 + \sigma^2 + \cdots + \sigma^2}{n^2} = \frac{\sigma^2}{n}.$$

If we are sampling from a population with unknown distribution, either finite or infinite, the sampling distribution of $\bar{X}$ will still be approximately normal with mean μ and variance σ^2/n provided that the sample size is large. This amazing result is an immediate consequence of the following theorem, called the **Central Limit Theorem**. The proof is outlined in Exercise 11 on page 199.

Theorem 6.14 (Central Limit Theorem)   *If $\bar{X}$ is the mean of a random sample of size n taken from a population with mean μ and finite variance σ^2, then the limiting form of the distribution of*

$$Z = \frac{\bar{X} - \mu}{\sigma/\sqrt{n}},$$

as $n \to \infty$, is the standard normal distribution $n(z; 0, 1)$.

The normal approximation for $\bar{X}$ will generally be good if $n \geq 30$ regardless of the shape of the population. If $n < 30$, the approximation is good only if the population is not too different from a normal distribution and, as stated above, if the population is known to be normal, the sampling distribution of $\bar{X}$ will follow a normal distribution exactly, no matter how small the size of the samples.

Example 6.18   An electrical firm manufacturers light bulbs that have a length of life that is approximately normally distributed, with mean equal to 800 hours and a standard deviation of 40 hours. Find the probability that a random sample of 16 bulbs will have an average life of less than 775 hours.

Solution. The sampling distribution of $\bar{X}$ will be approximately normal, with $\mu_{\bar{x}} = 800$ and $\sigma_{\bar{x}} = 40/\sqrt{16} = 10$. The desired probability is given by the area of the shaded region in Figure 6.5.

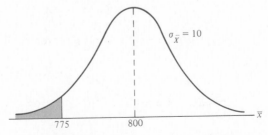

Figure 6.5   Area for Example 6.18.

Corresponding to $\bar{x} = 775$, we find that

$$z = \frac{775 - 800}{10} = -2.5,$$

and therefore

$$P(\bar{X} < 775) = P(Z < -2.5)$$
$$= 0.0062.$$

Example 6.19   Given the discrete uniform population

$$f(x) = \begin{cases} \frac{1}{4}, & x = 0, 1, 2, 3 \\ 0, & \text{elsewhere,} \end{cases}$$

find the probability that a random sample of size 36, selected with replacement, will yield a sample mean greater than 1.4 but less than 1.8 if the mean is measured to the nearest tenth.

Solution. Calculating the mean and variance of the uniform distribution by means of the formulas in Theorem 4.1, we find that

$$\mu = \frac{0 + 1 + 2 + 3}{4} = \frac{3}{2}$$

and

$$\sigma^2 = \frac{(0 - \frac{3}{2})^2 + (1 - \frac{3}{2})^2 + (2 - \frac{3}{2})^2 + (3 - \frac{3}{2})^2}{4}$$

$$= \frac{5}{4}.$$

The sampling distribution of $\bar{X}$ may be approximated by the normal distribution with mean $\mu_{\bar{x}} = 3/2$ and variance $\sigma_{\bar{X}}^2 = \sigma^2/n = 5/144$. Taking the square root, we find the standard deviation to be $\sigma_{\bar{x}} = 0.186$. The probability that $\bar{X}$ is greater than 1.4 but less than 1.8 is given by the area of the shaded region in Figure 6.6.

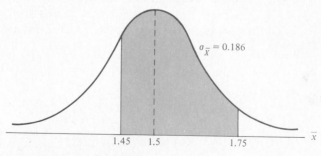

Figure 6.6   Area for Example 6.19.

The z values corresponding to $\bar{x}_1 = 1.45$ and $\bar{x}_2 = 1.75$ are

$$z_1 = \frac{1.45 - 1.5}{0.186} = -0.27$$

and

$$z_2 = \frac{1.75 - 1.5}{0.186} = 1.34.$$

Therefore,

$$
\begin{aligned}
P(1.4 < \bar{X} < 1.8) &\simeq P(-0.27 < Z < 1.34) \\
&= P(Z < 1.34) - P(Z < -0.27) \\
&= 0.9099 - 0.3936 \\
&= 0.5163.
\end{aligned}
$$

Suppose that we now have two populations, the first with mean μ_1 and variance σ_1^2, and the second with mean μ_2 and variance σ_2^2. Let the statistic $\bar{X}_1$ represent the mean of a random sample of size n_1 selected from the first population, and the statistic $\bar{X}_2$ represent the mean of a random sample selected from the second population, independent of the sample from the first population. What can we say about the sampling distribution of the difference $\bar{X}_1 - \bar{X}_2$ for repeated samples of size n_1 and n_2? According to Theorem 6.14, the variables $\bar{X}_1$ and $\bar{X}_2$ are both approximately normally distributed with means μ_1 and μ_2 and variances σ_1^2/n_1 and σ_2^2/n_2, respectively. This approximation improves as n_1 and n_2 increase. By choosing independent samples from the two populations the variables $\bar{X}_1$ and $\bar{X}_2$ will be independent, and then using Theorem 6.11, with $a_1 = 1$ and $a_2 = -1$, we can conclude that $\bar{X}_1 - \bar{X}_2$ is approximately normally distributed with mean

$$\mu_{\bar{X}_1 - \bar{X}_2} = \mu_{\bar{X}_1} - \mu_{\bar{X}_2} = \mu_1 - \mu_2$$

and variance

$$\sigma_{\bar{X}_1 - \bar{X}_2}^2 = \sigma_{\bar{X}_1}^2 + \sigma_{\bar{X}_2}^2 = \frac{\sigma_1^2}{n_1} + \frac{\sigma_2^2}{n_2}.$$

Theorem 6.15 *If independent samples of size n_1 and n_2 are drawn at random from two populations, discrete or continuous, with means μ_1 and μ_2 and variances σ_1^2 and σ_2^2, respectively, then the sampling distribution of the differences of means, $\bar{X}_1 - \bar{X}_2$, is approximately normally distributed with mean and variance given by*

$$\mu_{\bar{X}_1 - \bar{X}_2} = \mu_1 - \mu_2 \qquad and \qquad \sigma_{\bar{X}_1 - \bar{X}_2}^2 = \frac{\sigma_1^2}{n_1} + \frac{\sigma_2^2}{n_2}.$$

Hence

$$Z = \frac{(\bar{X}_1 - \bar{X}_2) - (\mu_1 - \mu_2)}{\sqrt{(\sigma_1^2/n_1) + (\sigma_2^2/n_2)}}$$

is approximately a standard normal variable.

If both n_1 and n_2 are greater than or equal to 30, the normal approximation for the distribution of $\bar{X}_1 - \bar{X}_2$ is very good regardless of the shapes of the two populations. However, even when n_1 and n_2 are less than 30, the normal approximation is reasonably good except when the populations are decidedly non-normal. Of course if both populations are normal, then $\bar{X}_1 - \bar{X}_2$ has a normal distribution no matter what the sizes are of n_1 and n_2.

Example 6.20   A sample of size $n_1 = 5$ is drawn at random from a population that is normally distributed with mean $\mu_1 = 50$ and variance $\sigma_1^2 = 9$, and the sample mean $\bar{x}_1$ is recorded. A second random sample of size $n_2 = 4$ is selected, independent of the first sample, from a different population that is also normally distributed, with mean $\mu_2 = 40$ and variance $\sigma_2^2 = 4$, and the sample mean $\bar{x}_2$ is recorded. Find $P(\bar{X}_1 - \bar{X}_2 < 8.2)$.

Solution. From the sampling distribution of $\bar{X}_1 - \bar{X}_2$ we know that the distribution is normal with mean

$$\mu_{\bar{X}_1 - \bar{X}_2} = \mu_1 - \mu_2$$
$$= 50 - 40 = 10$$

and variance

$$\sigma_{\bar{X}_1 - \bar{X}_2}^2 = \frac{\sigma_1^2}{n_1} + \frac{\sigma_2^2}{n_2}$$
$$= \frac{9}{5} + \frac{4}{4} = 2.8.$$

The desired probability is given by the area of the shaded region in Figure 6.7. Corresponding to the value $\bar{x}_1 - \bar{x}_2 = 8.2$, we find that

$$z = \frac{8.2 - 10}{\sqrt{2.8}} = -1.08,$$

and hence

$$P(\bar{X}_1 - \bar{X}_2 < 8.2) = P(Z < -1.08)$$
$$= 0.1401.$$

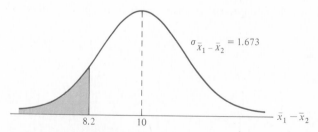

Figure 6.7   Area for Example 6.20.

Example 6.21   The television picture tubes of manufacturer A have a mean lifetime of 6.5 years and a standard deviation of 0.9 year, while those of manufacturer B have a mean lifetime of 6.0 years and a standard deviation of 0.8 year. What is the probability that a random sample of 36 tubes from manufacturer A will have a mean lifetime that is at least 1 year more than the mean lifetime of a sample of 49 tubes from manufacturer B?

Solution. We are given the following information:

Population 1	Population 2
$\mu_1 = 6.5$	$\mu_2 = 6.0$
$\sigma_1 = 0.9$	$\sigma_2 = 0.8$
$n_1 = 36$	$n_2 = 49$

If we use Theorem 6.15, the sampling distribution of $\bar{X}_1 - \bar{X}_2$ will be approximately normal and will have a mean and standard deviation given by

$$\mu_{\bar{X}_1 - \bar{X}_2} = 6.5 - 6.0 = 0.5$$

and

$$\sigma_{\bar{X}_1 - \bar{X}_2} = \sqrt{\frac{0.81}{36} + \frac{0.64}{49}} = 0.189.$$

The probability that the mean of 36 tubes from manufacturer A will be at least 1 year longer than the mean of 49 tubes from manufacturer B is given by the area of the shaded region in Figure 6.8. Corresponding to the value $\bar{x}_1 - \bar{x}_2 = 1.0$, we find that

$$z = \frac{1.0 - 0.5}{0.189} = 2.65,$$

and hence

$$P(\bar{X}_1 - \bar{X}_2 \geq 1.0) = P(Z > 2.65)$$
$$= 1 - P(Z < 2.65)$$
$$= 1 - 0.9960$$
$$= 0.0040.$$

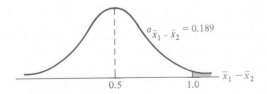

Figure 6.8   Area for Example 6.21.

Exercises

1. If all possible samples of size 16 are drawn from a normal population with mean equal to 50 and standard deviation equal to 5, what is the probability that a sample mean $\bar{X}$ will fall in the interval from $\mu_{\bar{X}} - 1.9\sigma_{\bar{X}}$ to $\mu_{\bar{X}} - 0.4\sigma_{\bar{X}}$? Assume that the sample means can be measured to any degree of accuracy.

2. Given the discrete uniform population

$$f(x) = \begin{cases} \frac{1}{3}, & x = 2, 4, 6 \\ 0, & \text{elsewhere}, \end{cases}$$

find the probability that a random sample of size 54, selected with replacement, will yield a sample mean greater than 4.1 but less than 4.4. Assume the means to be measured to the nearest tenth.

3. A certain type of thread is manufactured with a mean tensile strength of 78.3 kilograms and a standard deviation of 5.6 kilograms. Assuming that the population is infinite, how is the standard error of the mean changed when the sample size is
   (a) increased from 64 to 196?
   (b) decreased from 784 to 49?

4. If the standard error of the mean for the sampling distribution of random samples of size 36 from a large or infinite population is 2, how large must the size of the sample become if the standard error is to be reduced to 1.2?

5. A soft-drink machine is being regulated so that the amount of drink dispensed averages 240 milliliters with a standard deviation of 15 milliliters. Periodically, the machine is checked by taking a sample of 40 drinks and computing the average content. If the mean of the 40 drinks is a value within the interval $\mu_{\bar{X}} \pm 2\sigma_{\bar{X}}$, the machine is thought to be operating satisfactorily; otherwise, adjustments are made. In Section 6.5, the company official found the mean of 40 drinks to be $\bar{x} = 236$ milliliters and concluded that the machine needed no adjustment. Was this a reasonable decision?

6. The heights of 1000 students are approximately normally distributed with a mean of 174.5 centimeters and a standard deviation of 6.9 centi-

meters. If 200 random samples of size 25 are drawn from this population and the means recorded to the nearest tenth of a centimeter, determine
   (a) the mean and standard error of the sampling distribution of $\bar{X}$;
   (b) the number of sample means that fall between 172.5 and 175.8 centimeters inclusive;
   (c) the number of sample means falling below 172.0 centimeters.

7. The random variable X, representing the number of cherries in a cherry puff, has the following probability distribution:

x	4	5	6	7
$P(X = x)$	0.2	0.4	0.3	0.1

   (a) Find the mean μ and the variance σ^2 of X.
   (b) Find the mean $\mu_{\bar{X}}$ and the variance $\sigma_{\bar{X}}^2$ of the mean $\bar{X}$ for random samples of 36 cherry puffs.
   (c) Find the probability that the average number of cherries in 36 cherry puffs will be less than 5.5.

8. If a certain machine makes electrical resistors having a mean resistance of 40 ohms and a standard deviation of 2 ohms, what is the probability that a random sample of 36 of these resistors will have a combined resistance of more than 1458 ohms?

9. The average life of a manufacturer's blender is 5 years, with a standard deviation of 1 year. Assuming that the lives of these blenders follow approximately a normal distribution, find
   (a) the probability that the mean life of a random sample of 9 such blenders falls between 4.4 and 5.2 years;
   (b) the value of $\bar{x}$ to the right of which 15% of the means computed from random samples of size 9 would fall.

10. The amount of time that a bank teller spends on a customer is a random variable with a mean $\mu = 3.2$ minutes and a standard deviation $\sigma = 1.6$

minutes. If a random sample of 64 customers is observed, find the probability that their mean time at the teller's counter is
(a) at most 2.7 minutes;
(b) more than 3.5 minutes;
(c) at least 3.2 minutes but less than 3.4 minutes.

11. Central Limit Theorem
(a) Using Theorems 6.8, 6.9, and 6.10, show that

$$M_{(\bar{X}-\mu)/(\sigma/\sqrt{n})}(t) = e^{-\mu\sqrt{n}t/\sigma}\left[M_X\left(\frac{t}{\sigma\sqrt{n}}\right)\right]^n,$$

where $\bar{X}$ is the mean of a random sample of size n from a population $f(x)$ with mean μ and variance σ^2, and hence

$$\ln M_{(\bar{X}-\mu)/\sqrt{n}}(t) = \frac{-\mu\sqrt{n}t}{6} + n \ln M_X\left(\frac{t}{\sigma\sqrt{n}}\right).$$

(b) Use the result of Exercise 22 on page 180 to expand $M_X(t/\sigma\sqrt{n})$ as an infinite series in powers of t. We can then write $M_X(t/\sigma\sqrt{n}) = 1 + v$, where v is an infinite series.

(c) Assuming n sufficiently large, expand $\ln(1 + v)$ in a Maclaurin series and then show that

$$\lim_{n\to\infty} \ln M_{(\bar{X}-\mu)/(\sigma/\sqrt{n})}(t) = \frac{t^2}{2}$$

and hence

$$\lim_{n\to\infty} M_{(\bar{X}-\mu)/(\sigma/\sqrt{n})}(t) = e^{t^2/2}.$$

12. A random sample of size 25 is taken from a normal population having a mean of 80 and a standard deviation of 5. A second random sample of size 36 is taken from a different normal population having a mean of 75 and a standard deviation of 3. Find the probability that the sample mean computed from the 25 measurements will exceed the sample mean computed from the 36 measurements by at least 3.4 but less than 5.9. Assume the means to be measured to the nearest tenth.

13. The distribution of heights of a certain breed of terrier dogs has a mean height of 72 centimeters and a standard deviation of 10 centimeters, whereas the distribution of heights of a certain breed of poodles has a mean height of 28 centimeters with a standard deviation of 5 centimeters. Assuming that the sample means can be measured to any degree of accuracy, find the probability that the sample mean for a random sample of heights of 64 terriers exceeds the sample mean for a random sample of heights of 100 poodles by at most 44.2 centimeters.

14. The mean score for freshmen on an aptitude test, at a certain college, is 540, with a standard deviation of 50. What is the probability that two groups of students selected at random, consisting of 32 and 50 students, respectively, will differ in their mean scores by
(a) more than 20 points?
(b) an amount between 5 and 10 points?
Assume the means to be measured to any degree of accuracy.

15. Let $\bar{X}_1$ represent the mean of a random sample of size n_1, selected with replacement, from the discrete population

x	2	3	7
$f(x)$	$\frac{1}{3}$	$\frac{1}{3}$	$\frac{1}{3}$

and let $\bar{X}_2$ represent the mean of a random sample of size n_2, selected with replacement, from the discrete population

x	1	3
$f(x)$	$\frac{2}{3}$	$\frac{1}{3}$

If independent samples of size $n_1 = 125$ and $n_2 = 100$ are drawn with replacement, what is the probability that $\bar{X}_1 - \bar{X}_2$ will be greater than 1.84 but less than 2.63? Assume that the sample means can be measured to any degree of accuracy.

6.7 *Sampling Distribution of* $(n - 1)S^2/\sigma^2$

If a random sample of size n is drawn from a normal population with mean μ and variance σ^2, and the sample variance s^2 is computed, we obtain a value of the statistic S^2. The sampling distribution of S^2 has little practical application in statistics. Instead, we shall consider the distribution of the statistic $(n - 1)S^2/\sigma^2$.

By the addition and subtraction of the sample mean $\bar{X}$, it is easy to see that

$$\sum_{i=1}^{n} (X_i - \mu)^2 = \sum_{i=1}^{n} [(X_i - \bar{X}) + (\bar{X} - \mu)]^2$$

$$= \sum_{i=1}^{n} (X_i - \bar{X})^2 + \sum_{i=1}^{n} (\bar{X} - \mu)^2 + 2(\bar{X} - \mu) \sum_{i=1}^{n} (X_i - \bar{X})$$

$$= \sum_{i=1}^{n} (X_i - \bar{X})^2 + n(\bar{X} - \mu)^2.$$

Dividing each term of the equality by σ^2 and substituting $(n - 1)S^2$ for $\sum_{i=1}^{n} (X_i - \bar{X})^2$, we obtain

$$\frac{\sum_{i=1}^{n} (X_i - \mu)^2}{\sigma^2} = \frac{(n - 1)S^2}{\sigma^2} + \frac{(\bar{X} - \mu)^2}{\sigma^2/n}.$$

Now, according to the corollary of Theorem 6.12 we know that $\sum_{i=1}^{n} (X_i - \mu)^2/\sigma^2$ is a chi-square random variable with n degrees of freedom. The second term on the right of the equality is the square of a standard normal variable, since $\bar{X}$ is a normal random variable with mean $\mu_{\bar{X}} = \mu$ and variance $\sigma_{\bar{X}}^2 = \sigma^2/n$. Therefore, we may conclude from Example 6.5 that $(\bar{X} - \mu)^2/(\sigma^2/n)$ is a chi-square random variable with 1 degree of freedom. Using advanced techniques beyond the scope of this book, one can also show that the two chi-square variables $\sum_{i=1}^{n} (X_i - \mu)^2/\sigma^2$ and $(\bar{X} - \mu)^2/(\sigma^2/n)$ are independent.

Owing to the reproductive property of independent chi-square random variables, established in Theorem 6.12, it would seem reasonable to assume that $(n - 1)S^2/\sigma^2$ is also a chi-square random variable with $v = n - 1$ degrees of freedom. We state this result, without formal proof, in the following theorem.

Theorem 6.16   *If S^2 is the variance of a random sample of size n taken from a normal population having the variance σ^2, then the statistic*

$$X^2 = \frac{(n-1)S^2}{\sigma^2}$$

has a chi-square distribution with $v = n - 1$ degrees of freedom.

The values of the random variable X^2 are calculated from each sample by the formula

$$\chi^2 = \frac{(n-1)s^2}{\sigma^2}.$$

The probability that a random sample produces a χ^2 value greater than some specified value is equal to the area under the curve to the right of this value. It is customary to let χ_α^2 represent the χ^2 value above which we find an area of α. This is illustrated by the shaded region in Figure 6.9.

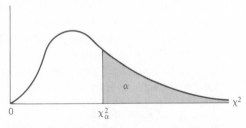

Figure 6.9   Tabulated values of the chi-square distribution.

Table A.5 gives values of χ_α^2 for various values of α and v. The areas, α, are the column headings; the degrees of freedom, v, are given in the left column; and the table entries are the χ^2 values. Hence the χ^2 value with 7 degrees of freedom, leaving an area of 0.05 to the right, is $\chi_{0.05}^2 = 14.067$. Owing to lack of symmetry, we must also use the tables to find $\chi_{0.95}^2 = 2.167$ for $v = 7$.

Exactly 95% of a chi-square distribution with $n - 1$ degrees of freedom lies between $\chi_{0.975}^2$ and $\chi_{0.025}^2$. A χ^2 value falling to the right of $\chi_{0.025}^2$ is not likely to occur unless our assumed value of σ^2 is too small. Similarly, a χ^2 value falling to the left of $\chi_{0.975}^2$ is unlikely unless our assumed value of σ^2 is too large. In other words, it is possible to have a χ^2 value to the left of $\chi_{0.975}^2$ or to the right of $\chi_{0.025}^2$ when σ^2 is correct, but if this should occur, it is more probable that the assumed value of σ^2 is in error.

Example 6.22   A manufacturer of car batteries guarantees that his batteries will last, on the average, 3 years with a standard deviation of 1 year. If five of these batteries have lifetimes of 1.9, 2.4, 3.0, 3.5, and 4.2 years, is the manufacturer still convinced that his batteries have a standard deviation of 1 year?

Solution. We first find the sample variance:

$$s^2 = \frac{(5)(48.26) - (15)^2}{(5)(4)} = 0.815.$$

Then

$$\chi^2 = \frac{(4)(0.815)}{1} = 3.26$$

is a value from a chi-square distribution with 4 degrees of freedom. Since 95% of the χ^2 values with 4 degrees of freedom fall between 0.484 and 11.143, the computed value with $\sigma^2 = 1$ is reasonable, and therefore the manufacturer has no reason to suspect that the standard deviation is other than 1 year.

6.8 t Distribution

Most of the time we are not fortunate enough to know the variance of the population from which we select our random samples. For samples of size $n \geq 30$, a good estimate of σ^2 is provided by calculating a value of S^2. What then happens to our statistic $(\bar{X} - \mu)/(\sigma/\sqrt{n})$ of Theorem 6.14 if we replace σ^2 by S^2? As long as S^2 provides a good estimate of σ^2 and does not vary much from sample to sample, which is usually the case for $n \geq 30$, the distribution of the statistic $(\bar{X} - \mu)/(S/\sqrt{n})$ is still approximately distributed as a standard normal variable.

If the sample size is small ($n < 30$), the values of S^2 fluctuate considerably from sample to sample (see Exercise 7 on page 211) and the distribution of the random variable $(\bar{X} - \mu)/(S/\sqrt{n})$ is no longer a standard normal distribution. We are now dealing with the distribution of a statistic that we shall call T, where

$$T = \frac{\bar{X} - \mu}{S/\sqrt{n}}.$$

In deriving the sampling distribution of T, we shall assume our random sample was selected from a normal population. We can then write

$$T = \frac{(\bar{X} - \mu)/(\sigma/\sqrt{n})}{\sqrt{S^2/\sigma^2}} = \frac{Z}{\sqrt{V/(n-1)}},$$

where

$$Z = \frac{\bar{X} - \mu}{\sigma/\sqrt{n}}$$

has the standard normal distribution, and

$$V = \frac{(n-1)S^2}{\sigma^2}$$

has a chi-square distribution with $v = n - 1$ degrees of freedom. In sampling from normal populations, one can show that $\bar{X}$ and S^2 are independent, and consequently so are Z and V. We are now in a position to derive the distribution of T.

Theorem 6.17   *Let Z be a standard normal random variable and V a chi-square random variable with v degrees of freedom. If Z and V are independent, then the distribution of the random variable T, where*

$$T = \frac{Z}{\sqrt{V/v}},$$

is given by

$$h(t) = \frac{\Gamma[(v + 1)/2]}{\Gamma(v/2)\sqrt{\pi v}} \left(1 + \frac{t^2}{v}\right)^{-(v+1)/2}, \qquad -\infty < t < \infty.$$

*This is known as the **t distribution** with v degrees of freedom.*

Proof. Since Z and V are independent random variables, their joint probability distribution is given by the product of the distribution of Z and V. That is,

$$f(z, v) = \begin{cases} \dfrac{1}{\sqrt{2\pi}} e^{-z^2/2} \dfrac{1}{2^{v/2}\Gamma(v/2)} v^{v/2 - 1} e^{-v/2}, & -\infty < z < \infty, \\ & 0 < v < \infty \\ 0, & \text{elsewhere.} \end{cases}$$

Let us define a second random variable $U = V$. The inverse solutions of $t = z/\sqrt{v/v}$ and $u = v$ are $z = t\sqrt{u}/\sqrt{v}$ and $v = u$, from which we obtain

$$J = \begin{vmatrix} \sqrt{u}/\sqrt{v} & t/2\sqrt{uv} \\ 0 & 1 \end{vmatrix} = \frac{\sqrt{u}}{\sqrt{v}}.$$

The transformation is one to one, mapping the points $\{(z, v)| -\infty < z < \infty, 0 < v < \infty\}$ into the set $\{(t, u)| -\infty < t < \infty, 0 < u < \infty\}$. Using Theorem 6.4, we find the joint probability distribution of T and U to be

$$g(t, u) = \begin{cases} \dfrac{1}{\sqrt{2\pi}\, 2^{v/2}\Gamma(v/2)} u^{v/2 - 1} e^{-\{(u/2)[1 + (t^2/v)]\}} \dfrac{\sqrt{u}}{\sqrt{v}}, & -\infty < t < \infty, \\ & 0 < u < \infty \\ 0, & \text{elsewhere.} \end{cases}$$

Integrating out u, we find that the distribution of T is given by

$$h(t) = \int_0^\infty g(t, u)\, du$$

$$= \int_0^\infty \frac{1}{\sqrt{2\pi v}\, 2^{v/2}\Gamma(v/2)} u^{[(v + 1)/2] - 1} e^{-\{(u/2)[1 + (t^2/v)]\}}\, du.$$

Let us substitute $z = u(1 + t^2/v)/2$ and $du = dz/(1 + t^2/v)$ to give

$$h(t) = \frac{1}{\sqrt{2\pi v}\, 2^{v/2}\Gamma(v/2)} \int_0^\infty \left(\frac{2z}{1 + t^2/v}\right)^{[(v+1)/2]-1} e^{-z} \left(\frac{2}{1 + t^2/v}\right) dz$$

$$= \frac{1}{\Gamma(v/2)\sqrt{\pi v}} \left(1 + \frac{t^2}{v}\right)^{-[(v+1)/2]} \int_0^\infty z^{[(v+1)/2]-1} e^{-z}\, dz$$

$$= \frac{\Gamma[(v+1)/2]}{\Gamma(v/2)\sqrt{\pi v}} \left(1 + \frac{t^2}{v}\right)^{-(v+1)/2}, \qquad -\infty < t < \infty.$$

The probability distribution of T was first published in 1908 in a paper by W. S. Gosset. At the time, Gosset was employed by an Irish brewery that disallowed publication of research by members of its staff. To circumvent this restriction, he published his work secretly under the name "Student." Consequently, the distribution of T is usually called the **Student t distribution**, or simply the **t distribution**. In deriving the equation of this distribution, Gosset assumed that the samples were selected from a normal population. Although this would seem to be a very restrictive assumption, it can be shown that nonnormal populations possessing bell-shaped distributions will still provide values of T that approximate the t distribution very closely.

The distribution of T is similar to the distribution of Z in that they both are symmetric about a mean of zero. Both distributions are bell-shaped, but the t distribution is more variable, owing to the fact that the T values depend on the fluctuations of two quantities, $\bar{X}$ and S^2, whereas the Z values depend only on the changes of $\bar{X}$ from sample to sample. The distribution of T differs from that of Z in that the variance of T depends on the sample size n and is always greater than 1. Only when the sample size $n \to \infty$ will the two distributions become the same. In Figure 6.10 we show the relationship between a standard normal distribution ($v = \infty$) and t distributions with 2 and 5 degrees of freedom.

The probability that a random sample produces a value $t = (\bar{x} - \mu)/(s/\sqrt{n})$ falling between any two specified values is equal to the area under the curve of the t distribution between the two ordinates corresponding to the specified

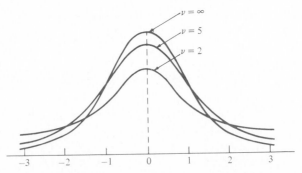

Figure 6.10   The t distribution curves for $v = 2$, 5, and ∞.

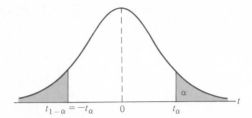

Figure 6.11   Symmetry property of the *t* distribution.

values. It would be a tedious task to attempt to set up separate tables giving the areas between every conceivable pair of ordinates for all values of $n \le 30$. Table A.4 gives only those *t* values above which we find a specified area α, where α is 0.1, 0.05, 0.025, 0.01, or 0.005. This table is set up differently from the table of normal-curve areas in that the areas are now the column headings and the entries are the *t* values. The left column gives the degrees of freedom. It is customary to let t_α represent the *t* value above which we find an area equal to α. Hence the *t* value with 10 degrees of freedom leaving an area of 0.025 to the right is $t = 2.228$. Since the *t* distribution is symmetric about a mean of zero, we have $t_{1-\alpha} = -t_\alpha$; that is, the *t* value leaving an area of $1 - \alpha$ to the right and therefore an area of α to the left is equal to the negative *t* value that leaves an area of α in the right tail of the distribution (see Figure 6.11). That is, $t_{0.95} = -t_{0.05}$, $t_{0.99} = -t_{0.01}$, and so forth.

Example 6.23   The *t* value with $v = 14$ degrees of freedom that leaves an area of 0.025 to the left, and therefore an area of 0.975 to the right is

$$t_{0.975} = -t_{0.025} = -2.145.$$

Example 6.24   Find $P(-t_{0.025} < T < t_{0.05})$.

Solution. Since $t_{0.05}$ leaves an area of 0.05 to the right, and $-t_{0.025}$ leaves an area of 0.025 to the left, we find a total area of

$$1 - 0.05 - 0.025 = 0.925$$

between $-t_{0.025}$ and $t_{0.05}$. Hence

$$P(-t_{0.025} < T < t_{0.05}) = 0.925.$$

Example 6.25   Find *k* such that $P(k < T < -1.761) = 0.045$, for a random sample of size 15 selected from a normal distribution.

Solution. From Table A.4 we note that 1.761 corresponds to $t_{0.05}$ when $v = 14$. Therefore, $-t_{0.05} = -1.761$. Since *k* in the original probability statement is to the left of $-t_{0.05} = -1.761$, let $k = -t_\alpha$. Then, from Figure 6.12, we have

$$0.045 = 0.05 - \alpha$$

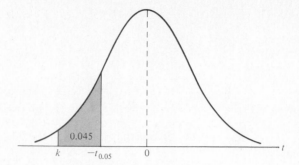

Figure 6.12 The t values for Example 6.25.

or

$$\alpha = 0.005.$$

Hence, from Table A.4 with $v = 14$,

$$k = -t_{0.005} = -2.977$$

and

$$P(-2.977 < T < -1.761) = 0.045.$$

Exactly 95% of the values of a t distribution with $v = n - 1$ degrees of freedom lie between $-t_{0.025}$ and $t_{0.025}$. Of course, there are other t values that contain 95% of the distribution such as $-t_{0.02}$ and $t_{0.03}$, but these values do not appear in Table A.4, and furthermore the shortest possible interval is obtained by choosing t values that leave exactly the same area in the two tails of our distribution. A t value that falls below $-t_{0.025}$ or above $t_{0.025}$ would tend to make us believe that either a very rare event has taken place or perhaps our assumption about μ is in error. Should this happen, we shall make the latter decision and claim that our assumed value of μ is in error. In fact, a t value falling below $-t_{0.01}$ or above $t_{0.01}$ would provide even stronger evidence that our assumed value of μ should be rejected. General procedures for testing claims concerning the value of the parameter μ will be treated in Chapter 8.

Example 6.26   A manufacturer of light bulbs claims that his bulbs will burn on the average 500 hours. To maintain this average, he tests 25 bulbs each month. If the computed t value falls between $-t_{0.05}$ and $t_{0.05}$, he is satisfied with his claim. What conclusion should he draw from a sample that has a mean $\bar{x} = 518$ hours and a standard deviation $s = 40$ hours? Assume the distribution of burning times to be approximately normal.

Solution. From Table A.4 we find that $t_{0.05} = 1.711$ for 24 degrees of freedom. Therefore, the manufacturer is satisfied with his claim if a sample of 25 bulbs yields a t value between -1.711 and 1.711. If $\mu = 500$, then

$$t = \frac{518 - 500}{40/\sqrt{25}} = 2.25,$$

a value well above 1.711. The probability of obtaining a t value, with $v = 24$, equal to or greater than 2.25 is approximately 0.02. If $\mu > 500$, the value of t computed from the sample would be more reasonable. Hence the manufacturer is likely to conclude that his bulbs are a better product than he thought.

6.9 F Distribution

One of the most important distributions in applied statistics is the F distribution. The statistic F is defined to be the ratio of two independent chi-square random variables, each divided by their degrees of freedom. Hence we can write

$$F = \frac{U/v_1}{V/v_2},$$

where U and V are independent random variables having chi-square distributions with v_1 and v_2 degrees of freedom, respectively. We shall now derive the sampling distribution of F.

Theorem 6.18 *Let U and V be two independent random variables having chi-square distributions with v_1 and v_2 degrees of freedom, respectively. Then the distribution of the random variable*

$$F = \frac{U/v_1}{V/v_2}$$

is given by

$$h(f) = \begin{cases} \dfrac{\Gamma[(v_1 + v_2)/2](v_1/v_2)^{v_1/2}}{\Gamma(v_1/2)\Gamma(v_2/2)} \dfrac{f^{v_1/2 - 1}}{(1 + v_1 f/v_2)^{(v_1 + v_2)/2}}, & 0 < f < \infty \\ 0, & \text{elsewhere.} \end{cases}$$

*This is known as the **F distribution** with v_1 and v_2 degrees of freedom.*

Proof. The joint probability distribution of the independent random variables U and V is given by

$$\phi(u, v) = r(u)s(v),$$

where $r(u)$ and $s(v)$ represent the distributions of U and V, respectively. Hence

$$\phi(u, v) = \frac{1}{2^{v_1/2}\Gamma(v_1/2)} u^{v_1/2 - 1}e^{-u/2} \frac{1}{2^{v_2/2}\Gamma(v_2/2)} v^{v_2/2 - 1}e^{-v/2}$$

$$= \begin{cases} \dfrac{1}{2^{(v_1 + v_2)/2}\Gamma(v_1/2)\Gamma(v_2/2)} u^{v_1/2 - 1}v^{v_2/2 - 1}e^{-(u+v)/2}, & \begin{array}{l} 0 < u < \infty, \\ 0 < v < \infty \end{array} \\ 0, & \text{elsewhere.} \end{cases}$$

Let us define a second random variable $W = V$. The inverse solutions of $f = (u/v_1)/(v/v_2)$ and $w = v$ are $u = (v_1/v_2)fw$ and $v = w$, from which we obtain

$$J = \begin{vmatrix} (v_1/v_2)w & (v_1/v_2)f \\ 0 & 1 \end{vmatrix} = \frac{v_1}{v_2} w.$$

The transformation is one to one, mapping the points $\{(u, v)|0 < u < \infty, 0 < v < \infty\}$ into the set $\{(f, w)|0 < f < w, 0 < w < \infty\}$. Using Theorem 6.4, we find that the joint probability distribution of F and W is

$$g(f, w) = \begin{cases} \dfrac{1}{2^{(v_1 + v_2)/2}\Gamma(v_1/2)\Gamma(v_2/2)} \left(\dfrac{v_1 fw}{v_2}\right)^{v_1/2 - 1} w^{v_2/2 - 1}e^{-(w/2)[(v_1 f/v_2) + 1]}\dfrac{v_1 w}{v_2}, \\ \qquad 0 < f < \infty, 0 < w < \infty \\ 0, \qquad \text{elsewhere.} \end{cases}$$

The distribution of F is then given by the marginal distribution

$$h(f) = \int_0^\infty g(f, w) \, dw$$

$$= \frac{(v_1/v_2)^{v_1/2}f^{v_1/2 - 1}}{2^{(v_1 + v_2)/2}\Gamma(v_1/2)\Gamma(v_2/2)} \int_0^\infty w^{[(v_1 + v_2)/2] - 1}e^{-(w/2)[(v_1 f/v_2) + 1]} \, dw.$$

Substituting $z = (w/2)[(v_1 f/v_2) + 1]$ and $dw = [2/(v_1 f/v_2 + 1)] \, dz$, we obtain

$$h(f) = \frac{(v_1/v_2)^{v_1/2}f^{v_1/2 - 1}}{2^{(v_1 + v_2)/2}\Gamma(v_1/2)\Gamma(v_2/2)} \int_0^\infty \left(\frac{2z}{v_1 f/v_2 + 1}\right)^{(v_1 + v_2)/2 - 1} e^{-z}\frac{2}{v_1 f/v_2 + 1} \, dz$$

$$= \frac{(v_1/v_2)^{v_1/2}f^{v_1/2 - 1}}{\Gamma(v_1/2)\Gamma(v_2/2)(1 + v_1 f/v_2)^{(v_1 + v_2)/2}} \int_0^\infty z^{(v_1 + v_2)/2 - 1}e^{-z} \, dz$$

$$= \begin{cases} \dfrac{\Gamma[(v_1 + v_2)/2](v_1/v_2)^{v_1/2}}{\Gamma(v_1/2)\Gamma(v_2/2)} \dfrac{f^{v_1/2 - 1}}{(1 + v_1 f/v_2)^{(v_1 + v_2)/2}}, & 0 < f < \infty \\ 0, & \text{elsewhere.} \end{cases}$$

The number of degrees of freedom associated with the chi-square random variable appearing in the numerator of F is always stated first, followed by the number of degrees of freedom associated with the chi-square random variable appearing in the denominator. Thus the curve of the F distribution depends not only on the two parameters v_1 and v_2 but also on the order in which we

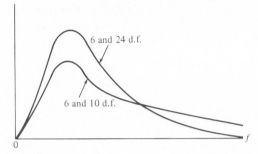

Figure 6.13 Typical F distributions.

state them. Once these two values are given, we can identify the curve. Typical F curves are shown in Figure 6.13.

Let f_α be the f value above which we find an area equal to α. This is illustrated by the shaded region in Figure 6.14. Table A.6 gives values of f_α only for $\alpha = 0.05$ and $\alpha = 0.01$ for various combinations of the degrees of freedom v_1 and v_2. Hence the f value with 6 and 10 degrees of freedom, leaving an area of 0.05 to the right, is $f_{0.05} = 3.22$. By means of the following theorem, Table A.6 can also be used to find values of $f_{0.95}$ and $f_{0.99}$. The proof is left for the reader.

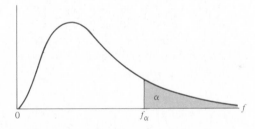

Figure 6.14 Tabulated values of the F distribution.

Theorem 6.19   *Writing $f_\alpha(v_1, v_2)$ for f_α with v_1 and v_2 degrees of freedom, we obtain*

$$f_{1-\alpha}(v_1, v_2) = \frac{1}{f_\alpha(v_2, v_1)}.$$

Thus the f value with 6 and 10 degrees of freedom, leaving an area of 0.95 to the right, is

$$f_{0.95}(6, 10) = \frac{1}{f_{0.05}(10, 6)} = \frac{1}{4.06} = 0.246.$$

Suppose that random samples of size n_1 and n_2 are selected from two normal populations with variances σ_1^2 and σ_2^2, respectively. From Theorem 6.16, we know that

$$X_1^2 = \frac{(n_1 - 1)S_1^2}{\sigma_1^2}$$

and

$$X_2^2 = \frac{(n_2 - 1)S_2^2}{\sigma_2^2}$$

are random variables having chi-square distributions with $v_1 = n_1 - 1$ and $v_2 = n_2 - 1$ degrees of freedom. Furthermore, since the samples are selected at random, we are dealing with independent random variables and then using Theorem 6.18, with $X_1^2 = U$ and $X_2^2 = V$, we obtain the following result.

Theorem 6.20   *If S_1^2 and S_2^2 are the variances of independent random samples of size n_1 and n_2 taken from normal populations with variances σ_1^2 and σ_2^2, respectively, then*

$$F = \frac{S_1^2/\sigma_1^2}{S_2^2/\sigma_2^2} = \frac{\sigma_2^2 S_1^2}{\sigma_1^2 S_2^2}$$

has an F distribution with $v_1 = n_1 - 1$ and $v_2 = n_2 - 1$ degrees of freedom.

In Chapters 7 and 8 we shall use Theorem 6.20 to make inferences concerning the variances of two normal populations. The F distribution is applied primarily, however, in the analysis-of-variance procedures of Chapters 11 through 13, where we wish to test the equality of several means simultaneously.

Exercises

1. For a chi-square distribution find
   (a) $\chi_{0.025}^2$ when $v = 15$;
   (b) $\chi_{0.01}^2$ when $v = 7$;
   (c) $\chi_{0.05}^2$ when $v = 24$.

2. For a chi-square distribution find the following:
   (a) $\chi_{0.005}^2$ when $v = 5$;
   (b) $\chi_{0.05}^2$ when $v = 19$;
   (c) $\chi_{0.01}^2$ when $v = 12$.

3. For a chi-square distribution find χ_α^2 such that
   (a) $P(X^2 < \chi_\alpha^2) = 0.99$ when $v = 4$;
   (b) $P(X^2 > \chi_\alpha^2) = 0.025$ when $v = 19$;
   (c) $P(37.652 < X^2 < \chi_\alpha^2) = 0.045$ when $v = 25$.

4. For a chi-square distribution find χ_α^2 such that
   (a) $P(X^2 > \chi_\alpha^2) = 0.01$ when $v = 21$;
   (b) $P(X^2 < \chi_\alpha^2) = 0.95$ when $v = 6$;
   (c) $P(\chi_\alpha^2 < X^2 < 23.209) = 0.015$ when $v = 10$.

5. Find the probability that a random sample of 25 observations, from a normal population with variance $\sigma^2 = 6$, will have a variance s^2
   (a) greater than 9.1;
   (b) between 3.462 and 10.745.
   Assume the sample variances to be continuous measurements.

6. The scores on a placement test given to college

freshmen for the past five years are approximately normally distributed with a mean $\mu = 74$ and a variance $\sigma^2 = 8$. Would you still consider $\sigma^2 = 8$ to be a valid value of the variance if a random sample of 20 students who take this placement test this year obtain a value of $s^2 = 20$?

7. Show that the variance of S^2 for random samples of size n from a normal population decreases as n becomes large.

HINT: First find the variance of $(n - 1)S^2/\sigma^2$.

8. (a) Find $t_{0.025}$ when $v = 14$.
   (b) Find $-t_{0.01}$ when $v = 10$.
   (c) Find $t_{0.995}$ when $v = 7$.
9. (a) Find $P(T < 2.365)$ when $v = 7$.
   (b) Find $P(T > 1.318)$ when $v = 24$.
   (c) Find $P(-1.356 < T < 2.179)$ when $v = 12$.
   (d) Find $P(T > -2.567)$ when $v = 17$.
10. (a) Find $P(-t_{0.005} < T < t_{0.01})$.
    (b) Find $P(T > -t_{0.025})$.
11. Given a random sample of size 24 from a normal distribution, find k such that
    (a) $P(-2.069 < T < k) = 0.965$;
    (b) $P(k < T < 2.807) = 0.095$;
    (c) $P(-k < T < k) = 0.90$.
12. A manufacturing firm claims that the batteries used in their electronic games will last an average of 30 hours. To maintain this average, 16 batteries are tested each month. If the computed t value falls between $-t_{0.025}$ and $t_{0.025}$, the firm is satisfied with its claim. What conclusion should the

firm draw from a sample that has a mean $\bar{x} = 27.5$ hours and a standard deviation $s = 5$ hours? Assume the distribution of battery lives to be approximately normal.

13. A normal population with unknown variance has a mean of 20. Is one likely to obtain a random sample of size 9 from this population with a mean of 24 and a standard deviation of 4.1? If not, what conclusion would you draw?

14. A cigarette manufacturer claims that his cigarettes have an average nicotine content of 1.83 milligrams. If a random sample of 8 cigarettes of this type have nicotine contents of 2.0, 1.7, 2.1, 1.9, 2.2, 2.1, 2.0, and 1.6 milligrams, would you agree with the manufacturer's claim?

15. For an F distribution find
    (a) $f_{0.05}$ with $v_1 = 7$ and $v_2 = 15$;
    (b) $f_{0.05}$ with $v_1 = 15$ and $v_2 = 7$;
    (c) $f_{0.01}$ with $v_1 = 24$ and $v_2 = 19$;
    (d) $f_{0.95}$ with $v_1 = 19$ and $v_2 = 24$;
    (e) $f_{0.99}$ with $v_1 = 28$ and $v_2 = 12$.

16. If S_1^2 and S_2^2 represent the variances of independent random samples of size $n_1 = 25$ and $n_2 = 31$, taken from normal populations with variances $\sigma_1^2 = 10$ and $\sigma_2^2 = 15$, respectively, find the $P(S_1^2/S_2^2 > 1.26)$.

17. If S_1^2 and S_2^2 represent the variances of independent random samples of size $n_1 = 8$ and $n_2 = 12$, taken from normal populations with equal variances, find the $P(S_1^2/S_2^2 < 4.89)$.

7

Estimation Theory

7.1 Statistical Inference

The theory of **statistical inference** consists of those methods by which one makes inferences or generalizations about a population. The trend of today is to distinguish between the **classical method** of estimating a population parameter, whereby inferences are based strictly on information obtained from a random sample selected from the population, and the **Bayesian method**, which utilizes prior subjective knowledge about the probability distribution of the unknown parameters in conjunction with the information provided by the sample data. Throughout most of this chapter we shall use classical methods to estimate unknown population parameters such as the mean, proportion, and the variance by computing statistics from random samples and applying the theory of sampling distributions, much of which was covered in Chapter 6. For completeness, a brief discussion of the Bayesian approach to statistical decision theory is presented in Sections 7.10 and 7.11.

Statistical inference may be divided into two major areas: **estimation** and **tests of hypotheses**. We treat these two areas separately, dealing with the theory of estimation in this chapter and the theory of hypothesis testing in Chapter 8. To distinguish clearly between the two areas, consider the following examples. A candidate for public office may wish to estimate the true proportion of voters favoring him by obtaining the opinions from a random sample of 100 eligible voters. The fraction of voters in the sample favoring the candidate could be used as an estimate of the true proportion in the population of voters. A knowledge of the sampling distribution of a proportion enables one to establish the degree of accuracy of our estimate. This problem falls in the area of estimation.

Now consider the case in which a housewife is interested in finding out whether brand A floor wax is more scuff-resistant than brand B floor wax. She might hypothesize that brand A is better than brand B and, after proper testing, accept or reject this hypothesis. In this example we do not attempt to estimate a parameter, but instead we try to arrive at a correct decision about a prestated hypothesis. Once again we are dependent on sampling theory to provide us with some measure of accuracy for our decision.

7.2 *Classical Methods of Estimation*

A **point estimate** of some population parameter θ is a single value $\hat{\theta}$ of a statistic $\hat{\Theta}$. For example, the value $\bar{x}$ of the statistic $\bar{X}$, computed from a sample of size n, is a point estimate of the population parameter μ. Similarly, $\hat{p} = x/n$, is a point estimate of the true proportion p for a binomial experiment.

The statistic that one uses to obtain a point estimate is called an **estimator** or a **decision function**. Hence the decision function S^2, which is a function of the random sample, is an estimator of σ^2 and the estimate s^2 is the "action" taken. Different samples will generally lead to different actions or estimates.

Definition 7.1   *The set of all possible actions that can be taken in an estimation problem is called the* **action space** *or* **decision space**.

An estimator is not expected to estimate the population parameter without error. We do not expect $\bar{X}$ to estimate μ exactly, but we certainly hope that it is not too far off. For a particular sample it is possible to obtain a closer estimate of μ by using the sample median $\tilde{X}$ as an estimator. Consider, for instance, a sample consisting of the values 2, 5, and 11 from a population whose mean is 4 but supposedly unknown. We would estimate μ to be $\bar{x} = 6$, using the sample mean as our estimate, or $\tilde{x} = 5$, using the sample median as our estimate. In this case the estimator $\tilde{X}$ produces an estimate closer to the true parameter than that of the estimator $\bar{X}$. On the other hand, if our random sample contains the values 2, 6, and 7, then $\bar{x} = 5$ and $\tilde{x} = 6$, so that $\bar{X}$ is now the better estimator. Not knowing the true value of μ, we must decide in advance whether to use $\bar{X}$ or $\tilde{X}$ as our estimator.

What are the desirable properties of a "good" decision function that would influence us to choose one estimator rather than another? Let $\hat{\Theta}$ be an estimator whose value $\hat{\theta}$ is a point estimate of some unknown population parameter θ. Certainly, we would like the sampling distribution of $\hat{\Theta}$ to have a mean equal to the parameter estimated. An estimator possessing this property is said to be **unbiased**.

Definition 7.2   *A statistic $\hat{\Theta}$ is said to be an **unbiased** estimator of the parameter θ if*

$$\mu_{\hat{\Theta}} = E(\hat{\Theta}) = \theta.$$

Example 7.1   Show that S^2 is an unbiased estimator of the parameter σ^2.

Solution. Let us write

$$\sum_{i=1}^{n} (X_i - \bar{X})^2 = \sum_{i=1}^{n} [(X_i - \mu) - (\bar{X} - \mu)]^2$$

$$= \sum_{i=1}^{n} (X_i - \mu)^2 - 2(\bar{X} - \mu)\sum_{i=1}^{n} (X_i - \mu) + n(\bar{X} - \mu)^2$$

$$= \sum_{i=1}^{n} (X_i - \mu)^2 - n(\bar{X} - \mu)^2.$$

Now

$$E(S^2) = E\left[\frac{\sum_{i=1}^{n} (X_i - \bar{X})^2}{n-1}\right]$$

$$= \frac{1}{n-1}\left[\sum_{i=1}^{n} E(X_i - \mu)^2 - nE(\bar{X} - \mu)^2\right]$$

$$= \frac{1}{n-1}\left(\sum_{i=1}^{n} \sigma_{X_i}^2 - n\sigma_{\bar{X}}^2\right).$$

However,

$$\sigma_{X_i}^2 = \sigma^2 \qquad \text{for } i = 1, 2, \ldots, n$$

and

$$\sigma_{\bar{X}}^2 = \frac{\sigma^2}{n}.$$

Therefore,

$$E(S^2) = \frac{1}{n-1}\left(n\sigma^2 - n\frac{\sigma^2}{n}\right) = \sigma^2.$$

Although S^2 is an unbiased estimator of σ^2, S, on the other hand, is a biased estimator of σ with the bias becoming insignificant for large samples.

If $\hat{\Theta}_1$ and $\hat{\Theta}_2$ are two unbiased estimators of the same population parameter θ, we would choose the estimator whose sampling distribution has the smaller variance. Hence, if $\sigma^2_{\hat{\Theta}_1} < \sigma^2_{\hat{\Theta}_2}$, we say that $\hat{\Theta}_1$ is a **more efficient estimator** of θ than $\hat{\Theta}_2$.

Definition 7.3   *If we consider all possible unbiased estimators of some parameter θ, the one with the smallest variance is called the **most efficient estimator** of θ.*

In Figure 7.1 we illustrate the sampling distributions of 3 different estimators $\hat{\Theta}_1$, $\hat{\Theta}_2$, and $\hat{\Theta}_3$, all estimating θ. It is clear that only $\hat{\Theta}_1$ and $\hat{\Theta}_2$ are unbiased, since their distributions are centered at θ. The estimator $\hat{\Theta}_1$ has a smaller variance than $\hat{\Theta}_2$ and is therefore more efficient. Hence our choice for an estimator of θ, among the three considered, would be $\hat{\Theta}_1$.

For normal populations one can show that both $\bar{X}$ and $\tilde{X}$ are unbiased estimators of the population mean μ, but the variance of $\bar{X}$ is smaller than the variance of $\tilde{X}$. Thus both estimates $\bar{x}$ and $\tilde{x}$ will, on the average, equal the population mean μ, but $\bar{x}$ is likely to be closer to μ for a given sample, and thus $\bar{X}$ is more efficient than $\tilde{X}$.

Even the most efficient unbiased estimator is unlikely to estimate the population parameter exactly. It is true that our accuracy increases with large samples, but where is still no reason why we should expect a point estimate from a given sample to be exactly equal to the population parameter it is supposed to estimate. Perhaps it would be more desirable to determine an interval within which we would expect to find the value of the parameter. Such an interval is called an **interval estimate**.

An interval estimate of a population parameter θ is an interval of the form $\hat{\theta}_L < \theta < \hat{\theta}_U$, where $\hat{\theta}_L$ and $\hat{\theta}_U$ depend on the value of the statistic $\hat{\Theta}$ for a particular sample and also on the sampling distribution of $\hat{\Theta}$. Thus a random sample of SAT verbal scores for students of the entering freshman class might produce an interval from 530 to 550 within which we expect to find the true average of all SAT verbal scores for the freshman class. The values of the end points, 530 and 550, will depend on the computed sample mean $\bar{x}$ and the sampling distribution of $\bar{X}$. As the sample size increases, we know that $\sigma^2_{\bar{X}} = \sigma^2/n$ decreases, and consequently our estimate is likely to be closer to the parameter μ, resulting in a shorter interval. Thus the interval estimate indicates, by its length, the accuracy of the point estimate.

Since different samples will generally yield different values of $\hat{\Theta}$ and, therefore, different values of $\hat{\theta}_L$ and $\hat{\theta}_U$, these end points of the interval are values of

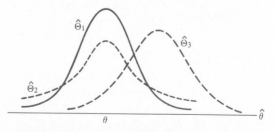

Figure 7.1   Sampling distributions of different estimators of θ.

corresponding random variables $\hat{\Theta}_L$ and $\hat{\Theta}_U$. From the sampling distribution of $\hat{\Theta}$ we shall be able to determine $\hat{\theta}_L$ and $\hat{\theta}_U$ such that the $P(\hat{\Theta}_L < \theta < \hat{\Theta}_U)$ is equal to any positive fractional value we care to specify. If, for instance, we find $\hat{\theta}_L$ and $\hat{\theta}_U$ such that

$$P(\hat{\Theta}_L < \theta < \hat{\Theta}_U) = 1 - \alpha,$$

for $0 < \alpha < 1$, then we have a probability of $1 - \alpha$ of selecting a random sample that will produce an interval containing θ. The interval $\hat{\theta}_L < \theta < \hat{\theta}_U$, computed from the selected sample, is then called a $(1 - \alpha)100\%$ **confidence interval**, the fraction $1 - \alpha$ is called the **confidence coefficient** or the **degree of confidence**, and the end points, $\hat{\theta}_L$ and $\hat{\theta}_U$, are called the lower and upper **confidence limits**. Thus, when $\alpha = 0.05$, we have a 95% confidence interval, and when $\alpha = 0.01$, we obtain a wider 99% confidence interval. The wider the confidence interval is, the more confident we can be that the given interval contains the unknown parameter. Of course, it is better to be 95% confident that the average life of a certain television transistor is between 6 and 7 years than to be 99% confident that it is between 3 and 10 years. Ideally, we prefer a short interval with a high degree of confidence. Sometimes, restrictions on the size of our sample prevent us from achieving short intervals without sacrificing some of our degree of confidence.

7.3 *Estimating the Mean*

A point estimator of the population mean μ is given by the statistic $\bar{X}$. The sampling distribution of $\bar{X}$ is centered at μ and in most applications the variance is smaller than that of any other estimator. Thus the sample mean $\bar{x}$ will be used as a point estimate for the population mean μ. Recall that $\sigma_{\bar{X}}^2 = \sigma^2/n$, so that a large sample will yield a value of $\bar{X}$ that comes from a sampling distribution with a small variance. Hence $\bar{x}$ is likely to be a very accurate estimate of μ when n is large.

Let us now consider the interval estimate of μ. If our sample is selected from a normal population or, failing this, if n is sufficiently large, we can establish a confidence interval for μ by considering the sampling distribution of $\bar{X}$. According to the central limit theorem, we can expect the sampling distribution of $\bar{X}$ to be approximately normally distributed with mean $\mu_{\bar{X}} = \mu$ and standard deviation $\sigma_{\bar{X}} = \sigma/\sqrt{n}$. Writing $z_{\alpha/2}$ for the z value above which we find an area of $\alpha/2$, we can see from Figure 7.2 that

$$P(-z_{\alpha/2} < Z < z_{\alpha/2}) = 1 - \alpha,$$

where

$$Z = \frac{\bar{X} - \mu}{\sigma/\sqrt{n}}.$$

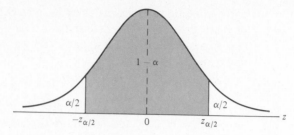

Figure 7.2 $P(-z_{\alpha/2} < Z < z_{\alpha/2}) = 1 - \alpha.$

Hence

$$P\left(-z_{\alpha/2} < \frac{\bar{X} - \mu}{\sigma/\sqrt{n}} < z_{\alpha/2}\right) = 1 - \alpha.$$

Multiplying each term in the inequality by $\sigma/\sqrt{n}$, and then subtracting $\bar{X}$ from each term and multiplying by -1 (reversing the sense of the inequalities), we obtain

$$P\left(\bar{X} - z_{\alpha/2} \frac{\sigma}{\sqrt{n}} < \mu < \bar{X} + z_{\alpha/2} \frac{\sigma}{\sqrt{n}}\right) = 1 - \alpha.$$

A random sample of size n is selected from a population whose variance σ^2 is known and the mean $\bar{x}$ is computed to give the following $(1 - \alpha)100\%$ confidence interval.

Confidence Interval for μ; σ Known   *If $\bar{x}$ is the mean of a random sample of size n from a population with known variance σ^2, a $(1 - \alpha)100\%$ confidence interval for μ is given by*

$$\bar{x} - z_{\alpha/2} \frac{\sigma}{\sqrt{n}} < \mu < \bar{x} + z_{\alpha/2} \frac{\sigma}{\sqrt{n}},$$

where $z_{\alpha/2}$ is the z value leaving an area of $\alpha/2$ to the right.

For small samples selected from nonnormal populations, we cannot expect our degree of confidence to be accurate. However, for samples of size $n \geq 30$, regardless of the shape of most populations, sampling theory guarantees good results.

Clearly, the values of the random variables $\hat{\Theta}_L$ and $\hat{\Theta}_U$, defined in Section 7.2, are the confidence limits

$$\hat{\theta}_L = \bar{x} - z_{\alpha/2} \frac{\sigma}{\sqrt{n}} \quad \text{and} \quad \hat{\theta}_U = \bar{x} + z_{\alpha/2} \frac{\sigma}{\sqrt{n}}.$$

Different samples will yield different values of $\bar{x}$ and therefore produce different interval estimates of the parameter μ as shown in Figure 7.3. The circular

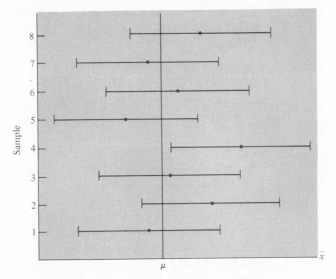

Figure 7.3 Interval estimates of μ for different samples.

dots at the center of each interval indicate the position of the point estimate $\bar{x}$ for each random sample. Most of the intervals are seen to contain μ, but not in every case. Note that all of these intervals are of the same width, since their widths depend only on the choice of $z_{\alpha/2}$ once $\bar{x}$ is determined. The larger the value we choose for $z_{\alpha/2}$, the wider we make all the intervals and the more confident we can be that the particular sample selected will produce an interval that contains the unknown parameter μ.

To compute a $(1 - \alpha)100\%$ confidence interval for μ, we have assumed that σ is known. Since this is generally not the case, we shall replace σ by the sample standard deviation s, provided that $n \geq 30$.

Example 7.2 The mean and standard deviation for the quality point averages of a random sample of 36 college seniors are calculated to be 2.6 and 0.3, respectively. Find the 95% and 99% confidence intervals for the mean of the entire senior class.

Solution. The point estimate of μ is $\bar{x} = 2.6$. Since the sample size is large, the standard deviation σ can be approximated by $s = 0.3$. The z value, leaving an area of 0.025 to the right and therefore an area of 0.975 to the left, is $z_{0.025} = 1.96$ (Table A.3). Hence the 95% confidence interval is

$$2.6 - (1.96)\left(\frac{0.3}{\sqrt{36}}\right) < \mu < 2.6 + (1.96)\left(\frac{0.3}{\sqrt{36}}\right),$$

which reduces to

$$2.50 < \mu < 2.70.$$

To find a 99% confidence interval, we find the z value leaving an area of 0.005 to the right and 0.995 to the left. Therefore, using Table A.3 again, $z_{0.005} = 2.575$, and the 99% confidence interval is

$$2.6 - (2.575)\left(\frac{0.3}{\sqrt{36}}\right) < \mu < 2.6 + (2.575)\left(\frac{0.3}{\sqrt{36}}\right),$$

or simply

$$2.47 < \mu < 2.73.$$

We now see that a longer interval is required to estimate μ with a higher degree of accuracy.

The $(1 - \alpha)100\%$ confidence interval provides an estimate of the accuracy of our point estimate. If μ is actually the center value of the interval, then $\bar{x}$ estimates μ without error. Most of the time, however, $\bar{x}$ will not be exactly equal to μ and the point estimate is in error. The size of this error will be the absolute value of the difference between μ and $\bar{x}$, and we can be $(1 - \alpha)100\%$ confident that this difference will not exceed $z_{\alpha/2}\, \sigma/\sqrt{n}$. We can readily see this if we draw a diagram of a hypothetical confidence interval as in Figure 7.4.

Figure 7.4   Error in estimating μ by $\bar{x}$.

Theorem 7.1   *If $\bar{x}$ is used as an estimate of μ, we can then be $(1 - \alpha)100\%$ confident that the error will not exceed $z_{\alpha/2}\, \sigma/\sqrt{n}$.*

In Example 7.2 we are 95% confident that the sample mean $\bar{x} = 2.6$ differs from the true mean μ by an amount less than 0.1 and 99% confident that the difference is less than 0.13.

Frequently, we wish to know how large a sample is necessary to ensure that the error in estimating μ will be less than a specified amount e. By Theorem 7.1 this means that we must choose n such that $z_{\alpha/2}\, \sigma/\sqrt{n} = e$. Solving this equation gives the following formula for n.

Theorem 7.2   *If $\bar{x}$ is used as an estimate of μ, we can be $(1 - \alpha)100\%$ confident that the error will not exceed a specified amount e when the sample size is*

$$n = \left(\frac{z_{\alpha/2}\, \sigma}{e}\right)^2.$$

When solving for the sample size, n, all fractional values are rounded up to the next whole number. By adhering to this principle, we can be sure that our degree of confidence never falls below $(1 - \alpha)100\%$.

Strictly speaking, the formula in Theorem 7.2 is applicable only if we know the variance of the population from which we are to select our sample. Lacking this information, we could take a preliminary sample of size $n \geq 30$ to provide an estimate of σ. Then, using s as an approximation for σ in Theorem 7.2, we could determine approximately how many observations are needed to provide the desired degree of accuracy.

Example 7.3   How large a sample is required in Example 7.2 if we want to be 95% confident that our estimate of μ is off by less than 0.05?

Solution. The sample standard deviation $s = 0.3$ obtained from the preliminary sample of size 36 will be used for σ. Then, by Theorem 7.2,

$$n = \left[\frac{(1.96)(0.3)}{0.05} \right]^2 = 138.3.$$

Therefore, we can be 95% confident that a random sample of size 139 will provide an estimate $\bar{x}$ differing from μ by an amount less than 0.05.

Frequently, we are attempting to estimate the mean of a population when the variance is unknown and it is impossible to obtain a sample of size $n \geq 30$. Cost can often be a factor that limits our sample size. As long as our population is approximately bell-shaped, confidence intervals can be computed when σ^2 is unknown and the sample size is small by using the sampling distribution of T, where

$$T = \frac{\bar{X} - \mu}{S/\sqrt{n}}.$$

The procedure is the same as for large samples except that we use the t distribution in place of the standard normal.

Referring to Figure 7.5, we can assert that

$$P(-t_{\alpha/2} < T < t_{\alpha/2}) = 1 - \alpha,$$

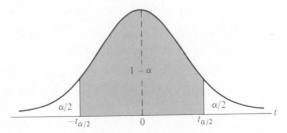

Figure 7.5 $P(-t_{\alpha/2} < T < t_{\alpha/2}) = 1 - \alpha.$

where $t_{\alpha/2}$ is the t value with $n - 1$ degrees of freedom, above which we find an area of $\alpha/2$. Because of symmetry, an equal area of $\alpha/2$ will fall to the left of $-t_{\alpha/2}$. Substituting for T, we write

$$P\left(-t_{\alpha/2} < \frac{\bar{X} - \mu}{S/\sqrt{n}} < t_{\alpha/2}\right) = 1 - \alpha.$$

Multiplying each term in the inequality by $S/\sqrt{n}$, and then subtracting $\bar{X}$ from each term and multiplying by -1, we obtain

$$P\left(\bar{X} - t_{\alpha/2}\frac{S}{\sqrt{n}} < \mu < \bar{X} + t_{\alpha/2}\frac{S}{\sqrt{n}}\right) = 1 - \alpha.$$

For our particular random sample of size n, the mean $\bar{x}$ and standard deviation s are computed and the following $(1 - \alpha)100\%$ confidence interval for μ is obtained.

Small-Sample Confidence Interval for μ; σ Unknown   *If $\bar{x}$ and s are the mean and standard deviation of a random sample of size $n < 30$ from an approximate normal population with unknown variance σ^2, a $(1 - \alpha)100\%$ confidence interval for μ is given by*

$$\bar{x} - t_{\alpha/2}\frac{s}{\sqrt{n}} < \mu < \bar{x} + t_{\alpha/2}\frac{s}{\sqrt{n}},$$

where $t_{\alpha/2}$ is the t value with $v = n - 1$ degrees of freedom, leaving an area of $\alpha/2$ to the right.

Example 5.4   The contents of 7 similar containers of sulfuric acid are 9.8, 10.2, 10.4, 9.8, 10.0, 10.2, and 9.6 liters. Find a 95% confidence interval for the mean of all such containers, assuming an approximate normal distribution.

Solution. The sample mean and standard deviation for the given data are

$$\bar{x} = 10.0 \quad \text{and} \quad s = 0.283.$$

Using Table A.4, we find $t_{0.025} = 2.447$ for $v = 6$ degrees of freedom. Hence the 95% confidence interval for μ is

$$10.0 - (2.477)\left(\frac{0.283}{\sqrt{7}}\right) < \mu < 10.0 + (2.447)\left(\frac{0.283}{\sqrt{7}}\right),$$

which reduces to

$$9.74 < \mu < 10.26.$$

7.4 Tolerance Limits

In Section 7.2 we defined a confidence interval for a parameter θ to be an interval of the form $\hat{\theta}_L < \theta < \hat{\theta}_U$, where $\hat{\theta}_L$ and $\hat{\theta}_U$ depend on the value of the statistic $\hat{\Theta}$ for a particular sample and also on the sampling distribution of $\hat{\Theta}$. That is, we computed confidence limits so that a specified fraction of the intervals computed from all possible samples of the same size contain the population parameter θ. For example, when all possible samples of the same size n are selected from a normal distribution, 95% of the intervals determined by the confidence limits $\bar{x} \pm 1.96\sigma/\sqrt{n}$ will contain the parameter μ. This fact was demonstrated in Figure 7.3. Hence we are 95% confident that the interval $\bar{x} \pm 1.96\sigma/\sqrt{n}$, computed from a particular sample, will contain the parameter μ.

Industrial engineers often find it more important to establish a confidence interval not containing the single value μ, but rather a confidence interval that will contain a fixed proportion or percentage of the measurements corresponding to some characteristic of a manufactured product. Such intervals are called **tolerance intervals** and their end points are called **tolerance limits**. One would expect a 95% tolerance interval that contains 0.90 or 90% of the measurements to be wider than the corresponding 95% confidence interval containing only the single parameter μ.

If we can assume from past experience that the measurements of our product are normally distributed with known mean μ and known standard deviation σ, the tolerance limits containing a specified proportion $1 - \alpha$ of the measurements are of the form $\mu \pm z_{\alpha/2}\,\sigma$, where $z_{\alpha/2}$ is determined from the table of normal curve areas. For a tolerance interval containing 0.90 of the measurements, $z_{0.05} = 1.645$. Unfortunately, in most industrial processes the values of μ and σ are seldom known. Nevertheless, tolerance limits can be defined by substituting the sample estimates $\bar{x}$ and s for μ and σ, respectively, and forming the interval $\bar{x} \pm ks$, where k is obtained from Table A.7.

Tolerance Limits *For a normal distribution of measurements with unknown mean μ and unknown standard deviation σ, **tolerance limits** are given by $\bar{x} \pm ks$, where k is determined so that one can assert with $100\gamma\%$ confidence that the given limits contain **at least** the proportion $1 - \alpha$ of the measurements.*

Table A.7 gives values of k for $1 - \alpha = 0.90, 0.95, 0.99$; $\gamma = 0.95, 0.99$; and for selected values of n from 2 to 1000.

Example 7.5 A machine is producing metal pieces that are cylindrical in shape. A sample of these pieces is taken and the diameters are found to be 1.01, 0.97, 1.03, 1.04,

0.99, 0.98, 0.99, 1.01, and 1.03 centimeters. Find the 99% tolerance limits that will contain 95% of the metal pieces produced by this machine, assuming an approximate normal distribution.

Solution. The sample mean and standard deviation for the given data are

$$\bar{x} = 1.0056 \qquad \text{and} \qquad s = 0.0245.$$

From Table A.7 for $n = 9$, $\gamma = 0.99$, and $1 - \alpha = 0.95$, we find $k = 4.550$. Hence the 99% tolerance limits are given by

$$1.0056 \pm (4.550)(0.0245),$$

or 0.894 and 1.117. That is, we are 99% confident that the tolerance interval from 0.894 to 1.117 will contain 95% of the metal pieces produced by this machine. It is interesting to note that the corresponding 99% confidence interval for μ (see Exercise 13 on page 225) has a lower limit of 0.978 and an upper limit of 1.033, verifying our earlier statement that a tolerance interval must necessarily be longer than a confidence interval with the same degree of confidence.

Exercises

1. Let us define $S'^2 = \sum\limits_{i=1}^{n} (X_i - \bar{X})^2/n$. Show that $E(S'^2) = [(n-1)/n]\sigma^2$, and hence S'^2 is a biased estimator for σ^2.

2. If X is a binomial random variable, show that

   (a) $\hat{P} = \dfrac{X}{n}$ is an unbiased estimator of p;

   (b) $P' = \dfrac{X + \sqrt{n}/2}{n + \sqrt{n}}$ is a biased estimator of p.

3. Show that the estimator P' of Exercise 2(b) becomes unbiased as $n \to \infty$.

4. An electrical firm manufactures light bulbs that have a length of life that is approximately normally distributed with a standard deviation of 40 hours. If a sample of 30 bulbs has an average life of 780 hours, find a 96% confidence interval for the population mean of all bulbs produced by this firm.

5. A soft-drink machine is regulated so that the amount of drink dispensed is approximately normally distributed with a standard deviation equal to 0.15 deciliters. Find a 95% confidence interval for the mean of all drinks dispensed by this machine if a random sample of 36 drinks has an average content of 2.25 deciliters.

6. The heights of a random sample of 50 college students showed a mean of 174.5 centimeters and a standard deviation of 6.9 centimeters.

   (a) Construct a 98% confidence interval for the mean height of all college students.

   (b) What can we assert with 98% confidence about the possible size of our error if we estimate the mean height of all college students to be 174.5 centimeters?

7. A random sample of 100 automobile owners shows that an automobile is driven on the average 23,500 kilometers per year, in the state of Virginia, with a standard deviation of 3900 kilometers.

   (a) Construct a 99% confidence interval for the average number of kilometers an automobile is driven annually in Virginia.

(b) What can we assert with 99% confidence about the possible size of our error if we estimate the average number of kilometers driven by car owners in Virginia as 23,500 kilometers per year?

8. How large a sample is needed in Exercise 4 if we wish to be 96% confident that our sample mean will be within 10 hours of the true mean?

9. How large a sample is needed in Exercise 5 if we wish to be 95% confident that our sample mean will be within 0.09 deciliter of the true mean?

10. An efficiency expert wishes to determine the average time that it takes to drill three holes in a certain metal clamp. How large a sample will he need to be 95% confident that his sample mean will be within 15 seconds of the true mean? Assume that it is known from previous studies that $\sigma = 40$ seconds.

11. A UCLA researcher claims that the life span of mice can be extended by as much as 25% when the calories in their food are reduced by approximately 40% from the time they are weaned. The restricted diets are enriched to normal levels by vitamins and protein. Assuming that it is known from previous studies that $\sigma = 5.8$ months, how many mice should be included in our sample if we wish to be 99% confident that the mean life span of the sample will be within 2 months of the population mean for all mice subjected to this reduced diet?

12. Regular consumption of presweetened cereals contributes to tooth decay, heart disease, and other degenerative diseases according to studies conducted by Dr. W. H. Bowen of the National Institutes of Health and Dr. J. Yudben, Professor of Nutrition and Dietetics at the University of London. In a random sample of 20 similar single servings of Alpha-Bits the average sugar content was 11.3 grams with a standard deviation of 2.45 grams. Assuming that the sugar contents are normally distributed, construct a 95% confidence interval for the mean sugar content for single servings of Alpha-Bits.

13. A machine is producing metal pieces that are cylindrical in shape. A sample of pieces is taken and the diameters are 1.01, 0.97, 1.03, 1.04, 0.99, 0.98, 0.99, 1.01, and 1.03 centimeters. Find a 99% confidence interval for the mean diameter of pieces

from this machine, assuming an approximate normal distribution.

14. A random sample of 8 cigarettes of a certain brand has an average nicotine content of 2.6 milligrams and a standard deviation of 0.9 milligram. Construct a 99% confidence interval for the true average nicotine content of this particular brand of cigarettes, assuming the distribution of nicotine contents to be approximately normal.

15. A random sample of 12 shearing pins are taken in a study of the Rockwell hardness of the head on the pin. Measurements on the Rockwell hardness were made for each of the 12, yielding an average value of 48.50 with a sample standard deviation of 1.5. Assuming the measurements to be normally distributed, construct a 90% confidence interval for the mean Rockwell hardness.

16. A random sample of 12 graduates of a certain secretarial school typed an average of 79.3 words per minute with a standard deviation of 7.8 words per minute. Assuming a normal distribution for the number of words typed per minute, find a 95% confidence interval for the average number of words typed by all graduates of this school.

17. A random sample of 25 cigarettes of a certain brand has an average nicotine content of 1.3 milligrams and a standard deviation of 0.17 milligram. Find the 95% tolerance limits that will contain 90% of the nicotine contents for this brand of cigarettes, assuming the measurements to be normally distributed.

18. The following measurements were recorded for the drying time, in hours, of a certain brand of latex paint:

3.4	2.5	4.8	2.9	3.6
2.8	3.3	5.6	3.7	2.8
4.4	4.0	5.2	3.0	4.8

Assuming that the measurements represent a random sample from a normal population, find the 99% tolerance limits that will contain 95% of the drying times.

19. Referring to Exercise 7, construct a 99% tolerance interval containing 99% of the miles driven by automobiles annually in Virginia. Assume the distribution of measurements to be approximately normal.

20. Referring to Exercise 15, construct a 95% interval containing 90% of the measurements.

7.5 *Estimating the Difference Between Two Means*

If we have two populations with means μ_1 and μ_2 and variances σ_1^2 and σ_2^2, respectively, a point estimator of the difference between μ_1 and μ_2 is given by the statistic $\bar{X}_1 - \bar{X}_2$. Therefore, to obtain a point estimate of $\mu_1 - \mu_2$, we shall select two independent random samples, one from each population, of size n_1 and n_2, and compute the difference, $\bar{x} - \bar{x}_2$, of the sample means.

If our independent samples are selected from normal populations, or failing this, if n_1 and n_2 are both at least 30, we can establish a confidence interval for $\mu_1 - \mu_2$ by considering the sampling distribution of $\bar{X}_1 - \bar{X}_2$.

According to Theorem 6.15, we can expect the sampling distribution of $\bar{X}_1 - \bar{X}_2$ to be approximately normally distributed with mean $\mu_{\bar{X}_1 - \bar{X}_2} = \mu_1 - \mu_2$ and standard deviation $\sigma_{\bar{X}_1 - \bar{X}_2} = \sqrt{(\sigma_1^2/n_1) + (\sigma_2^2/n_2)}$. Therefore, we can assert with a probability of $1 - \alpha$ that the standard normal variable

$$Z = \frac{(\bar{X}_1 - \bar{X}_2) - (\mu_1 - \mu_2)}{\sqrt{(\sigma_1^2/n_1) + (\sigma_2^2/n_2)}}$$

will fall between $-z_{\alpha/2}$ and $z_{\alpha/2}$. Referring once again to Figure 7.2, we write

$$P(-z_{\alpha/2} < Z < z_{\alpha/2}) = 1 - \alpha.$$

Substituting for Z, we state equivalently that

$$P\left[-z_{\alpha/2} < \frac{(\bar{X}_1 - \bar{X}_2) - (\mu_1 - \mu_2)}{\sqrt{(\sigma_1^2/n_1) + (\sigma_2^2/n_2)}} < z_{\alpha/2}\right] = 1 - \alpha,$$

which leads to the following $(1 - \alpha)100\%$ confidence interval for $\mu_1 - \mu_2$.

Confidence Interval for $\mu_1 - \mu_2$; σ_1^2 and σ_2^2 Known   *If $\bar{x}_1$ and $\bar{x}_2$ are the means of independent random samples of size n_1 and n_2 from populations with known variances σ_1^2 and σ_2^2, respectively, a $(1 - \alpha)100\%$ confidence interval for $\mu_1 - \mu_2$ is given by*

$$(\bar{x}_1 - \bar{x}_2) - z_{\alpha/2}\sqrt{\frac{\sigma_1^2}{n_1} + \frac{\sigma_2^2}{n_2}} < \mu_1 - \mu_2 < (\bar{x}_1 - \bar{x}_2) + z_{\alpha/2}\sqrt{\frac{\sigma_1^2}{n_1} + \frac{\sigma_2^2}{n_2}},$$

where $z_{\alpha/2}$ is the z value leaving an area of $\alpha/2$ to the right.

The degree of confidence is exact when samples are selected from normal populations. For nonnormal populations we obtain an approximate confidence interval that is very good when both n_1 and $n_2 \geq 30$. As before, if σ_1^2

and σ_2^2 are unknown and our samples are sufficiently large, we may replace σ_1^2 by s_1^2 and σ_2^2 by s_2^2 without appreciably affecting the confidence interval.

Example 7.6 A standardized chemistry test was given to 50 girls and 75 boys. The girls made an average grade of 76 with a standard deviation of 6, while the boys made an average grade of 82 with a standard deviation of 8. Find a 96% confidence interval for the difference $\mu_1 - \mu_2$, where μ_1 is the mean score of all boys and μ_2 is the mean score of all girls who might take this test.

Solution. The point estimate of $\mu_1 - \mu_2$ is $\bar{x}_1 - \bar{x}_2 = 82 - 76 = 6$. Since n_1 and n_2 are both large, we can substitute $s_1 = 8$ for σ_1 and $s_2 = 6$ for σ_2. Using $\alpha = 0.04$, we find $z_{0.02} = 2.05$ from Table A.3. Hence substitution in the above formula, the 96% confidence interval is

$$6 - 2.05 \sqrt{\frac{64}{75} + \frac{36}{50}} < \mu_1 - \mu_2 < 6 + 2.05 \sqrt{\frac{64}{75} + \frac{36}{50}}$$

or simply

$$3.43 < \mu_1 - \mu_2 < 8.57.$$

This procedure for estimating the difference between two means is applicable if σ_1^2 and σ_2^2 are known or can be estimated from large samples. If the sample sizes are small, we must again resort to the t distribution to provide confidence intervals that are valid when the populations are approximately normally distributed.

Let us now assume that σ_1^2 and σ_2^2 are unknown and that n_1 and n_2 are small (< 30). If $\sigma_1^2 = \sigma_2^2 = \sigma^2$, we obtain a standard normal variable in the form

$$Z = \frac{(\bar{X}_1 - \bar{X}_2) - (\mu_1 - \mu_2)}{\sqrt{\sigma^2[(1/n_1) + (1/n_2)]}}.$$

According to Theorem 6.16 the two random variables $(n_1 - 1)S_1^2/\sigma^2$ and $(n_2 - 1)S_2^2/\sigma^2$ have chi-square distributions with $n_1 - 1$ and $n_2 - 1$ degrees of freedom, respectively. Furthermore, they are independent chi-square variables, since the random samples were selected independently. Consequently, their sum

$$V = \frac{(n_1 - 1)S_1^2}{\sigma^2} + \frac{(n_2 - 1)S_2^2}{\sigma^2} = \frac{(n_1 - 1)S_1^2 + (n_2 - 1)S_2^2}{\sigma^2}$$

has a chi-square distribution with $v = n_1 + n_2 - 2$ degrees of freedom.

Since the preceding expressions for Z and V can be shown to be independent, it follows from Theorem 6.17 that the statistic

$$T = \frac{(\bar{X}_1 - \bar{X}_2) - (\mu_1 - \mu_2)}{\sqrt{\sigma^2[(1/n_1) + (1/n_2)]}} \Bigg/ \sqrt{\frac{(n_1 - 1)S_1^2 + (n_2 - 1)S_2^2}{\sigma^2(n_1 + n_2 - 2)}},$$

which has the t distribution with $v = n_1 + n_2 - 2$ degrees of freedom.

A point estimate of the unknown common variance σ^2 can be obtained by pooling the sample variances. Denoting the pooled estimator by S_p^2, we write

$$S_p^2 = \frac{(n_1 - 1)S_1^2 + (n_2 - 1)S_2^2}{n_1 + n_2 - 2}.$$

Substituting S_p^2 in the T statistic, we obtain the less cumbersome form

$$T = \frac{(\bar{X}_1 - \bar{X}_2) - (\mu_1 - \mu_2)}{S_p\sqrt{(1/n_1) + (1/n_2)}}.$$

Using the statistic T, we have

$$P(-t_{\alpha/2} < T < t_{\alpha/2}) = 1 - \alpha,$$

where $t_{\alpha/2}$ is the t value with $n_1 + n_2 - 2$ degrees of freedom, above which we find an area of $\alpha/2$. Substituting for T in the inequality, we write

$$P\left[-t_{\alpha/2} < \frac{(\bar{X}_1 - \bar{X}_2) - (\mu_1 - \mu_2)}{S_p\sqrt{(1/n_1) + (1/n_2)}} < t_{\alpha/2} \right] = 1 - \alpha.$$

After performing the usual mathematical manipulations, the difference of the sample means $\bar{x}_1 - \bar{x}_2$ and the pooled variance

$$s_p^2 = \frac{(n_1 - 1)s_1^2 + (n_2 - 1)s_2^2}{n_1 + n_2 - 2}$$

are computed and then the following $(1 - \alpha)100\%$ confidence interval for $\mu_1 - \mu_2$ is obtained.

Small-Sample Confidence Interval for $\mu_1 - \mu_2$; $\sigma_1^2 = \sigma_2^2$ but Unknown   *If $\bar{x}_1$ and $\bar{x}_2$ are the means of small independent random samples of size n_1 and n_2, respectively, from approximate normal populations with unknown but equal variances, a $(1 - \alpha)100\%$ confidence interval for $\mu_1 - \mu_2$ is given by*

$$(\bar{x}_1 - \bar{x}_2) - t_{\alpha/2}\, s_p \sqrt{\frac{1}{n_1} + \frac{1}{n_2}} < \mu_1 - \mu_2 < (\bar{x}_1 - \bar{x}_2) + t_{\alpha/2}\, s_p \sqrt{\frac{1}{n_1} + \frac{1}{n_2}},$$

where s_p is the pooled estimate of the population standard deviation and $t_{\alpha/2}$ is the t value with $\nu = n_1 + n_2 - 2$ degrees of freedom, leaving an area of $\alpha/2$ to the right.

Example 7.7   In the article "*Macroinvertebrate Community Structure as an Indicator of Acid Mine Pollution*" published in the *Journal of Environmental Pollution* (Vol. 6, 1974), we are given a report on an investigation undertaken in Cane Creek, Alabama, to determine the relationship between selected physiochemical parameters and different measures of macroinvertebrate community structure. One facet of the investigation was an evaluation of the effectiveness of a numerical species diversity index to indicate aquatic degradation due to acid mine drainage. Conceptually, a high index of macroinvertebrate species diversity should indicate an unstressed aquatic system, while a low diversity index should indicate a stressed aquatic system.

Two independent sampling stations were chosen for this study, one located downstream from the acid mine discharge point and the other located upstream. For 12 monthly samples collected at the downstream station the species diversity index had a mean value $\bar{x}_1 = 3.11$ and a standard deviation $s_1 = 0.771$, while 10 monthly samples collected at the upstream station had a mean index value $\bar{x}_2 = 2.04$ and a standard deviation $s_2 = 0.448$. Find a 90% confidence interval for the difference between the population means for the two locations, assuming that the populations are approximately normally distributed with equal variances.

Solution. Let μ_1 and μ_2 represent the population means, respectively, for the species diversity index at the downstream and upstream stations. We wish to find a 90% confidence interval for $\mu_1 - \mu_2$. Our point estimate of $\mu_1 - \mu_2$ is $\bar{x}_1 - \bar{x}_2 = 3.11 - 2.04 = 1.07$. The pooled estimate, s_p^2, of the common variance, σ^2, is

$$s_p^2 = \frac{(n_1 - 1)s_1^2 + (n_2 - 1)s_2^2}{n_1 + n_2 - 2}$$

$$= \frac{(11)(0.771^2) + (9)(0.448^2)}{12 + 10 - 2} = 0.417.$$

Taking the square root, we obtain $s_p = 0.646$. Using $\alpha = 0.1$, we find in Table A.4 that $t_{0.05} = 1.725$ for $v = n_1 + n_2 - 2 = 20$ degrees of freedom. Therefore, the 90% confidence interval for $\mu_1 - \mu_2$ is

$$1.07 - (1.725)(0.646)\sqrt{\tfrac{1}{12} + \tfrac{1}{10}} < \mu_1 - \mu_2 < 1.07 + (1.725)(0.646)\sqrt{\tfrac{1}{12} + \tfrac{1}{10}},$$

which simplifies to

$$0.593 < \mu_1 - \mu_2 < 1.547.$$

Hence we are 90% confident that the interval from 0.593 to 1.547 contains the difference of the population means for values of the species diversity index at the two stations. The fact that both confidence limits are positive indicates that, on the average, the index for the station located downstream from the discharge point is greater than the index for the station located upstream.

The procedure for constructing confidence intervals for $\mu_1 - \mu_2$ from small samples assumes the populations to be normal and the population variances to be equal. Slight departures from either of these assumptions do not seriously alter the degree of confidence for our interval. A procedure will be presented in Chapter 8 for testing the equality of two unknown population variances based on the information provided by the sample variances. If the population variances are considerably different, we still obtain good results when the populations are normal, provided that $n_1 = n_2$. Therefore, in a planned experiment, one should make every effort to equalize the size of the samples.

Let us now consider the problem of finding an interval estimate of $\mu_1 - \mu_2$ for small samples when the unknown population variances are not likely to be equal, and it is impossible to select samples of equal size. The statistic most often used in this case is

$$T' = \frac{(\bar{X}_1 - \bar{X}_2) - (\mu_1 - \mu_2)}{\sqrt{(S_1^2/n_1) + (S_2^2/n_2)}},$$

which has approximately a t distribution with v degrees of freedom, where

$$v = \frac{(s_1^2/n_1 + s_2^2/n_2)^2}{[(s_1^2/n_1)^2/(n_1 - 1)] + [(s_2^2/n_2)^2/(n_2 - 1)]}.$$

Since v is seldom an integer, we round it off to the nearest whole number.

Using the statistic T', we write

$$P(-t_{\alpha/2} < T' < t_{\alpha/2}) \simeq 1 - \alpha,$$

where $t_{\alpha/2}$ is the value of the t distribution with v degrees of freedom, above which we find an area of $\alpha/2$. Substituting for T' in the inequality, and following the exact steps as before, we state the final result.

Small-Sample Confidence Interval for $\mu_1 - \mu_2$; $\sigma_1^2 \neq \sigma_2^2$ and Unknown   *If $\bar{x}_1$ and s_1^2, and $\bar{x}_2$ and s_2^2, are the means and variances of small independent samples of size n_1 and n_2, respectively, from approximate normal distributions with unknown and unequal variances, an approximate $(1 - \alpha)100\%$ confidence interval for $\mu_1 - \mu_2$ is given by*

$$(\bar{x}_1 - \bar{x}_2) - t_{\alpha/2}\sqrt{\frac{s_1^2}{n_1} + \frac{s_2^2}{n_2}} < \mu_1 - \mu_2 < (\bar{x}_1 - \bar{x}_2) + t_{\alpha/2}\sqrt{\frac{s_1^2}{n_1} + \frac{s_2^2}{n_2}},$$

where $t_{\alpha/2}$ is the t value with

$$v = \frac{(s_1^2/n_1 + s_2^2/n_2)^2}{[(s_1^2/n_1)^2/(n_1 - 1)] + [(s_2^2/n_2)^2/(n_2 - 1)]}$$

degrees of freedom, leaving an area $\alpha/2$ to the right.

Example 7.8   A study on the "*Nutrient Retention and Macroinvertebrate Community Response to Sewage Stress in a Stream Ecosystem*" was conducted by the Department of Zoology at the Virginia Polytechnic Institute and State University in 1980 to estimate the difference in the amount of the chemical orthophosphorus measured at two different stations on the James River. Orthophosphorus is measured in milligrams per liter. Fifteen samples were collected from station 1 and 12 samples were obtained from station 2. The 15 samples from station 1 had an average orthophosphorus content of 3.84 milligrams per liter and a standard deviation of 3.07 milligrams per liter, while the

12 samples from station 2 had an average content of 1.49 milligrams per liter and a standard deviation of 0.80 milligram per liter. Find a 95% confidence interval for the difference in the true average orthophosphorus contents at these two stations, assuming that the observations came from normal populations with different variances.

Solution. For station 1 we have $\bar{x}_1 = 3.84$, $s_1 = 3.07$, and $n_1 = 15$. For station 2, $\bar{x}_2 = 1.49$, $s_2 = 0.80$, and $n_2 = 12$. We wish to find a 95% confidence interval for $\mu_1 - \mu_2$. Since the population variances are assumed to be unequal and our sample sizes are not the same, we can only find an approximate 95% confidence interval based on the t distribution with v degrees of freedom, where

$$v = \frac{(3.07^2/15 + 0.80^2/12)^2}{[(3.07^2/15)^2/14] + [(0.80^2/12)^2/11]}$$

$$= 16.3 \simeq 16.$$

Our point estimate of $\mu_1 - \mu_2$ is $\bar{x}_1 - \bar{x}_2 = 3.84 - 1.49 = 2.35$. Using $\alpha = 0.05$, we find in Table A.4 that $t_{0.025} = 2.120$ for $v = 16$ degrees of freedom. Therefore, the 95% confidence interval for $\mu_1 - \mu_2$ is

$$2.35 - 2.120\sqrt{\frac{3.07^2}{15} + \frac{0.80^2}{12}} < \mu_1 - \mu_2 < 2.35 + 2.120\sqrt{\frac{3.07^2}{15} + \frac{0.80^2}{12}}$$

which simplifies to

$$0.60 < \mu_1 - \mu_2 < 4.10.$$

Hence we are 95% confident that the interval from 0.60 to 4.10 milligrams per liter contains the difference of the true average orthophosphorus contents for these two locations.

We conclude this section by considering estimation procedures for the difference of two means when the samples are not independent and the variances of the two populations are not necessarily equal. This will be true if the observations in the two samples occur in pairs so that the two observations are related. For instance, if we run a test on a new diet using 15 individuals, the weights before and after completion of the test form our two samples. Observations in the two samples made on the same individual are related and hence form a pair. To determine if the diet is effective, we must consider the differences $d_1, d_2, \ldots, d_n$ of paired observations. These differences are the values of a random sample $D_1, D_2, \ldots, D_n$ from a population of differences that we shall assume to be normally distributed with mean $\mu_D = \mu_1 - \mu_2$ and variance σ_D^2. We estimate σ_D^2 by s_d^2, the variance of the differences that constitute our sample. The point estimator of μ_D is given by $\bar{D}$.

Since the variance of the statistic $\bar{D}$ for paired observations is less than the variance of $\bar{X}_1 - \bar{X}_2$ for independent random samples, any statistical inference concerning $\mu_1 - \mu_2$ based on $\bar{D}$ will be more sensitive. For this reason it is often desirable to plan an experiment in which pairing has been purposely introduced. This may be achieved by recording the X_1 and X_2 measurements for each of n different subjects, whether they be individuals, animals, or plants. Thus the random variables might represent the weights of each of n individuals before and after a controlled diet experiment. We could also choose n pairs of subjects with each pair having a similar characteristic such as IQ, age, breed, and so on; then for each pair one member is selected at random to yield a value of X_1 leaving the other member to provide the value of X_2. In this case X_1 and X_2 might represent the grades obtained by two individuals of equal IQ when one of the individuals is assigned at random to a class using the conventional lecture approach while the other individual is assigned to a class using programmed materials.

A $(1 - \alpha)100\%$ confidence interval for μ_D can be established by writing

$$P(-t_{\alpha/2} < T < t_{\alpha/2}) = 1 - \alpha,$$

where

$$T = \frac{\bar{D} - \mu_D}{S_d/\sqrt{n}}$$

and $t_{\alpha/2}$, as before, is a value of the t distribution with $n - 1$ degrees of freedom.

It is now a routine procedure to replace T, by its definition, in the inequality above and carry out the mathematical steps that lead to the following $(1 - \alpha)100\%$ confidence interval for $\mu_1 - \mu_2 = \mu_D$.

Confidence Interval for $\mu_D = \mu_1 - \mu_2$ for Paired Observations   *If $\bar{d}$ and s_d are the mean and standard deviation of the normally distributed differences of n random pairs of measurements, a $(1 - \alpha)100\%$ confidence interval for $\mu_D = \mu_1 - \mu_2$ is*

$$\bar{d} - t_{\alpha/2} \frac{s_d}{\sqrt{n}} < \mu_D < \bar{d} + t_{\alpha/2} \frac{s_d}{\sqrt{n}},$$

where $t_{\alpha/2}$ is the t value with $v = n - 1$ degrees of freedom, leaving an area of $\alpha/2$ to the right.

Example 7.9   In the article "*Essential Elements in Fresh and Canned Tomatoes,*" published in the *Journal of Food Science* (Vol. 46, 1981), the contents of essential elements were determined in fresh and canned tomatoes by atomic absorption spectrophotometry. The copper contents in fresh tomatoes compared to the copper contents for the same tomatoes after being canned were recorded as follows:

Pair	Fresh Tomatoes	Canned Tomatoes	d_i
1	0.066	0.085	0.019
2	0.079	0.088	0.009

Pair	Fresh Tomatoes	Canned Tomatoes	d_i
3	0.069	0.091	0.022
4	0.076	0.096	0.020
5	0.071	0.093	0.022
6	0.087	0.095	0.008
7	0.071	0.079	0.008
8	0.073	0.078	0.005
9	0.067	0.065	−0.002
10	0.062	0.068	0.006

Find a 98% confidence interval for the true difference in the mean copper contents of fresh and canned tomatoes assuming the distribution of differences to be normal.

Solution. We wish to find a 98% confidence interval for $\mu_1 - \mu_2$, where μ_1 and μ_2 represent the true average copper contents for canned tomatoes and fresh tomatoes, respectively. Since the observations are paired, $\mu_1 - \mu_2 = \mu_D$. The point estimate of μ_D is given by $\bar{d} = 0.0117$. The standard deviation, s_d, of the sample differences is

$$s_d = \sqrt{\frac{n \sum_{i=1}^{n} d_i^2 - \left(\sum_{i=1}^{n} d_i \right)^2}{n(n-1)}} = \sqrt{\frac{(10)(0.002003) - (0.117)^2}{(10)(9)}}$$

$$= 0.0084.$$

Using $\alpha = 0.02$, we find in Table A.4 that $t_{0.01} = 2.821$ for $v = n - 1 = 9$ degrees of freedom. Therefore, the 98% confidence interval is

$$0.0117 - (2.821)\left(\frac{0.0084}{\sqrt{10}}\right) < \mu_D < 0.0117 + (2.821)\left(\frac{0.0084}{\sqrt{10}}\right)$$

or simply

$$0.0042 < \mu_D < 0.0192$$

from which we conclude that there are significantly higher amounts of copper in canned tomatoes than in fresh tomatoes.

Exercises

1. A random sample of size $n_1 = 25$ taken from a normal population with a standard deviation $\sigma_1 = 5$ has a mean $\bar{x}_1 = 80$. A second random sample of size $n_2 = 36$, taken from a different normal population with a standard deviation $\sigma_2 = 3$, has a mean $\bar{x}_2 = 75$. Find a 94% confidence interval for $\mu_1 - \mu_2$.

2. Two kinds of thread are being compared for strength. Fifty pieces of each type of thread are tested under similar conditions. Brand A had an average tensile strength of 78.3 kilograms with a standard deviation of 5.6 kilograms, while brand B had an average tensile strength of 87.2 kilograms with a standard deviation of 6.3 kilograms. Construct a 95% confidence interval for the difference of the population means.

3. A study was made to determine if a certain metal treatment has any effect on the amount of metal removed in a pickling operation. A random sample of 100 pieces was immersed in a bath for 24 hours without the treatment, yielding an average of 12.2 millimeters of metal removed and a sample standard deviation of 1.1 millimeters. A second sample of 200 pieces was exposed to the treatment followed by the 24-hour immersion in the bath, resulting in an average removal of 9.1 millimeters of metal with a sample standard deviation of 0.9 millimeter. Compute a 98% confidence interval estimate for the difference between the population means. Does the treatment appear to reduce the mean amount of metal removed?

4. In a batch chemical process, two catalysts are being compared for their effect on the output of the process reaction. A sample of 12 batches was prepared using catalyst 1 and a sample of 10 batches was obtained using catalyst 2. The 12 batches for which catalyst 1 was used gave an average yield of 85 with a sample standard deviation of 4, while the average for the second sample gave an average of 81 and a sample standard deviation of 5. Find a 90% confidence interval for the difference between the population means, assuming that the populations are approximately normally distributed with equal variances.

5. Students may choose between a 3-semester-hour course in physics without labs and a 4-semester-hour course with labs. The final written examination is the same for each section. If 12 students in the section with labs made an average examination grade of 84 with a standard deviation of 4, and 18 students in the section without labs made an average grade of 77 with a standard deviation of 6, find a 99% confidence interval for the difference between the average grades for the two courses. Assume the populations to be approximately normally distributed with equal variances.

6. In a study conducted at the Virginia Polytechnic Institute and State University in 1983 on the development of ectomycorrhizal, a symbiotic relationship between the roots of trees and a fungus in which minerals are transferred from the fungus to the trees and sugars from the trees to the fungus, 20 northern red oak seedlings with the fungus *Pisolithus tinctorus* were grown in a greenhouse. All seedlings were planted in the same type of soil and received the same amount of sunshine and water. Half received no nitrogen at planting time to serve as a control and the other half received 368 ppm of nitrogen in the form $NaNO_3$. The stem weights, recorded in grams, at the end of 140 days were recorded as follows:

No Nitrogen	Nitrogen
0.32	0.26
0.53	0.43
0.28	0.47
0.37	0.49
0.47	0.52
0.43	0.75
0.36	0.79
0.42	0.86
0.38	0.62
0.43	0.46

Construct a 95% confidence interval for the difference in the mean stem weights between seedlings that receive no nitrogen and those that receive 368 ppm of nitrogen. Assume the populations to be normally distributed with equal variances.

7. The following data, recorded in days, represent the length of time to recovery for patients randomly treated with one of two medications to clear up severe bladder infections:

Medication 1	Medication 2
$n_1 = 14$	$n_2 = 16$
$\bar{x}_1 = 17$	$\bar{x}_2 = 19$
$s_1^2 = 1.5$	$s_2^2 = 1.8$

Find a 99% confidence interval for the difference $\mu_2 - \mu_1$ in the mean recovery time for the two medications, assuming normal populations with equal variances.

8. An experiment reported in *Popular Science*, in 1981, compared fuel economies for two types of similarly equipped diesel mini-trucks. Let us suppose that 12 Volkswagen and 10 Toyota trucks are used in 90-kilometer per hour steady-speed tests. If the 12 Volkswagen trucks average 16 kilometers per liter with a standard deviation of 1.0 kilometer per liter and the 10 Toyota trucks average 11 kilometers per liter with a standard deviation of 0.8 kilometer per liter, construct a 90% confidence interval for the difference between the average kilometers per liter of these two mini-trucks. Assume that the distances per liter for each truck model are approximately normally distributed with equal variances.

9. A taxi company is trying to decide whether to purchase brand A or brand B tires for its fleet of taxis. To estimate the difference in the two brands, an experiment is conducted using 12 of each brand. The tires are run until they wear out. The results are

$$\text{Brand } A: \bar{x}_1 = 36{,}300 \text{ kilometers,}$$
$$s_1 = 5000 \text{ kilometers.}$$

$$\text{Brand } B: \bar{x}_2 = 38{,}100 \text{ kilometers,}$$
$$s_2 = 6100 \text{ kilometers.}$$

Compute a 95% confidence interval for $\mu_1 - \mu_2$, assuming the populations to be approximately normally distributed.

10. The following data represent the running times of films produced by two motion-picture companies.

	Time (minutes)						
Company I	103	94	110	87	98		
Company II	97	82	123	92	175	88	118

Compute a 90% confidence interval for the difference between the average running times of films produced by the two companies. Assume that the running time differences are approximately normally distributed.

11. The government awarded grants to the agricultural departments of 9 universities to test the yield capabilities of two new varieties of wheat. Each variety was planted on plots of equal area at each university and the yields, in kilograms per plot, recorded as follows:

	University								
	1	2	3	4	5	6	7	8	9
Variety 1	38	23	35	41	44	29	37	31	38
Variety 2	45	25	31	38	50	33	36	40	43

Find a 95% confidence interval for the mean difference between the yields of the two varieties, assuming the differences of yields to be approximately normally distributed. Explain why pairing is necessary in this problem.

12. Referring to Exercise 9, find a 99% confidence interval for $\mu_1 - \mu_2$ if a tire from each company is assigned at random to the rear wheels of 8 taxis and the following distance, in kilometers, recorded:

Taxi	Brand A	Brand B
1	34,400	36,700
2	45,500	46,800
3	36,700	37,700
4	32,000	31,100
5	48,400	47,800
6	32,800	36,400
7	38,100	38,900
8	30,100	31,500

Assume that the differences of the distances are approximately normally distributed.

13. It is claimed that a new diet will reduce a person's weight by 4.5 kilograms on the average in a period of 2 weeks. The weights of 7 women who followed this diet were recorded before and after a 2-week period.

	Woman						
	1	2	3	4	5	6	7
Weight before	58.5	60.3	61.7	69.0	64.0	62.6	56.7
Weight after	60.0	54.9	58.1	62.1	58.5	59.9	54.4

Test a manufacturer's claim by computing a 95% confidence interval for the mean difference in the weight. Assume the differences of weights to be approximately normally distributed.

14. A health spa claims that a new exercise program will reduce a person's waist size by 2 centimeters on the average over a 5-day period. The waist sizes of 6 men who participated in this exercise program are recorded before and after the 5-day period in the following table:

	Man					
	1	2	3	4	5	6
Waist size before	90.4	95.5	98.7	115.9	104.0	85.6
Waist size after	91.7	93.9	97.4	112.8	101.3	84.0

By computing a 95% confidence interval for the mean reduction in waist size, determine whether the health spa's claim is valid. Assume the distribution of differences of waist sizes before and after the program to be approximately normal.

15. The study "*Evaluation of Prescribed Burning in Relation to Available Deer Browse*" was undertaken at the Virginia Polytechnic Institute and State University in 1964 to determine if fire can be used as a viable management tool to increase the amount of forage available to deer during the critical months in late winter and early spring. Calcium is a required element for plants and animals. The amount taken up and stored in the plant is closely correlated to the amount present in the soil. It was hypothesized that a fire may change the calcium levels present in the soil and thus affect the amount available to the deer. A large tract of land in the Fishburn Forest was selected for a prescribed burn. Soil samples were taken from 12 plots of equal area just prior to the burn on May 20, 1964, and analyzed for calcium. On July 16, 1964, postburn calcium levels were analyzed from the same plots. These values, in kilograms per plot, are presented in the following table:

Plot	Calcium Levels (kg/plot)	
	Preburn	Postburn
1	50	9
2	50	18
3	82	45
4	64	18
5	82	18
6	73	9
7	77	32
8	54	9
9	23	18
10	45	9
11	36	9
12	54	9

Construct a 95% confidence interval for the mean difference in the calcium level that is present in the soil prior to and after the prescribed burn. Assume the distribution of differences of calcium levels to be approximately normal.

16. In the article "*A Study of a Modified Membrane Filter Technique for the Enumeration of Stressed Fecal Coliforms in Urban Runoff*" (1977), the Department of Civil Engineering at the Virginia Polytechnic Institute and State University compared a modified (M-5 hr) assay technique for recovering fecal coliforms in stormwater runoff from an urban area to a most probable number (MPN) technique. A total of 12 runoff samples were collected between May 25 and June 28, 1977, and analyzed by the two techniques. The fecal coliform counts per 100 ml are recorded in the following table:

Sample	MPN Count	M-5 hr Count
1	2300	2010
2	1200	930
3	450	400
4	210	436
5	270	4100
6	450	2090
7	154	219
8	179	169
9	192	194

Sample	MPN Count	M-5 hr Count
10	230	174
11	340	274
12	194	183

Construct a 90% confidence interval for the difference in the mean fecal coliform counts between the M-5 hr and the MPN techniques. Assume that the count differences are approximately normally distributed.

7.6 Estimating a Proportion

A point estimator of the proportion p in a binomial experiment is given by the statistic $\hat{P} = X/n$, where X represents the number of successes in n trials. Therefore, the sample proportion $\hat{p} = x/n$ will be used as the point estimate of the parameter p.

If the unknown proportion p is not expected to be too close to zero or 1, we can establish a confidence interval for p by considering the sampling distribution of $\hat{P}$. Designating a failure in each binomial trial by the value 0 and a success by the value 1, the number of successes, x, can be interpreted as the sum of n values consisting only of zeros and ones, and $\hat{p}$ is just the sample mean of these n values. Hence, by the Central Limit Theorem, for n sufficiently large, $\hat{P}$ is approximately normally distributed with mean

$$\mu_{\hat{P}} = E(\hat{P}) = E\left(\frac{X}{n}\right) = \frac{np}{n} = p$$

and variance

$$\sigma_{\hat{P}}^2 = \sigma_{X/n}^2 = \frac{\sigma_X^2}{n^2} = \frac{npq}{n^2} = \frac{pq}{n}.$$

Therefore, we can assert that

$$P(-z_{\alpha/2} < Z < z_{\alpha/2}) = 1 - \alpha,$$

where

$$Z = \frac{\hat{P} - p}{\sqrt{pq/n}}$$

and $z_{\alpha/2}$ is the value of the standard normal curve above which we find an area of $\alpha/2$. Substituting for Z, we write

$$P\left(-z_{\alpha/2} < \frac{\hat{P} - p}{\sqrt{pq/n}} < z_{\alpha/2}\right) = 1 - \alpha.$$

Multiplying each term of the inequality by $\sqrt{pq/n}$, and then subtracting $\hat{P}$ and multiplying by -1, we obtain

$$P\left(\hat{P} - z_{\alpha/2}\sqrt{\frac{pq}{n}} < p < \hat{P} + z_{\alpha/2}\sqrt{\frac{pq}{n}}\right) = 1 - \alpha.$$

It is difficult to manipulate the inequalities so as to obtain a random interval whose end points are independent of p, the unknown parameter. When n is large, very little error is introduced by substituting the point estimate $\hat{p} = x/n$ for the p under the radical sign. Then we can write

$$P\left(\hat{P} - z_{\alpha/2}\sqrt{\frac{\hat{p}\hat{q}}{n}} < p < \hat{P} + z_{\alpha/2}\sqrt{\frac{\hat{p}\hat{q}}{n}}\right) \simeq 1 - \alpha.$$

For our particular random sample of size n, the sample proportion $\hat{p} = x/n$ is computed, and the following approximate $(1 - \alpha)100\%$ confidence interval for p is obtained.

Large-Sample Confidence Interval for p   *If $\hat{p}$ is the proportion of successes in a random sample of size n, and $\hat{q} = 1 - \hat{p}$, an approximate $(1 - \alpha)100\%$ confidence interval for the binomial parameter p is given by*

$$\hat{p} - z_{\alpha/2}\sqrt{\frac{\hat{p}\hat{q}}{n}} < p < \hat{p} + z_{\alpha/2}\sqrt{\frac{\hat{p}\hat{q}}{n}},$$

where $z_{\alpha/2}$ is the z value leaving an area of $\alpha/2$ to the right.

When n is small and the unknown proportion p is believed to be close to 0 or to 1, the confidence-interval procedure established here is unreliable and, therefore, should not be used. To be on the safe side, one should require both $n\hat{p}$ and $n\hat{q}$ to be greater than or equal to 5. The method for finding a confidence interval for the binomial parameter p is also applicable when the binomial distribution is being used to approximate the hypergeometric distribution, that is, when n is small relative to N, as illustrated in Example 7.10.

Example 7.10   In a random sample of $n = 500$ families owning television sets in the city of Hamilton, Canada, it was found that $x = 340$ subscribed to HBO. Find a 95% confidence interval for the actual proportion of families in this city who subscribe to HBO.

Solution. The point estimate of p is $\hat{p} = 340/500 = 0.68$. Using Table A.3, we find $z_{0.025} = 1.96$. Therefore, the 95% confidence interval for p is

$$0.68 - 1.96\sqrt{\frac{(0.68)(0.32)}{500}} < p < 0.68 + 1.96\sqrt{\frac{(0.68)(0.32)}{500}},$$

which simplifies to

$$0.64 < p < 0.72.$$

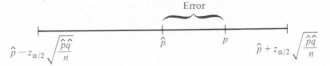

Figure 7.6 Error in estimating p by $\hat{p}$.

If p is the center value of a $(1 - \alpha)100\%$ confidence interval, then $\hat{p}$ estimates p without error. Most of the time, however, $\hat{p}$ will not be exactly equal to p and the point estimate is in error. The size of this error will be the positive difference that separates p and $\hat{p}$, and we can be $(1 - \alpha)100\%$ confident that this difference will not exceed $z_{\alpha/2}\sqrt{\hat{p}\hat{q}/n}$. We can readily see this if we draw a diagram of a typical confidence interval as in Figure 7.6.

Theorem 7.3 *If $\hat{p}$ is used as an estimate of p, we can be $(1 - \alpha)100\%$ confident that the error will not exceed $z_{\alpha/2}\sqrt{\hat{p}\hat{q}/n}$.*

In Example 7.10 we are 95% confident that the sample proportion $\hat{p} = 0.68$ differs from the true proportion p by an amount not exceeding 0.04.

Let us now determine how large a sample is necessary to ensure that the error in estimating p will be less than a specified amount e. By Theorem 7.3, this means we must choose n such that $z_{\alpha/2}\sqrt{\hat{p}\hat{q}/n} = e$.

Theorem 7.4 *If $\hat{p}$ is used as an estimate of p, we can be $(1 - \alpha)100\%$ confident that the error will be less than a specified amount e when the sample size is approximately*

$$n = \frac{z_{\alpha/2}^2\,\hat{p}\hat{q}}{e^2}.$$

Theorem 7.4 is somewhat misleading in that we must use $\hat{p}$ to determine the sample size n, but $\hat{p}$ is computed from the sample. If a crude estimate of p can be made without taking a sample, we could use this value for $\hat{p}$ and then determine n. Lacking such an estimate, we could take a preliminary sample of size $n \geq 30$ to provide an estimate of p. Then, using Theorem 7.4 we could determine approximately how many observations are needed to provide the desired degree of accuracy. Once again, all fractional values of n are rounded up to the next whole number.

Example 7.11 How large a sample is required in Example 7.10 if we want to be 95% confident that our estimate of p is within 0.02?

Solution. Let us treat the 500 families as a preliminary sample providing an estimate $\hat{p} = 0.68$. Then, by Theorem 7.4,

$$n = \frac{(1.96)^2(0.68)(0.32)}{(0.02)^2} = 2090.$$

Therefore, if we base our estimate of p on a random sample of size 2090, we can be 95% confident that our sample proportion will not differ from the true proportion by more than 0.02.

Occasionally, it will be impractical to obtain an estimate of p to be used in determining the sample size for a specified degree of confidence. If this happens, an upper bound for n is established by noting that $\hat{p}\hat{q} = \hat{p}(1 - \hat{p})$, which must be at most equal to $1/4$, since $\hat{p}$ must lie between 0 and 1. This fact may be verified by completing the square. Hence

$$\hat{p}(1 - \hat{p}) = -(\hat{p}^2 - \hat{p}) = \tfrac{1}{4} - (\hat{p}^2 - \hat{p} + \tfrac{1}{4})$$
$$= \tfrac{1}{4} - (\hat{p} - \tfrac{1}{2})^2,$$

which is always less than $1/4$ except when $\hat{p} = 1/2$ and then $\hat{p}\hat{q} = 1/4$. Therefore, if we substitute $\hat{p} = 1/2$ into the formula for n in Theorem 7.4, when, in fact, p actually differs from $1/2$, then n will turn out to be larger than necessary for the specified degree of confidence and as a result our degree of confidence will increase.

Theorem 7.5   *If $\hat{p}$ is used as an estimate of p, we can be* **at least** *$(1 - \alpha)100\%$ confident that the error will not exceed a specified amount e when the sample size is*

$$n = \frac{z_{\alpha/2}^2}{4e^2}.$$

Example 7.12   How large a sample is required in Example 7.10 if we want to be at least 95% confident that our estimate of p is within 0.02?

Solution. Unlike Example 7.11, we shall now assume that no preliminary sample has been taken to provide an estimate of p. Consequently, we can be at least 95% confident that our sample proportion will not differ from the true proportion by more than 0.02 if we choose a sample of size

$$n = \frac{(1.96)^2}{(4)(0.02)^2} = 2401.$$

Comparing the results of Example 7.11 and 7.12, we see that information concerning p, provided by a preliminary sample, or perhaps from past experience, enables us to choose a smaller sample while maintaining our required degree of accuracy.

7.7 Estimating the Difference Between Two Proportions

Consider the problem in which we wish to estimate the difference between two binomial parameters p_1 and p_2. For example, we might let p_1 be the proportion of smokers with lung cancer and p_2 the proportion of nonsmokers with lung cancer. Our problem, then, is to estimate the difference between these two proportions. First, we select independent random samples of size n_1 and n_2 from the two binomial populations with means $n_1 p_1$ and $n_2 p_2$ and variances $n_1 p_1 q_1$ and $n_2 p_2 q_2$, respectively, then determine the numbers x_1 and x_2 of people in each sample with lung cancer, and form the proportions $\hat{p}_1 = x_1/n_1$ and $\hat{p}_2 = x_2/n_2$. A point estimator of the difference between the two proportions, $p_1 - p_2$, is given by the statistic $\hat{P}_1 - \hat{P}_2$. Therefore, the difference of the sample proportions, $\hat{p}_1 - \hat{p}_2$, will be used as the point estimate of $p_1 - p_2$.

A confidence interval for $p_1 - p_2$ can be established by considering the sampling distribution of $\hat{P}_1 - \hat{P}_2$. From Section 7.6 we know that $\hat{P}_1$ and $\hat{P}_2$ are each approximately normally distributed, with means p_1 and p_2 and variances $p_1 q_1/n_1$ and $p_2 q_2/n_2$, respectively. By choosing independent samples from the two populations, the variables $\hat{P}_1$ and $\hat{P}_2$ will be independent, and then by the reproductive property of the normal distribution established in Theorem 6.11, we conclude that $\hat{P}_1 - \hat{P}_2$ is approximately normally distributed with mean

$$\mu_{\hat{P}_1 - \hat{P}_2} = p_1 - p_2$$

and variance

$$\sigma^2_{\hat{P}_1 - \hat{P}_2} = \frac{p_1 q_1}{n_1} + \frac{p_2 q_2}{n_2}.$$

Therefore, we can assert that

$$P(-z_{\alpha/2} < Z < z_{\alpha/2}) = 1 - \alpha,$$

where

$$Z = \frac{(\hat{P}_1 - \hat{P}_2) - (p_1 - p_2)}{\sqrt{(p_1 q_1/n_1) + (p_2 q_2/n_2)}}$$

and $z_{\alpha/2}$ is a value of the standard normal curve above which we find an area of $\alpha/2$. Substituting for Z, we write

$$P\left[-z_{\alpha/2} < \frac{(\hat{P}_1 - \hat{P}_2) - (p_1 - p_2)}{\sqrt{(p_1 q_1/n_1) + (p_2 q_2/n_2)}} < z_{\alpha/2}\right] = 1 - \alpha.$$

After performing the usual mathematical manipulations, we replace p_1, p_2, q_1, and q_2 under the radical sign by their estimates $\hat{p}_1 = x_1/n_1$, $\hat{p}_2 = x_2/n_2$, $\hat{q}_1 = 1 - \hat{p}_1$, and $\hat{q}_2 = 1 - \hat{p}_2$, provided that $n_1 \hat{p}_1$, $n_1 \hat{q}_1$, $n_2 \hat{p}_2$, and $n_2 \hat{q}_2$ are all greater than or equal to 5, and the following approximate $(1 - \alpha)100\%$ confidence interval for $p_1 - p_2$ is obtained.

Large-Sample Confidence Interval for $p_1 - p_2$ *If $\hat{p}_1$ and $\hat{p}_2$ are the proportion of successes in random samples of size n_1 and n_2, respectively, $\hat{q}_1 = 1 - \hat{p}_1$ and $\hat{q}_2 = 1 - \hat{p}_2$, an approximate $(1 - \alpha)100\%$ confidence interval for the difference of two binomial parameters, $p_1 - p_2$, is given by*

$$(\hat{p}_1 - \hat{p}_2) - z_{\alpha/2}\sqrt{\frac{\hat{p}_1 \hat{q}_1}{n_1} + \frac{\hat{p}_2 \hat{q}_2}{n_2}} < p_1 - p_2 < (\hat{p}_1 - \hat{p}_2) + z_{\alpha/2}\sqrt{\frac{\hat{p}_1 \hat{q}_1}{n_1} + \frac{\hat{p}_2 \hat{q}_2}{n_2}},$$

where $z_{\alpha/2}$ is the z value leaving an area of $\alpha/2$ to the right.

Example 7.13   A certain change in a manufacturing procedure for component parts is being considered. Samples are taken using both the existing and the new procedure in order to determine if the new procedure results in an improvement. If 75 of 1500 items from the existing procedure were found to be defective and 80 of 2000 items from the new procedure were found to be defective, find a 90% confidence interval for the true difference in the fraction of defectives between the existing and the new process.

Solution. Let p_1 and p_2 be the true proportions of defectives for the existing and new procedures, respectively. Hence $\hat{p}_1 = 75/1500 = 0.05$ and $\hat{p} = 80/2000 = 0.04$, and the point estimate of $p_1 - p_2$ is $\hat{p}_1 - \hat{p}_2 = 0.05 - 0.04 = 0.01$. Using Table A.3, we find $z_{0.05} = 1.645$. Therefore, substituting into this formula we obtain the 90% confidence interval

$$0.01 - 1.645\sqrt{\frac{(0.05)(0.95)}{1500} + \frac{(0.04)(0.96)}{2000}} < p_1 - p_2$$

$$< 0.01 + 1.645\sqrt{\frac{(0.05)(0.95)}{1500} + \frac{(0.04)(0.96)}{2000}},$$

which simplifies to

$$-0.0017 < p_1 - p_2 < 0.0217.$$

Since the interval contains the value 0, there is no reason to believe that the new procedure produced a significant decrease in the proportion of defectives over the existing method.

Exercises

1. (a) A random sample of 200 voters is selected and 114 are found to support an annexation suit. Find the 96% confidence interval for the fraction of the voting population favoring the suit.
   (b) What can we assert with 96% confidence about the possible size of our error if we estimate the fraction of voters favoring the annexation suit to be 0.57?

2. (a) A random sample of 500 cigarette smokers is selected and 86 are found to have a preference for brand X. Find the 90% confidence interval for the fraction of the population of cigarette smokers who prefer brand X.
   (b) What can we assert with 90% confidence about the possible size of our error if we estimate the fraction of cigarette smokers who prefer brand X to be 0.172?

3. In a random sample of 1000 homes in a certain city, it is found that 228 are heated by oil. Find the 99% confidence interval for the proportion of homes in this city that are heated by oil.

4. Compute a 98% confidence interval for the proportion of defective items in a process when it is found that a sample of size 100 yields 8 defectives.

5. A new-rocket-launching system is being considered for deployment of small, short-range rockets. The existing system has $p = 0.8$ as the probability of a successful launch. A sample of 40 experimental launches is made with the new system and 34 are successful.
   (a) Construct a 95% confidence interval for p.
   (b) Would you conclude that the new system is better?

6. A geneticist is interested in the proportion of African makes that have a certain minor blood disorder. In a random sample of 100 African males, 24 are found to be afflicted.
   (a) Compute a 99% confidence interval for the proportion of African males that have this blood disorder.
   (b) What can we assert with 99% confidence about the possible size of our error if we esti-

mate the proportion of African males with this blood disorder to be 0.24?

7. (a) According to a report in the *Roanoke Times & World-News*, August 20, 1981, approximately 2/3 of the 1600 adults polled by telephone said they think the space shuttle program is a good investment for the country. Find a 95% confidence interval for the proportion of American adults who think the space shuttle program is a good investment for the country.
   (b) What can we assert with 95% confidence about the possible size of our error if we estimate the proportion of American adults who think the space shuttle program is a good investment to be 2/3?

8. In the newspaper article referred to in Exercise 7, 32% of the 1600 adults polled said the U.S. space program should emphasize scientific exporation. How large a sample of adults is needed in the poll if one wishes to be 95% confident that the estimated percentage will be within 2% of the true percentage?

9. How large a sample is needed in Exercise 1 if we wish to be 96% confident that our sample proportion will be within 0.02 of the true fraction of the voting population?

10. How large a sample is needed in Exercise 3 if we wish to be 99% confident that our sample proportion will be within 0.05 of the true proportion of homes in this city that are heated by oil?

11. How large a sample is needed in Exercise 4 if we wish to be 98% confident that our sample proportion will be within 0.05 of the true proportion defective?

12. A study is to be made to estimate the percentage of citizens in a town who favor having their water fluoridated. How large a sample is needed if one wishes to be at least 95% confident that our estimate is within 1% of the true percentage?

13. According to Dr. Memory Elvin-Lewis, head of the microbiology department at Washington University School of Dental Medicine in St. Louis, a

couple of cups of either green or oolong tea each day will provide sufficient fluoride to protect your teeth from decay. People who do not like tea and who live in unfluoridated areas should ask their local governments to consider having their water fluoridated. How large a sample is needed to estimate the percentage of citizens in a certain town who favor having their water fluoridated if one wishes to be at least 99% confident that the estimate is within 1% of the true percentage?

14. A study is to be made to estimate the proportion of residents in a certain city and its suburbs who favor the construction of a nuclear power plant. How large a sample is needed if one wishes to be at least 95% confident that the estimate is within 0.04 of the true proportion of residents in this city and its suburbs that favor the construction of the nuclear power plant?

15. A certain geneticist is interested in the proportion of males and females in the population that have a certain minor blood disorder. In a random sample of 1000 males, 250 are found to be afflicted, whereas 275 of 1000 females tested appear to have the disorder. Compute a 95% confidence interval for the difference between the proportion of males and females that have the blood disorder.

16. A cigarette-manufacturing firm claims that its brand A line of cigarettes outsells its brand B line by 8%. If it is found that 42 of 200 smokers prefer brand A and 18 of 150 smokers prefer brand B, compute a 94% confidence interval for the difference between the proportions of sales of the 2 brands and decide if the 8% difference is a valid claim.

17. A clinical trial is conducted to determine if a certain type of inoculation has an effect on the incidence of a certain disease. A sample of 1000 rats was kept in a controlled environment for a period of 1 year and 500 of the rats were given the inoculation. Of the group not given the drug, there were 120 incidences of the disease, while 98 of the inoculated group contracted it. If we call p_1 the probability of incidence of the disease in uninoculated rats and p_2 the probability of incidence after receiving the drug, compute a 90% confidence interval for $p_1 - p_2$.

18. An anthropologist is interested in the proportion of individuals in two Indian tribes with double occipital hair whorls. Suppose that independent samples are taken from each of the two tribes, and it is found that 24 of 100 Indians from tribe A and 36 of 120 Indians from tribe B possess this characteristic. Construct a 95% confidence interval for the difference $p_B - p_A$ between the proportions of these two tribes with occipital hair whorls.

19. In a study to estimate the proportion of residents in a certain city and its suburbs who favor the construction of a nuclear power plant, it is found that 168 of 400 urban residents favor the construction while only 145 of 500 suburban residents are in favor. Find a 95% confidence interval for the difference between the proportion of urban and suburban residents who favor construction of the nuclear plant.

7.8 *Estimating the Variance*

If a sample of size n is drawn from a normal population with variance σ^2, and the sample variance s^2 is computed, we obtain a value of the statistic S^2. This computed sample variance will be used as a point estimate of σ^2. Hence the statistic S^2 is called an estimator of σ^2.

An interval estimate of σ^2 can be established by using the statistic

$$X^2 = \frac{(n-1)S^2}{\sigma^2}.$$

According to Theorem 6.16, the statistic X^2 has a chi-square distribution with $n-1$ degrees of freedom when samples are chosen from a normal population.

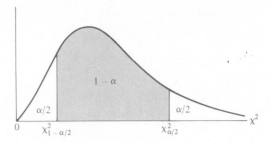

Figure 7.7 $P(\chi^2_{1-\alpha/2} < X^2 < \chi^2_{\alpha/2}) = 1 - \alpha.$

We may write (see Figure 7.7)

$$P(\chi^2_{1-\alpha/2} < X^2 < \chi^2_{\alpha/2}) = 1 - \alpha,$$

where $\chi^2_{1-\alpha/2}$ and $\chi^2_{\alpha/2}$ are values of the chi-square distribution with $n - 1$ degrees of freedom, leaving areas of $1 - \alpha/2$ and $\alpha/2$, respectively, to the right. Substituting for X^2, we write

$$P\left[\chi^2_{1-\alpha/2} < \frac{(n-1)S^2}{\sigma^2} < \chi^2_{\alpha/2}\right] = 1 - \alpha.$$

Dividing each term in the inequality by $(n - 1)S^2$, and then inverting each term (thereby changing the sense of the inequalities), we obtain

$$P\left[\frac{(n-1)S^2}{\chi^2_{\alpha/2}} < \sigma^2 < \frac{(n-1)S^2}{\chi^2_{1-\alpha/2}}\right] = 1 - \alpha.$$

For our particular random sample of size n, the sample variance s^2 is computed, and the following $(1 - \alpha)100\%$ confidence interval for σ^2 is obtained.

Confidence Interval for σ^2   *If s^2 is the variance of a random sample of size n from a normal population, a $(1 - \alpha)100\%$ confidence interval for σ^2 is given by*

$$\frac{(n-1)s^2}{\chi^2_{\alpha/2}} < \sigma^2 < \frac{(n-1)s^2}{\chi^2_{1-\alpha/2}},$$

where $\chi^2_{\alpha/2}$ and $\chi^2_{1-\alpha/2}$ are χ^2 values with $v = n - 1$ degrees of freedom, leaving areas of $\alpha/2$ and $1 - \alpha/2$, respectively, to the right.

A $(1 - \alpha)100\%$ confidence interval for σ is obtained by taking the square root of each endpoint of the interval for σ^2.

Example 7.14   The following are the weights, in decagrams, of 10 packages of grass seed distributed by a certain company: 46.4, 46.1, 45.8, 47.0, 46.1, 45.9, 45.8, 46.9, 45.2, and 46.0. Find a 95% confidence interval for the variance of all such packages of grass seed distributed by this company, assuming a normal population.

Solution. First we find

$$s^2 = \frac{n \sum\limits_{i=1}^{n} x_i^2 - \left(\sum\limits_{i=1}^{n} x_i\right)^2}{n(n-1)} = \frac{(10)(21{,}273.12) - (461.2)^2}{(10)(9)} = 0.286.$$

To obtain a 95% confidence interval, we choose $\alpha = 0.05$. Then, using Table A.5 with $v = 9$ degrees of freedom we find $\chi^2_{0.025} = 19.023$ and $\chi^2_{0.975} = 2.700$. Therefore, the 95% confidence interval for σ^2 is given by

$$\frac{(9)(0.286)}{19.023} < \sigma^2 < \frac{(9)(0.286)}{2.700},$$

or simply

$$0.135 < \sigma^2 < 0.953.$$

7.9 *Estimating the Ratio of Two Variances*

A point estimate of the ratio of two population variances σ_1^2/σ_2^2 is given by the ratio s_1^2/s_2^2 of the sample variances. Hence the statistic S_1^2/S_2^2 is called an estimator of σ_1^2/σ_2^2.

If σ_1^2 and σ_2^2 are the variances of normal populations, we can establish an interval estimate of σ_1^2/σ_2^2 by using the statistic

$$F = \frac{\sigma_2^2 S_1^2}{\sigma_1^2 S_2^2}.$$

According to Theorem 6.20 the random variable F has an F distribution with $v_1 = n_1 - 1$ and $v_2 = n_2 - 1$ degrees of freedom. Therefore, we may write (see Figure 7.8)

$$P[f_{1-\alpha/2}(v_1, v_2) < F < f_{\alpha/2}(v_1, v_2)] = 1 - \alpha,$$

where $f_{1-\alpha/2}(v_1, v_2)$ and $f_{\alpha/2}(v_1, v_2)$ are the values of the F distribution with v_1 and v_2 degrees of freedom, leaving areas of $1 - \alpha/2$ and $\alpha/2$, respectively, to the

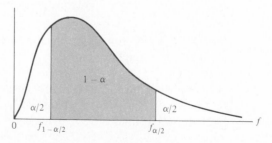

Figure 7.8 $P[f_{1-\alpha/2}(v_1, v_2) < F < f_{\alpha/2}(v_1, v_2)] = 1 - \alpha.$

right. Substituting for F, we write

$$P\left[f_{1-\alpha/2}(v_1, v_2) < \frac{\sigma_2^2 S_1^2}{\sigma_1^2 S_2^2} < f_{\alpha/2}(v_1, v_2)\right] = 1 - \alpha.$$

Multiplying each term in the inequality by S_2^2/S_1^2, and then inverting each term (again changing the sense of the inequalities), we obtain

$$P\left[\frac{S_1^2}{S_2^2}\frac{1}{f_{\alpha/2}(v_1, v_2)} < \frac{\sigma_1^2}{\sigma_2^2} < \frac{S_1^2}{S_2^2}\frac{1}{f_{1-\alpha/2}(v_1, v_2)}\right] = 1 - \alpha.$$

The results of Theorem 6.19 enable us to replace the quantity $f_{1-\alpha/2}(v_1, v_2)$ by $1/f_{\alpha/2}(v_2, v_1)$. Therefore,

$$P\left[\frac{S_1^2}{S_2^2}\frac{1}{f_{\alpha/2}(v_1, v_2)} < \frac{\sigma_1^2}{\sigma_2^2} < \frac{S_1^2}{S_2^2}f_{\alpha/2}(v_2, v_1)\right] = 1 - \alpha.$$

For any two independent random samples of size n_1 and n_2 selected from two normal populations, the ratio of the sample variances, s_1^2/s_2^2, is computed and the following $(1 - \alpha)100\%$ confidence interval for σ_1^2/σ_2^2 is obtained.

Confidence Interval for σ_1^2/σ_2^2 *If s_1^2 and s_2^2 are the variances of independent samples of size n_1 and n_2, respectively, from normal populations, then a $(1 - \alpha)100\%$ confidence interval for σ_1^2/σ_2^2 is*

$$\frac{s_1^2}{s_2^2}\frac{1}{f_{\alpha/2}(v_1, v_2)} < \frac{\sigma_1^2}{\sigma_2^2} < \frac{s_1^2}{s_2^2}f_{\alpha/2}(v_2, v_1),$$

where $f_{\alpha/2}(v_1, v_2)$ is an f value with $v_1 = n_1 - 1$ and $v_2 = n_2 - 1$ degrees of freedom leaving an area of $\alpha/2$ to the right, and $f_{\alpha/2}(v_2, v_1)$ is a similar f value with $v_2 = n_2 - 1$ and $v_1 = n_1 - 1$ degrees of freedom.

As in the previous section, a $(1 - \alpha)100\%$ confidence interval for σ_1/σ_2 is obtained by taking the square root of each endpoint of the interval for σ_1^2/σ_2^2.

Example 7.15 A confidence interval for the difference in the mean orthophosphorus contents, measured in milligrams per liter, at two stations on the James River was constructed in Example 7.8 on page 230 by assuming the normal population variance to be unequal. Justify this assumption by constructing a 98% confidence interval for σ_1^2/σ_2^2 and for σ_1/σ_2, where σ_1^2 and σ_2^2 are the variances of the populations of orthophosphorus contents at station 1 and station 2, respectively.

Solution. From Example 7.8 we have $n_1 = 15$, $n_2 = 12$, $s_1 = 3.07$, and $s_2 = 0.80$. For a 98% confidence interval, $\alpha = 0.02$. Interpolating in Table A.6, we find $f_{0.01}(14, 11) \simeq 4.30$ and $f_{0.01}(11, 14) \simeq 3.87$. Therefore, the 98% confidence interval for σ_1^2/σ_2^2 is

$$\frac{3.07^2}{0.80^2}\left(\frac{1}{4.30}\right) < \frac{\sigma_1^2}{\sigma_2^2} < \frac{3.07^2}{0.80^2}(3.87),$$

which simplifies to

$$3.425 < \frac{\sigma_1^2}{\sigma_2^2} < 56.991.$$

Taking square roots of the confidence limits, we find that a 98% confidence interval for σ_1/σ_2 is

$$1.851 < \frac{\sigma_1}{\sigma_2} < 7.549.$$

Since this interval does not allow for the possibility of σ_1/σ_2 being equal to 1, we were correct in assuming $\sigma_1 \neq \sigma_2$ or $\sigma_1^2 \neq \sigma_2^2$ in Example 7.8.

Exercises

1. A manufacturer of car batteries claims that his batteries will last, on average, 3 years with a variance of 1 year. If 5 of these batteries have lifetimes of 1.9, 2.4, 3.0, 3.5, and 4.2 years, construct a 95% confidence interval for σ^2 and decide if the manufacturer's claim that $\sigma^2 = 1$ is valid. Assume the population of battery lives to be approximately normally distributed.

2. A random sample of 20 students obtained a mean of $\bar{x} = 72$ and a variance of $s^2 = 16$ on a college placement test in mathematics. Assuming the scores to be normally distributed, construct a 98% confidence interval for σ^2.

3. Construct a 95% confidence interval for σ in Exercise 12 on page 225.

4. Construct a 99% confidence interval for σ^2 in Exercise 13 on page 225.

5. Construct a 99% confidence interval for σ in Exercise 14 on page 225.

6. Construct a 90% confidence interval for σ^2 in Exercise 15 on page 225.

7. Construct a 98% confidence interval for σ_1/σ_2 in Exercise 8 on page 235, where σ_1 and σ_2 are, respectively, the standard deviations for the distances obtained per liter of fuel by the Volkswagen and Toyota mini-trucks.

8. Construct a 90% confidence interval for σ_1^2/σ_2^2 in Exercise 9 on page 235. Were we justified in assuming that $\sigma_1^2 = \sigma_2^2$ when we constructed our confidence interval for $\mu_1 - \mu_2$?

9. Construct a 90% confidence interval for σ_1^2/σ_2^2 in Exercise 10 on page 235. Should we have assumed $\sigma_1^2 = \sigma_2^2$ in constructing our confidence interval for $\mu_{II} - \mu_I$?

7.10 Bayesian Methods of Estimation

The classical methods of estimation that we have studied so far are based solely on information provided by the random sample. These methods essentially interpret probabilities as relative frequencies. For example, in arriving at a 95% confidence interval for μ, we interpret the statement $P(-1.96 < Z < 1.96) = 0.95$ to mean that 95% of the time in repeated experiments Z will fall between -1.96 and 1.96. Probabilities of this type that can be interpreted in the frequency sense will be referred to as **objective probabilities**. The Bayesian approach to statistical methods of estimation combines sample information with other available prior information that may appear to be pertinent.

Consider the problem of finding a point estimate of the parameter θ for the population $f(x; \theta)$. The classical approach would be to take a random sample of size n and substitute the information provided by the sample into the appropriate estimator or decision function. Thus in the case of a binomial population $b(x; n, p)$ our estimate of p, the proportion of successes, would be $\hat{p} = x/n$. Now, suppose that additional information is given about θ, namely, that it is known to vary according to some probability distribution $f(\theta)$, often called a **prior distribution**, with **prior mean** μ_0 and **prior variance** σ_0^2. That is, we are now assuming θ to be a value of a random variable Θ with probability distribution $f(\theta)$ and we wish to estimate the particular value θ for the population from which we selected our random sample. The probabilities associated with this prior distribution are called **subjective probabilities**, in that they measure a person's *degree of belief* in the location of the parameter. The person uses his or her own experience and knowledge as the basis for arriving at the subjective probabilities given by the prior distribution. Bayesian techniques use the prior distribution $f(\theta)$ along with the joint distribution of the sample, $f(x_1, x_2, \ldots, x_n; \theta)$, to compute the **posterior distribution** $f(\theta \mid x_1, x_2, \ldots, x_n)$. The posterior distribution consists of information from both the subjective prior distribution and the objective sampling distribution and expresses our degree of belief in the location of the parameter θ after we have observed the sample.

Let us write $f(x_1, x_2, \ldots, x_n \mid \theta)$ instead of $f(x_1, x_2, \ldots, x_n; \theta)$ for the joint probability distribution of our sample whenever we wish to indicate that the parameter is also a random variable. The joint distribution of the sample X_1, $X_2, \ldots, X_n$ and the parameter Θ is then

$$f(x_1, x_2, \ldots, x_n, \theta) = f(x_1, x_2, \ldots, x_n \mid \theta) f(\theta),$$

from which we readily obtain the marginal distribution

$$g(x_1, x_2, \ldots, x_n) = \begin{cases} \sum_\theta f(x_1, x_2, \ldots, x_n, \theta) & \text{(discrete case)} \\ \int_{-\infty}^{\infty} f(x_1, x_2, \ldots, x_n, \theta)\, d\theta & \text{(continuous case).} \end{cases}$$

Hence the posterior distribution may be written

$$f(\theta \mid x_1, x_2, \ldots, x_n) = \frac{f(x_1, x_2, \ldots, x_n, \theta)}{g(x_1, x_2, \ldots, x_n)}.$$

Definition 7.4   *The mean of the posterior distribution $f(\theta \mid x_1, x_2, \ldots, x_n)$, denoted by θ^*, is called the **Bayes estimate** for θ.*

Example 7.16 Let us assume that the prior distribution for the proportion p of defectives produced by a machine is

p	0.1	0.2
$f(p)$	0.6	0.4

Find the Bayes estimate for the proportion of defectives being produced by this machine if a random sample of size 2 yields 1 defective.

Solution. Let X be the number of defectives in our sample. Then the probability distribution for our sample is

$$f(x \mid p) = b(x; n, p) = \binom{2}{x} p^x q^{2-x}, \qquad x = 0, 1, 2.$$

If $p = 0.1$, the probability that the random sample of size 2 yields 1 defective is found to be

$$f(1 \mid 0.1) = b(1; 2, 0.1) = \binom{2}{1}(0.1)(0.9) = 0.18.$$

Similarly, when $p = 0.2$, we find that

$$f(1 \mid 0.2) = b(1; 2, 0.2) = \binom{2}{1}(0.2)(0.8) = 0.32.$$

From the fact that $f(x, p) = f(x \mid p) f(p)$, we can set up the table

p	0.1	0.2
$f(1, p)$	0.108	0.128

from which we get

$$g(1) = \sum_p f(1, p) = 0.236.$$

We obtain the posterior distribution for the proportion of defectives, p, when $x = 1$, from the formula $f(p \mid x) = f(x, p)/g(x)$. Hence we have

p	0.1	0.2
$f(p \mid x = 1)$	0.458	0.542

from which we get the Bayes estimate of p, denoted by p^*, to be

$$p^* = (0.1)(0.458) + (0.2)(0.542) = 0.1542.$$

Note that the Bayes estimate is much smaller than the 1/2 value given by the classical estimate $\hat{p} = x/n$.

Example 7.17   Repeat Example 7.16 using the uniform prior distribution $f(p) = 1, 0 < p < 1$.

Solution. As before, we have

$$f(x \mid p) = \binom{2}{x} p^x q^{2-x}, \qquad x = 0, 1, 2.$$

From the fact that $f(x, p) = f(x \mid p) f(p)$, we can write

$$f(1, p) = \binom{2}{1} pq$$

$$= 2p(1 - p), \qquad 0 < p < 1$$

and then

$$g(1) = \int_0^1 2p(1 - p) \, dp = \tfrac{1}{3}.$$

The posterior distribution for the proportion of defectives, p, when $x = 1$, is then

$$f(p \mid x = 1) = \frac{2p(1 - p)}{\tfrac{1}{3}}$$

$$= 6p(1 - p), \qquad 0 < p < 1,$$

from which we get the Bayes estimate of p to be

$$p^* = 6 \int_0^1 p^2(1 - p) \, dp = \tfrac{1}{2}.$$

In this case we see that the Bayes estimate p^* and the classical estimate $\hat{p}$ are equivalent.

A $(1 - \alpha)100\%$ **Bayesian interval** for the parameter θ can be constructed by finding an interval centered at the posterior mean that contains $(1 - \alpha)100\%$ of the posterior probability.

Definition 7.5   *The interval $a < \theta < b$ will be called a $(1 - \alpha)100\%$ **Bayes interval** for θ if*

$$\int_{\theta^*}^b f(\theta \mid x_1, x_2, \ldots, x_n) \, d\theta = \int_a^{\theta^*} f(\theta \mid x_1, x_2, \ldots, x_n) \, d\theta = \frac{1 - \alpha}{2}.$$

Bayesian methods of estimation concerning the mean μ of a normal population are based on the following theorem.

Theorem 7.6   *If $\bar{x}$ is the mean of a random sample of size n from a normal population with known variance σ^2, and the prior distribution of the population mean is a normal distribution with mean μ_0 and variance σ_0^2, then the posterior distribution of the* *population mean is also a normal distribution with mean μ^* and standard deviation σ^*, where*

$$\mu^* = \frac{n\bar{x}\sigma_0^2 + \mu_0\sigma^2}{n\sigma_0^2 + \sigma^2} \qquad and \qquad \sigma^* = \sqrt{\frac{\sigma_0^2\sigma^2}{n\sigma_0^2 + \sigma^2}}.$$

Proof. Multiplying the density of our sample

$$f(x_1, x_2, \ldots, x_n \mid \mu) = \frac{1}{(2\pi)^{n/2} \cdot \sigma^n} \exp\left[-\left(\frac{1}{2}\right)\sum_{i=1}^{n}\left(\frac{x_i - \mu}{\sigma}\right)^2\right],$$

for $-\infty < x_i < \infty$ and $i = 1, 2, \ldots, n$ by our prior

$$f(\mu) = \frac{1}{\sqrt{2\pi}\,\sigma_0}\, e^{-(1/2)[(\mu - \mu_0)/\sigma_0]^2}, \qquad -\infty < \mu < \infty,$$

we obtain the joint density of the random sample and the mean of the population from which the sample is selected. That is,

$$f(x_1, x_2, \ldots, x_n, \mu) = \frac{1}{(2\pi)^{(n+1)/2} \cdot \sigma^n\sigma_0}$$

$$\times \exp\left\{-\left(\frac{1}{2}\right)\left[\sum_{i=1}^{n}\left(\frac{x_i - \mu}{\sigma}\right)^2 + \left(\frac{\mu - \mu_0}{\sigma_0}\right)^2\right]\right\}.$$

In Section 6.7 we established the identity

$$\sum_{i=1}^{n}(x_i - \mu)^2 = \sum_{i=1}^{n}(x_i - \bar{x})^2 + n(\bar{x} - \mu)^2,$$

which enables us to write

$$f(x_1, x_2, \ldots, x_n, \mu) = \frac{1}{(2\pi)^{(n+1)/2} \cdot \sigma^n\sigma_0} \exp\left[-\left(\frac{1}{2}\right)\sum_{i=1}^{n}\left(\frac{x_i - \bar{x}}{\sigma}\right)^2\right]$$

$$\times\, e^{-(1/2)\{n[(\bar{x} - \mu)/\sigma]^2 + [(\mu - \mu_0)/\sigma_0]^2\}}.$$

Completing the square in the second exponent, we can write the joint density of the random sample and the population mean in the form

$$f(x_1, x_2, \ldots, x_n, \mu) = K e^{-(1/2)[(\mu - \mu^*)/\sigma^*]^2},$$

where

$$\mu^* = \frac{n\bar{x}\sigma_0^2 + \mu_0\sigma^2}{n\sigma_0^2 + \sigma^2},$$

$$\sigma^* = \sqrt{\frac{\sigma_0^2\sigma^2}{n\sigma_0^2 + \sigma^2}},$$

and K is a function of the sample values and the known parameters. The marginal distribution of the sample is then

$$g(x_1, x_2, \ldots, x_n) = K\sqrt{2\pi}\,\sigma^* \int_{-\infty}^{\infty} \frac{1}{\sqrt{2\pi}\,\sigma^*} e^{-(1/2)[(\mu-\mu^*)/\sigma^*]^2}\,d\mu$$

$$= K\sqrt{2\pi}\,\sigma^*,$$

and the posterior distribution is

$$f(\mu \mid x_1, x_2, \ldots, x_n) = \frac{f(x_1, x_2, \ldots, x_n, \mu)}{g(x_1, x_2, \ldots, x_n)}$$

$$= \frac{1}{\sqrt{2\pi}\,\sigma^*} e^{-(1/2)[(\mu-\mu^*)/\sigma^*]^2}, \qquad -\infty < \mu < \infty,$$

which is easily identified as a normal distribution with mean μ^* and standard deviation σ^* where μ^* and σ^* are defined above.

The Central Limit Theorem allows us to use Theorem 7.6 also when we select random samples of size $n \geq 30$ from nonnormal populations, and when the prior distribution of the mean is approximately normal.

To compute μ^* and σ^* by the formulas of Theorem 7.6, we have assumed that σ^2 is known. Since this is generally not the case, we shall replace σ^2 by the sample variance s^2 whenever $n \geq 30$. The posterior mean μ^* is the Bayes estimate of the population mean μ, and a $(1-\alpha)100\%$ **Bayesian interval** for μ can be constructed by computing the interval

$$\mu^* - z_{\alpha/2}\,\sigma^* < \mu < \mu^* + z_{\alpha/2}\,\sigma^*,$$

which is centered at the posterior mean and contains $(1-\alpha)100\%$ of the posterior probability.

Example 7.18 An electrical firm manufactures light bulbs that have a length of life that is approximately normally distributed with a standard deviation of 100 hours. Prior experience leads us to believe that μ is a value of a normal random variable with a mean $\mu_0 = 800$ hours and a standard deviation $\sigma_0 = 10$ hours. If a random sample of 25 bulbs has an average life of 780 hours, find a 95% Bayesian interval for μ.

Solution. According to Theorem 7.6, the posterior distribution of the mean is also a normal distribution with mean

$$\mu^* = \frac{(25)(780)(10)^2 + (800)(100)^2}{(25)(10)^2 + (100)^2} = 796$$

and standard deviation

$$\sigma^* = \sqrt{\frac{(10)^2(100)^2}{(25)(10)^2 + (100)^2}} = \sqrt{80}.$$

The 95% Bayesian interval for μ is then given by

$$796 - 1.96\sqrt{80} < \mu < 796 + 1.96\sqrt{80}$$

or

$$778.5 < \mu < 813.5.$$

By ignoring the prior information about μ in Example 7.18, we could proceed as in Section 7.3 and construct the classical 95% confidence interval

$$780 - (1.96)\left(\frac{100}{\sqrt{25}}\right) < \mu < 780 + (1.96)\left(\frac{100}{\sqrt{25}}\right)$$

or

$$740.8 < \mu < 819.2,$$

which is seen to be wider than the corresponding Bayesian interval.

7.11 Decision Theory

In our discussion of the classical approach to point estimation, we adopted the criterion that selects the decision function that is most efficient. That is, we choose from all possible unbiased estimators the one with the smallest variance as our "best" estimator. In *decision theory* we also take into account the rewards for making correct decisions and the penalties for making incorrect decisions. This leads to a new criterion that chooses the decision function $\hat{\Theta}$ that penalizes us the least when the action taken is incorrect. It is convenient now to introduce a **loss function** whose values depend on the true value of the parameter θ and the action $\hat{\theta}$. This is usually written in functional notation as $L(\hat{\Theta}; \theta)$. In many decision-making problems it is desirable to use a loss function of the form

$$L(\hat{\Theta}; \theta) = |\hat{\Theta} - \theta|$$

or perhaps

$$L(\hat{\Theta}; \theta) = (\hat{\Theta} - \theta)^2$$

in arriving at a choice between two or more decision functions.

Since θ is unknown, it must be assumed that it can equal any of several possible values. The set of all possible values that θ can assume is called the **parameter space**. For each possible value of θ in the parameter space, the loss function will vary from sample to sample. We define the **risk function** for the decision function $\hat{\Theta}$ to be the expected value of the loss function when the value of the parameter is θ and denote this function by $R(\hat{\Theta}; \theta)$. Hence we have

$$R(\hat{\Theta}; \theta) = E[L(\hat{\Theta}; \theta)].$$

One method of arriving at a choice between $\hat{\Theta}_1$ and $\hat{\Theta}_2$ as an estimator for θ would be to apply the **minimax criterion**. Essentially, we determine the maximum value of $R(\hat{\Theta}_1; \theta)$ and the maximum value of $R(\hat{\Theta}_2; \theta)$ in the parameter space and then choose the decision function that provided the minimum of these two maximum risks.

Example 7.19 According to the minimax criterion, is $\bar{X}$ or $\tilde{X}$ a better estimator of the mean μ of a normal population with known variance σ^2, based on a random sample of size n when the loss function is of the form $L(\hat{\Theta}; \theta) = (\hat{\Theta} - \theta)^2$?

Solution. The loss function corresponding to $\bar{X}$ is given by

$$L(\bar{X}; \mu) = (\bar{X} - \mu)^2.$$

Hence the risk function is

$$R(\bar{X}; \mu) = E[(\bar{X} - \mu)^2] = \frac{\sigma^2}{n}$$

for every μ in the parameter space. Similarly, one can show that the risk function corresponding to $\tilde{X}$ is given by

$$R(\tilde{X}; \mu) = E[(\tilde{X} - \mu)^2] \simeq \frac{\pi\sigma^2}{2n}$$

for every μ in the parameter space. In view of the fact that $\sigma^2/n < \pi\sigma^2/2n$, the minimax criterion selects $\bar{X}$ rather than $\tilde{X}$ as the better estimator for μ.

In some practical situations we may have additional information concerning the unknown parameter θ. For example, suppose that we wish to estimate the binomial parameter p, the proportion of defectives produced by a machine during a certain day when we know that p varies from day to day. If we can write down the prior distribution $f(p)$, then it is possible to determine the expected value of the risk function for each decision function. The expected risk corresponding to the estimator $\hat{P} = X/n$, often referred to as the **Bayes risk**, is written $B(\hat{P}) = E[R(\hat{P}; P)]$, where we are now treating the true proportion of defectives as a random variable. In general, when the unknown parameter is treated as a random variable with a prior distribution given by $f(\theta)$, the Bayes risk in estimating θ by means of the estimator $\hat{\Theta}$ is given by

$$B(\hat{\Theta}) = E[R(\hat{\Theta}; \Theta)] = \begin{cases} \displaystyle\sum_i R(\hat{\Theta}; \theta_i)f(\theta_i) & \text{(discrete case)} \\ \displaystyle\int_{-\infty}^{\infty} R(\hat{\Theta}; \theta)f(\theta)\, d\theta & \text{(continuous case).} \end{cases}$$

The decision function $\hat{\Theta}$ that minimizes $B(\hat{\Theta})$ is called the **Bayes estimator** of θ. We shall make no attempt in this book to derive a Bayes estimator, but instead we shall employ the Bayes risk to establish a criterion for choosing between two estimators.

Bayes' Criterion   *Suppose that the parameter θ is a value of the random variable Θ and $f(\theta)$ is the value of its probability distribution at θ. If $\hat{\Theta}_1$ and $\hat{\Theta}_2$ are two estimators of θ and $B(\hat{\Theta}_1) < B(\hat{\Theta}_2)$, then $\hat{\Theta}_1$ is selected as the better estimator for θ.*

The foregoing discussion on decision theory might better be understood if one considers the following two examples.

Example 7.20   Suppose that a friend has three similar coins except for the fact that the first one has two heads, the second one has two tails, and the third one is honest. We wish to estimate which coin our friend is flipping on the basis of two flips of the coin. Let θ be the number of heads on the coin. Consider two decision functions $\hat{\Theta}_1$ and $\hat{\Theta}_2$, where $\hat{\Theta}_1$ is the estimator that assigns to θ the number of heads that occur when the coin is flipped twice and $\hat{\Theta}_2$ is the estimator that assigns the value of 1 to θ no matter what the experiment yields. If the loss function is of the form $L(\hat{\Theta}; \theta) = (\hat{\Theta} - \theta)^2$, which estimator is better according to the minimax procedure?

Solution. For the estimator $\hat{\Theta}_1$, the loss function assumes the values $L(\hat{\theta}_1; \theta) = (\hat{\theta}_1 - \theta)^2$, where $\hat{\theta}_1$ may be 0, 1, or 2, depending on the true value of θ. Clearly, if $\theta = 0$ or 2, both flips will yield all tails or all heads and our decision will be a correct one. Hence $L(0; 0) = 0$ and $L(2; 2) = 0$, from which one may easily conclude that $R(\hat{\Theta}_1; 0) = 0$ and $R(\hat{\Theta}_1; 2) = 0$. However, when $\theta = 1$ we could obtain 0, 1, or 2 heads in the two flips with probabilities 1/4, 1/2, and 1/4, respectively. In this case we have $L(0; 1) = 1$, $L(1; 1) = 0$, and $L(2; 1) = 1$, from which we find that

$$R(\hat{\Theta}_1; 1) = 1 \times \tfrac{1}{4} + 0 \times \tfrac{1}{2} + 1 \times \tfrac{1}{4} = \tfrac{1}{2}.$$

For the estimator $\hat{\Theta}_2$, the loss function assumes values given by $L(\hat{\theta}_2; \theta) = (\hat{\theta}_2 - \theta)^2 = (1 - \theta)^2$. Hence $L(1; 0) = 1$, $L(1; 1) = 0$, and $L(1; 2) = 1$, and the corresponding risks are $R(\hat{\Theta}_2; 0) = 1$, $R(\hat{\Theta}_2; 1) = 0$, and $R(\hat{\Theta}_2; 2) = 1$. Since the maximum risk is 1/2 for the estimator $\hat{\Theta}_1$ compared to a maximum risk of 1 for $\hat{\Theta}_2$, the minimax criterion selects $\hat{\Theta}_1$ as the better of the two estimators.

Example 7.21   Referring to Example 7.20, let us suppose that our friend flips the honest coin 80% of the time and the other two coins each about 10% of the time. Does the Bayes criterion select $\hat{\Theta}_1$ or $\hat{\Theta}_2$ as the better estimator?

Solution. The parameter Θ may now be treated as a random variable with the following probability distribution:

θ	0	1	2
$f(\theta)$	0.1	0.8	0.1

For the estimator $\hat{\Theta}_1$, the Bayes risk is

$$B(\hat{\Theta}_1) = R(\hat{\Theta}_1; 0)f(0) + R(\hat{\Theta}_1; 1)f(1) + R(\hat{\Theta}_1; 2)f(2)$$
$$= (0)(0.1) + \tfrac{1}{2}(0.8) + (0)(0.1) = 0.4.$$

Similarly, for the estimator $\hat{\Theta}_2$, we have

$$B(\hat{\Theta}_2) = R(\hat{\Theta}_2; 0)f(0) + R(\hat{\Theta}_2; 1)f(1) + R(\hat{\Theta}_2; 2)f(2)$$
$$= (1)(0.1) + (0)(0.8) + (1)(0.1) = 0.2.$$

Since $B(\hat{\Theta}_2) < B(\hat{\Theta}_1)$, the Bayes criterion selects $\hat{\Theta}_2$ as the better estimator for the parameter θ.

Exercises

1. Estimate the proportion of defectives being produced by the machine in Example 7.16 on page 250 if the random sample of size 2 yields 2 defectives.

2. Let us assume that the prior distribution for the proportion p of drinks from a vending machine that overflow is

p	0.05	0.10	0.15
$f(p)$	0.3	0.5	0.2

   If 2 of the next 9 drinks from this machine overflow, find
   (a) the posterior distribution for the proportion p;
   (b) the Bayes estimate of p.

3. Repeat Exercise 2 when 1 of the next 4 drinks overflows and the uniform prior distribution is

$$f(p) = 10, \qquad 0.05 < p < 0.15.$$

4. The developer of a new condominium complex claims that 3 out of 5 buyers will prefer a two-bedroom unit, while his banker claims that it would be more correct to say that 7 out of 10 buyers will prefer a two-bedroom unit. In previous predictions of this type the banker has been twice as reliable as the developer. If 12 of the next 15 condominiums sold in this complex are two-bedroom units, find
   (a) the posterior probabilities associated with the claims of the developer and banker;
   (b) a point estimate of the proportion of buyers who prefer a two-bedroom unit.

5. The burn time for the first stage of a rocket is a normal random variable with a standard deviation of 0.8 minute. Assume a normal prior distribution for μ with a mean of 8 minutes and a standard deviation of 0.2 minute. If 10 of these rockets are fired and the first stage has an average burn time of 9 minutes, find a 95% Bayesian interval for μ.

6. The daily profit from a cigarette vending machine placed in a restaurant is a value of a normal random variable with unknown mean μ and variance σ^2. Of course, the mean will vary somewhat from restaurant to restaurant, and the distributor feels that these average daily profits can best be described by a normal distribution with

mean $\mu_0 = \$8.00$ and standard deviation $\sigma_0 = \$0.40$. If one of these cigarette machines, placed in a certain restaurant, showed an average daily profit of $\bar{x} = \$6.75$ during the first 30 days with a standard deviation of $s = \$1.20$, find

(a) a Bayes estimate of the true average daily profit for this restaurant;

(b) a 96% Bayesian interval of μ for this restaurant;

(c) the probability that the average daily profits from the machine in this restaurant is between $6.59 and $7.12.

7. The mathematics department of a large university is designing a placement test to be given to the incoming freshman classes. Members of the department feel that the average grade for this test will vary from one freshman class to another. This variation of the average class grade is expressed subjectively by a normal distribution with mean $\mu_0 = 72$ and variance $\sigma_0^2 = 5.76$.

(a) What prior probability does the department assign to the actual average grade being somewhere between 71.8 and 73.4 for next year's freshman class?

(b) If the test is tried on a random sample of 100 freshman students from the next incoming freshman class resulting in an average grade of 70 with a variance of 64, construct a 95% Bayesian interval for μ.

(c) What posterior probability should the department assign to the event of part (a)?

8. Suppose that in Example 7.18 the electrical firm does not have enough prior information regarding the population mean length of life to be able to assume a normal distribution for μ. The firm believes, however, that μ is surely between 770 and 830 hours and it is felt that a more realistic Bayesian approach would be to assume the prior distribution

$$f(\mu) = \tfrac{1}{60}, \qquad 770 < \mu < 830.$$

If a random sample of 25 bulbs gives an average life of 780 hours, follow the steps of the proof for Theorem 7.6 to find the posterior distribution $f(\mu \mid x_1, x_2, \ldots, x_{25})$.

9. Suppose that the time to failure T of a certain

hinge is an exponential random variable with probability density

$$f(t) = \theta e^{-\theta t}, \qquad t > 0.$$

From prior experience we are led to believe that θ is a value of an exponential random variable with probability density

$$f(\theta) = 2e^{-2\theta}, \qquad \theta > 0.$$

If we have a sample of n observations on T, show that the posterior distribution of Θ is a gamma distribution with parameters

$$\alpha = n + 1 \quad \text{and} \quad \beta = 1 \Big/ \Big(\sum_{i=1}^{n} t_i + 2 \Big).$$

10. We wish to estimate the binomial parameter p by the decision function $\hat{P}$, the proportion of successes in a bonomial experiment consisting of n trials. Find $R(\hat{P}; p)$ when the loss function is of the form $L(\hat{P}; p) = (\hat{P} - p)^2$.

11. Suppose that an urn contains three balls, of which θ are red and the remainder black, where θ can vary from 0 to 3. We wish to estimate θ by selecting two balls in succession without replacement. Let $\hat{\Theta}_1$ be the decision function that assigns to θ the value 0 if neither ball is red, the value 1 if the first ball only is red, the value 2 if the second ball only is red, and the value 3 if both balls are red. Using a loss function of the form $L(\hat{\Theta}_1; \theta) = |\hat{\Theta}_1 - \theta|$, find $R(\hat{\Theta}_1; \theta)$.

12. In Exercise 11 consider the estimator $\hat{\Theta}_2 = X(X + 1)/2$, where X is the number of red balls in our sample. Find $R(\hat{\Theta}_2; \theta)$.

13. Use the minimax criterion to determine whether the estimator $\hat{\Theta}_1$ of Exercise 11 or the estimator $\hat{\Theta}_2$ of Exercise 12 is the better estimator.

14. Use the Bayes criterion to determine whether the estimator $\hat{\Theta}_1$ of Exercise 11 or the estimator $\hat{\Theta}_2$ of Exercise 12 is the better estimator, given the following additional information:

θ	0	1	2	3
$f(\theta)$	0.1	0.5	0.1	0.3

8

Tests of Hypotheses

8.1 Statistical Hypotheses

Often the problem confronting us is not so much the estimation of a population parameter as discussed in Chapter 7, but rather the formulation of a set of rules that leads to a decision culminating in the acceptance or rejection of some statement or hypothesis about the population. For example, a medical researcher may decide on the basis of experimental evidence whether or not coffee drinking increases the risk of cancer in humans; an engineer might have to decide on the basis of sample data whether there is a difference between the accuracy of two kinds of gauges; or a sociologist might wish to collect appropriate data to enable her to decide whether blood type and eye color of an individual are independent variables. Procedures that lead to the acceptance or rejection of statistical hypotheses such as these comprise a major area of statistical inference. First, let us define precisely what we mean by a **statistical hypothesis**.

Definition 8.1   *A **statistical hypothesis** is an assertion or conjecture concerning one or more populations.*

The truth or falsity of a statistical hypothesis is never known with absolute certainty unless we examine the entire population. This, of course, would be impractical in most situations. Instead, we take a random sample from the population of interest and use the information contained in this sample to decide whether the hypothesis is likely to be true or false. Evidence from the sample that is inconsistent with the stated hypothesis leads to a rejection of the hypothesis, whereas evidence supporting the hypothesis leads to its acceptance. We should make it clear at this point that the acceptance of a statistical hypothesis is a result of insufficient evidence to reject it and does not necessarily imply that it is true. For example, in tossing a coin 100 times we might test the hypothesis that the coin is balanced. In terms of population parameters, we are testing the hypothesis that the proportion of heads is $p = 0.5$ if the coin were tossed indefinitely. An outcome of 48 heads would not be surprising if the coin is balanced. Such a result would surely support the hypothesis $p = 0.5$. One might argue that such an occurrence is also consistent with the hypothesis that $p = 0.45$. Thus, in accepting the hypothesis, the only thing we can be reasonably certain about is that the true proportion of heads is not a great deal different from one half. If the 100 trials had resulted in only 35 heads, we would then have evidence to support the rejection of our hypothesis. In view of the fact that the probability of obtaining 35 or fewer heads in 100 tosses of a balanced coin is approximately 0.002, either a very rare event has occurred or we are right in concluding that $p \neq 0.5$.

Although we shall use the terms *accept* and *reject* frequently throughout this chapter, it is important to understand that **the rejection of a hypothesis is to conclude that it is false, while the acceptance of a hypothesis merely implies that we have insufficient evidence to believe otherwise**. Because of this terminology, the statistician or experimenter will often choose to state the hypothesis in a form that hopefully will be rejected. If the medical researcher wishes to prove that coffee drinking increases the risk of cancer, she would assume that there is no increase in the risk and then attempt to reject this contention. Similarly, to support the claim that one kind of gauge is more accurate than another, the engineer tests the hypothesis that there is no difference in the accuracy of the two kinds of gauges.

In the past hypotheses that were formulated with the hope that they would be rejected led to the use of the term **null hypothesis**. Today this term is applied to any hypothesis we wish to test and is denoted by H_0. The rejection of H_0 leads to the acceptance of an **alternative hypothesis**, denoted by H_1. A null hypothesis concerning a population parameter will always be stated so as to specify an exact value of the parameter, whereas the alternative hypothesis allows for the possibility of several values. Hence, if H_0 is the null hypothesis $p = 0.5$ for a binomial population, the alternative hypothesis H_1 would be one of the following: $p > 0.5$, $p < 0.5$, or $p \neq 0.5$.

8.2 *Testing a Statistical Hypothesis*

To illustrate the concepts used in testing a statistical hypothesis about a population, consider the following example. A certain type of cold vaccine is known to be only 25% effective after a period of 2 years. To determine if a new and somewhat more expensive vaccine is superior in providing protection against the same virus for a longer period of time, suppose that 20 people are chosen at random and inoculated. In an actual study of this type the participants receiving the new vaccine might number several thousand. The number 20 is being used here only to demonstrate the basic steps in carrying out a statistical test. If more than 8 of those receiving the new vaccine surpass the 2-year period without contracting the virus, the new vaccine will be considered superior to the one presently in use. The requirement that the number exceed 8 is somewhat arbitrary but appears reasonable in that it represents a modest gain over the 5 people that could be expected to receive protection if the 20 people had been inoculated with the vaccine already in use. We are essentially testing the null hypothesis that the new vaccine is equally effective after a period of 2 years as the one now commonly used against the alternative hypothesis that the new vaccine is in fact superior. This is equivalent to testing the hypothesis that the binomial parameter for the probability of a success on a given trial is $p = 1/4$ against the alternative that $p > 1/4$. This is usually written as follows:

$$H_0: \quad p = 1/4,$$
$$H_1: \quad p > 1/4.$$

The statistic on which we base our decision is X, the number of individuals in our test group who receive protection from the new vaccine for a period of at least 2 years. The possible values of X, from 0 to 20, are divided into two groups: those numbers less than or equal to 8 and those greater than 8. All possible scores greater than 8 constitute the **critical region**, and all possible scores less than or equal to 8 determine the **acceptance region**. The last number that we observe in passing from the acceptance region into the critical region is called the **critical value**. In our illustration the critical value is the number 8. Therefore, if $x > 8$, we reject H_0 in favor of the alternative hypothesis H_1. If $x \le 8$, we accept H_0. This decision criterion is illustrated in Figure 8.1.

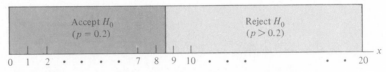

Figure 8.1 Decision criterion for testing $p = 0.2$ versus $p > 0.2$.

The decision procedure just described could lead to either of two wrong conclusions. For instance, the new vaccine may be no better than the one now in use and, for this particular randomly selected group of individuals, more than 8 surpass the 2-year period without contracting the virus. We would be committing an error by rejecting H_0 in favor of H_1 when, in fact, H_0 is true. Such an error is called a **type I error**.

Definition 8.2 *Rejection of the null hypothesis when it is true is called a* **type I error**.

Table 8.1 Possible Situations in Testing a Statistical Hypothesis

	H_0 Is True	H_0 Is False
Accept H_0	Correct decision	Type II error
Reject H_0	Type I error	Correct decision

A second kind of error is committed if 8 or fewer of the group surpass the 2-year period successfully and we conclude that the new vaccine is no better when it actually is better. In this case we would accept H_0 when it is false. This is called a **type II error**.

Definition 8.3   *Acceptance of the null hypothesis when it is false is called a* **type II error**.

In testing any statistical hypothesis, there are four possible situations that determine whether our decision is correct or in error. These four situations are summarized in Table 8.1.

The probability of committing a type I error, also called the **level of significance**, is denoted by the Greek letter α. In our illustration, a type I error will occur when more than 8 individuals surpass the 2-year period without contracting the virus using a new vaccine that is actually equivalent to the one in use. Hence, if X is the number of individuals who remain free of the virus for at least 2 years,

$$\alpha = P(\text{type I error})$$
$$= P(X > 8 \text{ when } p = \tfrac{1}{4})$$
$$= \sum_{x=9}^{20} b(x; 20, \tfrac{1}{4})$$
$$= 1 - \sum_{x=0}^{8} b(x; 20, \tfrac{1}{4})$$
$$= 1 - 0.9591$$
$$= 0.0409.$$

We say that the null hypothesis, $p = 1/4$, is being tested at the $\alpha = 0.0409$ level of significance. Sometimes the level of significance is called the **size** of the critical region. A critical region of size 0.0409 is very small and therefore it is unlikely that a type I error will be committed. Consequently, it would be most unusual for more than 8 individuals to remain immune to a virus for a 2-year period using a new vaccine that is essentially equivalent to the one now on the market.

The probability of committing a type II error, denoted by β, is impossible to compute unless we have a specific alternative hypothesis. If we test the null hypothesis that $p = 1/4$ against the alternative hypothesis that $p = 1/2$, then we are able to compute the probability of accepting H_0 when it is false. We simply find the probability of obtaining 8 or fewer in the group that surpass the 2-year period when $p = 1/2$. In this case

$$\beta = P(\text{type II error})$$
$$= P(X \leq 8 \text{ when } p = \tfrac{1}{2})$$
$$= \sum_{x=0}^{8} b(x; 20, \tfrac{1}{2})$$
$$= 0.2517.$$

This is a rather high probability indicating a poor test procedure. It is quite likely that we shall reject the new vaccine when, in fact, it is superior to that now in use. Ideally, we like to use a test procedure for which both the type I and type II errors are small.

It is possible that the director of the testing program is willing to make a type II error if the more expensive vaccine is not significantly superior. The only time he wishes to guard against the type II error is when the true value of p is at least 0.7. If $p = 0.7$, this test procedure gives

$$\beta = P(\text{type II error})$$
$$= P(X \leq 8 \text{ when } p = 0.7)$$
$$= \sum_{x=0}^{8} b(x; 20, 0.7)$$
$$= 0.0051.$$

With such a small probability of committing a type II error, it is extremely unlikely that the new vaccine would be rejected when it is 70% effective after a period of 2 years. As the alternative hypothesis approaches unity, the value of β diminishes to zero.

Let us assume that the director of the testing program is unwilling to commit a type II error when the alternative hypothesis $p = 1/2$ is true even though we have found the probability of such an error to be $\beta = 0.2517$. A reduction in β is always possible by increasing the size of the critical region. For example, consider what happens to the values of α and β when we change our critical value to 7 so that all scores greater than 7 fall in the critical region and those less than or equal to 7 fall in the acceptance region. Now, in testing $p = 1/4$ against the alternative hypothesis that $p = 1/2$, we find that

$$\alpha = \sum_{x=8}^{20} b(x; 20, \tfrac{1}{4})$$
$$= 1 - \sum_{x=0}^{7} b(x; 20, \tfrac{1}{4})$$
$$= 1 - 0.8982$$
$$= 0.1018$$

and

$$\beta = \sum_{x=0}^{7} b(x; 20, \tfrac{1}{2})$$

$$= 0.1316.$$

By adopting a new decision procedure, we have reduced the probability of committing a type II error at the expense of increasing the probability of committing a type I error. For a fixed sample size, a decrease in the probability of one error will usually result in an increase in the probability of the other error. Fortunately, the probability of committing both types of error can be reduced by increasing the sample size. Consider the same problem using a random sample of 100 individuals. If more than 36 of the group surpass the 2-year period, we reject the null hypothesis that $p = 1/4$ and accept the alternative hypothesis that $p > 1/4$. The critical value is now 36. All possible scores above 36 constitute the critical region and all possible scores less than or equal to 36 fall in the acceptance region.

To determine the probability of committing a type I error, we shall use the normal curve approximation with

$$\mu = np = (100)(\tfrac{1}{4}) = 25$$

and

$$\sigma = \sqrt{npq} = \sqrt{(100)(\tfrac{1}{4})(\tfrac{3}{4})} = 4.33.$$

Referring to Figure 8.2, we need the area under the normal curve to the right of $x = 36.5$. The corresponding z value is

$$z = \frac{36.5 - 25}{4.33} = 2.66.$$

From Table A.3 we find that

$$\alpha = P(\text{type I error})$$
$$= P(X > 36 \text{ when } p = \tfrac{1}{4})$$
$$\simeq P(Z > 2.66)$$
$$= 1 - P(Z < 2.66)$$
$$= 1 - 0.9961$$
$$= 0.0039.$$

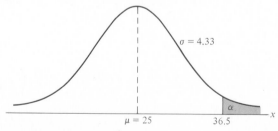

Figure 8.2 Probability of a type I error.

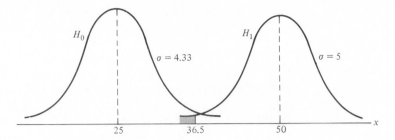

Figure 8.3 Probability of a type II error.

If H_0 is false and the true value of H_1 is $p = 1/2$, we can determine the probability of a type II error using the normal curve approximation with

$$\mu = np = (100)(\tfrac{1}{2}) = 50$$

and

$$\sigma = \sqrt{npq} = \sqrt{(100)(\tfrac{1}{2})(\tfrac{1}{2})} = 5.$$

The probability of falling in the acceptance region when H_1 is true is given by the area of the shaded region to the left of $x = 36.5$ in Figure 8.3. The z value corresponding to $x = 36.5$ is

$$z = \frac{36.5 - 50}{5} = -2.7.$$

Therefore,

$$
\begin{aligned}
\beta &= P(\text{type II error}) \\
&= P(X \le 36 \text{ when } p = \tfrac{1}{2}) \\
&\simeq P(Z < -2.7) \\
&= 0.0035.
\end{aligned}
$$

Obviously, the type I and type II errors will rarely occur if the experiment consists of 100 individuals.

The concepts discussed here for a discrete population can equally well be applied to continuous populations. Consider the null hypothesis that the average weight of male students in a certain college is 68 kilograms against the alternative hypothesis that it is unequal to 68. That is, we wish to test

$$
\begin{aligned}
H_0&: \quad \mu = 68, \\
H_1&: \quad \mu \ne 68.
\end{aligned}
$$

The alternative hypothesis allows for the possibility that $\mu < 68$ or $\mu > 68$.

A sample mean that falls close to the hypothesized value of 68 would be considered evidence in favor of H_0. On the other hand, a sample mean that is considerably less than or more than 68 would be evidence inconsistent with H_0 and therefore favoring H_1. A critical region might arbitrarily be chosen to

Reject H_0 ($\mu \neq 68$)	Accept H_0 ($\mu = 68$)	Reject H_0 ($\mu \neq 68$)

67   68   69 $\bar{x}$

Figure 8.4   Decision criterion for testing $\mu = 68$ versus $\mu \neq 68$.

be the two intervals $\bar{x} < 67$ and $\bar{x} > 69$. The acceptance region will then be the interval $67 \leq \bar{x} \leq 69$. This decision criterion is illustrated in Figure 8.4.

Let us now use the decision criterion of Figure 8.4 to calculate the probabilities of committing type I and type II errors when testing the null hypothesis that $\mu = 68$ kilograms against the alternative that $\mu \neq 68$ kilograms for the continuous population of student's weights.

Assume the standard deviation of the population of weights to be $\sigma = 3.6$. For large samples we may substitute s for σ if no other estimate of σ is available. Our decision statistic, based on a random sample of size $n = 36$, will be $\bar{X}$, the most efficient estimator of μ. From the Central Limit Theorem, we know that the sampling distribution of $\bar{X}$ is approximately normally distributed with standard deviation $\sigma_{\bar{x}} = \sigma/\sqrt{n} = 3.6/6 = 0.6$.

The probability of committing a type I error, or the level of significance of our test, is equal to the sum of the areas that have been shaded in each tail of the distribution in Figure 8.5. Therefore,

$$\alpha = P(\bar{X} < 67 \text{ when } \mu = 68) + P(\bar{X} > 69 \text{ when } \mu = 68).$$

The z values corresponding to $\bar{x}_1 = 67$ and $\bar{x}_2 = 69$ when H_0 is true are

$$z_1 = \frac{67 - 68}{0.6} = -1.67$$

and

$$z_2 = \frac{69 - 68}{0.6} = 1.67.$$

Therefore,

$$\alpha = P(Z < -1.67) + P(Z > 1.67)$$
$$= 2P(Z < -1.67)$$
$$= 0.0950.$$

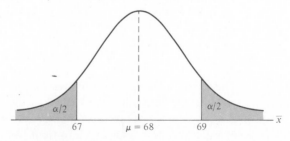

Figure 8.5   Critical region for testing $\mu = 68$ versus $\mu \neq 68$.

Thus 9.5% of all samples of size 36 would lead us to reject $\mu = 68$ kilograms when it is true. To reduce α, we have a choice of increasing the sample size or widening the acceptance region. Suppose that we increase the sample size to $n = 64$. Then $\sigma_{\bar{x}} = 3.6/8 = 0.45$. Now

$$z_1 = \frac{67 - 68}{0.45} = -2.22$$

and

$$z_2 = \frac{69 - 68}{0.45} = 2.22.$$

Hence

$$\alpha = P(Z < -2.22) + P(Z > 2.22)$$
$$= 2P(Z < -2.22)$$
$$= 0.0264.$$

The reduction in α is not sufficient by itself to guarantee a good testing procedure. We must evaluate β for various alternative hypotheses that we feel should be accepted if true. Therefore, if it is important to reject H_0 when the true mean is some value $\mu \geq 70$ or $\mu \leq 66$, then the probability of committing a type II error should be computed and examined for the alternatives $\mu = 66$ and $\mu = 70$. Because of symmetry, it is only necessary to consider the probability of accepting the null hypothesis that $\mu = 68$ when the alternative $\mu = 70$ is true. A type II error will result when the sample mean $\bar{x}$ falls between 67 and 69 when H_1 is true. Therefore, referring to Figure 8.6, we find that

$$\beta = P(67 \leq \bar{X} \leq 69 \text{ when } \mu = 70).$$

The z values corresponding to $\bar{x}_1 = 67$ and $\bar{x}_2 = 69$ when H_1 is true are

$$z_1 = \frac{67 - 70}{0.45} = -6.67$$

and

$$z_2 = \frac{69 - 70}{0.45} = -2.22.$$

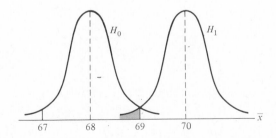

Figure 8.6   Type II error for testing $\mu = 68$ versus $\mu = 70$.

Therefore,

$$\beta = P(-6.67 < Z < -2.22)$$
$$= P(Z < -2.22) - P(Z < -6.67)$$
$$= 0.0132 - 0.0000$$
$$= 0.0132.$$

If the true value of μ is the alternative $\mu = 66$, the value of β will again be 0.0132. For all possible values of $\mu < 66$ or $\mu > 70$, the value of β will be even smaller when $n = 64$, and consequently there would be little chance of accepting H_0 when it is false.

The probability of committing a type II error increases rapidly when the true value of μ approaches, but is not equal to, the hypothesized value. Of course, this is usually the situation where we do not mind making a type II error. For example, if the alternative hypothesis $\mu = 68.5$ is true, we do not mind committing a type II error by concluding that the true answer is $\mu = 68$. The probability of making such an error will be high when $n = 64$. Referring to Figure 8.7, we have

$$\beta = P(67 \le \bar{X} \le 69 \text{ when } \mu = 68.5).$$

The z values corresponding to $\bar{x}_1 = 67$ and $\bar{x}_2 = 69$ when $\mu = 68.5$ are

$$z_1 = \frac{67 - 68.5}{0.45} = -3.33$$

and

$$z_2 = \frac{69 - 68.5}{0.45} = 1.11.$$

Therefore,

$$\beta = P(-3.33 < Z < 1.11)$$
$$= P(Z < 1.11) - P(Z < -3.33)$$
$$= 0.8665 - 0.0004$$
$$= 0.8661.$$

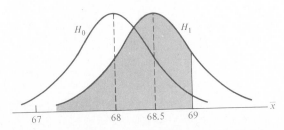

Figure 8.7 Type II error for testing $\mu = 68$ versus $\mu = 68.5$.

The preceding examples illustrate the following important properties:

1. The type I error and type II error are related. A decrease in the probability of one generally results in an increase in the probability of the other.
2. The size of the critical region, and therefore the probability of committing a type I error, can always be reduced by adjusting the critical value(s).
3. An increase in the sample size n will reduce α and β simultaneously.
4. If the null hypothesis is false, β is a maximum when the true value of a parameter approaches the hypothesized value. The greater the distance between the true value and the hypothesized value, the smaller β will be.

8.3 One-Tailed and Two-Tailed Tests

A test of any statistical hypothesis, where the alternative is **one-sided** such as

$$H_0: \quad \theta = \theta_0,$$
$$H_1: \quad \theta > \theta_0,$$

or perhaps

$$H_0: \quad \theta = \theta_0,$$
$$H_1: \quad \theta < \theta_0,$$

is called a **one-tailed test**. The critical region for the alternative hypothesis $\theta > \theta_0$ lies entirely in the right tail of the distribution, while the critical region for the alternative hypothesis $\theta < \theta_0$ lies entirely in the left tail. In a sense, the inequality symbol points in the direction where the critical region lies. A one-tailed test was used in the vaccine experiment of Section 8.2 to test the hypothesis $p = 1/4$ against the one-sided alternative $p > 1/4$ for the binomial distribution.

A test of any statistical hypothesis where the alternative is **two-sided**, such as

$$H_0: \quad \theta = \theta_0,$$
$$H_1: \quad \theta \neq \theta_0,$$

is called a **two-tailed test**, since the critical region is split into two parts having equal probabilities placed in each tail of the distribution of the test statistic. The alternative hypothesis $\theta \neq \theta_0$ states that either $\theta < \theta_0$ or $\theta > \theta_0$. A two-tailed test was used to test the null hypothesis that $\mu = 68$ kilograms against the two-sided alternative $\mu \neq 68$ kilograms for the continuous population of student weights in Section 8.2.

The null hypothesis, H_0, will always be stated using the equality sign so as to specify a single value. In this way the probability of committing a type I error can be controlled. Whether one sets up a one-tailed or a two-tailed test will depend on the conclusion to be drawn if H_0 is rejected. The location of the critical region can be determined only after H_1 has been stated. For example, in testing a new drug, one sets up the hypothesis that it is no better than similar drugs now on the market and tests this against the alternative hypothesis that the new drug is superior. Such an alternative hypothesis will result in a one-tailed test with the critical region in the right tail. However, if we wish to compare a new teaching technique with the conventional classroom procedure, the alternative hypothesis should allow for the new approach to be either inferior or superior to the conventional procedure. Hence the test is two-tailed with the critical region divided equally so as to fall in the extreme left and right tails of the distribution of our statistic.

Certain guidelines are desirable in determining which hypothesis should be stated as H_0 and which should be stated as H_1. First, read the problem carefully and determine the claim that you want to test. Should the claim suggest a simple direction such as *more than, less than, superior to, inferior to,* and so on, then H_1 will be stated using the inequality symbol ($<$ or $>$) corresponding to the suggested direction. If, for example, in testing a new drug we think that *more than* 30% of the people will be helped, we immediately write H_1: $p > 0.3$ and then the null hypothesis is written H_0: $p = 0.3$. Should the claim suggest a compound direction (equality as well as direction) such as *at least, equal to or greater, at most, no more than,* and so on, then this entire compound direction ($\leq$ or $\geq$) is expressed as H_0, but using only the equality sign, and H_1 is given by the opposite direction. Finally, if no direction whatsoever is suggested by the claim, then H_1 is stated using the *not equal* symbol ($\neq$).

Example 8.1   The manufacturer of a certain brand of cigarettes claims that the average nicotine content does not exceed 2.5 milligrams. State the null and alternative hypotheses to be used in testing this claim and determine where the critical region is located.

Solution. The manufacturer's claim should be rejected only if μ is greater than 2.5 milligrams and should be accepted if μ is less than or equal to 2.5 milligrams. Since the null hypothesis always specifies a single value of the parameter, we test

$$H_0: \quad \mu = 2.5,$$
$$H_1: \quad \mu > 2.5.$$

Although we have stated the null hypothesis with an equal sign, it is understood to include any value not specified by the alternative hypothesis. Consequently, the acceptance of H_0 does not imply that μ is exactly equal to 2.5 milligrams but rather that we do not have sufficient evidence favoring H_1. Since we have a one-tailed test, the greater than symbol indicated that the critical region lies entirely in the right tail of the distribution of our statistic $\bar{X}$.

Example 8.2   A real estate agent claims that 60% of all private residences being built today are 3-bedroom homes. To test this claim, a large sample of new residences is inspected; the proportion of these homes with 3 bedrooms is recorded and used as our test statistic. State the null and alternative hypotheses to be used in this test and determine the location of the critical region.

Solution. If the test statistic is substantially higher or lower than $p = 0.6$, we would reject the agent's claim. Hence we should make the test

$$H_0: \quad p = 0.6,$$
$$H_1: \quad p \neq 0.6.$$

The alternative hypothesis implies a two-tailed test with the critical region divided equally in both tails of the distribution of $\hat{P}$, our test statistic.

In testing hypotheses in which the test statistic is discrete, the critical region may be chosen arbitrarily and its size determined. If the size α is too large, it can be reduced by making an adjustment in the critical value. It may be necessary to increase the sample size to offset the increase that automatically occurs in β. In testing hypotheses in which the test statistic is continuous, it is customary to choose the value of α to be 0.05 or 0.01 and then find the critical region. For example, in a two-tailed test at the 0.05 level of significance, the critical values for a statistic having a standard normal distribution will be $-z_{0.025} = -1.96$ and $z_{0.025} = 1.96$. In terms of z values, the critical region of size 0.05 will be $z < -1.96$ and $z > 1.96$. An observed value of our test statistic is said to be **significant** if the null hypothesis is rejected at the specified level of significance.

The computer printout for a statistical package designed to test hypotheses will most likely include the probability of the test statistic taking on a value at least as extreme as the observed value when H_0 is true. This probability, denoted by P, is called a **P value**. It represents the lowest level of significance at which the observed value of our test statistic is significant. In the case of a one-tailed test in which the critical region lies entirely in the right tail of a standard normal distribution, the P value is just the area under the curve to the right of the z value computed from the available data. The smaller the P value, the less likely is one to observe such an extreme value and the more significant is the result. Therefore, if $P = 0.03$ appears on the printout, the observed value of the test statistic is significant for all $\alpha \geq 0.03$ and therefore it is certainly significant at the 0.05 level but not at the 0.01 level.

In the remaining sections of this chapter we consider several special tests of hypotheses that are frequently used by statisticians. The steps for testing a hypothesis concerning a population parameter θ against some alternative hypothesis may be summarized as follows:

1. State the null hypothesis H_0 that $\theta = \theta_0$.
2. Choose an appropriate alternative hypothesis H_1 from one of the alternatives $\theta < \theta_0$. $\theta > \theta_0$, or $\theta \neq \theta_0$.

3. Choose a significance level of size α.
4. Select the appropriate test statistic and establish the critical region. If the decision is to be based on a P value it is not necessary to state the critical region.
5. Compute the value of the test statistic from the sample data.
6. Decision: Reject H_0 if the test statistic has a value in the critical region or if the computed P value is less than or equal to the desired significance level α; otherwise accept H_0.

Exercises

1. Suppose that an allergist wishes to test the hypothesis that at least 30% of the public is allergic to some cheese products. Explain how the allergist could commit
   (a) a type I error;
   (b) a type II error.

2. A sociologist is concerned about the effectiveness of a training course designed to get more drivers to use seat belts in automobiles.
   (a) What hypothesis is she testing if she commits a type I error by erroneously concluding that the training course is ineffective?
   (b) What hypothesis is she testing if she commits a type II error by erroneously concluding that the training course is effective?

3. A large manufacturing firm is being charged with discrimination in its hiring practices.
   (a) What hypothesis is being tested if a jury commits a type I error by finding the firm guilty?
   (b) What hypothesis is being tested if a jury commits a type II error by finding the firm guilty?

4. The proportion of adults living in a small town who are college graduates is estimated to be $p = 0.3$. To test this hypothesis, a random sample of 15 adults is selected. If the number of college graduates in our sample is anywhere from 2 to 7, we shall accept the null hypothesis that $p = 0.3$; otherwise, we shall conclude that $p \neq 0.3$.
   (a) Evaluate α assuming $p = 0.3$.
   (b) Evaluate β for the alternative $p = 0.2$ and $p = 0.4$.
   (c) Is this a good test procedure?

5. Repeat Exercise 4 when 200 adults are selected and the acceptance region is defined to be $48 \leq x \leq 72$, where x is the number of college graduates in our sample.

6. The proportion of families buying milk from company A in a certain city is believed to be $p = 0.6$. If a random sample of 10 families shows that 3 or less buy milk from company A, we shall reject the hypothesis that $p = 0.6$ in favor of the alternative $p < 0.6$.
   (a) Find the probability of committing a type I error if the true proportion is $p = 0.6$.
   (b) Find the probability of committing a type II error for the alternatives $p = 0.3$, $p = 0.4$, and $p = 0.5$.

7. Repeat Exercise 6 when 50 families are selected and the critical region is defined to be $x \leq 24$, where x is the number of families in our sample that buy milk from company A.

8. A dry cleaning establishment claims that a new spot remover will remove no more than 70% of the spots to which it is applied. To check this claim, the spot remover will be used on 12 spots chosen at random. If fewer than 11 of the spots are removed, we shall accept the null hypothesis that $p = 0.7$; otherwise we conclude that $p > 0.7$.
   (a) Evaluate α, assuming that $p = 0.7$.
   (b) Evaluate β for the alternative $p = 0.9$.

9. Repeat Exercise 8 when 100 spots are treated and the critical region is defined to be $x > 82$, where x is the number of spots removed.

10. In the publication *Relief from Arthritis* by Thorsons Publishers, Ltd. (1979), John E. Croft claims

that over 40% of the sufferers from osteoarthritis received measurable relief from an ingredient produced by a particular species of mussel found off the coast of New Zealand. To test this claim, the mussel extract is to be given to a group of 7 osteoarthritic patients. If 3 or more of the patients receive relief, we shall accept the null hypothesis that $p = 0.4$; otherwise, we conclude that $p < 0.4$.

(a) Evaluate α, assuming that $p = 0.4$.

(b) Evaluate β for the alternative $p = 0.3$.

11. Repeat Exercise 10 when 70 patients are given the mussel extract and the critical region is defined to be $x < 24$, where x is the number of osteoarthritic patients who receive relief.

12. A random sample of 400 voters in a certain city are asked if they favor an additional 4% gasoline sales tax to provide badly needed revenues for street repairs. If more than 220 but fewer than 260 favor the sales tax, we shall conclude that 60% of the voters are for it.

(a) Find the probability of committing a type I error if 60% of the voters favor the increased tax.

(b) What is the probability of committing a type II error using this test procedure if actually only 48% of the voters are in favor of the additional gasoline tax?

13. Suppose, in Exercise 12, we conclude that 60% of the voters favor the gasoline sales tax if more than 214 but fewer than 266 voters in our sample favor it. Show that this new acceptance region results in a smaller value for α at the expense of increasing β.

14. A manufacturer has developed a new fishing line, which he claims has a mean breaking strength of 15 kilograms with a standard deviation of 0.5 kilogram. To test the hypothesis that $\mu = 15$ kilograms against the alternative that $\mu < 15$ kilograms, a random sample of 50 lines will be tested. The critical region is defined to be $\bar{x} < 14.9$.

(a) Find the probability of committing a type I error when H_0 is true.

(b) Evaluate β for the alternatives $\mu = 14.8$ and $\mu = 14.9$ kilograms.

15. A soft-drink machine at the Longhorn Steak House is regulated so that the amount of drink dispensed is approximately normally distributed with a mean of 200 milliliters and a standard deviation of 15 milliliters. The machine is checked periodically by taking a sample of 9 drinks and computing the average content. If $\bar{x}$ falls in the interval $191 < \bar{x} < 209$, the machine is thought to be operating satisfactorily; otherwise, we conclude that $\mu \neq 200$ milliliters.

(a) Find the probability of committing a type I error when $\mu = 200$ milliliters.

(b) Find the probability of committing a type II error when $\mu = 215$ milliliters.

16. Repeat Exercise 15 for samples of size $n = 25$. Use the same critical region.

17. A new cure has been developed for a certain type of cement that results in a compressive strength of 5000 kilograms per square centimeter and a standard deviation of 120. To test the hypothesis that $\mu = 5000$ against the alternative that $\mu < 5000$, a random sample of 50 pieces of cement is tested. The critical region is defined to be $\bar{x} < 4970$.

(a) Find the probability of committing a type I error when H_0 is true.

(b) Evaluate β for the alternatives $\mu = 4970$ and $\mu = 4960$.

18. If we plot the probabilities of accepting H_0 corresponding to various alternatives for μ (including the value specified by H_0) and connect all the points by a smooth curve, we obtain the **operating characteristic curve** of the test criterion, or simply the **OC curve**. Note that the probability of accepting H_0 when it is true is simply $1 - \alpha$. Operating characteristic curves are widely used in industrial applications to provide a visual display of the merits of the test criterion. With reference to Exercise 15, find the probabilities of accepting H_0 for the following 9 values of μ and plot the OC curve: 184, 188, 192, 196, 200, 204, 208, 212, and 216.

19. State the null and alternative hypotheses to be used in testing the following claims and determine generally where the critical region is located:

(a) The mean snowfall at Lake George during the month of February is 21.8 centimeters.

(b) No more than 20% of the faculty at the local university contributed to the annual giving fund.

(c) On the average, children attend schools within 6.2 kilometers of their homes in suburban St. Louis.

(d) At least 70% of next year's new cars will be in the compact and subcompact category.

(e) The proportion of voters favoring the incumbent in the upcoming election is 0.58.

(f) The average rib-eye steak at the Longhorn Steak House is at least 340 grams.

20. State the null and alternative hypotheses to be used in testing the following claims and determine generally where the critical region is located:

(a) At most, 20% of next year's wheat crop will be exported to the Soviet Union.

(b) On the average, American housewives drink 3 cups of coffee per day.

(c) The proportion of graduates in Virginia this year majoring in the social sciences is at least 0.15.

(d) The average donation to the American Lung Association is no more than $10.

(e) Residents in suburban Richmond commute, on the average, 15 kilometers to their place of employment.

8.4 *Tests Concerning Means*

Consider the problem of testing the hypothesis that the mean μ of a population, with known variance σ^2, equals a specified value μ_0 against the two-sided alternative that the mean is not equal to μ_0; that is, we shall test

$$H_0: \quad \mu = \mu_0,$$
$$H_1: \quad \mu \neq \mu_0.$$

The appropriate statistic on which we base our decision is the random variable $\bar{X}$. From the Central Limit Theorem we already know that for most populations the sampling distribution of $\bar{X}$ is approximately normally distributed with mean $\mu_{\bar{X}} = \mu$ and variance $\sigma_{\bar{X}}^2 = \sigma^2/n$, where μ and σ^2 are the mean and variance of the population from which we select a random sample of size n. If we use a significance level of α, it is possible to find two critical values, a and b, such that the interval $a \leq \bar{x} \leq b$ defines the acceptance region and the two tails of the distribution, $\bar{x} < a$ and $\bar{x} > b$, constitute the critical region.

The critical region is usually stated in terms of z values by means of the transformation

$$z = \frac{\bar{x} - \mu_0}{\sigma/\sqrt{n}}.$$

Hence, for an α level of significance, the critical values of the random variable Z that correspond to a and b are shown in Figure 8.8 to be

$$-z_{\alpha/2} = \frac{a - \mu_0}{\sigma/\sqrt{n}}$$

and

$$z_{\alpha/2} = \frac{b - \mu_0}{\sigma/\sqrt{n}}.$$

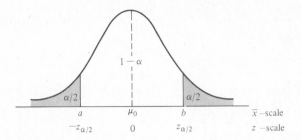

Figure 8.8 Critical region for the alternative hypothesis $\mu \neq \mu_0$.

From the population we select a random sample of size n and compute the sample mean $\bar{x}$. If $\bar{x}$ falls in the acceptance region, $a \leq \bar{x} \leq b$, then

$$z = \frac{\bar{x} - \mu_0}{\sigma/\sqrt{n}}$$

will fall in the region $-z_{\alpha/2} \leq z \leq z_{\alpha/2}$, and we conclude that $\mu = \mu_0$; otherwise, we reject H_0 and accept the alternative hypothesis that $\mu \neq \mu_0$. Since the population variance is seldom known, we may replace σ by s in our computations provided that s is computed from a sample of size $n \geq 30$.

The two-tailed test procedure just described is equivalent to finding a $(1 - \alpha)100\%$ confidence interval for μ and accepting H_0 if μ_0 lies in the interval. If μ_0 lies outside the interval, we reject H_0 in favor of the alternative hypothesis H_1. Consequently, when one makes inferences about the mean μ from a population with known variance σ^2, whether it be by the construction of a confidence interval or through the testing of a statistical hypothesis, the same value, $z = (\bar{x} - \mu)/(\sigma/\sqrt{n})$, is employed.

In general, if one uses an appropriate z or t value in Chapter 7 to construct a confidence interval for a population mean μ, or perhaps for the difference $\mu_1 - \mu_2$ of two population means, we can also use that same z or t value to test the hypothesis that $\mu = \mu_0$ or $\mu_1 - \mu_2 = d_0$ against an appropriate alternative. Of course, all the underlying assumptions made in Chapter 7 relative to the use of a given statistic apply to the tests described here. This essentially means that all our samples are selected either from approximately normal populations or are of size $n \geq 30$, in which case we can refer to the Central Limit Theorem to justify using a normal test statistic.

In Table 8.2 we list the values of the statistics used to test specified hypotheses H_0 concerning means and give the corresponding critical regions for one- and two-sided alternative hypotheses H_1. Several of these tests are illustrated in the following examples.

Example 8.3 A random sample of 100 recorded deaths in the United States during the past year showed an average life span of 71.8 years, with a standard deviation of 8.9 years. Does this seem to indicate that the average life span today is greater than 70 years? Use a 0.05 level of significance.

Table 8.2 Tests Concerning Means

H_0	Value of Test Statistic	H_1	Critical Region
$\mu = \mu_0$	$z = \dfrac{\bar{x} - \mu_0}{\sigma/\sqrt{n}}$; σ known or $n \geq 30$	$\mu < \mu_0$ $\mu > \mu_0$ $\mu \neq \mu_0$	$z < -z_\alpha$ $z > z_\alpha$ $z < -z_{\alpha/2}$ and $z > z_{\alpha/2}$
$\mu = \mu_0$	$t = \dfrac{\bar{x} - \mu_0}{s/\sqrt{n}}$; $v = n - 1$, σ unknown and $n < 30$	$\mu < \mu_0$ $\mu > \mu_0$ $\mu \neq \mu_0$	$t < -t_\alpha$ $t > t_\alpha$ $t < -t_{\alpha/2}$ and $t > t_{\alpha/2}$
$\mu_1 - \mu_2 = d_0$	$z = \dfrac{(\bar{x}_1 - \bar{x}_2) - d_0}{\sqrt{(\sigma_1^2/n_1) + (\sigma_2^2/n_2)}}$; σ_1 and σ_2 known $n \geq 30$	$\mu_1 - \mu_2 < d_0$ $\mu_1 - \mu_2 > d_0$ $\mu_1 - \mu_2 \neq d_0$	$z < -z_\alpha$ $z > z_\alpha$ $z < -z_{\alpha/2}$ and $z > z_{\alpha/2}$
$\mu_1 - \mu_2 = d_0$	$t = \dfrac{(\bar{x}_1 - \bar{x}_2) - d_0}{s_p\sqrt{(1/n_1) + (1/n_2)}}$; $v = n_1 + n_2 - 2$, $\sigma_1 = \sigma_2$ but unknown, $s_p^2 = \dfrac{(n_1 - 1)s_1^2 + (n_2 - 1)s_2^2}{n_1 + n_2 - 2}$	$\mu_1 - \mu_2 < d_0$ $\mu_1 - \mu_2 > d_0$ $\mu_1 - \mu_2 \neq d_0$	$t < -t_\alpha$ $t > t_\alpha$ $t < -t_{\alpha/2}$ and $t > t_{\alpha/2}$
$\mu_1 - \mu_2 = d_0$	$t' = \dfrac{(\bar{x}_1 - \bar{x}_2) - d_0}{\sqrt{(s_1^2/n_1) + (s_2^2/n_2)}}$; $v = \dfrac{(s_1^2/n_1 + s_2^2/n_2)^2}{\dfrac{(s_1^2/n_1)^2}{n_1 - 1} + \dfrac{(s_2^2/n_2)^2}{n_2 - 1}}$; $\sigma_1 \neq \sigma_2$ and unknown	$\mu_1 - \mu_2 < d_0$ $\mu_1 - \mu_2 > d_0$ $\mu_1 - \mu_2 \neq d_0$	$t' < -t_\alpha$ $t' > t_\alpha$ $t' < -t_{\alpha/2}$ and $t' > t_{\alpha/2}$
$\mu_D = d_0$	$t = \dfrac{\bar{d} - d_0}{s_d/\sqrt{n}}$; $v = n - 1$, paired observations	$\mu_D < d_0$ $\mu_D > d_0$ $\mu_D \neq d_0$	$t < -t_\alpha$ $t > t_\alpha$ $t < -t_{\alpha/2}$ and $t > t_{\alpha/2}$

Solution. Following the six-step procedure outlined in Section 8.3, we have

1. H_0: $\mu = 70$ years.
2. H_1: $\mu > 70$ years.
3. $\alpha = 0.05$.
4. Critical region: $z > 1.645$, where

$$z = \frac{\bar{x} - \mu_0}{\sigma/\sqrt{n}}.$$

5. Computations: $\bar{x} = 71.8$ years, $\sigma \simeq s = 8.9$ years, and

$$z = \frac{71.8 - 70}{8.9/\sqrt{100}} = 2.02.$$

6. Decision: Reject H_0 and conclude that the average life span today is greater than 70 years.

In Example 8.3 the P value corresponding to $z = 2.02$ is given by the area of the shaded region in Figure 8.9. Using Table A.3, we have

$$P = P(Z > 2.02) = 0.0217$$

and since the desired significance level of 0.05 exceeds this P value, we reject H_0.

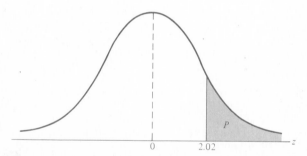

Figure 8.9   *P* value for Example 8.3.

Example 8.4   A manufacturer of sports equipment has developed a new synthetic fishing line that he claims has a mean breaking strength of 8 kilograms with a standard deviation of 0.5 kilogram. Test the hypothesis that $\mu = 8$ kilograms against the alternative that $\mu \neq 8$ kilograms if a random sample of 50 lines is tested and found to have a mean breaking strength of 7.8 kilograms. Use a 0.01 level of significance.

Solution. Using the six-step procedure, we have

1. H_0: $\mu = 8$ kilograms.
2. H_1: $\mu \neq 8$ kilograms.
3. $\alpha = 0.01$.

4. Critical region: $z < -2.575$ and $z > 2.575$, where

$$z = \frac{\bar{x} - \mu_0}{\sigma/\sqrt{n}}.$$

5. Computations: $\bar{x} = 7.8$ kilograms, $n = 50$, and hence

$$z = \frac{7.8 - 8}{0.5/\sqrt{50}} = -2.83.$$

6. Decision: Reject H_0 and conclude that the average breaking strength is not equal to 8 but is, in fact, less than 8 kilograms.

Since the test in Example 8.4 is two-tailed, the desired P value is twice the area of the shaded region in Figure 8.10 to the left of $z = -2.83$. Therefore, using Table A.3, we have

$$P = P(|Z| > 2.83) = 2P(Z < -2.83) = 0.0046,$$

which allows us to reject the null hypothesis that $\mu = 8$ kilograms at the 0.01 level of significance.

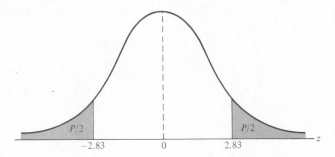

Figure 8.10   *P* Value for Example 8.4.

Example 8.5   The *Edison Electric Institute* has published figures on the annual number of kilowatt hours expended by various home appliances. It is claimed that a vacuum cleaner expends an average of 46 kilowatt hours per year. If a random sample of 12 homes included in a planned study indicates that vacuum cleaners expend an average of 42 kilowatt hours per year with a standard deviation of 11.9 kilowatt hours, does this suggest at the 0.05 level of significance that vacuum cleaners expend, on the average, less than 46 kilowatt hours annually? Assume the population of kilowatt hours to be normal.

Solution

1. H_0: $\mu = 46$ kilowatt hours.
2. H_1: $\mu < 46$ kilowatt hours.
3. $\alpha = 0.05$.

4. Critical region: $t < -1.796$, where

$$t = \frac{\bar{x} - \mu_0}{s/\sqrt{n}}$$

with $v = 11$ degrees of freedom.
5. Computations: $\bar{x} = 42$ kilowatt hours, $s = 11.9$ kilowatt hours, and $n = 12$. Hence

$$t = \frac{42 - 46}{11.9/\sqrt{12}} = -1.16.$$

6. Decision: Accept H_0 and conclude that the average number of kilowatt hours expended annually by home vacuum cleaners is not significantly less than 46.

Example 8.6 An experiment was performed to compare the abrasive wear of two different laminated materials. Twelve pieces of material 1 were tested, by exposing each piece to a machine measuring wear. Ten pieces of material 2 were similarly tested. In each case, the depth of wear was observed. The samples of material 1 gave an average (coded) wear of 85 units with a standard deviation of 4, while the samples of material 2 gave an average of 81 and a standard deviation of 5. Can we conclude at the 0.05 level of significance that the abrasive wear of material 1 exceeds that of material 2 by more than 2 units? Assume the populations to be approximately normal with equal variances.

Solution. Let μ_1 and μ_2 represent the population means of the abrasive wear for material 1 and material 2, respectively. Using the six-step procedure, we have
1. H_0: $\mu_1 - \mu_2 = 2$.
2. H_1: $\mu_1 - \mu_2 > 2$.
3. $\alpha = 0.05$.
4. Critical region: $t > 1.725$, where

$$t = \frac{(\bar{x}_1 - \bar{x}_2) - d_0}{s_p \sqrt{\dfrac{1}{n_1} + \dfrac{1}{n_2}}}$$

with $v = 20$ degrees of freedom.
5. Computations:

$$\bar{x}_1 = 85, \qquad s_1 = 4, \qquad n_1 = 12,$$
$$\bar{x}_2 = 81, \qquad s_2 = 5, \qquad n_2 = 10.$$

Hence

$$s_p = \sqrt{\frac{(11)(16) + (9)(25)}{12 + 10 - 2}} = 4.478$$

$$t = \frac{(85 - 81) - 2}{4.478 \sqrt{(1/12) + (1/10)}} = 1.04.$$

6. Decision: Accept H_0 and conclude that the abrasive wear of material 1 does not exceed that of material 2 by more than 2 units.

Example 8.7 In the paper "*Influence of Physical Restraint and Restraint-Facilitating Drugs on Blood Measurements of White-Tailed Deer and Other Selected Mammals,*" Virginia Polytechnic Institute and State University (1976), J. A. Wesson examined the influence of the drug *succinyl-choline* on the circulation levels of androgens in the blood. Blood samples from wild, free-ranging deer were obtained via the jugular vein immediately after an intramuscular injection of succinyl-choline using darts and a capture gun. Deer were bled again approximately 30 minutes after the injection and then released. The levels of androgens at time of capture and 30 minutes later, measured in nanograms per milliliter (ng/ml), for 15 deer are as follows:

Deer	Time of Injection	Androgen (ng/ml) 30 Minutes After Injection	d_i
1	2.76	7.02	4.26
2	5.18	3.10	−2.08
3	2.68	5.44	2.76
4	3.05	3.99	0.94
5	4.10	5.21	1.11
6	7.05	10.26	3.21
7	6.60	13.91	7.31
8	4.79	18.53	13.74
9	7.39	7.91	0.52
10	7.30	4.85	−2.45
11	11.78	11.10	−0.68
12	3.90	3.74	−0.16
13	26.00	94.03	68.03
14	67.48	94.03	26.55
15	17.04	41.70	24.66

Assuming the populations of androgen at time of injection and 30 minutes later are normally distributed, test at the 0.05 level of significance whether the androgen concentrations are altered after 30 minutes of restraint.

Solution. Let μ_1 and μ_2 be the average androgen concentration at the time of injection and 30 minutes later, respectively. We proceed as follows:

1. H_0: $\mu_1 = \mu_2$ or $\mu_D = \mu_1 - \mu_2 = 0$.
2. H_1: $\mu_1 \neq \mu_2$ or $\mu_D = \mu_1 - \mu_2 \neq 0$.
3. $\alpha = 0.05$.

4. Critical region: $t < -2.145$ and $t > 2.145$, where

$$t = \frac{\bar{d} - d_0}{s_d/\sqrt{n}}$$

with $\nu = 14$ degrees of freedom.

5. Computations: The sample mean and standard deviation for the d_i's are

$$\bar{d} = 9.848 \qquad \text{and} \qquad s_d = 18.474.$$

Therefore,

$$t = \frac{9.848 - 0}{18.474/\sqrt{15}} = 2.06.$$

6. Decision: Accept H_0 and conclude that there is no difference in circulating levels of androgen at time of immobilization and 30 minutes later.

8.5 *Choice of Sample Size for Testing Means*

The significance level for testing a statistical hypothesis is normally controlled by the experimenter, while β, or the **power** of the test defined by $1 - \beta$, is controlled by using the proper sample size. In this section we shall discuss the choice of sample size for tests involving one and two population means. For situations in which the normal distribution is used and the population variance or variances are known, it is a simple matter to determine the sample size necessary to attain the desired power.

Suppose that we wish to test the hypothesis

$$H_0: \quad \mu = \mu_0,$$
$$H_1: \quad \mu > \mu_0,$$

with a significance level α when the variance σ^2 is known. For a specific alternative, say $\mu = \mu_0 + \delta$, the power of our test is shown in Figure 8.11 to be

$$1 - \beta = P(\bar{X} > a \text{ when } \mu = \mu_0 + \delta).$$

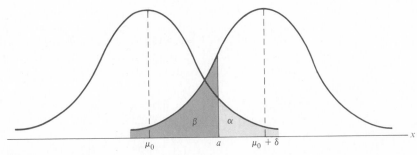

Figure 8.11   Testing $\mu = \mu_0$ versus $\mu = \mu_0 + \delta$.

Therefore,

$$\beta = P(\bar{X} < a \text{ when } \mu = \mu_0 + \delta)$$

$$= P\left[\frac{\bar{X} - (\mu_0 + \delta)}{\sigma/\sqrt{n}} < \frac{a - (\mu_0 + \delta)}{\sigma/\sqrt{n}} \text{ when } \mu = \mu_0 + \delta\right].$$

Under the alternative hypothesis $\mu = \mu_0 + \delta$, the statistic

$$\frac{\bar{X} - (\mu_0 + \delta)}{\sigma/\sqrt{n}}$$

is the standard normal variable Z. Therefore,

$$\beta = P\left(Z < \frac{a - \mu_0}{\sigma/\sqrt{n}} - \frac{\delta}{\sigma/\sqrt{n}}\right)$$

$$= P\left(Z < z_\alpha - \frac{\delta}{\sigma/\sqrt{n}}\right),$$

from which we conclude that

$$-z_\beta = z_\alpha - \frac{\delta\sqrt{n}}{\sigma}$$

and hence

$$n = \frac{(z_\alpha + z_\beta)^2 \sigma^2}{\delta^2},$$

a result that is also true when the alternative hypothesis is $\mu < \mu_0$.

In the case of a two-tailed test we obtain the power $1 - \beta$ for a specified alternative when

$$n \simeq \frac{(z_{\alpha/2} + z_\beta)^2 \sigma^2}{\delta^2}.$$

Example 8.8   Suppose that we wish to test the hypothesis

$$H_0: \quad \mu = 68 \text{ kilograms,}$$
$$H_1: \quad \mu > 68 \text{ kilograms,}$$

for the weights of male students at a certain college using an $\alpha = 0.05$ level of significance when it is known that $\sigma = 5$. Find the sample size required if the power of our test is to be 0.95 when the true mean is 69 kilograms.

Solution. Since $\alpha = \beta = 0.05$, we have $z_\alpha = z_\beta = 1.645$. For the alternative $\mu = 69$, we take $\delta = 1$ and then

$$n = \frac{(1.645 + 1.645)^2(25)}{1} = 270.6.$$

Therefore, it requires 271 observations if the test is to reject the null hypothesis 95% of the time when, in fact, μ is as large as 69 kilograms.

A similar procedure can be used to determine the sample size $n = n_1 = n_2$ required for a specific power of the test in which two population means are being compared. For example, suppose that we wish to test the hypothesis

$$H_0: \quad \mu_1 - \mu_2 = d_0,$$
$$H_1: \quad \mu_1 - \mu_2 \neq d_0,$$

when σ_1 and σ_2 are known. For a specific alternative, say $\mu_1 - \mu_2 = d_0 + \delta$, the power of our test is shown in Figure 8.12 to be

$$1 - \beta = P(|\bar{X}_1 - \bar{X}_2| > a \text{ when } \mu_1 - \mu_2 = d_0 + \delta).$$

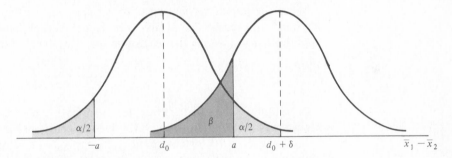

Figure 8.12   Testing $\mu_1 - \mu_2 = d_0$ versus $\mu_1 - \mu_2 = d_0 + \delta$.

Therefore,

$$\beta = P(-a < \bar{X}_1 - \bar{X}_2 < a \text{ when } \mu_1 - \mu_2 = d_0 + \delta)$$

$$= P\left[\frac{-a - (d_0 + \delta)}{\sqrt{(\sigma_1^2 + \sigma_2^2)/n}} < \frac{\bar{X}_1 - \bar{X}_2 - (d_0 + \delta)}{\sqrt{(\sigma_1^2 + \sigma_2^2)/n}}\right.$$

$$\left. < \frac{a - (d_0 + \delta)}{\sqrt{(\sigma_1^2 + \sigma_2^2)/n}} \text{ when } \mu_1 - \mu_2 = d_0 + \delta\right].$$

Under the alternative hypothesis $\mu_1 - \mu_2 = d_0 + \delta$, the statistic

$$\frac{\bar{X}_1 - \bar{X}_2 - (d_0 + \delta)}{\sqrt{(\sigma_1^2 + \sigma_2^2)/n}}$$

is the standard normal variable Z. Now, writing

$$-z_{\alpha/2} = \frac{-a - d_0}{\sqrt{(\sigma_1^2 + \sigma_2^2)/n}} \quad \text{and} \quad z_{\alpha/2} = \frac{a - d_0}{\sqrt{(\sigma_1^2 + \sigma_2^2)/n}},$$

we have

$$\beta = P\left[-z_{\alpha/2} - \frac{\delta}{\sqrt{(\sigma_1^2 + \sigma_2^2)/n}} < Z < z_{\alpha/2} - \frac{\delta}{\sqrt{(\sigma_1^2 + \sigma_2^2)/n}}\right],$$

from which we conclude that

$$-z_\beta \simeq z_{\alpha/2} - \frac{\delta}{\sqrt{(\sigma_1^2 + \sigma_2^2)/n}}$$

and hence

$$n \simeq \frac{(z_{\alpha/2} + z_\beta)^2(\sigma_1^2 + \sigma_2^2)}{\delta^2}.$$

For the one-tailed test, the expression for the required sample size when $n = n_1 = n_2$ is given by

$$n = \frac{(z_\alpha + z_\beta)^2(\sigma_1^2 + \sigma_2^2)}{\delta^2}.$$

When the population variance (or variances in the two-sample situation) is unknown, the choice of sample size is not straightforward. In testing the hypothesis $\mu = \mu_0$ when the true value is $\mu = \mu_0 + \delta$, the statistic

$$\frac{\bar{X} - (\mu_0 + \delta)}{S/\sqrt{n}}$$

does not follow the t distribution, as one might expect, but instead follows the **noncentral t distribution**. However, tables or charts based on the noncentral t distribution do exist for determining the appropriate sample size if some estimate of σ is available or if δ is a multiple of σ. Table A.8 gives the sample sizes needed to control the values of α and β for various values of

$$\Delta = \frac{|\delta|}{\sigma} = \frac{|\mu - \mu_0|}{\sigma}$$

for both one-tailed and two-tailed tests. In the case of the two-sample t test in which the variance are unknown but assumed equal, we obtain the sample sizes $n = n_1 = n_2$ needed to control the values of α and β for various values of

$$\Delta = \frac{|\delta|}{\sigma} = \frac{|\mu_1 - \mu_2 - d_0|}{\sigma}$$

from Table A.9.

Example 8.9   In comparing the performance of two catalysts on the effect of a reaction yield, a two-sample t test is to be conducted with $\alpha = 0.05$. The variances in the yields are considered to be the same for the two catalysts. How large a sample for each catalyst is needed to test the hypothesis

$$H_0: \quad \mu_1 = \mu_2,$$
$$H_1: \quad \mu_1 \neq \mu_2,$$

if it is essential to detect a difference of 0.8σ between the catalysts with probability 0.9?

Solution. From Table A.9, with $\alpha = 0.05$ for a two-tailed test, $\beta = 0.1$, and

$$\Delta = \frac{|0.8\sigma|}{\sigma} = 0.8,$$

we find the required sample size to be $n = 34$.

Exercises

1. An electrical firm manufactures light bulbs that have a length of life that is approximately normally distributed with a mean of 800 hours and a standard deviation of 40 hours. Test the hypothesis that $\mu = 800$ hours against the alternative $\mu \neq 800$ hours if a random sample of 30 bulbs has an average life of 788 hours. Use a 0.04 level of significance.

2. A random sample of 36 drinks from a soft-drink machine has an average content of 21.9 deciliters, with a standard deviation of 1.42 deciliters. Test the hypothesis that $\mu = 22.2$ deciliters against the alternative hypothesis, $\mu < 22.2$, at the 0.05 level of significance.

3. In a research report by Richard H. Weindruch of the UCLA Medical School, it is claimed that mice with an average lifespan of 32 months will live to be about 40 months old when 40% of the calories in their food are replaced by vitamins and protein. Is there any reason to believe that $\mu < 40$ if 64 mice that are placed on this diet have an average life of 38 months with a standard deviation of 5.8 months? Use a 0.025 level of significance.

4. The average height of females in the freshman class of a certain college has been 162.5 centimeters with a standard deviation of 6.9 centimeters. Is there reason to believe that there has been a change in the average height if a random sample of 50 females in the present freshman class has an average height of 165.2 centimeters? Use a 0.02 level of significance.

5. It is claimed that an automobile is driven on the average less than 20,000 kilometers per year. To test this claim, a random sample of 100 automobile owners are asked to keep a record of the kilometers they travel. Would you agree with this claim if the random sample showed an average of 23,500 kilometers and a standard deviation of 3900 kilometers? Use a 0.01 level of significance.

6. The *Edison Electric Institute* has published figures on the annual hours of use of various home appliances. It is claimed that a trash compactor is run an average of 125 hours per year. If a random sample of 49 homes equipped with trash compactors indicates an annual average use of 126.9 hours with a standard deviation of 8.4 hours, does this suggest at the 0.05 level of significance that trash compactors are used, on the average, more than 125 hours per year?

7. Test the hypothesis that the average content of containers of a particular lubricant is 10 liters if the contents of a random sample of 10 containers are 10.2, 9.7, 10.1, 10.3, 10.1, 9.8, 9.9, 10.4, 10.3, and 9.8 liters. Use a 0.01 level of significance and assume that the distribution of contents is normal.

8. According to *Dietary Goals for the United States* (1977), high sodium intake may be related to ulcers, stomach cancer, and migraine headaches. The human requirement for salt is only 220 milligrams per day, which is surpassed in most single servings of ready-to-eat cereals. If a random sample of 20 similar servings of Special K has a mean sodium content of 244 milligrams of sodium and a standard deviation of 24.5 milligrams, does this suggest at the 0.05 level of significance that the average sodium content for single servings of Special K is greater than 220 milligrams? Assume the distribution of sodium contents to be normal.

9. A random sample of 8 cigarettes of a certain brand has an average nicotine content of 4.2 milligrams and a standard deviation of 1.4 milligrams. Is this in line with the manufacturer's claim that the average nocotine content does not exceed 3.5 milligrams? Use a 0.01 level of significance and

assume the distribution of nicotine contents to be normal.

10. Last year the employees of the city sanitation department donated an average of $10.00 to the volunteer rescue squad. Test the hypothesis at the 0.01 level of significance that the average contribution this year is still $10.00 if a random sample of 12 employees showed an average donation of $10.90 with a standard deviation of $1.75. Assume that the donations are approximately normally distributed.

11. Past experience indicates that the time for high school seniors to complete a standardized test is a normal random variable with a mean of 35 minutes. If a random sample of 20 high school seniors took an average of 33.1 minutes to complete this test with a standard deviation of 4.3 minutes, test the hypothesis at the 0.025 level of significance that $\mu = 35$ minutes against the alternative that $\mu < 35$ minutes.

12. A random sample of size $n_1 = 25$, taken from a normal population with a standard deviation $\sigma_1 = 5.2$, has a mean $\bar{x}_1 = 81$. A second random sample of size $n_2 = 36$, taken from a different normal population with a standard deviation $\sigma_2 = 3.4$, has a mean $\bar{x}_2 = 76$. Test the hypothesis, at the 0.06 level of significance, that $\mu_1 = \mu_2$ against the alternative $\mu_1 \neq \mu_2$.

13. A manufacturer claims that the average tensile strength of thread A exceeds the average tensile strength of thread B by at least 12 kilograms. To test his claim, 50 pieces of each type of thread are tested under similar conditions. Type A thread had an average tensile strength of 86.7 kilograms with a standard deviation of 6.28 kilograms, while type B thread had an average tensile strength of 77.8 kilograms with a standard deviation of 5.61 kilograms. Test the manufacturer's claim using a 0.05 level of significance.

14. A study was made to estimate the difference in salaries of college professors in the private and state colleges of North Carolina. A random sample of 100 professors in private colleges showed an average 9-month salary of $32,000 with a standard deviation of $13,000. A random sample of 200 professors in state colleges showed an average salary of $32,900 with a standard deviation of $1400. Test the hypothesis that the

average salary for professors teaching in state colleges does not exceed the average salary for professors teaching in private colleges by more than $500. Use a 0.01 level of significance.

15. A study is made to see if increasing the substrate concentration has an appreciable effect on the velocity of a chemical reaction. With a substrate concentration of 1.5 moles per liter, the reaction was run 15 times with an average velocity of 7.5 micromoles per 30 minutes and a standard deviation of 1.5. With a substrate concentration of 2.0 miles per liter, 12 runs were made, yielding an average velocity of 8.8 micromoles per 30 minutes and a sample standard deviation of 1.2. Is there any reason to believe that this increase in substrate concentration causes an increase in the mean velocity by more than 0.5 micromole per 30 minutes? Use a 0.01 level of significance and assume the populations to be approximately normally distributed with equal variances.

16. A study was made to determine if the subject matter in a physics course is better understood when a lab constitutes part of the course. Students were allowed to choose between a 3-semester-hour course without labs and a 4-semester-hour course with labs. In the section with labs 11 students made an average grade of 85 with a standard deviation of 4.7, and in the section without labs 17 students made an average grade of 79 with a standard deviation of 6.1. Would you say that the laboratory course increases the average grade by as much as 8 points? Use a 0.01 level of significance and assume the populations to be approximately normally distributed with equal variances.

17. To find out whether a new serum will arrest leukemia, 9 mice, which have all reached an advanced stage of the disease, are selected. Five mice receive the treatment and 4 do not. The survival times, in years, from the time the experiment commenced are as follows:

Treatment	2.1	5.3	1.4	4.6	0.9
No treatment	1.9	0.5	2.8	3.1	

At the 0.05 level of significance can the serum be said to be effective? Assume the two distributions to be normally distributed with equal variances.

18. A large automobile manufacturing company is trying to decide whether to purchase brand A or brand B tires for its new models. To help arrive at a decision, an experiment is conducted using 12 of each brand. The tires are run until they wear out. The results are

Brand A: $\bar{x}_1 = 37,900$ kilometers,
$s_1 = 5100$ kilometers

Brand B: $\bar{x}_2 = 39,800$ kilometers,
$s_2 = 5900$ kilometers.

Test the hypothesis at the 0.05 level of significance that there is no difference in the 2 brands of tires. Assume the populations to be approximately normally distributed.

19. In Exercise 8 on page 235, test the hypothesis that Volkswagen mini-trucks, on the average, exceed similarly equipped Toyota mini-trucks by 4 kilometers per liter. Use a 0.10 level of significance.

20. A UCLA researcher claims that the average life span of mice can be extended by as much as 8 months when the calories in their food are reduced by approximately 40% from the time they are weaned. The restricted diets are enriched to normal levels by vitamins and protein. Suppose that a random sample of 10 mice are fed a normal diet and live an average life span of 32.1 months with a standard deviation of 3.2 months, while a random sample of 15 mice are fed the restricted diet and live an average life span of 37.6 months with a standard deviation of 2.8 months. Test the hypothesis at the 0.05 level of significance that the average life span of mice on this restricted diet is increased by 8 months against the alternative that the increase is less than 8 months. Assume the distributions of life spans for the regular and restricted diets are approximately normal with equal variances.

21. The following data represent the running times of films produced by 2 motion-picture companies:

	Time (minutes)					
Company 1	102	86	98	109	92	
Company 2	81	165	97	134	92	87	114

Test the hypothesis that the average running time of films produced by company 2 exceeds the average running time of films produced by company 1 by 10 minutes against the one-sided alternative that the difference is more than 10 minutes. Use a 0.1 level of significance and assume the distributions of times to be approximately normal with unequal variances.

22. In the study *"Interrelationships Between Stress, Dietary Intake, and Plasma Ascorbic Acid During Pregnancy"* conducted at the Virginia Polytechnic Institute and State University in May 1983, the plasma ascorbic acid levels of pregnant women were compared for smokers versus nonsmokers. Thirty-two women in the last three months of pregnancy, free of major health disorders, and ranging in age from 15 to 32 years were selected for the study. Prior to the collection of 20 ml of blood, the participants were told to avoid breakfast, forego their vitamin supplements, and avoid foods high in ascorbic acid content. From the blood samples, the following plasma ascorbic acid values of each subject were determined in milligrams per 100 milliliters:

Plasma Ascorbic Acid Values		
Nonsmokers		Smokers
0.97	1.16	0.48
0.72	0.86	0.71
1.00	0.85	0.98
0.81	0.58	0.68
0.62	0.57	1.18
1.32	0.64	1.36
1.24	0.98	0.78
0.99	1.09	1.64
0.90	0.92	
0.74	0.78	
0.88	1.24	
0.94	1.18	

Is there sufficient evidence at the 0.05 level of significance to conclude that there is a difference between plasma ascorbic acid levels of smokers and nonsmokers? Assume that the two sets of data came from normal populations with unequal variances.

23. A study on the *"Nutrient Retention and Macro-invertebrate Community Response to Sewage Stress*

in a Stream Ecosystem" was conducted by the Department of Zoology at the Virginia Polytechnic Institute and State University in 1980 to determine if there is a significant difference in the density of organisms at two different stations located on Cedar Run, a secondary stream located in the Roanoke River drainage basin. Sewage from a sewage treatment plant and overflow from the Federal Mogul Corporation settling pond enter the stream near its head waters. The following data give the density measurements, in number of organisms per square meter, at the two different collecting stations:

Number of Organisms per Square Meter			
Station 1		Station 2	
5030	4980	2800	2810
13,700	11,910	4670	1330
10,730	8130	6890	3320
11,400	26,850	7720	1230
860	17,660	7030	2130
2200	22,800	7330	2190
4250	1130		
15,040	1690		

Can we conclude, at the 0.05 level of significance, that the average densities at the two stations are equal? Assume that the observations come from normal populations with different variances.

24. Five samples of a ferrous-type substance are to be used to determine if there is a difference between a laboratory chemical analysis and an X-ray fluorescence analysis of the iron content. Each sample was split into two subsamples and the two types of analysis were applied. Following are the coded data showing the iron content analysis:

Analysis	Sample				
	1	2	3	4	5
X-ray	2.0	2.0	2.3	2.1	2.4
Chemical	2.2	1.9	2.5	2.3	2.4

Assuming that the populations are normal, test at the 0.05 level of significance whether the two methods of analysis give, on the average, the same result.

25. A taxi company is trying to decide whether the use of radial tires instead of regular belted tires improves fuel economy. Twelve cars were equipped with radial tires and driven over a prescribed test course. Without changing drivers, the same cars were then equipped with regular belted tires and driven once again over the test course. The gasoline consumption, in kilometers per liter, was recorded as follows:

Car	Kilometers per Liter	
	Radial Tires	Belted Tires
1	4.2	4.1
2	4.7	4.9
3	6.6	6.2
4	7.0	6.9
5	6.7	6.8
6	4.5	4.4
7	5.7	5.7
8	6.0	5.8
9	7.4	6.9
10	4.9	4.7
11	6.1	6.0
12	5.2	4.9

At the 0.025 level of significance, can we conclude that cars equipped with radial tires give better fuel economy than those equipped with belted tires? Assume the populations to be normally distributed.

26. In Exercise 13 on page 235, use the t distribution to test the hypothesis, at the 0.05 level of significance, that the diet reduces a person's weight by 4.5 kilograms on the average against the alternative hypothesis that the mean difference in weight is less than 4.5 kilograms.

27. According to the article *"Practice and Fatigue Effects on the Programming of a Coincident Timing Response,"* published in the *Journal of Human Movement Studies* in 1976, practice under fatigued conditions distorts mechanisms which govern performance. An experiment was conducted using 15 college males who were trained to make a continuous horizontal right to left arm movement from a microswitch to a barrier, knocking over the barrier coincident with the arrival of a clock sweephand to the 6 o'clock position. The absolute value of the difference between the time, in milliseconds, that it took to knock over the barrier and the time for the sweephand to reach the 6 o'clock position (500 msec) was recorded. Each participant performed the task five times under prefatigue and postfatigue conditions and the sums of the absolute differences for the five performances were recorded as follows:

Subject	Absolute Time Differences (msec)	
	Prefatigue	Postfatigue
1	158	91
2	92	59
3	65	215
4	98	226
5	33	223
6	89	91
7	148	92
8	58	177
9	142	134
10	117	116
11	74	153
12	66	219
13	109	143
14	57	164
15	85	100

An increase in the mean absolute time differences when the task is performed under postfatigue conditions would support the claim that practice under fatigued conditions distorts mechanisms that govern performance. Assuming the populations to be normally distributed, test this claim at the 0.05 level of significance.

28. In the study "*Comparison of Sorbic Acid in Country Ham Before and After Storage*" conducted by the Department of Human Nutrition and Foods at the Virginia Polytechnic Institute and State University in 1983, the following data on the comparison of sorbic acid residuals in parts per million in ham immediately after dipping in a sorbate solution and after 60 days of storage were

recorded:

Slice	Sorbic Acid Residuals in Ham	
	Before Storage	After Storage
1	224	116
2	270	96
3	400	239
4	444	329
5	590	437
6	660	597
7	1400	689
8	680	576

Assuming the populations to be normally distributed, is there sufficient evidence, at the 0.05 level of significance, to say that the length of storage influences sorbic acid residual concentrations?

29. How large a sample is required in Exercise 2 if the power of our test is to be 0.90 when the true mean is 21.3? Assume that $\sigma = 1.42$.

30. If the distribution of lifespans in Exercise 3 is approximately normal, how large a sample is required in order that the probability of committing a type II error be 0.1 when the true mean is 35.9 months? Assume that $\sigma = 5.8$ months.

31. How large a sample is required in Exercise 4 if the power of our test is to be 0.95 when the true average height differs from 162.5 by 3.1 centimeters?

32. How large should the samples be in Exercise 13 if the power of our test is to be 0.95 when the true difference between thread types A and B is 8 kilograms?

33. How large a sample is required in Exercise 9 if the power of our test is to be 0.8 when the true nicotine content exceeds the hypothesized value by 1.2σ?

34. On testing

$$H_0: \quad \mu = 14,$$
$$H_1: \quad \mu \neq 14,$$

an $\alpha = 0.05$ level t test is being considered. What sample size is necessary in order that the probability is 0.1 of falsely accepting H_0 when the true population mean differs from 14 by 0.5? From a preliminary sample we estimate σ to be 1.25.

8.6 *Tests Concerning Proportions*

Tests of hypotheses concerning proportions are required in many areas. The politician is certainly interested in knowing what fraction of the voters will favor him in the next election. All manufacturing firms are concerned about the proportion of defectives when a shipment is made. The gambler depends on a knowledge of the proportion of outcomes that he considers favorable.

We shall consider the problem of testing the hypothesis that the proportion of successes in a binomial experiment equals some specified value. That is, we are testing the null hypothesis H_0 that $p = p_0$, where p is the parameter of the binomial distribution. The alternative hypothesis may be one of the usual one-sided or two-sided alternatives: $p < p_0$, $p > p_0$, or $p \neq p_0$.

The appropriate statistic on which we base our decision criterion is the binomial random variable X, although we could just as well use the statistic $\hat{P} = X/n$. Values of X that are far from the mean $\mu = np_0$ will lead to the rejection of the null hypothesis. Because X is a discrete binomial variable, it is unlikely that a critical region can be established whose size is *exactly* equal to a prespecified value of α. For this reason it is preferable, in dealing with small samples, to base our decisions on P values. To test the hypothesis

$$H_0: \quad p = p_0,$$
$$H_1: \quad p < p_0,$$

we use the binomial distribution to compute the P value

$$P = P(X \leq x \text{ when } p = p_0).$$

The value x is the number of successes in our sample of size n. If this P value is less than or equal to α, our test is significant at the α level and we reject H_0 in favor of H_1. Similarly, to test the hypothesis

$$H_0: \quad p = p_0,$$
$$H_1: \quad p > p_0,$$

at the α level of significance, we compute

$$P = P(X \geq x \text{ when } p = p_0)$$

and reject H_0 in favor of H_1 if this P value is less than or equal to α. Finally, to test the hypothesis

$$H_0: \quad p = p_0,$$
$$H_1: \quad p \neq p_0,$$

at the α level of significance, we compute

$$P = 2P(X \leq x \text{ when } p = p_0)$$

if $x < np_0$ or

$$P = 2P(X \geq x \text{ when } p = p_0)$$

if $x > np_0$ and reject H_0 in favor of H_1 if the computed P value is less than or equal to α.

The steps for testing a null hypothesis about a proportion against various alternatives using the binomial probabilities of Table A.1 are as follows:

Testing a Proportion; Small Samples

1. H_0: $p = p_0$.
2. H_1: *Alternatives are $p < p_0, p > p_0$, or $p \neq p_0$.*
3. *Choose a level of significance equal to α.*
4. *Test statistic: Binomial variable X with $p = p_0$.*
5. *Computations: Find x, the number of successes and compute the appropriate P value.*
6. *Decision: Reject H_0 if the computed P value is less than or equal to α.*

Example 8.10   A builder claims that heat pumps are installed in 70% of all homes being constructed today in the city of Richmond. Would you agree with this claim if a random survey of new homes in this city shows that 8 out of 15 had heat pumps installed? Use a 0.10 level of significance.

Solution

1. H_0: $p = 0.7$.
2. H_1: $p \neq 0.7$.
3. $\alpha = 0.10$.
4. Test statistic: Binomial variable X with $p = 0.7$ and $n = 15$.
5. Computations: $x = 8$ and $np_0 = (15)(0.7) = 10.5$. Therefore, from Table A.1, the computed P value is

$$P = 2P(X \leq 8 \text{ when } p = 0.7)$$

$$= 2 \sum_{x=0}^{8} b(x; 15, 0.7)$$

$$= 0.2622 > 0.10.$$

6. Decision: Accept H_0 and conclude that there is insufficient reason to doubt the builder's claim.

In Section 5.4 we saw that binomial probabilities were obtainable from the actual binomial formula or from Table A.1 when n is small. For large n, approximation procedures are required. When the hypothesized value p_0 is very close to 0 or 1, the Poisson distribution, with parameter $\mu = np_0$, may be used. However, the normal-curve approximation, with parameters $\mu = np_0$ and $\sigma^2 = np_0 q_0$, is usually preferred for large n and is very accurate as long as p_0

is not extremely close to 0 or to 1. If we use the normal approximation, the z **value for testing $p = p_0$** is given by

$$z = \frac{x - np_0}{\sqrt{np_0 q_0}},$$

which is a value of the standard normal variable Z. Hence, for a two-tailed test at the α level of significance, the critical region is $z < -z_{\alpha/2}$ and $z > z_{\alpha/2}$. For the one-sided alternative $p < p_0$, the critical region is $z < -z_\alpha$, and for the alternative $p > p_0$, the critical region is $z > z_\alpha$.

Example 8.11 A commonly prescribed drug on the market for relieving nervous tension is believed to be only 60% effective. Experimental results with a new drug administered to a random sample of 100 adults who were suffering from nervous tension showed that 70 received relief. Is this sufficient evidence to conclude that the new drug is superior to the one commonly prescribed? Use a 0.05 level of significance.

Solution

1. H_0: $p = 0.6$.
2. H_1: $p > 0.6$.
3. $\alpha = 0.05$.
4. Critical region: $z > 1.645$.
5. Computations: $x = 70, n = 100, np_0 = (100)(0.6) = 60$, and

$$z = \frac{70 - 60}{\sqrt{(100)(0.6)(0.4)}} = 2.04.$$

6. Decision: Reject H_0 and conclude that the new drug is superior.

8.7 Testing the Difference Between Two Proportions

Situations often arise where we wish to test the hypothesis that two proportions are equal. For example, we might try to prove that the proportion of doctors who are pediatricians in one state is equal to the proportion of pediatricians in another state. A person may decide to give up smoking only if he or she is convinced that the proportion of smokers with lung cancer exceeds the proportion of nonsmokers with lung cancer.

In general, we wish to test the null hypothesis that two proportions, or binomial parameters, are equal. That is, we are testing $p_1 = p_2$ against one of the alternatives $p_1 < p_2$, $p_1 > p_2$, or $p_1 \neq p_2$. Of course, this is equivalent to testing the null hypothesis that $p_1 - p_2 = 0$ against one of the alternatives $p_1 - p_2 < 0$, $p_1 - p_2 > 0$, or $p_1 - p_2 \neq 0$. The statistic on which we base our decision is the random variable $\hat{P}_1 - \hat{P}_2$. Independent samples of size n_1 and n_2 are selected at random from two binomial populations and the proportion of successes $\hat{P}_1$ and $\hat{P}_2$ for the two samples are computed.

In our construction of confidence intervals for p_1 and p_2 we noted, for n sufficiently large, that the point estimator $\hat{P}_1$ and $\hat{P}_2$ was approximately normally distributed with mean

$$\mu_{\hat{P}_1 - \hat{P}_2} = p_1 - p_2$$

and variance

$$\sigma^2_{\hat{P}_1 - \hat{P}_2} = \frac{p_1 q_1}{n_1} + \frac{p_2 q_2}{n_2}.$$

Therefore, our acceptance and critical regions can be established by using the standard normal variable

$$Z = \frac{(\hat{P}_1 - \hat{P}_2) - (p_1 - p_2)}{\sqrt{(p_1 q_1/n_1) + (p_2 q_2/n_2)}}.$$

When H_0 is true, we can substitute $p_1 = p_2 = p$ and $q_1 = q_2 = q$ (where p and q are the common values) in the preceding formula for Z to give the form

$$Z = \frac{\hat{P}_1 - \hat{P}_2}{\sqrt{pq[(1/n_1) + (1/n_2)]}}.$$

To compute a value of Z, however, we must estimate the parameters p and q that appear in the radical. Upon pooling the data from both samples, the **pooled estimate of the proportion p** is

$$\hat{p} = \frac{x_1 + x_2}{n_1 + n_2},$$

where x_1 and x_2 are the number of successes in each of the two samples. Substituting $\hat{p}$ for p and $\hat{q} = 1 - \hat{p}$ for q, the z **value for testing $p_1 = p_2$** is determined from the formula

$$z = \frac{\hat{p}_1 - \hat{p}_2}{\sqrt{\hat{p}\hat{q}[(1/n_1) + (1/n_2)]}}.$$

The critical regions for the appropriate alternative hypotheses are set up as before using critical points of the standard normal curve. Hence, for the alternative $p_1 \neq p_2$ at the α level of significance, the critical region is $z < -z_{\alpha/2}$ and $z > z_{\alpha/2}$. For a test where the alternative is $p_1 < p_2$, the critical region is $z < -z_{\alpha}$, and when the alternative is $p_1 > p_2$, the critical region is $z > z_{\alpha}$.

Example 8.12   A vote is to be taken among the residents of a town and the surrounding county to determine whether a proposed chemical plant should be constructed. The construction site is within the town limits and for this reason many voters in the county feel that the proposal will pass because of the large proportion of town voters who favor the construction. To determine if there is a significant difference in the proportion of town voters and county voters favoring the proposal, a poll is taken. If 120 of 200 town voters favor the proposal and 240 of 500 county residents favor it, would you agree that the proportion of town voters favoring the proposal is higher than the proportion of county voters? Use a 0.025 level of significance.

Solution. Let p_1 and p_2 be the true proportion of voters in the town and county, respectively, favoring the proposal. We now follow the six-step procedure:

1. H_0: $p_1 = p_2$.
2. H_1: $p_1 > p_2$.
3. $\alpha = 0.025$.
4. Critical region: $z > 1.96$.
5. Computations:

$$\hat{p}_1 = \frac{x_1}{n_1} = \frac{120}{200} = 0.60$$

$$\hat{p}_2 = \frac{x_2}{n_2} = \frac{240}{500} = 0.48$$

$$\hat{p} = \frac{x_1 + x_2}{n_1 + n_2} = \frac{120 + 240}{200 + 500} = 0.51.$$

Therefore,

$$z = \frac{0.60 - 0.48}{\sqrt{(0.51)(0.49)[(1/200) + (1/500)]}} = 2.9.$$

6. Decision: Reject H_0 and agree that the proportion of town voters favoring the proposal is higher than the proportion of county voters.

Exercises

1. A distributor of cigarettes claims that 20% of the smokers in Miami prefer Kent cigarettes. To test this claim, 20 cigarette smokers are selected at random and asked what brand they prefer. If 6 of the 20 named Kent as their preference, what conclusion do we draw? Use a 0.05 level of significance.

2. Suppose that, in the past, 40% of all adults favored capital punishment. Do we have reason to

believe that the proportion of adults favoring capital punishment today has increased if, in a random sample of 15 adults, 8 favor capital punishment? Use a 0.05 level of significance.

3. A coin is tossed 20 times resulting in 5 heads. Is this sufficient evidence to reject the hypothesis at the 0.03 level of significance that the coin is balanced in favor of the alternative that heads occur less than 50% of the time?

4. It is believed that at least 60% of the residents in a certain area favor an annexation suit by a neighboring city. What conclusion would you draw if only 110 in a sample of 200 voters favor the suit? Use a 0.04 level of significance.

5. A fuel oil company claims that one-fifth of the homes in a certain city are heated by oil. Do we have reason to doubt this claim if, in a random sample of 1000 homes in this city, it is found that 236 are heated by oil? Use a 0.01 level of significance.

6. At a certain college it is estimated that at most 25% of the students ride bicycles to class. Does this seem to be a valid estimate if, in a random sample of 90 college students, 28 are found to ride bicycles to class? Use a 0.05 level of significance.

7. A new radar device is being considered for a certain defense missile system. The system is checked by experimenting with actual aircraft in which a *kill* or a *no kill* is simulated. If in 300 trials, 250 kills occur, accept or reject, at the 0.04 level of significance, the claim that the probability of a kill with the new system does not exceed the 0.8 probability of the existing device.

8. In a controlled laboratory experiment scientists at the University of Minnesota discovered that 25% of a certain strain of rats subjected to a 20% coffee bean diet and then force-fed a powerful cancer-causing chemical later developed cancerous tumors. Would we have reason to believe that the proportion of rats developing tumors when subjected to this diet has increased if the experiment were repeated and 16 of 48 rats developed tumors? Use a 0.05 level of significance.

9. In a study to estimate the proportion of residents in a certain city and its suburbs who favor the construction of a nuclear power plant, it is found that 63 of 100 urban residents favor the construction while only 59 of 125 suburban residents are in favor. Is there a significant difference between the proportion of urban and suburban residents who favor construction of the nuclear plant? Use a 0.04 level of significance.

10. In a study on the fertility of married women conducted by Martin O'Connell and Carolyn C. Rogers for the Census Bureau in 1979, two groups of childless wives aged 25 to 29 were selected at random and each wife was asked if she eventually planned to have a child. One group was selected from among those wives married less than two years and the other from among those wives married five years. Suppose that 240 of 300 wives married less than two years planned to have children some day compared to 288 of the 400 wives married five years. Can we conclude at the 0.025 level of significance that the proportion of wives married less than two years who planned to have children is significantly higher than the proportion of wives married five years?

11. A cigarette manufacturing firm distributes two brands of cigarettes. If it is found that 56 of 200 smokers prefer brand A and that 29 of 150 smokers prefer brand B, can we conclude at the 0.06 level of significance that brand A outsells brand B?

12. A geneticist is interested in the proportion of males and females in a population that have a certain minor blood disorder. In a random sample of 100 males, 31 are found to be afflicted, whereas only 24 of 100 females tested appear to have the disorder. Can we conclude at the 0.01 level of significance that the proportion of men in the population afflicted with this blood disorder is significantly greater than the proportion of women afflicted?

13. A study was made to determine whether more Italians than Americans prefer white champagne to pink champagne at weddings. Of the 300 Italians selected at random, 72 preferred white champagne, and of the 400 Americans selected, 70 preferred white champagne rather than pink. Can we conclude that a high proportion of Italians rather than Americans prefer white champagne at weddings? Use a 0.05 level of significance.

14. *z* value for testing $p_1 - p_2 = d_0$: To test the null hypothesis H_0 that $p_1 - p_2 = d_0$, where $d_0 \neq 0$, we base our decision on

$$z = \frac{\hat{p}_1 - \hat{p}_2 - d_0}{\sqrt{\dfrac{\hat{p}_1 \hat{q}_1}{n_1} + \dfrac{\hat{p}_2 \hat{q}_2}{n_2}}}, \ldots$$

which is a value of a random variable whose distribution approximates the standard normal distribution as long as n_1 and n_2 are both large.

(a) With reference to Example 8.12 on page 293, test the hypothesis that the percentage of town voters favoring the construction of the chemical plant will not exceed the percentage of county voters by more than 3%. Use a 0.025 level of significance.

(b) With reference to Exercise 11, test the hypoth-

esis at the 0.06 level of significance that brand A outsells brand B by 10% against the alternative hypothesis that the difference is less than 10%.

8.8 Tests Concerning Variances

In this section we are concerned with testing hypotheses concerning population variances or standard deviations. In other words, we are interested in testing hypotheses concerning the uniformity of a population or perhaps in comparing the uniformity of one population with that of a second population. We might, therefore, be interested in testing the hypothesis that the variability in the percentage of impurities in a certain kind of fruit preserve does not exceed some specified value or that the variability in the length of life for some brand of exterior house paint is equal to the variability in the length of life for some competitive brand.

Let us first consider the problem of testing the null hypothesis H_0 that the population variance σ^2 equals a specified value σ_0^2 against one of the usual alternatives $\sigma^2 < \sigma_0^2$, $\sigma^2 > \sigma_0^2$, or $\sigma^2 \neq \sigma_0^2$. The appropriate statistic on which we base our decision is the same chi-square statistic of Theorem 6.16 on page 000 that was used in Chapter 7 to construct a confidence interval for σ^2. Therefore, if we assume that the distribution of the population being sampled is at least approximately normal, the **chi-square value for testing $\sigma^2 = \sigma_0^2$** is given by

$$\chi^2 = \frac{(n-1)s^2}{\sigma_0^2},$$

where n is the sample size, s^2 is the sample variance, and σ_0^2 is the value of σ^2 given by the null hypothesis. If H_0 is true, χ^2 is a value of the chi-square distribution with $v = n - 1$ degrees of freedom. Hence, for a two-tailed test at the α level of significance, the critical region is $\chi^2 < \chi_{1-\alpha/2}^2$ and $\chi^2 > \chi_{\alpha/2}^2$. For the one-sided alternative $\sigma^2 < \sigma_0^2$, the critical region is $\chi^2 < \chi_{1-\alpha}^2$, and for the one-sided alternative $\sigma^2 > \sigma_0^2$, the critical region is $\chi^2 > \chi_\alpha^2$.

To test a hypothesis about a population variance, we proceed by the same basic six steps outlined at the end of Section 8.3.

Example 8.13   A manufacturer of car batteries claims that the life of his batteries is approximately normally distributed with a standard deviation equal to 0.9 year. If a random sample of 10 of these batteries has a standard deviation of 1.2 years, do you think that $\sigma > 0.9$ year? Use a 0.05 level of significance.

Solution

1. H_0: $\sigma^2 = 0.81$.
2. H_1: $\sigma^2 > 0.81$.

3. $\alpha = 0.05$.

4. Critical region: From Figure 8.13, we see that the null hypothesis is rejected when $\chi^2 > 16.919$, where

$$\chi^2 = \frac{(n-1)s^2}{\sigma_0^2}$$

with $v = 9$ degrees of freedom.

5. Computations: $s^2 = 1.44$, $n = 10$, and

$$\chi^2 = \frac{(9)(1.44)}{0.81} = 16.0.$$

6. Decision: Accept H_0 and conclude that there is insufficient reason to doubt that the standard deviation is 0.9 year.

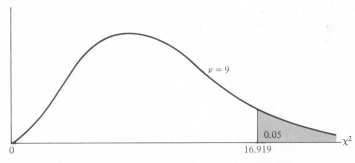

Figure 8.13 Critical region for the alternative hypothesis $\sigma^2 > 81$.

Now let us consider the problem of testing the equality of the variances σ_1^2 and σ_2^2 of two populations. That is, we shall test the null hypothesis H_0 that $\sigma_1^2 = \sigma_2^2$ against one of the usual alternatives $\sigma_1^2 < \sigma_2^2$, $\sigma_1^2 > \sigma_2^2$, or $\sigma_1^2 \neq \sigma_2^2$. For independent random samples of size n_1 and n_2, respectively, from the two populations, the *f* **value for testing** $\boldsymbol{\sigma_1^2 = \sigma_2^2}$ is the ratio

$$f = \frac{s_1^2}{s_2^2},$$

where s_1^2 and s_2^2 are the variances computed from the two samples. If the two populations are approximately normally distributed and the null hypothesis is true, according to Theorem 6.20 on page 210, the ratio $f = s_1^2/s_2^2$ is a value of the F distribution with $v_1 = n_1 - 1$ and $v_2 = n_2 - 1$ degrees of freedom. Therefore, the critical regions of size α corresponding to the one-sided alternatives $\sigma_1^2 < \sigma_2^2$ and $\sigma_1^2 > \sigma_2^2$ are, respectively, $f < f_{1-\alpha}(v_1, v_2)$ and $f > f_\alpha(v_1, v_2)$. For the two-sided alternative $\sigma_1^2 \neq \sigma_2^2$, the critical region is given by $f < f_{1-\alpha/2}(v_1, v_2)$ and $f > f_{\alpha/2}(v_1, v_2)$.

Example 8.14 In testing for the difference in the abrasive wear of the two materials in Example 8.6 on page 279, we assumed that the two unknown population variances are equal. Were we justified in making this assumption? Use a 0.10 level of significance.

Solution. Let σ_1^2 and σ_2^2 be the population variances for the abrasive wear of material 1 and material 2, respectively. Following the six-step procedure, we have

1. H_0: $\sigma_1^2 = \sigma_2^2$.
2. H_1: $\sigma_1^2 \neq \sigma_2^2$.
3. $\alpha = 0.10$.
4. Critical region: From Figure 8.14, we see that

$$f_{0.05}(11, 9) = 3.11$$

and, by using Theorem 6.19 on page 209,

$$f_{0.95}(11, 9) = \frac{1}{f_{0.05}(9, 11)} = 0.34.$$

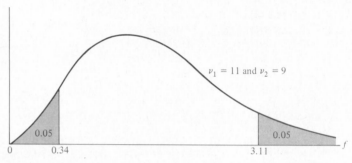

Figure 8.14   Critical region for the alternative $\sigma_1^2 \neq \sigma_2^2$.

Therefore, the null hypothesis is rejected when $f < 0.34$ or $f > 3.11$, where $f = s_1^2/s_2^2$ with $v_1 = 11$ and $v_2 = 9$ degrees of freedom.
5. Computations: $s_1^2 = 16$, $s_2^2 = 25$, and hence

$$f = \tfrac{16}{25} = 0.64.$$

6. Decision: Accept H_0 and conclude that we were justified in assuming the unknown variances equal in Example 8.6.

Exercises

---

1. The volume of containers of a particular lubricant is known to be normally distributed with a variance of 0.03 liter. Test the hypothesis that $\sigma^2 = 0.03$ against the alternative that $\sigma^2 \neq 0.03$ for the random sample of 10 containers in Exercise 7 on page 285. Use a 0.01 level of significance.

2. Past experience indicates that the time for high school seniors to complete a standardized test is a normal random variable with a standard deviation of 6 minutes. Test the hypothesis that $\sigma = 6$ against the alternative that $\sigma < 6$ if a random sample of 20 high school seniors has a standard deviation $s = 4.51$. Use a 0.05 level of significance.

3. The nicotine content of a certain brand of cigarettes is known to be normally distributed with a variance of 1.3 milligrams. Test the hypothesis that $\sigma^2 = 1.3$ against the alternative that $\sigma^2 \neq 1.3$ if a random sample of 8 of these cigarettes has a standard deviation $s = 1.8$. Use a 0.05 level of significance.

4. Past data indicate that the amount of money contributed by the working residents of a large city to a volunteer rescue squad is a normal random variable with a standard deviation of $1.40. It has been suggested that the contributions to the rescue squad from just the employees of the sanitation department are much more variable. If the contributions of a random sample of 12 employees from the sanitation department had a standard deviation of $1.75, can we conclude at the 0.01 level of significance that the standard deviation of the contributions of all sanitation workers is greater than that of all workers living in this city?

5. A soft-drink dispensing machine is said to be out of control if the variance of the contents exceeds 1.15 deciliters. If a random sample of 25 drinks from this machine has a variance of 2.03 deciliters, does this indicate at the 0.05 level of significance that the machine is out of control? Assume that the contents are approximately normally distributed.

6. **Large-sample test of $\sigma^2 = \sigma_0^2$:** When $n \geq 30$ we can test the null hypothesis that $\sigma^2 = \sigma_0^2$, or $\sigma = \sigma_0$, by computing

$$z = \frac{s - \sigma_0}{\sigma_0/\sqrt{2n}},$$

which is a value of a random variable whose sampling distribution is approximately the standard normal distribution.

(a) With reference to Example 8.3 on page 275, test at the 0.05 level of significance whether $\sigma = 7.5$ years against the alternative that $\sigma \neq 7.5$ years.

(b) It is suspected that the variance of the distribution of distances in kilometers achieved per 5 liters of fuel by a new automobile model equipped with a diesel engine is less than the variance of the distribution of distances achieved by the same model equipped with a six-cylinder gasoline engine, which is known to be $\sigma^2 = 6.25$. If 72 test runs in the diesel model have a variance of 4.41, can we conclude at the 0.02 level of significance that the variance of the distances achieved by the diesel model is less than that of the gasoline model?

7. A study is conducted to compare the length of time between men and women to assemble a certain product. Past experience indicates that the distribution of times for both men and women are approximately normal but the variance of the times for women is less than that for men. A random sample of times for 11 men and 14 women produced the following data:

Men	Women
$n_1 = 11$	$n_2 = 14$
$s_1 = 6.1$	$s_2 = 5.3$

Test the hypothesis that $\sigma_1^2 = \sigma_2^2$ against the alternative that $\sigma_1^2 > \sigma_2^2$. Use a 0.01 level of significance.

8. In Exercise 23 on page 287, test the hypothesis at the 0.02 level of significance that $\sigma_1^2 = \sigma_2^2$ against the alternative that $\sigma_1^2 \neq \sigma_2^2$, where σ_1^2 and σ_2^2 are the variances for the number of organisms per square meter at the two different locations on Cedar Run.

9. With reference to Exercise 18 on page 287, test the hypothesis that $\sigma_1 = \sigma_2$ against the alternative that $\sigma_1 < \sigma_2$, where σ_1 and σ_2 are the standard deviations of the distances obtained by brand A and brand B tires, respectively. Use a 0.05 level of significance.

10. With reference to Exercise 21 on page 287, test the hypothesis that $\sigma_1^2 = \sigma_2^2$ against the alternative that $\sigma_1^2 \neq \sigma_2^2$, where σ_1^2 and σ_2^2 are the variances for the running times of films produced by company 1 and company 2, respectively. Use a 0.10 level of significance.

11. Two types of instruments for measuring the amount of sulfur monoxide in the atmosphere are being compared in an air-pollution experiment. It is desired to determine whether the two types of instruments yield measurements having the same variability. The following readings were recorded for the two instruments:

Sulfur Monoxide	
Instrument A	Instrument B
0.86	0.87
0.82	0.74
0.75	0.63

Sulfur Monoxide	
Instrument A	Instrument B
0.61	0.55
0.89	0.76
0.64	0.70
0.81	0.69
0.68	0.57
0.65	0.53

Assuming the populations of measurements to be approximately normally distributed, test the hypothesis that $\sigma_A = \sigma_B$ against the alternative that $\sigma_A \neq \sigma_B$. Use a 0.02 level of significance.

8.9 Goodness-of-Fit Test

Throughout this chapter we have been concerned with the testing of statistical hypotheses about single population parameters such as μ, σ^2, and p. Now we shall consider a test to determine if a population has a specified theoretical distribution. The test is based on how good a fit we have between the frequency of occurrence of observations in an observed sample and the expected frequencies obtained from the hypothesized distribution.

To illustrate, consider the tossing of a die. We hypothesize that the die is honest, which is equivalent to testing the hypothesis that the distribution of outcomes is the discrete uniform distribution

$$f(x) = \tfrac{1}{6}, \qquad x = 1, 2, \ldots, 6.$$

Suppose that the die is tossed 120 times and each outcome is recorded. Theoretically, if the die is balanced, we would expect each face to occur 20 times. The results are given in Table 8.3. By comparing the observed frequencies with

Table 8.3 Observed and Expected Frequencies of 120 Tosses of a Die

	Faces					
	1	2	3	4	5	6
Observed	20	22	17	18	19	24
Expected	20	20	20	20	20	20

the corresponding expected frequencies, we must decide whether these discrepancies are likely to occur as a result of sampling fluctuations and the die is balanced or the die is not honest and the distribution of outcomes is not uniform. It is common practice to refer to each possible outcome of an experiment as a cell. Hence, in our illustration, we have 6 cells. The appropriate statistic on which we base our decision criterion for an experiment involving k cells is defined by the following theorem.

Theorem 8.1   *A **goodness-of-fit test** between observed and expected frequencies is based on the quantity*

$$\chi^2 = \sum_{i=1}^{k} \frac{(o_i - e_i)^2}{e_i},$$

where χ^2 is a value of a random variable whose sampling distribution is approximated very closely by the chi-square distribution with $v = k - 1$ degrees of freedom. The symbols o_i and e_i represent the observed and expected frequencies, respectively, for the ith cell.

The number of degrees of freedom associated with the chi-square distribution used here is equal to $k - 1$, since there are only $k - 1$ freely determined cell frequencies. That is, once $k - 1$ cell frequencies are determined, so is the frequency for the kth cell.

If the observed frequencies are close to the corresponding expected frequencies, the χ^2 value will be small, indicating a good fit. If the observed frequencies differ considerably from the expected frequencies, the χ^2 value will be large and the fit is poor. A good fit leads to the acceptance of H_0, whereas a poor fit leads to its rejection. The critical region will, therefore, fall in the right tail of the chi-square distribution. For a level of significance equal to α, we find the critical value χ_α^2 from Table A.5, and then $\chi^2 > \chi_\alpha^2$ constitutes the critical region. **The decision criterion described here should not be used unless each of the expected frequencies is at least equal to 5.** This restriction may require the combining of adjacent cells resulting in a reduction in the number of degrees of freedom.

From Table 8.3 we find the χ^2 value to be

$$\chi^2 = \frac{(20 - 20)^2}{20} + \frac{(22 - 20)^2}{20} + \frac{(17 - 20)^2}{20}$$
$$+ \frac{(18 - 20)^2}{20} + \frac{(19 - 20)^2}{20} + \frac{(24 - 20)^2}{20}$$
$$= 1.7.$$

Using Table A.5, we find $\chi_{0.05}^2 = 11.070$ for $v = 5$ degrees of freedom. Since 1.7 is less than the critical value, we fail to reject H_0 and conclude that the distribution is uniform. In other words, the die is balanced.

Table 8.4 Observed and Expected
Frequencies of Battery Lives Assuming
Normality

Class Boundaries	o_i	e_i
1.45–1.95	2 ⎫	0.5 ⎫
1.95–2.45	1 ⎬ 7	2.1 ⎬ 8.5
2.45–2.95	4 ⎭	5.9 ⎭
2.95–3.45	15	10.3
3.45–3.95	10	10.7
3.95–4.45	5 ⎫ 8	7.0 ⎫ 10.5
4.45–4.95	3 ⎭	3.5 ⎭

As a second illustration let us test the hypothesis that the frequency distribution of battery lives given in Table 2.4 on page 56 may be approximated by a normal distribution with mean $\mu = 3.5$ and standard deviation $\sigma = 0.7$. The expected frequencies for the 7 classes (cells), listed in Table 8.4, are obtained by computing the areas under the hypothesized normal curve that fall between the various class boundaries.

For example, the z values corresponding to the boundaries of the fourth class are

$$z_1 = \frac{2.95 - 3.5}{0.7} = -0.79$$

and

$$z_2 = \frac{3.45 - 3.5}{0.7} = -0.07.$$

From Table A.3 we find the area between $z_1 = -0.79$ and $z_2 = -0.07$ to be

$$\text{Area} = P(-0.79 < Z < -0.07)$$
$$= P(Z < -0.07) - P(Z < -0.79)$$
$$= 0.4721 - 0.2148$$
$$= 0.2573.$$

Hence the expected frequency for the fourth class is

$$e_4 = (0.2573)(40) = 10.3.$$

It is customary to round these frequencies to one decimal.

The expected frequency for the first class interval is obtained by using the total area under the normal curve to the left of the boundary 1.95. For the last class interval, we use the total area to the right of the boundary 4.45. All other expected frequencies are determined by the method described for the fourth class. Note that we have combined adjacent classes in Table 8.4, where the expected frequencies are less than 5. Consequently, the total number of intervals is reduced from 7 to 4 resulting in $v = 3$ degrees of freedom. The χ^2 value is then given by

$$\chi^2 = \frac{(7 - 8.5)^2}{8.5} + \frac{(15 - 10.3)^2}{10.3} + \frac{(10 - 10.7)^2}{10.7} + \frac{(8 - 10.5)^2}{10.5}$$

$$= 3.05.$$

Since the computed χ^2 value is less than $\chi^2_{0.05} = 7.815$ for 3 degrees of freedom, we have no reason to reject the null hypothesis and conclude that the normal distribution with $\mu = 3.5$ and $\sigma = 0.7$ provides a good fit for the distribution of battery lives.

8.10 Test for Independence

The chi-square test procedure discussed in Section 8.9 can also be used to test the hypothesis of independence of two variables of classification. Suppose that we wish to determine whether the opinions of the voting residents of the state of Illinois concerning a new tax reform are independent of their levels of income. A random sample of 1000 registered voters from the state of Illinois are classified as to whether they are in a low, medium, or high income bracket and whether or not they favor a new tax reform. The observed frequencies are presented in Table 8.5, which is known as a **contingency table**.

A contingency table with r rows and c columns is referred to as an $r \times c$ table ("$r \times c$" is read "r by c"). The row and column totals in Table 8.5 are called **marginal frequencies**. Our decision to accept or reject the null hypothesis, H_0, of independence between a voter's opinion concerning the new tax reform and his or her level of income is based upon how good a fit we have between the observed frequencies in each of the 6 cells of Table 8.5 and the

Table 8.5 2 × 3 Contingency Table

Tax Reform	Income Level			Total
	Low	Medium	High	
For	182	213	203	598
Against	154	138	110	402
Total	336	351	313	1000

frequencies that we would expect for each cell under the assumption that H_0 is true. To find these expected frequencies, let us define the following events:

> L: An individual selected is in the low-income level.
>
> M: An individual selected is in the medium-income level.
>
> H: An individual selected is in the high-income level.
>
> F: An individual selected is for the new tax reform.
>
> A: An individual selected is against the new tax reform.

By using the marginal frequencies, we can list the following probability estimates:

$$P(L) = \frac{336}{1000}, \qquad P(M) = \frac{351}{1000}, \qquad P(H) = \frac{313}{1000},$$

$$P(F) = \frac{598}{1000}, \qquad P(A) = \frac{402}{1000}.$$

Now, if H_0 is true and the two variables are independent, we should have

$$P(L \cap F) = P(L)P(F) = \left(\frac{336}{1000}\right)\left(\frac{598}{1000}\right),$$

$$P(L \cap A) = P(L)P(A) = \left(\frac{336}{1000}\right)\left(\frac{402}{1000}\right),$$

$$P(M \cap F) = P(M)P(F) = \left(\frac{351}{1000}\right)\left(\frac{598}{1000}\right),$$

$$P(M \cap A) = P(M)P(A) = \left(\frac{351}{1000}\right)\left(\frac{402}{1000}\right),$$

$$P(H \cap F) = P(H)P(F) = \left(\frac{313}{1000}\right)\left(\frac{598}{1000}\right),$$

$$P(H \cap A) = P(H)P(A) = \left(\frac{313}{1000}\right)\left(\frac{402}{1000}\right).$$

The expected frequencies are obtained by multiplying each cell probability by the total number of observations. As before, we round these frequencies to one decimal. Thus the expected number of low-income voters in our sample who favor the new tax reform is estimated to be

$$\left(\frac{336}{1000}\right)\left(\frac{598}{1000}\right)(1000) = \frac{(336)(598)}{1000} = 200.9$$

when H_0 is true. The general rule for obtaining the **expected frequency** of any cell is given by the following formula:

$$Expected\ frequency = \frac{(column\ total) \times (row\ total)}{grand\ total}.$$

Table 8.6 Observed and Expected Frequencies

Tax Reform	Income Levels			Total
	Low	Medium	High	
For	182(200.9)	213(209.9)	203(187.2)	598
Against	154(135.1)	138(141.1)	110(125.8)	402
Total	336	351	313	1000

The expected frequency for each cell is recorded in parentheses beside the actual observed value in Table 8.6. Note that the expected frequencies in any row or column add up to the appropriate marginal total. In our example we need to compute only the two expected frequencies in the top row of Table 8.6 and then find the others by subtraction. The number of degrees of freedom associated with the chi-square test used here is equal to the number of cell frequencies that may be filled in freely when we are given the marginal totals and the grand total, and in this illustration that number is 2. A simple formula providing the correct number of degrees of freedom is given by

$$v = (r - 1)(c - 1).$$

Hence, for our example, $v = (2 - 1)(3 - 1) = 2$ degrees of freedom.

To test the null hypothesis of independence, we use the following decision criterion:

Test for Independence *Calculate*

$$\chi^2 = \sum_i \frac{(o_i - e_i)^2}{e_i},$$

where the summation extends over all rc cells in the r × c contingency table. If $\chi^2 > \chi_\alpha^2$ with $v = (r - 1)(c - 1)$ degrees of freedom, reject the null hypothesis of independence at the α level of significance; otherwise, accept the null hypothesis.

Applying this criterion to our example, we find that

$$\chi^2 = \frac{(182 - 200.9)^2}{200.9} + \frac{(213 - 209.9)^2}{209.9} + \frac{(203 - 187.2)^2}{187.2}$$

$$+ \frac{(154 - 135.1)^2}{135.1} + \frac{(138 - 141.1)^2}{141.1} + \frac{(110 - 125.8)^2}{125.8}$$

$$= 7.85.$$

From Table A.5 we find that $\chi_{0.05}^2 = 5.991$ for $v = (2 - 1)(3 - 1) = 2$ degrees of freedom. The null hypothesis is rejected at the 0.05 level of significance, and we conclude that a voter's opinion concerning the new tax reform and his or her level of income are not independent.

It is important to remember that the statistic on which we base our decision has a distribution that is only approximated by the chi-square distribution. The computed χ^2 values depend on the cell frequencies and consequently are discrete. The continuous chi-square distribution seems to approximate the discrete sampling distribution of X^2 very well, provided that the number of degrees of freedom is greater than 1. In a 2×2 contingency table, where we have only 1 degree of freedom, a correction called **Yate's correction for continuity** is applied. The corrected formula then becomes

$$\chi^2 \text{ (corrected)} = \sum_i \frac{(|o_i - e_i| - 0.5)^2}{e_i}.$$

If the expected cell frequencies are large, the corrected and uncorrected results are almost the same. When the expected frequencies are between 5 and 10, Yates' correction should be applied. For expected frequencies less than 5, the Fisher–Irwin exact test should be used. A discussion of this test may be found in *Basic Concepts of Probability and Statistics* by Hodges and Lehmann (see the Bibliography). The Fisher–Irwin test may be avoided, however, by choosing a larger sample.

8.11 Test for Homogeneity

When we tested for independence in Section 8.10, a random sample of 1000 voters was selected and the row and column totals of our contingency table were determined by chance. Another type of problem for which the method of Section 8.10 applies is one in which either the row or column totals are predetermined. Suppose, for example, that we decide in advance to select 200 Democrats, 150 Republicans, and 150 Independents from the voters of the state of North Carolina and record whether they are for a proposed abortion law, against it, or undecided. The observed responses are given in Table 8.7.

Table 8.7 3×3 Contingency Table

Abortion Law	Political Affiliation			Total
	Democrat	Republican	Independent	
For	82	70	62	214
Against	93	62	67	222
Undecided	25	18	21	64
Total	200	150	150	500

Now, rather than test for independence, we test the hypothesis that the population proportions within each row are the same. That is, we test the hypothesis that the proportions of Democrats, Republicans, and Independents favoring the abortion law are the same; the proportions of each political affiliation against the law are the same; and the proportions of each political affiliation that are undecided are the same. We are basically interested in determining whether the three categories of voters are **homogeneous** with respect to their opinions concerning the proposed abortion law. Such a test is called a **test for homogeneity**.

Assuming homogeneity, we again find the expected cell frequencies by multiplying the corresponding row and column totals and then dividing by the grand total. The analysis then proceeds using the same chi-square statistic as before. We illustrate this process in the following example for the data of Table 8.7.

Example 8.15 Referring to the data of Table 8.7, test the hypothesis that the opinions concerning the proposed abortion law are the same within each political affiliation. Use a 0.05 level of significance.

Solution

1. H_0: For each opinion the proportions of Democrats, Republicans, and Independents are the same.
2. H_1: For at least one opinion the proportions of Democrats, Republicans, and Independents are not the same.
3. $\alpha = 0.05$.
4. Critical region: $\chi^2 > 9.488$ with $v = 4$ degrees of freedom.
5. Computations: Using the expected cell frequency formula on page 304, we need to compute the 4 cell frequencies. All other frequencies are found by subtraction. The observed and expected cell frequencies are displayed in Table 8.8.

Table 8.8 Observed and Expected Frequencies

Abortion Law	Political Affiliation			Total
	Democrat	Republican	Independent	
For	82(85.6)	70(64.2)	59(64.2)	214
Against	93(88.8)	62(66.6)	67(66.6)	222
Undecided	25(25.6)	18(19.2)	24(19.2)	64
Total	200	150	150	500

Now,

$$\chi^2 = \frac{(82 - 85.6)^2}{85.6} + \frac{(70 - 64.2)^2}{64.2} + \frac{(59 - 64.2)^2}{64.2}$$

$$+ \frac{(93 - 88.8)^2}{88.8} + \frac{(62 - 66.6)^2}{66.6} + \frac{(67 - 66.6)^2}{66.6}$$

$$+ \frac{(25 - 25.6)^2}{25.6} + \frac{(18 - 19.2)^2}{19.2} + \frac{(24 - 19.2)^2}{19.2}$$

$$= 2.90.$$

6. Decision: Accept H_0 and conclude that the proportions of Democrats, Republicans, and Independents are the same for each stated opinion. That is, the three political affiliations are homogeneous with respect to the opinions expressed by the voters.

8.12 *Testing for Several Proportions*

The chi-square statistic for testing for homogeneity is also applicable when testing the hypothesis that k binomial parameters have the same value. This is, therefore, an extension of the test presented in Section 8.7 for determining differences between two proportions to a test for determining differences among k proportions. Hence we are interested in testing the null hypothesis

$$H_0:   p_1 = p_2 = \cdots = p_k$$

against the alternative hypothesis, H_1, that the population proportions are *not all equal*. To perform this test, we first observe independent random samples of size $n_1, n_2, \ldots, n_k$ from the k populations and arrange the data as in the $2 \times k$ contingency table, Table 8.9.

If we depend on whether the sizes of the random samples were predetermined or occurred at random, the test procedure is identical to the test for homogeneity or the test for independence. Therefore, the expected cell

Table 8.9 k Independent Binomial Samples

	Sample			
	1	2	$\cdots$	k
Successes	x_1	x_2	$\cdots$	x_k
Failures	$n_1 - x_1$	$n_2 - x_2$	$\cdots$	$n_k - x_k$

frequencies are calculated as before and substituted together with the observed frequencies into the chi-square statistic

$$\chi^2 = \sum_i \frac{(o_i - e_i)^2}{e_i},$$

with

$$v = (2 - 1)(k - 1) = k - 1$$

degrees of freedom. By selecting the appropriate upper-tail critical region of the form $\chi^2 > \chi_\alpha^2$, one can now reach a decision concerning H_0.

Example 8.16   In a shop study, a set of data was collected to determine whether or not the proportion of defectives produced by workers was the same for the day, evening, or night shift worked. The following data were collected:

	Shift		
	Day	Evening	Night
Defectives	45	55	70
Nondefectives	905	890	870

Use a 0.025 level of significance to determine if the proportion of defectives is the same for all three shifts.

Solution. Let p_1, p_2, and p_3 represent the true proportion of defectives for the day, evening, and night shift, respectively. Using the six-step procedure, we have

1. H_0: $p_1 = p_2 = p_3$.
2. H_1: p_1, p_2, and p_3 are not all equal.
3. $\alpha = 0.025$.
4. Critical region: $\chi^2 > 7.378$ for $v = 2$ degrees of freedom.
5. Computations: Corresponding to the observed frequencies $o_1 = 45$ and $o_2 = 55$, we find

$$e_1 = \frac{(950)(170)}{2835} = 57.0 \quad \text{and} \quad e_2 = \frac{(945)(170)}{2835} = 56.7.$$

All other expected frequencies are found by subtraction and are displayed in Table 8.10.

Table 8.10 Observed and Expected Frequencies

		Shift		
	Day	Evening	Night	Total
Defectives	45(57.0)	55(56.7)	70(56.3)	170
Nondefectives	905(893.0)	890(888.3)	870(883.7)	2665
Total	950	945	940	2835

Now,

$$\chi^2 = \frac{(45 - 57.0)^2}{57.0} + \frac{(55 - 56.7)^2}{56.7} + \frac{(70 - 56.3)^2}{56.3}$$

$$+ \frac{(905 - 893.0)^2}{893.0} + \frac{(890 - 888.3)^2}{888.3} + \frac{(870 - 883.7)^2}{883.7}$$

$$= 6.29.$$

6. Decision: Accept H_0 and conclude that the proportion of defectives produced is about the same for all shifts.

Exercises

1. A die is tossed 180 times with the following results:

x	1	2	3	4	5	6
f	28	36	36	30	27	23

Is this a balanced die? Use a 0.01 level of significance.

2. In 100 tosses of a coin, 63 heads and 37 tails are observed. Is this a balanced coin? Use a 0.05 level of significance.

3. A machine is supposed to mix peanuts, hazelnuts, cashews, and pecans in the ratio 5 : 2 : 2 : 1. A can containing 500 of these mixed nuts was found to have 269 peanuts, 112 hazelnuts, 74 cashews, and 45 pecans. At the 0.05 level of significance, test the hypothesis that the machine is mixing the nuts in the ratio 5 : 2 : 2 : 1.

4. The grades in a statistics course for a particular semester were as follows:

Grade	A	B	C	D	F
f	14	18	32	20	16

Test the hypothesis, at the 0.05 level of significance, that the distribution of grades is uniform.

5. Three cards are drawn from an ordinary deck of playing cards, with replacement, and the number Y of spades is recorded. After repeating the experiment 64 times, the following outcomes were

recorded:

y	0	1	2	3
f	21	31	12	0

Test the hypothesis at the 0.01 level of significance that the recorded data may be fitted by the binomial distribution $b(y; 3, 1/4)$, $y = 0, 1, 2, 3$.

6. Three marbles are selected from an urn containing 5 red marbles and 3 green marbles. After recording the number X of red marbles, the marbles are replaced in the urn and the experiment repeated 112 times. The results obtained are as follows:

x	0	1	2	3
f	1	31	55	25

Test the hypothesis at the 0.05 level of significance that the recorded data may be fitted by the hypergeometric distribution $h(x; 8, 3, 5)$, $x = 0, 1, 2, 3$.

7. A coin is thrown until a head occurs and the number X of tosses recorded. After repeating the experiment 256 times, we obtained the following results:

x	1	2	3	4	5	6	7	8
f	136	60	34	12	9	1	3	1

Test the hypothesis at the 0.05 level of significance that the observed distribution of X may be fitted by the geometric distribution $g(x; 1/2)$, $x = 1, 2, 3, \ldots$.

8. Repeat Exercise 5 using a new set of data obtained by actually carrying out the described experiment 64 times.

9. Repeat Exercise 7 using a new set of data obtained by performing the described experiment 256 times.

10. In Exercise 1 on page 59, test the goodness of fit between the observed class frequencies and the corresponding expected frequencies of a normal distribution with $\mu = 65$ and $\sigma = 21$, using a 0.05 level of significance.

11. In Exercise 5 on page 60, test the goodness of fit between the observed class frequencies and the corresponding expected frequencies of a normal distribution with $\mu = 1.8$ and $\sigma = 0.4$, using a 0.01 level of significance.

12. In an experiment to study the dependence of hypertension on smoking habits, the following data were taken on 180 individuals:

	Nonsmokers	Moderate Smokers	Heavy Smokers
Hypertension	21	36	30
No hypertension	48	26	19

Test the hypothesis that the presence or absence of hypertension is independent of smoking habits. Use a 0.05 level of significance.

13. A random sample of 90 adults are classified according to sex and the number of hours they watch television during a week:

	Sex	
	Male	Female
Over 25 hours	15	29
Under 25 hours	27	19

Use a 0.01 level of significance and test the hypothesis that the time spent watching television is independent of whether the viewer is male or female.

14. A random sample of 200 married men, all retired, were classified according to education and number of children:

	Number of Children		
Education	0–1	2–3	Over 3
Elementary	14	37	32
Secondary	19	42	17
College	12	17	10

Test the hypothesis, at the 0.05 level of significance, that the size of a family is independent of the level of education attained by the father.

15. A criminologist conducted a survey to determine whether the incidence of certain types of crime varied from one part of a large city to another. The particular crimes of interest were assault, burglary, larceny, and homicide. The following table shows the numbers of crimes committed in four areas of the city during the past year:

District	Type of Crime			
	Assault	Burglary	Larceny	Homicide
1	162	118	451	18
2	310	196	996	25
3	258	193	458	10
4	280	175	390	19

Can we conclude from these data at the 0.01 level of significance that the occurrence of these types of crime is dependent upon the city district?

16. A college infirmary conducted an experiment to determine the degree of relief provided by three cough remedies. Each cough remedy was tried on 50 students and the following data recorded:

	Cough Remedy		
	NyQuil	Robitussin	Triaminic
No relief	11	13	9
Some relief	32	28	27
Total relief	7	9	14

Test the hypothesis, at the 0.05 level of significance, that the three cough remedies are equally effective.

17. To determine current attitudes about prayers in public schools, a survey was conducted in 4 Virginia counties. The following table gives the attitudes of 200 parents from Craig County, 150 parents from Giles County, 100 parents from Franklin County, and 100 parents from Montgomery County:

Attitude	County			
	Craig	Giles	Franklin	Montgomery
Favor	65	66	40	34
Oppose	42	30	33	42
No opinion	93	54	27	24

At the 0.01 level of significance, test for homogeneity of attitudes among the 4 counties concerning prayers in the public schools.

18. According to a new Johns Hopkins University study published in the *American Journal of Public Health*, widows survive longer than widowers. Consider the following survival data collected on 100 widows and 100 widowers following the death of a spouse:

Years Lived	Widow	Widower
Less than 5	25	39
5 to 10	42	40
More than 10	33	21

Can we conclude at the 0.05 level of significance that the proportions of widows and widowers are equal with respect to the different time periods that a spouse survives after the death of his or her mate?

19. The following responses concerning the standard of living at the time of an independent opinion poll of 1000 households versus one year earlier seems to be in agreement with the results of a study published in *Across the Board* (June 1981):

Period	Standard of Living			Total
	Somewhat Better	Same	Not as Good	
1980: Jan.	72	144	84	300
May	63	135	102	300
Sept.	47	100	53	200
1981: Jan.	40	105	55	200

Test the hypothesis at the 0.05 level of significance that the proportions of households within each standard of living category are the same for each of the 4 time periods.

20. A survey was conducted in Indiana, Kentucky, and Ohio to determine the attitude of voters concerning school busing. A poll of 200 voters from each of these states yielded the following results:

	Voter Attitude		
State	Support	Do Not Support	Undecided
Indiana	82	97	21
Kentucky	107	66	27
Ohio	93	74	33

At the 0.025 level of significance, test the null hypothesis that the proportions of voters within each attitude category are the same for each of the three states.

21. A survey was conducted in two Virginia cities to determine voter sentiment for two gubernatorial candidates in an upcoming election. Five hundred voters were randomly selected from each city and the following data were recorded:

	City	
Voter Sentiment	Richmond	Norfolk
Favor A	204	225
Favor B	211	198
Undecided	85	77

At the 0.05 level of significance, test the null hypothesis that proportions of voters favoring candidate A, candidate B, or undecided are the same for each city.

22. In a study to estimate the proportion of wives who regularly watch soap operas, it is found that 52 of 200 wives in Denver, 31 of 150 wives in Phoenix, and 37 of 150 wives in Rochester watch at least one soap opera. Use a 0.05 level of significance to test the hypothesis that there is no difference among the true proportions of wives who watch soap operas in these 3 cities.

23. If a can containing 500 nuts is selected at random from each of three different distributors of mixed nuts and there are, respectively, 345, 313, and 359 peanuts in each of the cans, can we conclude at the 0.01 level of significance that the mixed nuts of the three distributors contain equal proportions of peanuts?

24. A study was made to determine whether there is a difference between the proportions of parents in the states of Maryland, Virginia, Georgia, and Alabama who favor placing Bibles in the elementary schools. The responses of 100 parents selected at random in each of these states are recorded in the following table:

Preference	State			
	Maryland	Virginia	Georgia	Alabama
Yes	65	71	78	82
No	35	29	22	18

Can we conclude that the proportions of parents who favor placing Bibles in the schools are the same for these four states? Use a 0.025 level of significance.

9

Linear Regression and Correlation

9.1 Linear Regression

Often, in practice, one is called upon to solve problems involving sets of variables when it is known that there exists some inherent relationship among the variables. For example, in an industrial situation it may be known that the tar content in the outlet stream in a chemical process is related to the inlet temperature. It may be of interest to develop a method of prediction, that is, a procedure for estimating the tar content for various levels of the inlet temperature from experimental information. The statistical aspect of the problem then becomes one of arriving at the best estimate of the relationship between the variables.

For this example and most applications there is a clear distinction between the variables as far as their role in the experimental process is concerned. Quite often there is a single *dependent variable* or response Y, which is uncontrolled in the experiment. This response depends on one or more *independent* or **regressor variables** say, $x_1, x_2, \ldots, x_k$, which are measured with negligible error and indeed are often controlled in the experiment. Thus the independent variables $x_1, x_2, \ldots, x_k$ are *not* random variables and therefore have no distributional properties. In the example cited earlier, inlet temperature is the independent variable x and tar content is the response Y. The relationship, fitted to a set of experimental data, is characterized by a prediction equation called a **regression equation**. In the case of a single Y and a single x, the situation becomes a regression of Y on x. For k independent variables, we speak in terms of a regression of Y on x_1, x_2, ..., x_k. A chemical engineer may, in fact, be concerned with the amount of hydrogen lost from samples of a particular metal when the material is placed in storage. In this case there may be two inputs, storage time x_1 in hours and storage temperature x_2 in degrees centigrade. The response would then be hydrogen loss Y in parts per million.

In this chapter we shall deal with the topic of **simple linear regression**, treating only the case of a single regressor variable. For the case of more than one regressor variable, the reader is referred to Chapter 10. Let us denote a random sample of size n by the set $\{(x_i, y_i);\ i = 1, 2, ..., n\}$. If additional samples were taken using exactly the same values of x, we should expect the y values to vary. Hence the value y_i in the ordered pair (x_i, y_i) is a value of some random variable Y_i. For convenience we define $Y | x$ to be the random variable Y corresponding to a fixed value x and denote its mean and variance by $\mu_{Y|x}$ and $\sigma^2_{Y|x}$, respectively. Clearly then, if $x = x_i$, the symbol $Y | x_i$ represents the random variable Y_i with mean $\mu_{Y|x_i}$ and variance $\sigma^2_{Y|x_i}$.

The term **linear regression** implies that $\mu_{Y|x}$ is linearly related to x by the **population regression equation**

$$\mu_{Y|x} = \alpha + \beta x$$

where the **regression coefficients** α and β are parameters to be estimated from the sample data. Denoting their estimates by a and b, respectively, we can then estimate $\mu_{Y|x}$ by $\hat{y}$ from the **sample regression line**

$$\hat{y} = a + bx,$$

where the estimates a and b represent the y intercept and slope, respectively. The symbol $\hat{y}$ is used here to distinguish between the estimated or predicted value given by the sample regression line and an actual observed experimental value y for some value of x.

One of the more challenging problems confronting the water pollution control field is presented by the tanning industry. Tannery wastes are chemically complex. They are characterized by high values of biochemical oxygen demand, volatile solids, and other pollution measures. Consider the experimental data of Table 9.1, which was obtained from 33 samples of chemically treated waste in the study "*Chemical Treatment of Spent Vegetable Tan Liquor*" conducted at the Virginia Polytechnic Institute and State University in 1970. Readings on x, the percent reduction in total solids, and y, the percent reduction in chemical oxygen demand for the 33 samples were recorded.

The data of Table 9.1 have been plotted in Figure 9.1 to give a **scatter diagram**. From an inspection of this scatter diagram, it is seen that the points follow closely a straight line, indicating that the assumption of linearity between the two variables appears to be reasonable.

The sample regression line and a hypothetical true regression line have been

Table 9.1 Measures of Solids and Chemical Oxygen Demand

Solids Reduction (%) x	Chemical Oxygen Demand (%) y
3	5
7	11
11	21
15	16
18	16
27	28
29	27
30	25
30	35
31	30
31	40
32	32
33	34
33	32
34	34
36	37
36	38
36	34
37	36
38	38
39	37
39	36
39	45
40	39
41	41
42	40
42	44
43	37
44	44
45	46
46	46
47	49
50	51

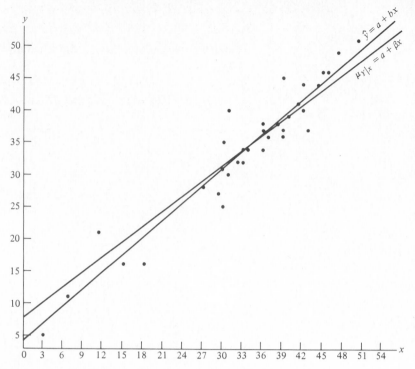

Figure 9.1   Scatter diagram with regression lines.

drawn on the scatter diagram of Figure 9.1. One expects that the agreement between the sample line and the unknown hypothetical line will be good when a large amount of data is available.

In the following section we shall develop procedures for finding estimates of the regression coefficients, α and β, in order that the regression equation can be used for predicting or estimating the mean response or an individual response for a specific value of the independent variable x.

9.2 Simple Linear Regression

In the case of simple linear regression where there is a single independent regressor variable x and a single dependent random variable Y, the data may be represented by the pairs of observations $\{(x_i, y_i); \ i = 1, 2, \ldots, n\}$. It is informative to use the concepts from the previous section to define each random variable $Y_i = Y \mid x_i$ by means of a **statistical model**. If we postulate that all means $\mu_{Y \mid x_i}$ fall on a straight line, each Y_i may be described by the **simple linear regression model**

$$Y_i = \mu_{Y \mid x_i} + E_i = \alpha + \beta x_i + E_i,$$

where the random error E_i, the model error, must necessarily have a mean of zero. Each observation (x_i, y_i) in our sample satisfies the equation

$$y_i = \alpha + \beta x_i + \varepsilon_i,$$

where ε_i is the value assumed by E_i when Y_i takes on the value y_i.

Similarly, using the estimated or fitted regression line

$$\hat{y} = a + bx,$$

each pair of observations satisfies the relation

$$y_i = a + bx_i + e_i,$$

where $e_i = y_i - \hat{y}_i$ is called a **residual** and describes the error in the fit of the model at the ith data point. The difference between e_i and ε_i is clearly shown in Figure 9.2.

We shall find a and b, the estimates of α and β, so that the sum of the squares of the residuals is a minimum. The residual sum of squares is often called the sum of squares of the errors about the regression line and denoted by SSE. This minimization procedure for estimating the parameters is called the **method of least squares**. Hence we shall find a and b so as to minimize

$$SSE = \sum_{i=1}^{n} e_i^2 = \sum_{i=1}^{n} (y_i - \hat{y}_i)^2 = \sum_{i=1}^{n} (y_i - a - bx_i)^2.$$

Differentiating SSE with respect to a and b, we have

$$\frac{\partial(SSE)}{\partial a} = -2 \sum_{i=1}^{n} (y_i - a - bx_i)$$

$$\frac{\partial(SSE)}{\partial b} = -2 \sum_{i=1}^{n} (y_i - a - bx_i)x_i.$$

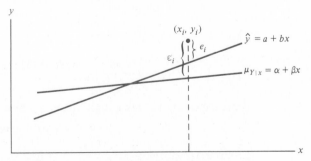

Figure 9.2 Comparing ε_i with the residual e_i.

Setting the partial derivatives equal to zero and rearranging the terms, we obtain the equations (called the **normal equations**)

$$na + b \sum_{i=1}^{n} x_i = \sum_{i=1}^{n} y_i$$

$$a \sum_{i=1}^{n} x_i + b \sum_{i=1}^{n} x_i^2 = \sum_{i=1}^{n} x_i y_i,$$

which may be solved simultaneously to yield computing formulas for a and b.

Estimating the Regression Coefficients   *Given the sample $\{(x_i, y_i); i = 1, 2, \ldots, n\}$, the least-squares estimates a and b of the regression coefficients α and β are computed from the formulas*

$$b = \frac{n \sum_{i=1}^{n} x_i y_i - \left(\sum_{i=1}^{n} x_i \right)\left(\sum_{i=1}^{n} y_i \right)}{n \sum_{i=1}^{n} x_i^2 - \left(\sum_{i=1}^{n} x_i \right)^2}$$

and

$$a = \frac{\sum_{i=1}^{n} y_i - b \sum_{i=1}^{n} x_i}{n}.$$

The calculations of a and b using the data of Table 9.1 are illustrated in the following example.

Example 9.1   Estimate the regression line for the pollution data of Table 9.1.

Solution. Using standard computer packages, we find that

$$\sum_{i=1}^{33} x_i = 1104, \qquad \sum_{i=1}^{33} y_i = 1124, \qquad \sum_{i=1}^{33} x_i y_i = 41{,}355, \qquad \sum_{i=1}^{33} x_i^2 = 41{,}086.$$

Therefore,

$$b = \frac{(33)(41{,}355) - (1104)(1124)}{(33)(41{,}086) - (1104)^2} = 0.903643$$

and

$$a = \frac{1124 - (0.903643)(1104)}{33} = 3.829640.$$

Thus the estimated regression line is given by

$$\hat{y} = 3.8296 + 0.9036x,$$

where the coefficients are rounded to four decimals. By substituting two values of x into this equation, say $x = 3$ and $x = 50$, we obtain the predicted values

$\hat{y} = 6.5$ and $\hat{y} = 49$. The sample regression line in Figure 9.1 was drawn by connecting these two points with a straight line.

Using the regression line of Example 9.1, we would predict a 31% reduction in the chemical oxygen demand when the reduction in the total solids is 30%. The 31% reduction in the chemical oxygen demand may be interpreted as an estimate of the population mean $\mu_{Y|30}$ or as an estimate of a new observation when the reduction in total solids is 30%. Such estimates, however, are subject to error. Even when the experiment is controlled so that the reduction in total solids is 30%, it is unlikely that we would measure a reduction in the chemical oxygen demand exactly equal to 31%. In fact, the original data recorded in Table 9.1 show that measurements of 25% and 35% were recorded for the reduction in oxygen demand when the reduction in total solids was kept at 30%.

Exercises

1. The study "*Development of LIFTEST, A Dynamic Technique to Assess Individual Capability to Lift Material*" was conducted at the Virginia Polytechnic Institute and State University in 1982 to determine if certain static arm strength measures have an influence on the "dynamic lift" characteristics of an individual. Twenty-five individuals were subjected to strength tests and then were asked to perform a weight-lifting test in which weight was dynamically lifted overhead. The data are as follows:

Individual	Arm Strength, x	Dynamic Lift, y
1	17.3	71.7
2	19.3	48.3
3	19.5	88.3
4	19.7	75.0
5	22.9	91.7
6	23.1	100.0
7	26.4	73.3
8	26.8	65.0
9	27.6	75.0
10	28.1	88.3
11	28.2	68.3
12	28.7	96.7

Individual	Arm Strength, x	Dynamic Lift, y
13	29.0	76.7
14	29.6	78.3
15	29.9	60.0
16	29.9	71.7
17	30.3	85.0
18	31.3	85.0
19	36.0	88.3
20	39.5	100.0
21	40.4	100.0
22	44.3	100.0
23	44.6	91.7
24	50.4	100.0
25	55.9	71.7

(a) Estimate α and β for the linear regression curve $\mu_{Y|x} = \alpha + \beta x$.

(b) Find a point estimate of $\mu_{Y|30}$.

2. The grades of a class of 9 students on a midterm report (x) and on the final examination (y) are as follows:

x	77	50	71	72	81	94	96	99	67
y	82	66	78	34	47	85	99	99	68

(a) Estimate the linear regression line.
(b) Estimate the final examination grade of a student who received a grade of 85 on the midterm report.

3. A study was made on the amount of converted sugar in a certain process at various temperatures. The data were coded and recorded as follows:

x, Temperature	y, Converted Sugar
1.0	8.1
1.1	7.8
1.2	8.5
1.3	9.8
1.4	9.5
1.5	8.9
1.6	8.6
1.7	10.2
1.8	9.3
1.9	9.2
2.0	10.5

(a) Estimate the linear regression line.
(b) Estimate the mean amount of converted sugar produced when the coded temperature is 1.75.

4. In a certain type of metal test specimen, the normal stress on a specimen is known to be functionally related to the shear resistance. The following is a set of coded experimental data on the two variables:

x, Normal Stress	y, Shear Resistance
26.8	26.5
25.4	27.3
28.9	24.2
23.6	27.1
27.7	23.6
23.9	25.9
24.7	26.3
28.1	22.5
26.9	21.7
27.4	21.4
22.6	25.8
25.6	24.9

(a) Estimate the regression line $\mu_{Y|x} = \alpha + \beta x$.
(b) Estimate the shear resistance for a normal stress of 24.5 kilograms per square centimeter.

5. The amounts of a chemical compound y, which dissolved in 100 grams of water at various temperatures, x, were recorded as follows:

x °C	y (grams)		
0	8	6	8
15	12	10	14
30	25	21	24
45	31	33	28
60	44	39	42
75	48	51	44

(a) Find the equation of the regression line.
(b) Graph the line on a scatter diagram.
(c) Estimate the amount of chemical that will dissolve in 100 grams of water at 50°C.

6. A mathematics placement test is given to all entering freshmen at a small college. A student who receives a grade below 35 is denied admission to the regular mathematics course and placed in a remedial class. The placement test scores and the final grades for 20 students who took the regular course were recorded as follows:

Placement Test	Course Grade	Placement Test	Course Grade
50	53	90	54
35	41	80	91
35	61	60	48
40	56	60	71
55	68	60	71
65	36	40	47
35	11	55	53
60	70	50	68
90	79	65	57
35	59	50	79

(a) Plot a scatter diagram.
(b) Find the equation of the regression line to predict course grades from placement test scores.
(c) Graph the line on the scatter diagram.

(d) If 60 is the minimum passing grade, below which placement test score should students in the future be denied admission to this course?

7. A study was made by a retail merchant to determine the relation between weekly advertising expenditures and sales. The following data were recorded:

Advertising Costs ($)	Sales ($)
40	385
20	400
25	395
20	365
30	475
50	440
40	490
20	420
50	560
40	525
25	480
50	510

(a) Plot a scatter diagram.
(b) Find the equation of the regression line to predict weekly sales from advertising expenditures.
(c) Estimate the weekly sales when advertising costs are $35.

8. The following data were collected to determine the relationship between high school rank in class and grade-point average at the end of the freshman year in college:

Grade-Point Average, y	Decile Rank, x	Grade-Point Average, y	Decile Rank, x
1.93	3	1.40	8
2.55	2	1.45	4
1.72	1	1.72	8
2.48	1	3.80	1
2.87	1	2.13	5
1.87	3	1.81	6
1.34	4	2.33	1
3.03	1	2.53	1
2.54	2	2.04	2
2.34	2	3.20	2

(a) Find the equation of the regression line to predict the grade-point average of college freshmen from high school rank in class.
(b) Predict the grade-point average for an entering freshman who ranks in the third decile of her graduating class.

9. In a study between the amount of rainfall and the quantity of air pollution removed, the following data were collected:

Daily Rainfall, x (0.01 centimeter)	Particulate Removed, y (micrograms per cubic meter)
4.3	126
4.5	121
5.9	116
5.6	118
6.1	114
5.2	118
3.8	132
2.1	141
7.5	108

(a) Find the equation of the regression line to predict the particulate removed from the amount of daily rainfall.
(b) Estimate the amount of particulate removed when the daily rainfall is $x = 4.8$ units.

10. The following data are the selling prices z of a certain make and model of used cars w years old:

w, Years	z, Dollars
1	6350
2	5695
2	5750
3	5395
5	4985
5	4895

(a) Fit a curve of the form $\mu_{Z|w} = \gamma \delta^w$ by means of the nonlinear sample regression equation $\hat{z} = cd^w$.
HINT: Write

$$\ln \hat{z} = \ln c + (\ln d)w$$
$$= a + bw$$

where $a = \ln c$ and $b = \ln d$, and then estimate a and b by the formulas on page 320 using the sample points $(w_i, \ln z_i)$.

(b) Estimate the selling price of such a car when it is 4 years old.

11. The pressure P of a gas corresponding to various volumes V was recorded as follows:

V (cm³)	50	60	70	90	100
P(kg/cm²)	64.7	51.3	40.5	25.9	7.8

The ideal gas law is given by the equation $PV^\gamma = C$, where γ and C are constants.

(a) Following the suggested procedure of Exercise 10, find the least squares estimates of γ and C from the given data.

(b) Estimate P when $V = 80$ cubic centimeters.

12. The following data, representing the average running speeds for various distances by 5 joggers all over 70 years of age, was published in the *American Scientist* (May-June 1981):

Distance, s (km)	1.6	3	5	10	42
Average speed, v (m/sec)	4.7	4.2	4.1	3.9	3.8

(a) Following the suggested procedure of Exercise 10, fit a curve of the form $\mu_{V|s} = \gamma s^\delta$ by means of the nonlinear sample regression equation $\hat{v} = cs^d$.

(b) Predict the average running speed for a jogger over the age of 70 who runs a distance of 10 kilometers.

9.3 Properties of the Least Squares Estimators

In addition to the assumptions that the error term in the model

$$Y_i = \alpha + \beta x_i + E_i$$

is a random variable with mean zero, suppose that we make the further assumption that each E_i has the same variance σ^2 and that $E_1, E_2, \ldots, E_n$ are independent from run to run in the experiment. With these assumptions on the E_i's, we have a procedure for finding the means and variances for the estimators of α and β.

It is important to remember that our values of a and b, based on a given sample of n observations, are only estimates of true parameters α and β. If the experiment is repeated over and over again, each time using the same fixed values of x, the resulting estimates of α and β will most likely differ from experiment to experiment. These different estimates may be thought of as values assumed by the random variables A and B.

Since the values of x remain fixed, the values of A and B depend on the variations in the values of y, or, more precisely, on the values of the random variables $Y_1, Y_2, \ldots, Y_n$. The distributional assumptions on the E_i's imply that the Y_i's, $i = 1, 2, \ldots, n$, are also independently distributed, each with mean $\mu_{Y|x_i} = \alpha + \beta x_i$ and equal variances σ^2; that is, $\sigma^2_{Y|x_i} = \sigma^2$ for $i = 1, 2, \ldots, n$.

Since the estimator

$$B = \frac{n \sum\limits_{i=1}^{n} x_i Y_i - \left(\sum\limits_{i=1}^{n} x_i \right) \left(\sum\limits_{i=1}^{n} Y_i \right)}{n \sum\limits_{i=1}^{n} x_i^2 - \left(\sum\limits_{i=1}^{n} x_i \right)^2}$$

$$= \frac{\sum\limits_{i=1}^{n} (x_i - \bar{x})(Y_i - \bar{Y})}{\sum\limits_{i=1}^{n} (x_i - \bar{x})^2}$$

$$= \frac{\sum\limits_{i=1}^{n} (x_i - \bar{x})Y_i}{\sum\limits_{i=1}^{n} (x_i - \bar{x})^2}$$

is a linear function of the random variables $Y_1, Y_2, \ldots, Y_n$ with coefficients

$$a_i = \frac{x_i - \bar{x}}{\sum\limits_{i=1}^{n} (x_i - \bar{x})^2}, \qquad i = 1, 2, \ldots, n,$$

we may conclude from Corollary 2 of Theorem 3.17 that

$$\mu_B = E(B) = \frac{\sum\limits_{i=1}^{n} (x_i - \bar{x})E(Y_i)}{\sum\limits_{i=1}^{n} (x_i - \bar{x})^2}$$

$$= \frac{\sum\limits_{i=1}^{n} (x_i - \bar{x})(\alpha + \beta x_i)}{\sum\limits_{i=1}^{n} (x_i - \bar{x})^2} = \beta$$

and then using Corollary 3 of Theorem 3.10,

$$\sigma_B^2 = \frac{\sum\limits_{i=1}^{n} (x_i - \bar{x})^2 \sigma_{Y_i}^2}{\left[\sum\limits_{i=1}^{n} (x_i - \bar{x})^2 \right]^2} = \frac{\sigma^2}{\sum\limits_{i=1}^{n} (x_i - \bar{x})^2}.$$

It can be shown (Exercise 1 on page 334) that the random variable A has the mean

$$\mu_A = \alpha$$

and variance

$$\sigma_A^2 = \frac{\displaystyle\sum_{i=1}^{n} x_i^2}{n \displaystyle\sum_{i=1}^{n} (x_i - \bar{x})^2} \, \sigma^2.$$

From the foregoing results, it becomes apparent that the least squares estimators for α and β are both unbiased estimators.

To be able to draw inferences on α and β, it becomes necessary to arrive at an estimate of the parameter σ^2 appearing in the two preceding variance formulas for A and B. The parameter σ^2, the model error variance, reflects random variation or experimental error variation, around the regression line. In deriving the estimator for σ^2, it is advantageous from a theoretical point of view, to introduce the notation

$$S_{xx} = \sum_{i=1}^{n} (x_i - \bar{x})^2 = \sum_{i=1}^{n} x_i^2 - \frac{\left(\sum_{i=1}^{n} x_i\right)^2}{n}$$

$$S_{yy} = \sum_{i=1}^{n} (y_i - \bar{y})^2 = \sum_{i=1}^{n} y_i^2 - \frac{\left(\sum_{i=1}^{n} y_i\right)^2}{n}$$

$$S_{xy} = \sum_{i=1}^{n} (x_i - \bar{x})(y_i - \bar{y}) = \sum_{i=1}^{n} x_i y_i - \frac{\left(\sum_{i=1}^{n} x_i\right)\left(\sum_{i=1}^{n} y_i\right)}{n}.$$

Now we may write the error sum of squares as follows:

$$SSE = \sum_{i=1}^{n} (y_i - a - bx_i)^2$$

$$= \sum_{i=1}^{n} [(y_i - \bar{y}) - b(x_i - \bar{x})]^2$$

$$= \sum_{i=1}^{n} (y_i - \bar{y})^2 - 2b \sum_{i=1}^{n} (x_i - \bar{x})(y_i - \bar{y}) + b^2 \sum_{i=1}^{n} (x_i - \bar{x})^2$$

$$= S_{yy} - 2bS_{xy} + b^2 S_{xx}$$

$$= S_{yy} - bS_{xy},$$

the final step following from the fact that $b = S_{xy}/S_{xx}$. We may now prove the following theorem.

Theorem 9.1   *An unbiased estimate of σ^2 is given by*

$$s^2 = \frac{SSE}{n-2} = \frac{S_{yy} - bS_{xy}}{n-2}.$$

Proof. Interpreting the error sum of squares as a random variable whose values will vary if the experiment were repeated several times, we may write

$$SSE = S_{YY} - BS_{xY}$$
$$= S_{YY} - B^2 S_{xx} \qquad \text{(since } S_{xY} = BS_{xx})$$
$$= \sum_{i=1}^{n} (Y_i - \bar{Y})^2 - B^2 \sum_{i=1}^{n} (x_i - \bar{x})^2$$
$$= \sum_{i=1}^{n} Y_i^2 - n\bar{Y}^2 - \left(\sum_{i=1}^{n} x_i^2 - n\bar{x}^2 \right) B^2.$$

Now, taking expected values, we have

$$E(SSE) = \sum_{i=1}^{n} E(Y_i^2) - nE(\bar{Y}^2) - \left(\sum_{i=1}^{n} x_i^2 - n\bar{x}^2 \right) E(B^2).$$

By Theorem 3.2 on page 85 we may substitute the quantities

$$E(Y_i^2) = \sigma_{Y_i}^2 + \mu_{Y_i}^2,$$
$$E(\bar{Y}^2) = \sigma_{\bar{Y}}^2 + \mu_{\bar{Y}}^2,$$
$$E(B^2) = \sigma_B^2 + \mu_B^2,$$

into the preceding equation to give

$$E(SSE) = \sum_{i=1}^{n} (\sigma_{Y_i}^2 + \mu_{Y_i}^2) - n(\sigma_{\bar{Y}}^2 + \mu_{\bar{Y}}^2) - \left(\sum_{i=1}^{n} x_i^2 - n\bar{x}^2 \right) (\sigma_B^2 + \mu_B^2)$$

$$= n\sigma^2 + \sum_{i=1}^{n} (\alpha + \beta x_i)^2 - n\left[\frac{\sigma^2}{n} + (\alpha + \beta\bar{x})^2 \right]$$

$$- \left(\sum_{i=1}^{n} x_i^2 - n\bar{x}^2 \right) \left(\frac{\sigma^2}{S_{xx}} + \beta^2 \right).$$

Setting $S_{xx} = \sum_{i=1}^{n} x_i^2 - n\bar{x}^2$ and simplifying, we obtain $E(SSE) = (n-2)\sigma^2$. Therefore,

$$E(S^2) = \frac{E(SSE)}{n-2} = \sigma^2$$

and s^2 is an unbiased estimate of σ^2.

9.4 *Inferences Concerning the Regression Coefficients*

Aside from merely estimating the linear relationship between x and Y for purposes of prediction, the experimenter may also be interested in drawing certain inferences about the slope and intercept. In order to allow for the testing of hypotheses and the construction of confidence intervals on α and β, one must be willing to make the further assumption that each E_i, $i = 1, 2, \ldots,$ n, is normally distributed. This assumption implies that $Y_1, Y_2, \ldots, Y_n$ are also normally distributed, each with probability distribution $n(y_i; \alpha + \beta x_i, \sigma)$. Since A and B are linear functions of independent normal variables, we may now deduce from Theorem 6.11 on page 177 that A and B are normally distributed with probability distributions $n(a; \alpha, \sigma_A)$ and $n(; \beta, \sigma_B)$ respectively.

It turns out that under the normality assumption, a result very much analogous to that given in Theorem 6.16 on page 201 allows us to conclude that $(n - 2)S^2/\sigma^2$ is a chi-square variable with $n - 2$ degrees of freedom, independent of the random variable B. Theorem 6.17 on page 203 then assures us that the statistic

$$T = \frac{(B - \beta)/(\sigma/\sqrt{S_{xx}})}{S/\sigma}$$

$$= \frac{B - \beta}{S/\sqrt{S_{xx}}}$$

has a t distribution with $n - 2$ degrees of freedom. The statistic T can be used to construct a $(1 - \alpha)100\%$ confidence interval for the coefficient β.

Confidence Interval for β   *A $(1 - \alpha)100\%$ confidence interval for the parameter β in the regression line $\mu_{Y|x} = \alpha + \beta x$ is*

$$b - \frac{t_{\alpha/2}\, s}{\sqrt{S_{xx}}} < \beta < b + \frac{t_{\alpha/2}\, s}{\sqrt{S_{xx}}},$$

where $t_{\alpha/2}$ is a value of the t distribution with $n - 2$ degrees of freedom.

Example 9.2   Find a 95% confidence interval for β in the regression line $\mu_{Y|x} = \alpha + \beta x$ based on the pollution data in Table 9.1.

Solution. In Example 9.1 we found that

$$\sum_{i=1}^{33} x_i = 1104, \qquad \sum_{i=1}^{33} x_i^2 = 41{,}086, \qquad \sum_{i=1}^{33} y_i = 1124, \qquad \sum_{i=1}^{33} x_i y_i = 41{,}355.$$

Referring to the data in Table 9.1, we now find $\sum_{i=1}^{33} y_i^2 = 41{,}998$. Therefore,

$$S_{xx} = 41{,}086 - \frac{(1104)^2}{33} = 4152.18,$$

$$S_{yy} = 41{,}998 - \frac{(1124)^2}{33} = 3713.88,$$

$$S_{xy} = 41{,}355 - \frac{(1104)(1124)}{33} = 3752.09.$$

Recall that $b = 0.903643$. Hence

$$s^2 = \frac{S_{yy} - bS_{xy}}{n-2}$$

$$= \frac{3713.88 - (0.903643)(3752.09)}{31} = 10.430004.$$

Therefore, taking the square root we obtain $s = 3.22960$. Using Table A.4, we find $t_{0.025} \simeq 2.045$ for 31 degrees of freedom. Therefore, a 95% confidence interval for β is given by

$$0.903643 - \frac{(2.045)(3.22960)}{\sqrt{4152.18}} < \beta < 0.903643 + \frac{(2.045)(3.22960)}{\sqrt{4152.18}}$$

which simplifies to

$$0.8011 < \beta < 1.0061.$$

To test the null hypothesis H_0 that $\beta = \beta_0$ against a suitable alternative, we again use the t distribution with $n-2$ degrees of freedom to establish a critical region and then base our decision on the value of

$$t = \frac{b - \beta_0}{s/\sqrt{S_{xx}}}.$$

The method is illustrated in the following example.

Example 9.3 Using the estimated value $b = 0.903643$ of Example 9.1, test the hypothesis that $\beta = 1.0$ at the 0.01 level of significance against the alternative that $\beta < 1.0$.

Solution

1. H_0: $\beta = 1.0$.
2. H_1: $\beta < 1.0$.
3. Choose a 0.01 level of significance.
4. Critical region: $t < -2.462$, approximately.

5. Computations:

$$t = \frac{0.903643 - 1.0}{3.22960/\sqrt{4152.18}} = -1.92.$$

6. Decision: Accept H_0 and conclude that β is not significantly less than 1.0 at the 0.01 level of significance.

Confidence intervals and hypothesis testing on the coefficient α may be established from the fact that A is also normally distributed. It is not difficult to show that

$$T = \frac{A - \alpha}{S\sqrt{\sum\limits_{i=1}^{n} x_i^2/nS_{xx}}}$$

has a t distribution with $n - 2$ degrees of freedom from which we may construct a $(-\alpha)100\%$ confidence interval for α.

Confidence Interval for α   *A $(1 - \alpha)100\%$ confidence interval for the parameter α in the regression line $\mu_{Y|x} = \alpha + \beta x$ is*

$$a - \frac{t_{\alpha/2}\, S\sqrt{\sum\limits_{i=1}^{n} x_i^2}}{\sqrt{nS_{xx}}} < \alpha < a + \frac{t_{\alpha/2}\, S\sqrt{\sum\limits_{i=1}^{n} x_i^2}}{\sqrt{nS_{xx}}},$$

where $t_{\alpha/2}$ is a value of the t distribution with $n - 2$ degrees of freedom.

Note that the symbol α is being used here in two totally unrelated ways, first as the level of significance and then as the intercept of the regression line.

Example 9.4   Find a 95% confidence interval for α in the regression line $\mu_{Y|x} = \alpha + \beta x$, based on the data in Table 9.1.

Solution. In Example 9.2 we found that $S_{xx} = 4152.18$ and $s = 3.22960$. From Example 9.1 we had $\sum\limits_{i=1}^{33} x_i^2 = 41{,}086$ and $a = 3.829640$. Using Table A.4, we find $t_{0.025} \simeq 2.045$ for 31 degrees of freedom. Therefore, a 95% confidence interval for α is given by

$$3.829640 - \frac{(2.045)(3.22960)\sqrt{41{,}086}}{\sqrt{(33)(4152.18)}} < \alpha < 3.829640 + \frac{(2.045)(3.22960)\sqrt{41{,}086}}{\sqrt{(33)(4152.18)}}$$

which simplifies to

$$0.2131 < \alpha < 7.4462.$$

To test the null hypothesis H_0 that $\alpha = \alpha_0$ against a suitable alternative, we can use the t distribution with $n - 2$ degrees of freedom to establish a critical region and then base our decision on the value of

$$t = \frac{a - \alpha_0}{s\sqrt{\sum\limits_{i=1}^{n} x_i^2/nS_{xx}}}.$$

Example 9.5 Using the estimated value $a = 3.829640$ in Example 9.1, test the hypothesis that $\alpha = 0$ at the 0.05 level of significance against the alternative that $\alpha \neq 0$.

Solution.

1. H_0: $\alpha = 0$.
2. H_1: $\alpha \neq 0$.
3. Choose a 0.05 level of significance.
4. Critical region: $t < -2.045$ and $t > 2.045$, approximately.
5. Computations:

$$t = \frac{3.829640 - 0}{3.22960\sqrt{41{,}086/(33)(4152.18)}} = 2.17.$$

6. Decision: Reject H_0 and conclude that $\alpha \neq 0$.

9.5 Prediction

The equation $\hat{y} = a + bx$ may be used to predict the **mean response** $\mu_{Y|x_0}$ at $x = x_0$, where x_0 is not necessarily one of the prechosen values, or it may be used to predict a single value y_0 of the variable Y_0 when $x = x_0$. We would expect the error of prediction to be higher in the case of a single predicted value than in the case where a mean is predicted. This, then, will affect the width of our intervals for the values being predicted.

Suppose that the experimenter wishes to construct a confidence interval for $\mu_{Y|x_0}$. We shall use the point estimator $\hat{Y}_0 = A + Bx_0$ to estimate $\mu_{Y|x_0} = \alpha + \beta x_0$. It can be shown that the sampling distribution of $\hat{Y}_0$ is normal with mean

$$\mu_{\hat{Y}_0} = E(\hat{Y}_0) = E(A + Bx_0) = \alpha + \beta x_0 = \mu_{Y|x_0}$$

and variance

$$\sigma_{\hat{Y}_0}^2 = \sigma_{A + Bx_0}^2 = \sigma_{\bar{Y} + B(x_0 - \bar{x})}^2$$

$$= \sigma^2\left[\frac{1}{n} + \frac{(x_0 - \bar{x})^2}{S_{xx}}\right],$$

the latter following from the fact that $\text{cov}(\bar{Y}, B) = 0$ (see Exercise 2 on page 335). Thus the $(1 - \alpha)100\%$ confidence interval on the mean response $\mu_{Y|x_0}$ can now be constructed from the statistic

$$T = \frac{\hat{Y}_0 - \mu_{Y|x_0}}{S\sqrt{(1/n) + [(x_0 - \bar{x})^2/S_{xx}]}},$$

which has a t distribution with $n - 2$ degrees of freedom.

Confidence Interval for $\mu_{Y|x_0}$ *A $(1 - \alpha)100\%$ confidence interval for the mean response $\mu_{Y|x_0}$ is given by*

$$\hat{y}_0 - t_{\alpha/2}\, s \sqrt{\frac{1}{n} + \frac{(x_0 - \bar{x})^2}{S_{xx}}} < \mu_{Y|x_0} < \hat{y}_0 + t_{\alpha/2}\, s \sqrt{\frac{1}{n} + \frac{(x_0 - \bar{x})^2}{S_{xx}}},$$

where $t_{\alpha/2}$ is a value of the t distribution with $n - 2$ degrees of freedom.

Example 9.6   Using the data of Table 9.1, construct 95% confidence limits for the mean response $\mu_{Y|x}$.

Solution. From the regression equation we find for $x_0 = 20$, say,

$$\hat{y}_0 = 3.829633 + (0.903643)(20) = 21.9025.$$

Previously, we had $\bar{x} = 33.4545$, $S_{xx} = 4152.18$, $s = 3.22960$, and $t_{0.025} \simeq 2.045$ for 31 degrees of freedom. Therefore, a 95% confidence interval for $\mu_{Y|20}$ is given by

$$21.9025 - (2.045)(3.22960)\sqrt{\frac{1}{33} + \frac{(20 - 33.4545)^2}{4152.18}} < \mu_{Y|20}$$

$$< 21.9025 + (2.045)(3.22960)\sqrt{\frac{1}{33} + \frac{(20 - 33.4545)^2}{4152.18}}$$

or simply

$$20.1071 < \mu_{Y|20} < 23.6979.$$

Repeating the previous calculations for each of several different values of x_0, one can obtain the corresponding confidence limits on each $\mu_{Y|x_0}$. Figure 9.3 displays the data points, the estimated regression line, and the upper and lower confidence limits on the mean of $Y | x$.

    Another type of interval that is often misinterpreted and confused with that given for $\mu_{Y|x}$ is the prediction interval on a future observed response. Actually, in many instances the prediction interval is more relevant to the scientist or engineer than the confidence interval on the mean. In the tar content–inlet temperature example, there would certainly be interest not only in estimating the mean tar content at a specific temperature but also in con-

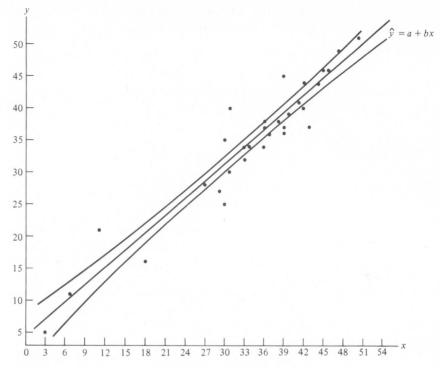

Figure 9.3 Confidence limits for the mean value of $Y \mid x$.

structing a confidence interval for predicting the actual amount of tar content at the given temperature for some future measurement.

To obtain a **prediction interval** for any single value y_0 of the variable Y_0, it is necessary to estimate the variance of the differences between the ordinates $\hat{y}_0$, obtained from the computed regression lines in repeated sampling when $x = x_0$, and the corresponding true ordinate y_0. We can think of the difference $\hat{y}_0 - y_0$ as a value of the random variable $\hat{Y}_0 - Y_0$, whose sampling distribution can be shown to be normal with mean

$$\mu_{\hat{Y}_0 - Y_0} = E(\hat{Y}_0 - Y_0)$$
$$= E[A + Bx_0 - (\alpha + \beta x_0 + E_0)]$$
$$= 0$$

and variance

$$\sigma^2_{\hat{Y}_0 - Y_0} = \sigma^2_{A + Bx_0 - E_0}$$
$$= \sigma^2_{\bar{Y} + B(x_0 - \bar{x}) - E_0}$$
$$= \sigma^2 \left[1 + \frac{1}{n} + \frac{(x_0 - \bar{x})^2}{S_{xx}} \right].$$

Thus the $(1 - \alpha)100\%$ prediction interval for a single predicted value y_0 can be constructed from the statistic

$$T = \frac{\hat{Y}_0 - Y_0}{S\sqrt{1 + (1/n) + [(x_0 - \bar{x})^2/S_{xx}]}},$$

which has a t distribution with $n - 2$ degrees of freedom.

Prediction Interval for y_0   *A $(1 - \alpha)100\%$ prediction interval for a single response y_0 is given by*

$$\hat{y}_0 - t_{\alpha/2}\, s\sqrt{1 + \frac{1}{n} + \frac{(x_0 - \bar{x})^2}{S_{xx}}} < y_0 < \hat{y}_0 + t_{\alpha/2}\, s\sqrt{1 + \frac{1}{n} + \frac{(x_0 - \bar{x})^2}{S_{xx}}},$$

where $t_{\alpha/2}$ is a value of the t distribution with $n - 2$ degrees of freedom.

Clearly, there is a distinction between the concept of a confidence interval and the prediction interval described above. The confidence interval interpretation is identical to that described for all confidence intervals on population parameters discussed throughout the book. Indeed, $\mu_{Y|x_0}$ is a population parameter. The computed prediction interval, however, represents an interval that has a probability equal to $1 - \alpha$ of containing not a parameter but a future value y_0 of the random variable Y_0.

Example 9.7   Using the data of Table 9.1, construct a 95% confidence interval for y_0 when $x_0 = 20$.

Solution. We have $n = 33$, $x_0 = 20$, $\bar{x} = 33.4545$, $\hat{y}_0 = 21.9025$, $S_{xx} = 4152.18$, $s = 3.22960$, and $t_{0.025} \simeq 2.045$ for 31 degrees of freedom. Therefore, a 95% prediction interval for y_0 is given by

$$21.9025 - (2.045)(3.22960)\sqrt{1 + \frac{1}{33} + \frac{(20 - 33.4545)^2}{4152.18}} < y_0$$

$$< 21.9025 + (2.045)(3.22960)\sqrt{1 + \frac{1}{33} + \frac{(20 - 33.4545)^2}{4152.18}},$$

which simplifies to

$$15.0583 < y_0 < 28.7467.$$

Exercises

1. Assuming that the E_i's are normal, independent with zero means and common variance σ^2, show that A, the least squares estimator of α in $\mu_{Y|x} = \alpha + \beta x$ is normally distributed with mean α and variance

$$\sigma_A^2 = \frac{\sum_{i=1}^{n} x_i^2}{n \sum_{i=1}^{n} (x_i - \bar{x})^2}\, \sigma^2.$$

2. For a simple linear regression model

$$Y_i = \alpha + \beta x_i + E_i, \qquad i = 1, 2, \ldots, n,$$

where the E_i's are independent and normally distributed with zero means and equal variances σ^2, show that $\bar{Y}$ and

$$B = \frac{\sum\limits_{i=1}^{n} (x_i - \bar{x}) Y_i}{\sum\limits_{i=1}^{n} (x_i - \bar{x})^2}$$

have zero covariance.

3. With reference to Exercise 1 on page 321,
   (a) evaluate s^2;
   (b) test the hypothesis that $\beta = 0$ against the alternative that $\beta \neq 0$ at the 0.05 level of significance and interpret the resulting decision.

4. With reference to Exercise 2 on page 321,
   (a) evaluate s^2;
   (b) construct a 95% confidence interval for α;
   (c) construct a 95% confidence interval for β.

5. With reference to Exercise 3 on page 322,
   (a) evaluate s^2;
   (b) construct a 95% confidence interval for α;
   (c) construct a 95% confidence interval for β.

6. With reference to Exercise 4 on page 322,
   (a) evaluate s^2;
   (b) construct a 99% confidence interval for α;
   (c) construct a 99% confidence interval for β.

7. With reference to Exercise 5 on page 322,
   (a) evaluate s^2;
   (b) construct a 99% confidence interval for α;
   (c) construct a 99% confidence interval for β.

8. Test the hypothesis that $\alpha = 10$ in Exercise 6 on page 322 against the alternative that $\alpha > 10$. Use a 0.05 level of significance.

9. Test the hypothesis that $\beta = 6$ in Exercise 7 on page 323 against the alternative that $\beta < 6$. Use a 0.025 level of significance.

10. Using the value of s^2 found in Exercise 4(a), construct a 95% confidence interval for $\mu_{Y|80}$ in Exercise 2 on page 321.

11. With reference to Exercise 4 on page 322, use the value of s^2 found in Exercise 6(a) to construct.
    (a) a 95% confidence interval for the mean shear resistance when $x = 24.5$;
    (b) a 95% prediction interval for a single predicted value of the shear resistance when $x = 24.5$.

12. Using the value of s^2 found in Exercise 5(a), graph the regression line and the 95% confidence bands for the mean response $\mu_{Y|x}$ for the data of Exercise 3 on page 322.

13. Using the value of s^2 found in Exercise 5(a), construct a 95% confidence interval for the amount of converted sugar corresponding to $x = 1.6$ in Exercise 3 on page 322.

14. With reference to Exercise 5 on page 322, use the value of s^2 found in Exercise 7(a) to construct
    (a) a 99% confidence interval for the average amount of chemical that will dissolve in 100 grams of water at 50°C;
    (b) a 99% prediction interval for the amount of chemical that will dissolve in 100 grams of water at 50°C.

15. With reference to Exercise 6 on page 322, use the value of s^2 found in Exercise 8 to construct
    (a) a 95% confidence interval for the average course grade of students who make a 35 on the placement test;
    (b) a 95% prediction interval for the course grade of a student who made a 35 on the placement test.

16. With reference to Exercise 7 on page 323, use the value of s^2 found in Exercise 9 to construct
    (a) a 95% confidence interval for the average weekly sales when $45 is spent on advertising;
    (b) a 95% prediction interval for the weekly sales when $45 is spent on advertising.

9.6 Choice of a Regression Model

Much of what has been presented to this point on regression involving a single independent variable depends on the assumption that the model chosen is correct, the presumption that $\mu_{Y|x}$ is related to x linearly in the parameters. Certainly, one would not expect the prediction of the response to be good if there are several independent variables, not considered in the model, that are affecting the response and are varying in the system. In addition, the prediction would certainly be inadequate if the true structure relating $\mu_{Y|x}$ to x is extremely nonlinear in the range of the variables considered.

Often the simple linear regression model is used even though it is known that the model is something other than linear or that the true structure is unknown. This approach is often sound, particularly when the range of x is narrow. Thus the model used becomes an approximating function that one hopes is an adequate representation of the true picture in the region of interest. One should note, however, the effect of an inadequate model on the results presented thus far. For example, if the true model, unknown to the experimenter, is linear in more than one x, say,

$$\mu_{Y|x_1, x_2} = \alpha + \beta x_1 + \gamma x_2,$$

then the ordinary least squares estimate $b = S_{xy}/S_{xx}$, calculated by only considering x_1 in the experiment, is, under general circumstances, a biased estimate of the coefficient β, the bias being a function of the additional coefficient γ (see Exercise 3 on page 344). Also, the estimate s^2 for σ^2 is biased due to the additional variable.

9.7 Analysis-of-Variance Approach

Often the problem of analyzing the quality of the estimated regression line is handled through an **analysis-of-variance** approach. This is merely a procedure whereby the total variation in the dependent variable is subdivided into meaningful components that are then observed and treated in a systematic fashion. The analysis of variance, discussed extensively in Chapter 11, is a powerful tool that is used in many applications.

Suppose that we have n experimental data points in the usual form (x_i, y_i) and that the regression line is estimated. In our estimation of σ^2 in Section 9.3 we established the identity

$$S_{yy} = bS_{xy} + SSE.$$

An alternative and perhaps more informative formulation is given by

$$\sum_{i=1}^{n} (y_i - \bar{y})^2 = \sum_{i=1}^{n} (\hat{y}_i - \bar{y})^2 + \sum_{i=1}^{n} (y_i - \hat{y}_i)^2.$$

So we have achieved a partitioning of the **total corrected sum of squares of** y into two components that should reflect particular meaning to the experimenter. We shall indicate this partitioning symbolically as

$$SST = SSR + SSE.$$

The first component on the right is called the **regression sum of squares** and it reflects the amount of variation in the y values **explained by the model**, in this case the postulated straight line. The second component is just the familiar error sum of squares, which reflects variation about the regression line. It can be shown that SSR/σ^2 and SSE/σ^2 are values of independent chi-square variables with 1 and $n - 2$ degrees of freedom, respectively, and then by Theorem 6.12 it follows that SST/σ^2 is also a value of a chi-square variable with $n - 1$ degrees of freedom.

Suppose that we are interested in testing the hypothesis

$$H_0: \quad \beta = 0,$$
$$H_1: \quad \beta \neq 0,$$

where the null hypothesis essentially says that the model is $\mu_{Y|x} = \alpha$ and the sample regression line has the form $\hat{y} = a$, where $a = \bar{y}$. That is, the variation in Y results from chance or random fluctuations which are independent of the values of x. To carry out such a test we compute

$$f = \frac{SSR/1}{SSE/(n - 2)} = \frac{SSR}{s^2}$$

and reject H_0 at the α level of significance when $f > f_\alpha(1, n - 2)$.

In practice, one first computes

$$SST = S_{yy}$$
$$SSR = bS_{xy}$$

and then, making use of the previous sum of squares identity, obtains

$$SSE = SST - SSR.$$

The computations are usually summarized by means of an **analysis-of-variance table** as indicated in Table 9.2. It is customary to refer to the various sum of squares divided by their respective degrees of freedom as the **mean squares**.

When the null hypothesis is rejected, that is, when the computed F statistic exceeds the critical value $f_\alpha(1, n - 2)$, we conclude that there is a significant amount of variation in the response accounted for by the postulated model, the straight-line function. If the F statistic is in the acceptance region, we

Table 9.2 Analysis of Variance for Testing $\beta = 0$

Source of Variation	Sum of Squares	Degrees of Freedom	Mean Square	Computed f
Regression	SSR	1	SSR	SSR/s^2
Error	SSE	$n-2$	$s^2 = \dfrac{SSE}{n-2}$	
Total	SST	$n-1$		

conclude that the data did not reflect sufficient evidence to support the model postulated.

In Section 9.4 a procedure was given whereby the statistic

$$T = \frac{B - \beta_0}{S/\sqrt{S_{xx}}}$$

was used to test the hypothesis

$$H_0: \quad \beta = \beta_0,$$
$$H_1: \quad \beta \neq \beta_0,$$

where T follows the t distribution with $n-2$ degrees of freedom. The hypothesis is rejected if $|t| > t_{\alpha/2}$ for an α level of significance. It is interesting to note that in the special case in which we are testing

$$H_0: \quad \beta = 0,$$
$$H_1: \quad \beta \neq 0,$$

the value of our T statistic becomes

$$t = \frac{b}{s/\sqrt{S_{xx}}}$$

and the hypothesis under consideration is identical to that being tested in Table 9.2. Namely, the null hypothesis states that the variation in the response is due merely to chance. The analysis of variance uses the F distribution rather than the t distribution, but **the two procedures are identical**. This we can see by writing

$$t^2 = \frac{b^2 S_{xx}}{s^2} = \frac{b S_{xy}}{s^2} = \frac{SSR}{s^2},$$

which is identical to the f value used in the analysis of variance. The basic relationship between the t distribution with v degrees of freedom and the F distribution with 1 and v degrees of freedom is given by

$$t_{\alpha/2}^2 = f_\alpha(1, v).$$

9.8 Test for Linearity of Regression

In certain types of experimental situations the researcher has the capability to obtain repeated observations on the response for each value of x. Although it is not necessary to have these repetitions in order to estimate α and β, nevertheless it does enable the experimenter to obtain quantitative information concerning the appropriateness of the model. In fact, if repeated observations have been generated, the experimenter can make a significance test to aid in determining whether or not the model is adequate.

Let us select a random sample of n observations using k distinct values of x, say $x_1, x_2, \ldots, x_k$, such that the sample contains n_1 observed values of the random variable Y_1 corresponding to x_1, n_2 observed values of Y_2 corresponding to $x_2, \ldots, n_k$ observed values of Y_k corresponding to x_k. Of necessity, $n = \sum\limits_{i=1}^{k} n_i$. We define

$$y_{ij} = \text{the } j\text{th value of the random variable } Y_i,$$

$$y_i. = T_i. = \sum_{j=1}^{n_i} y_{ij},$$

$$\bar{y}_i. = \frac{T_i.}{n_i}.$$

Hence, if $n_4 = 3$ measurements of Y are made corresponding to $x = x_4$, we would indicate these observations by $y_{41}, y_{42},$ and y_{43}. Then

$$T_4. = y_{41} + y_{42} + y_{43}.$$

The error sum of squares consists of two parts: the amount due to the variation between the values of Y within given values of x and a component that is normally called the **lack of fit** contribution. The first component reflects mere random variation or **pure experimental error**, while the second component is a measure of the systematic variation brought about by higher-order terms. In our case these are terms in x other than the linear or first-order contribution. Note that in choosing a linear model we are essentially assuming that this second component does not exist and hence our error sum of squares is completely due to random errors. If this should be the case, then $s^2 = SSE/(n-2)$ is an unbiased estimate of σ^2. However, if the model does not adequately fit the data, then the error sum of squares is inflated and produces a biased estimate of σ^2. Whether or not the model fits the data, an unbiased estimate of σ^2 can always be obtained when we have repeated observations simply by computing

$$s_i^2 = \frac{\sum\limits_{j=1}^{n_i} (y_{ij} - \bar{y}_{i.})^2}{n_i - 1}; \qquad i = 1, 2, \ldots, k,$$

for each of the k distinct values of x and then pooling these variances to give

$$s^2 = \frac{\sum\limits_{i=1}^{k} (n_i - 1)s_i^2}{n - k} = \frac{\sum\limits_{i=1}^{k} \sum\limits_{j=1}^{n_i} (y_{ij} - \bar{y}_{i.})^2}{n - k}.$$

The numerator of s^2 is a measure of the pure experimental error. A computational procedure for separating the error sum of squares into the two components representing pure error and lack of fit is as follows:

1. Compute the pure error sum of squares

$$\sum\limits_{i=1}^{k} \sum\limits_{j=1}^{n_i} (y_{ij} - \bar{y}_{i.})^2 = \sum\limits_{i=1}^{k} \sum\limits_{j=1}^{n_i} y_{ij}^2 - \sum\limits_{i=1}^{k} \frac{T_i^2}{n_i}.$$

This sum of squares has $n - k$ degrees of freedom associated with it and the resulting mean square is our unbiased estimate s^2 of σ^2.
2. Subtract the pure error sum of squares from the error sum of squares, SSE, thereby obtaining the sum of squares due to lack of fit. The degrees of freedom for lack of fit are also obtained by simply subtracting $(n - 2) - (n - k) = k - 2$.

The computations required for testing hypotheses in a regression problem with repeated measurements on the response may be summarized as shown in Table 9.3.

Figures 9.4 and 9.5 show a pictorial display of the sample points for the "correct model" and "incorrect model" situations. In Figure 9.4 where the

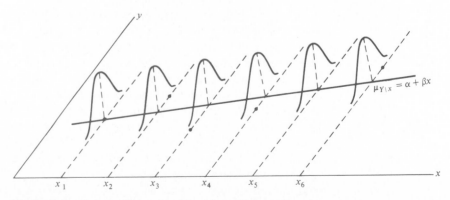

Figure 9.4   Correct linear model with no lack-of-fit component.

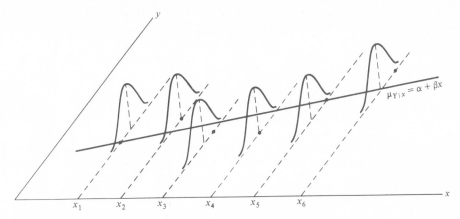

Figure 9.5 Incorrect linear model with lack-of-fit component.

$\mu_{Y|x}$ fall on a straight line, there is no lack of fit when a linear model is assumed so that the sample variation around the regression line is pure error resulting from the variation that occurs among repeated observations. In Figure 9.5 where the $\mu_{Y|x}$ clearly do not fall on a straight line, the lack of fit from erroneously choosing a linear model accounts for a large portion of the variation around the regression line in addition to the pure error.

The concept of lack of fit is extremely important in applications of regression analysis. In fact, the need to construct or design an experiment that will account for lack of fit becomes more critical as the problem and the underlying mechanism involved become more complicated. Surely, one cannot always be certain that his postulated structure, in this case the linear regression model, is correct or even an adequate representation. The following

Table 9.3 Analysis of Variance for Testing Linearity of Regression

Source of Variation	Sum of Squares	Degrees of Freedom	Mean Square	Computed f
Regression	SSR	1	SSR	$\dfrac{SSR}{s^2}$
Error	SSE	$n-2$		
Lack of fit	$\left\{\begin{array}{l} SSE - SSE(\text{pure}) \end{array}\right.$	$\left\{\begin{array}{l} k-2 \end{array}\right.$	$\dfrac{SSE - SSE(\text{pure})}{k-2}$	$\dfrac{SSE - SSE(\text{pure})}{s^2(k-2)}$
Pure error	$\left.\begin{array}{l} SSE(\text{pure}) \end{array}\right.$	$\left.\begin{array}{l} n-k \end{array}\right.$	$s^2 = \dfrac{SSE(\text{pure})}{n-k}$	
Total	SST	$n-1$		

example shows how the error sum of squares is partitioned into the two components representing pure error and lack of fit. The adequacy of the model is tested at the α level of significance by comparing the lack-of-fit mean square divided by s^2 with $f_\alpha(k - 2, n - k)$.

Example 9.8 Observations on the yield of a chemical reaction taken at various temperatures were recorded as follows:

$y(\%)$	$x(°C)$		$y(\%)$	$x(°C)$
77.4	150		88.9	250
76.7	150		89.2	250
78.2	150		89.7	250
84.1	200		94.8	300
84.5	200		94.7	300
83.7	200		95.9	300

Estimate the linear model $\mu_{Y|x} = \alpha + \beta x$ and test for lack of fit.

Solution. We have $n_1 = n_2 = n_3 = n_4 = 3$. Therefore,

$$S_{yy} = \sum_{i=1}^{4} \sum_{j=1}^{3} y_{ij}^2 - \frac{\left(\sum_{i=1}^{4} \sum_{j=1}^{3} y_{ij} \right)^2}{12}$$

$$= 90{,}265.5200 - 89{,}752.4033$$

$$= 513{,}1167$$

$$S_{xx} = \sum_{i=1}^{4} n_i x_i^2 - \frac{\left(\sum_{i=1}^{4} n_i x_i \right)^2}{12}$$

$$= 645{,}000 - 607{,}500$$

$$= 37{,}500$$

$$S_{xy} = \sum_{i=1}^{4} \sum_{j=1}^{3} x_i y_{ij} - \frac{\left(\sum_{i=1}^{4} n_i x_i \right)\left(\sum_{i=1}^{4} \sum_{j=1}^{3} y_{ij} \right)}{12}$$

$$= 237{,}875 - 233{,}505$$

$$= 4370$$

$$\bar{y} = 86.4833 \quad \text{and} \quad \bar{x} = 225.$$

The regression coefficients are then given by

$$b = \frac{4370}{37{,}500} = 0.1165$$

and

$$a = 86.4833 - (0.1165)(225) = 60.2708.$$

Hence our estimated regression line is

$$\hat{y} = 60.2708 + 0.1165x.$$

To test for lack of fit, we proceed in the usual manner:

1. H_0: the regression is linear in x.
2. H_1: the regression is nonlinear in x.
3. Choose of 0.05 level of significance.
4. Critical region: $f > 4.46$ with 2 and 8 degrees of freedom.
5. Computations: We have

$$SST = S_{yy} = 513.1167$$
$$SSR = bS_{xy} = (0.1165)(4370) = 509.1050$$
$$SSE = S_{yy} - bS_{xy} = 4.0117.$$

To compute the pure error sum of squares, we first write

$$x_1 = 150 \qquad T_1. = 232.3$$
$$x_2 = 200 \qquad T_2. = 252.3$$
$$x_3 = 250 \qquad T_3. = 267.8$$
$$x_4 = 300 \qquad T_4. = 285.4.$$

Therefore,

$$SSE(\text{pure}) = 90{,}265.52 - \frac{232.3^2 + 252.3^2 + 267.8^2 + 285.4^2}{3}$$

$$= 2.66.$$

These results and the remaining computations are exhibited in Table 9.4.

Table 9.4 Analysis of Variance on Yield-Temperature Data

Source of Variation	Sum of Squares	Degrees of Freedom	Mean Square	Computed f
Regression	509.1050	1	509.1050	1531.60
Error	4.0117	10		
Lack of fit	$\{$ 1.3517	$\{$ 2	0.6758	2.03
Pure error	2.6600	8	0.3324	
Total	513.1167	11		

6. Conclusion: The partitioning of the total variation in this manner reveals a significant variation accounted for by the linear model and an insignificant amount of variation due to lack of fit. Thus the experimental data do not seem to suggest the need to consider terms higher than first order in the model and the null hypothesis is accepted.

Exercises

1. (a) Find the least squares estimate for the parameter β in the linear equation $\mu_{Y|x} = \beta x$.
   (b) Estimate the regression line passing through the origin for the following data:

x	0.5	1.5	3.2	4.2	5.1	6.5
y	1.3	3.4	6.7	8.0	10.0	13.2

2. Suppose it is not known in Exercise 1 whether or not the true regression should pass through the origin. Estimate the general linear model $\mu_{Y|x} = \alpha + \beta x$ and test the hypothesis that $\alpha = 0$ at the 0.10 level of significance against the alternative that $\alpha \neq 0$.

3. Suppose an experimenter postulates a model of the type

   $$Y_i = \alpha + \beta x_{1i} + E_i, \qquad i = 1, 2, \ldots, n,$$

   when in fact, an additional variable, say x_2, also contributes linearly to the response. The true model is then given by

   $$Y_i = \alpha + \beta x_{1i} + \gamma x_{2i} + E_i, \qquad i = 1, 2, \ldots, n.$$

   Compute the expected value of the estimator

   $$B = \frac{\displaystyle\sum_{i=1}^{n} (x_{1i} - \bar{x}_1) Y_i}{\displaystyle\sum_{i=1}^{n} (x_{1i} - \bar{x}_1)^2}.$$

4. Use an analysis-of-variance approach to test the hypothesis that $\beta = 0$ against the alternative hypothesis $\beta \neq 0$ in Exercise 3 on page 322 at the 0.05 level of significance.

5. Organophosphate (OP) compounds are used to a large extent as pesticides. However, it is important to study their effect on species that are exposed to them. In the laboratory study "*Some Effects of Organophosphate Pesticides on Wildlife Species*," by the Department of Fisheries and Wildlife at the Virginia Polytechnic Institute and State University (1983), an experiment was conducted in which different dosages of a particular OP pesticide were administered to 5 groups of 5 mice (peromysius leucopus). The 25 mice were female of similar age and condition. One group received no chemical. The basic response (y) was a measure of activity in the brain. It was postulated that brain activity would decrease with an increase in OP dosage. The data are as follows:

Animal	Dose (mg/kg body weight), x	Activity (moles/liter minute), y
1	0.0	10.9
2	0.0	10.6
3	0.0	10.8
4	0.0	9.8
5	0.0	9.0
6	2.3	11.0
7	2.3	11.3
8	2.3	9.9
9	2.3	9.2
10	2.3	10.1
11	4.6	10.6
12	4.6	10.4

Animal	Dose (mg/kg body weight), x	Activity (moles/liter minute), y
13	4.6	8.8
14	4.6	11.1
15	4.6	8.4
16	9.2	9.7
17	9.2	7.8
18	9.2	9.0
19	9.2	8.2
20	9.2	2.3
21	18.4	2.9
22	18.4	2.2
23	18.4	3.4
24	18.4	5.4
25	18.4	8.2

(a) Using the model

$$Y_i = \alpha + \beta x_i + E_i, \qquad i = 1, 2, \ldots, 25,$$

find the least-squares estimates of α and β.

(b) Construct an analysis-of-variance table in which the lack of fit and pure error have been separated. Determine if the lack of fit is significant at the 0.05 level. Interpret the results.

6. Test for linearity of regression in Exercise 5 on page 322. Use a 0.05 level of significance.

7. Test for linearity of regression in Exercise 6 on page 322. Use a 0.05 level of significance.

8. The amounts of solids removed from a particular material when exposed to drying periods of different lengths are as follows:

x (hours)	y (grams)	
4.4	13.1	14.2
4.5	9.0	11.5
4.8	10.4	11.5
5.5	13.8	14.8
5.7	12.7	15.1
5.9	9.9	12.7
6.3	13.8	16.5
6.9	16.4	15.7
7.5	17.6	16.9
7.8	18.3	17.2

(a) Estimate the linear regression line.

(b) Test at the 0.05 level of significance whether the linear model is adequate.

9.9 Correlation

Up to this point we have assumed that the independent regressor variable x is a physical or scientific variable but not a random variable. In fact, in this context, x is often called a **mathematical variable**, which, in the sampling process, is measured with negligible error. In many applications of regression techniques it is more realistic to assume that both X and Y are random variables and the measurements $\{(x_i, y_i); i = 1, 2, \ldots, n\}$ are observations from a population having the joint density function $f(x, y)$. We shall consider the problem of measuring the relationship between the two variables X and Y. For example, if X and Y represent the length and circumference of a particular kind of bone in the adult body, we might conduct and anthropological study to determine whether large values of X are associated with large values of Y, and vice versa. On the other hand, if X represents the age of a used automobile and Y represents the retail book value of the automobile, we would expect large values of X to correspond to small values of Y and small values of X to correspond to large values of Y. **Correlation analysis** attempts to measure the strength of such relationships between two variables by means of a single number called a **correlation coefficient**.

In theory it is often assumed that the conditional distribution $f(y \mid x)$ of Y, for fixed values of X, is normal with mean $\mu_{Y|x} = \alpha + \beta x$ and variance $\sigma_{Y|x}^2 = \sigma^2$ and that X is likewise normally distributed with mean μ_X and variance σ_X^2. The joint density of X and Y is then given by

$$f(x, y) = n(y \mid x; \alpha + \beta x, \sigma) n(x; \mu_X, \sigma_X)$$

$$= \frac{1}{2\pi \sigma_X \sigma} \exp\left\{ -\left(\frac{1}{2}\right)\left[\left(\frac{y - (\alpha + \beta x)}{\sigma}\right)^2 + \left(\frac{x - \mu_X}{\sigma_X}\right)^2\right]\right\},$$

for $-\infty < x < \infty$ and $-\infty < y < \infty$.

Let us write the random variable Y in the form

$$Y = \alpha + \beta X + E,$$

where X is now a random variable independent of the random error E. Since the mean of the random error E is zero, it follows that

$$\mu_Y = \alpha + \beta \mu_X$$

and

$$\sigma_Y^2 = \sigma^2 + \beta^2 \sigma_X^2.$$

Substituting for α and σ^2 into the above expression for $f(x, y)$, we obtain the **bivariate normal distribution**

$$f(x, y) = \frac{1}{2\pi \sigma_X \sigma_Y \sqrt{1 - \rho^2}}$$

$$\times \exp\left\{ -\frac{1}{2(1 - \rho^2)}\left[\left(\frac{x - \mu_X}{\sigma_X}\right)^2 - 2\rho\left(\frac{x - \mu_X}{\sigma_X}\right)\left(\frac{y - \mu_Y}{\sigma_Y}\right) + \left(\frac{y - \mu_Y}{\sigma_Y}\right)^2\right]\right\},$$

for $-\infty < x < \infty$ and $-\infty < y < \infty$, where

$$\rho^2 = 1 - \frac{\sigma^2}{\sigma_Y^2} = \beta^2 \frac{\sigma_X^2}{\sigma_Y^2}.$$

The constant ρ (rho) is called the **population correlation coefficient** and plays a major role in many bivariate data analysis problems. It is important for the reader to understand the physical interpretation of this correlation coefficient and the distinction between correlation and regression. The term *regression* still has meaning here. In fact, the straight line given by $\mu_{Y|x} = \alpha + \beta x$ is still called the regression line as before, and the estimates of α and β are identical to those given in Section 9.2. The value of ρ is 0 when $\beta = 0$, which results when there essentially is no linear regression; that is, the regression line is horizontal and any knowledge of X is useless in predicting Y. Since $\sigma_Y^2 \geq \sigma^2$, we must have $\rho^2 \leq 1$ and hence $-1 \leq \rho \leq 1$. Values of $\rho = \pm 1$ only occur when $\sigma^2 = 0$, in which case we have a perfect linear relationship between the two variables. Thus a value of ρ equal to $+1$ implies a perfect linear relationship with a positive slope, while a value of ρ equal to -1 results from a perfect linear relationship with a negative slope. It might be said then that sample estimates of ρ close to unity in magnitude imply good correlation or **linear association** between X and Y, while values near zero indicate little or no correlation.

To obtain a sample estimate of ρ, recall from Section 9.3 that the error sum of squares is given by

$$SSE = S_{yy} - bS_{xy}.$$

Dividing both sides of this equation by S_{yy} and replacing S_{xy} by bS_{xx}, we obtain the relation

$$b^2 \frac{S_{xx}}{S_{yy}} = 1 - \frac{SSE}{S_{yy}}.$$

The value of $b^2 S_{xx}/S_{yy}$ is zero when $b = 0$, which will occur when the sample points show no linear relationship. Since $S_{yy} \geq SSE$, we conclude that $b^2 S_{xx}/S_{yy}$ must be between 0 and 1. Consequently, $b\sqrt{S_{xx}/S_{yy}}$ must range from -1 to $+1$, negative values corresponding to lines with negative slopes and positive values to lines with positive slopes. A value of -1 or $+1$ will occur when $SSE = 0$, but this is the case where all sample points lie in a straight line. Hence a perfect linear relationship exists between X and Y when $b\sqrt{S_{xx}/S_{yy}} = \pm 1$. Clearly, the quantity $b\sqrt{S_{xx}/S_{yy}}$, which we shall henceforth designate as r, can be used as an estimate of the population correlation coefficient ρ. It is customary to refer to the estimate r as the **Pearson product-moment correlation coefficient** or simply the **sample correlation coefficient**.

Correlation Coefficient   *The measure ρ of linear association between two variables X and Y is estimated by the* **sample correlation coefficient** r, *where*

$$r = b\sqrt{\frac{S_{xx}}{S_{yy}}} = \frac{S_{xy}}{\sqrt{S_{xx}S_{yy}}}.$$

For values of r between -1 and $+1$ we must be careful in our interpretation. For example, values of r equal to 0.3 and 0.6 only mean that we have two positive correlations, one somewhat stronger than the other. It is wrong to conclude that $r = 0.6$ indicates a linear relationship twice as good as that indicated by the value $r = 0.3$. On the other hand, if we write

$$r^2 = \frac{S_{xy}^2}{S_{xx}S_{yy}} = \frac{SSR}{S_{yy}},$$

then r^2, which is usually referred to as the **sample coefficient of determination**, represents the proportion of the variation of S_{yy} explained by the regression of Y on x, namely, SSR. That is, r^2 expresses the proportion of the total variation in the values of the variable Y that can be accounted for or explained by a linear relationship with the values of the random variable X. Thus a correlation of 0.6 means that 0.36 or 36% of the total variation of the values of Y in our sample is accounted for by a linear relationship with values of X.

Example 9.9   It is important that scientific researchers in the area of forest products be able to study correlation among the anatomy and mechanical properties of trees. According to the study *"Quantitative Anatomical Characteristics of Plantation Grown Loblolly Pine (Pinus Taeda L.) and Cottonwood (Populus Deltoides Bart. Ex Marsh.) and Their Relationships to Mechanical Properties"* conducted by the Department of Forestry and Forest Products at the Virginia Polytechnic Institute and State University (1983), an experiment in which 29 loblolly pines were randomly selected for investigation, yielded the following table on the specific gravity in grams/cm³ and the modulus of rupture in kilo pascals (KPa):

Specific Gravity, x (gms/cm³)	Modulus of Rupture, y (KPa)
0.414	29,186
0.383	29,266
0.399	26,215
0.402	30,162
0.442	38,867
0.422	37,831
0.466	44,576
0.500	46,097
0.514	59,698
0.530	67,705
0.569	66,088
0.558	78,486
0.577	89,869
0.572	77,369
0.548	67,095
0.581	85,156
0.557	69,571
0.550	84,160
0.531	73,466
0.550	78,610
0.556	67,657

Specific Gravity, x (gms/cm^3)	Modulus of Rupture, y (KPa)
0.523	74,017
0.602	87,291
0.569	86,836
0.544	82,540
0.557	81,699
0.530	82,096
0.547	75,657
0.585	80,490

Compute and interpret the sample correlation coefficient.

Solution. From the data we find

$$S_{xx} = 0.11273, \qquad S_{yy} = 11,807,324,786, \qquad S_{xy} = 34,422.75972.$$

Therefore,

$$r = \frac{34,422.75972}{\sqrt{(0.11273)(11,807,324,786)}} = 0.9435.$$

A correlation coefficient of 0.9435 indicates a good linear relationship between X and Y. Since $r^2 = 0.8902$, we can say that approximately 89% of the variation in the values of Y is accounted for by a linear relationship with X.

A test of the special hypothesis $\rho = 0$ versus an appropriate alternative is equivalent to testing $\beta = 0$ for the simple linear regression model and therefore the procedures of Section 9.7 using either the t distribution with $n - 2$ degrees of freedom or the F distribution with 1 and $n - 2$ degrees of freedom are applicable. However, if one wishes to avoid the analysis-of-variance procedure and compute only the sample correlation coefficient, it can easily be verified (see Exercise 3 on page 351) that the t value given by

$$t = \frac{b}{s/\sqrt{S_{xx}}} = \frac{\sqrt{SSR}}{s}$$

on page 338 can also be written as

$$t = \frac{r\sqrt{n - 2}}{\sqrt{1 - r^2}},$$

which, as before, is a value of the statistic T having a t distribution with $n - 2$ degrees of freedom.

Example 9.10 For the data of Example 9.9 test the hypothesis that there is no linear association among the variables.

Solution

1. H_0: $\rho = 0$.
2. H_1: $\rho \neq 0$.
3. $\alpha = 0.05$.
4. Critical region: $t < -2.052$ and $t > 2.052$.
5. Computations:

$$t = \frac{0.9435(\sqrt{27})}{\sqrt{1 - 0.9435^2}} = 14.79.$$

6. Decision: Reject the hypothesis of no linear association.

A test of the more general hypothesis $\rho = \rho_0$ against a suitable alternative is easily conducted from the sample information. If X and Y follow the bivariate normal distribution, the quantity

$$\frac{1}{2} \ln \left(\frac{1 + r}{1 - r} \right)$$

is a value of a random variable that follows approximately the normal distribution with mean $(1/2) \ln [(1 + \rho)/(1 - \rho)]$ and variance $1/(n - 3)$. Thus the test procedure is to compute

$$z = \frac{\sqrt{n - 3}}{2} \left[\ln \left(\frac{1 + r}{1 - r} \right) - \ln \left(\frac{1 + \rho_0}{1 - \rho_0} \right) \right]$$

$$= \frac{\sqrt{n - 3}}{2} \ln \left[\frac{(1 + r)(1 - \rho_0)}{(1 - r)(1 + \rho_0)} \right]$$

and compare with the critical points of the standard normal distribution.

Example 9.11   For the data of Example 9.9, test the null hypothesis that $\rho = 0.9$ against the alternative that $\rho \neq 0.9$. Use a 0.05 level of significance.

Solution

1. H_0: $\rho = 0.9$.
2. H_1: $\rho \neq 0.9$.
3. $\alpha = 0.05$.
4. Critical region: $z < -1.96$ and $z > 1.96$.
5. Computations:

$$z = \frac{\sqrt{26}}{2} \ln \left[\frac{(1 + 0.9435)0.1}{1 - 0.9435} \right] = 3.15.$$

6. Decision: H_0 is rejected. The population correlation coefficient differs significantly from the value of 0.9.

It should be pointed out that in correlation studies, as in linear regression problems, the results obtained are only as good as the model that is assumed. In the correlation techniques studied here, a bivariate normal density is

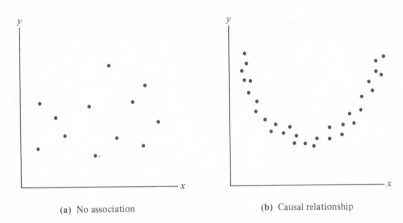

(a) No association   (b) Causal relationship

Figure 9.6   Scatter diagram showing zero correlation.

assumed for the variables X and Y, with the mean value of Y at each x value being linearly related to x. To observe the suitability of the linearity assumption, a preliminary plotting of the experimental data is often helpful. A value of the sample correlation coefficient close to zero will result from data that display a strictly random effect as in Figure 9.6(a), thus implying little or no causal relationship. It is important to remember that the correlation coefficient between two variables is a measure of their linear relationship, and that a value of $r = 0$ implies a lack of linearity and not a lack of association. Hence, if a strong quadratic relationship exists between X and Y as indicated in Figure 9.6(b), we shall still obtain a zero correlation indicating a nonlinear relationship.

Exercises

1. Compute and interpret the correlation coefficient for the following grades of 6 students selected at random:

Mathematics grade	70	92	80	74	65	83
English grade	74	84	63	87	78	90

2. Test the hypothesis that $\rho = 0$ in Exercise 1 against the alternative that $\rho \neq 0$. Use a 0.05 level of significance.

3. Show the necessary steps in converting the equation

$$t = \frac{b}{s/\sqrt{S_{xx}}}$$

to the equivalent form

$$t = \frac{r\sqrt{n-2}}{\sqrt{1-r^2}}.$$

4. The following data were obtained in a study of the relationship between the weight and chest size of infants at birth:

Weight (kg)	Chest Size (cm)
2.75	29.5
2.15	26.3
4.41	32.2
5.52	36.5

Weight (kg)	Chest Size (cm)
3.21	27.2
4.32	27.7
2.31	28.3
4.30	30.3
3.71	28.7

(a) Calculate r.

(b) Test the null hypothesis that $\rho = 0$ against the alternative that $\rho > 0$ at the 0.01 level of significance.

(c) What percentage of the variation in the infant chest sizes is explained by differences in weight?

5. With reference to Exercise 1 on page 321,
   (a) calculate r;
   (b) test the hypothesis that $\rho = 0$ against the alternative that $\rho \neq 0$ at the 0.05 level of significance.

6. With reference to Exercise 9 on page 323.
   (a) calculate r;
   (b) test the null hypothesis that $\rho = -0.5$ against the alternative that $\rho < -0.5$ at the 0.025 level of significance;
   (c) determine the percentage of the variation in the amount of particulate removed that is due to changes in the daily amount of rainfall.

7. The following data represent the chemistry grades for a random sample of 12 freshmen at a certain college along with their scores on an intelligence test administered while they were still seniors in high school:

Student	Test Score, x	Chemistry Grade, y
1	65	85
2	50	74
3	55	76
4	65	90
5	55	85
6	70	87
7	65	94
8	70	98
9	55	81
10	70	91
11	50	76
12	55	74

(a) Compute and interpret the sample correlation coefficient.

(b) Test the hypothesis that $\rho = 0.5$ against the alternative that $\rho > 0.5$ at the 0.01 level of significance.

10

Multiple Linear Regression

10.1 Introduction

In most research problems where regression analysis is applied, more than one independent variable is needed in the regression model. The complexity of most scientific mechanisms is such that in order to be able to predict an important response, a **multiple regression model** is needed. When this model is linear in the coefficients, it is called a **multiple linear regression model**. For the case of k independent variables $x_1, x_2, \ldots, x_k$, the mean of $Y \mid x_1, x_2, \ldots, x_k$ is given by the multiple linear regression model

$$\mu_{Y \mid x_1, x_2, \ldots, x_k} = \beta_0 + \beta_1 x_1 + \cdots + \beta_k x_k$$

and the estimated response is obtained from the sample regression equation

$$\hat{y} = b_0 + b_1 x_1 + \cdots + b_k x_k,$$

where each regression coefficient β_i is estimated by b_i from the sample data using the method of least squares. As in the case of a single independent variable, the multiple linear regression model can often be an adequate representation of a more complicated structure within certain ranges of the independent variables.

Similar least squares techniques can also be applied in estimating the coefficients when the linear model involves, say, powers and products of the independent variables. For example, when $k = 1$, the experimenter may feel that the means $\mu_{Y|x}$ do not fall on a straight line but are more appropriately described by the **polynomial regression model**

$$\mu_{Y|x} = \beta_0 + \beta_1 x + \beta_2 x^2 + \cdots + \beta_r x^r,$$

and the estimated response is obtained from the polynomial regression equation

$$\hat{y} = b_0 + b_1 x + b_2 x^2 + \cdots + b_r x^r.$$

Confusion arises occasionally when we speak of a polynomial model as a linear model. However, statisticians normally refer to a linear model as one in which the parameters occur linearly, regardless of how the independent variables enter the model. An example of a nonlinear model is the **exponential relationship** given by

$$\mu_{Y|x} = \alpha \beta^x,$$

which is estimated by the regression equation

$$\hat{y} = ab^x.$$

There are many phenomena in science and engineering that are inherently nonlinear in nature and, when the true structure is known, an attempt should certainly be made to fit the actual model. The literature on estimation by least squares of nonlinear models is voluminous and, except for Exercises 10, 11, and 12 on page 323, we do not attempt to cover the subject in this book. A student who wants a good account of some aspects of this subject should consult *Applied Regression Analysis* by Draper and Smith (see the Bibliography).

10.2 *Estimating the Coefficients*

In this section we shall obtain the least squares estimates of the parameters β_0, $\beta_1, \ldots, \beta_k$ by fitting the multiple linear regression model

$$\mu_{Y|x_1, x_2, \ldots, x_k} = \beta_0 + \beta_1 x_1 + \beta_2 x_2 + \cdots + \beta_k x_k$$

to the data points

$$\{(x_{1i}, x_{2i}, \ldots, x_{ki}, y_i); \quad i = 1, 2, \ldots, n \text{ and } n > k\},$$

where y_i is the observed response to the values $x_{1i}, x_{2i}, \ldots, x_{ki}$ of the k independent variables $x_1, x_2, \ldots, x_k$. Each observation $(x_{1i}, x_{2i}, \ldots, x_{ki}, y_i)$ satisfies the equation

$$y_i = \beta_0 + \beta_1 x_{1i} + \beta_2 x_{2i} + \cdots + \beta_k x_{ki} + \varepsilon_i$$

or

$$y_i = b_0 + b_1 x_{1i} + b_2 x_{2i} + \cdots + b_k x_{ki} + e_i,$$

where ε_i and e_i are the random error and residual, respectively, associated with the response y_i. In using the concept of least squares to arrive at estimates b_0, $b_1, \ldots, b_k$, we minimize the expression

$$SSE = \sum_{i=1}^{n} e_i^2 = \sum_{i=1}^{n} (y_i - b_0 - b_1 x_{1i} - b_2 x_{2i} - \cdots - b_k x_{ki})^2.$$

Differentiating SSE in turn with respect to b_0, b_1, b_2, $\ldots$, b_k, and equating to zero, we generate the set of $k + 1$ normal equations

$$nb_0 + b_1 \sum_{i=1}^{n} x_{1i} + b_2 \sum_{i=1}^{n} x_{2i} + \cdots + b_k \sum_{i=1}^{n} x_{ki} = \sum_{i=1}^{n} y_i$$

$$b_0 \sum_{i=1}^{n} x_{1i} + b_1 \sum_{i=1}^{n} x_{1i}^2 + b_2 \sum_{i=1}^{n} x_{1i} x_{2i} + \cdots + b_k \sum_{i=1}^{n} x_{1i} x_{ki} = \sum_{i=1}^{n} x_{1i} y_i$$

$$\vdots \qquad \vdots \qquad \vdots \qquad \vdots \qquad \vdots$$

$$b_0 \sum_{i=1}^{n} x_{ki} + b_1 \sum_{i=1}^{n} x_{ki} x_{1i} + b_2 \sum_{i=1}^{n} x_{ki} x_{2i} + \cdots + b_k \sum_{i=1}^{n} x_{ki}^2 = \sum_{i=1}^{n} x_{ki} y_i.$$

These equations can be solved for b_0, b_1, $b_2, \ldots, b_k$ by any appropriate method for solving systems of linear equations.

Example 10.1 In analytical chemistry, X-ray fluorescence analysis is a tool for estimating ingredient percentages in multicomponent mixtures. Often the estimation of concentrations depends heavily on the user's ability to fit suitable regression models. In the paper "*Corrections for Matrix Effects in X-Ray Fluorescence Analysis Using Multiple Regression Methods,*" published in *Analytical Chemistry* (Vol. 37, 1965), propellant slurries containing 4 ingredients were prepared. Concentrations of the components varied in the slurries to produce calibration type standards. The data are as follows:

y	x_1	x_2	x_3	x_4
0.5514	1.1240	0.8980	0.8219	0.9906
0.4426	0.9285	0.8872	0.9308	0.9944
0.5631	1.1214	0.8030	0.7668	1.1221
0.5624	1.1635	0.8706	0.9272	0.9832
0.4505	0.9415	0.8064	0.9026	1.1127
0.5290	1.0712	0.8404	0.8662	1.0836
0.4702	0.9561	0.8731	0.8206	1.0290
0.5001	1.0186	0.8431	0.8346	1.0591
0.4425	0.9039	0.8314	0.7596	1.0994

The response y_i is the measured concentration of an ingredient A. The measured value x_1 is the X-ray "intensity ratio" associated with ingredient A and the values x_2, x_3, and x_4 are intensity ratios for the additional components in the slurry. As a result of enhancement and absorption effects, response y is best predicted after regressing it against intensity values associated with *all* components. Thus the model is given by

$$\mu_{Y|x_1, x_2, x_3, x_4} = \beta_0 + \beta_1 x_1 + \beta_2 x_2 + \beta_3 x_3 + \beta_4 x_4 .$$

Fit this multiple linear regression model to the given data and then estimate the concentration of ingredient A for a mixture whose X-ray intensity ratios are, respectively, $x_1 = 1.091$, $x_2 = 0.855$, $x_3 = 0.758$, and $x_4 = 1.005$.

Solution. From the given data we find $n = 9$ and

$$\sum_{i=1}^{9} x_{1i} = 9.2287 \qquad \sum_{i=1}^{9} x_{2i} = 7.6532 \qquad \sum_{i=1}^{9} x_{3i} = 7.6303$$

$$\sum_{i=1}^{9} x_{4i} = 9.4741 \qquad \sum_{i=1}^{9} x_{1i}^2 = 9.5394 \qquad \sum_{i=1}^{9} x_{2i}^2 = 6.5172$$

$$\sum_{i=1}^{9} x_{3i}^2 = 6.5015 \qquad \sum_{i=1}^{9} x_{4i}^2 = 9.9974 \qquad \sum_{i=1}^{9} x_{1i} x_{2i} = 7.8510$$

$$\sum_{i=1}^{9} x_{1i} x_{3i} = 7.8257 \qquad \sum_{i=1}^{9} x_{1i} x_{4i} = 9.7037 \qquad \sum_{i=1}^{9} x_{2i} x_{3i} = 6.4943$$

$$\sum_{i=1}^{9} x_{2i} x_{4i} = 8.0421 \qquad \sum_{i=1}^{9} x_{3i} x_{4i} = 8.0182 \qquad \sum_{i=1}^{9} y_i = 4.5118$$

$$\sum_{i=1}^{9} x_{1i} y_i = 4.6663 \qquad \sum_{i=1}^{9} x_{2i} y_i = 3.8375 \qquad \sum_{i=1}^{9} x_{3i} y_i = 3.8226$$

$$\sum_{i=1}^{9} x_{4i} y_i = 4.7456 .$$

Inserting these values into the normal equations above, we obtain

$$9b_0 + 9.2287b_1 + 7.6532b_2 + 7.6303b_3 + 9.4741b_4 = 4.5118$$
$$9.2287b_0 + 9.5394b_1 + 7.8510b_2 + 7.8257b_3 + 9.7037b_4 = 4.6663$$
$$7.6532b_0 + 7.8510b_1 + 6.5172b_2 + 6.9434b_3 + 8.0421b_4 = 3.8375$$
$$7.6303b_0 + 7.8257b_1 + 6.4943b_2 + 6.5015b_3 + 8.0182b_4 = 3.8226$$
$$9.4741b_0 + 9.7031b_1 + 8.0421b_2 + 8.0182b_3 + 9.9974b_4 = 4.7456 .$$

The solution of this set of equations yields the unique estimates

$$b_0 = -0.3004, \qquad b_1 = 0.5387, \qquad b_2 = 0.1770,$$
$$b_3 = -0.0704, \qquad b_4 = 0.1506 .$$

Therefore, our regression equation is

$$\hat{y} = -0.3004 + 0.5387x_1 + 0.1770x_2 - 0.0704x_3 + 0.1506x_4 .$$

For a mixture whose X-ray intensities are $x_1 = 1.091$, $x_2 = 0.855$, $x_3 = 0.758$, and $x_4 = 1.005$, the estimated concentration of component A is given by

$$y = -0.3004 + (0.5387)(1.091) + (0.1770)(0.855)$$
$$-(0.0704)(0.758) + (0.1506)(1.005)$$
$$= 0.5366.$$

Now suppose that we wish to fit the polynomial equation

$$\mu_{Y|x} = \beta_0 + \beta_1 x + \beta_2 x^2 + \cdots + \beta_r x^r$$

to the n pairs of observations $\{(x_i, y_i); i = 1, 2, \ldots, n\}$. Each observation, y_i, satisfies the equation

$$y_i = \beta_0 + \beta_1 x_i + \beta_2 x_i^2 + \cdots \beta_r x_i^r + \varepsilon_i$$

or

$$y_i = b_0 + b_1 x_i + b_2 x_i^2 + \cdots b_r x_i^r + e_i,$$

where r is the degree of the polynomial, and ε_i and e_i are again the random error and residual associated with the response y_i. Here, the number of pairs, n, must be at least as large as $r + 1$, the number of parameters to be estimated. Notice that the polynomial model can be considered a special case of the more general multiple linear regression model, where we set $x_1 = x$, $x_2 = x^2$, ..., $x_r = x^r$. The normal equations assume the form

$$nb_0 + b_1 \sum_{i=1}^{n} x_i + b_2 \sum_{i=1}^{n} x_i^2 + \cdots + b_r \sum_{i=1}^{n} x_i^r = \sum_{i=1}^{n} y_i$$

$$b_0 \sum_{i=1}^{n} x_i + b_1 \sum_{i=1}^{n} x_i^2 + b_2 \sum_{i=1}^{n} x_i^3 + \cdots + b_r \sum_{i=1}^{n} x_i^{r+1} = \sum_{i=1}^{n} x_i y_i$$

$$\vdots \qquad \vdots \qquad \vdots \qquad \vdots \qquad \vdots$$

$$b_0 \sum_{i=1}^{n} x_i^r + b_1 \sum_{i=1}^{n} x_i^{n+1} + b_2 \sum_{i=1}^{n} x_i^{r+2} + \cdots + b_r \sum_{i=1}^{n} x_i^{2r} = \sum_{i=1}^{n} x_i^r y_i,$$

which are solved as before for $b_0, b_1, \ldots, b_r$.

Example 10.2 Given the data

x	0	1	2	3	4	5	6	7	8	9
y	9.1	7.3	3.2	4.6	4.8	2.9	5.7	7.1	8.8	10.2

fit a regression curve of the form $\mu_{Y|x} = \beta_0 + \beta_1 x + \beta_2 x^2$ and then estimate $\mu_{Y|2}$.

Solution. From the data given, we find

$$\sum_{i=1}^{10} x_i = 45 \qquad \sum_{i=1}^{10} x_i^2 = 297 \qquad \sum_{i=1}^{10} x_i^3 = 2025 \qquad \sum_{i=1}^{10} x_i^4 = 15{,}332$$

$$\sum_{i=1}^{10} y_i = 53.7 \qquad \sum_{i=1}^{10} x_i y_i = 307.3 \qquad \sum_{i=1}^{10} x_i^2 y_i = 2153.3 \qquad n = 10.$$

Solving the normal equations

$$\begin{aligned}
10b_0 + 45b_1 + 297b_2 &= 53.7 \\
45b_0 + 297b_1 + 2025b_2 &= 307.3 \\
297b_0 + 2025b_1 + 15{,}332b_2 &= 2153.3,
\end{aligned}$$

we obtain

$$b_0 = 8.697, \qquad b_1 = -2.341, \qquad b_2 = 0.288.$$

Therefore,

$$\hat{y} = 8.697 - 2.341x + 0.288x^2.$$

When $x = 2$ our estimate of $\mu_{Y|2}$ is

$$\begin{aligned}
\hat{y} &= 8.697 - (2.341)(2) + (0.288)(2^2) \\
&= 5.2.
\end{aligned}$$

10.3 *Estimating the Coefficients Using Matrices*

In fitting a multiple linear regression model, particularly when the number of variables exceeds two, a knowledge of matrix theory can facilitate the mathematical manipulations considerably. Suppose that the experimenter has k independent variables $x_1, x_2, \ldots, x_k$ and n observations $y_1, y_2, \ldots, y_n$, each of which can be expressed by the equation

$$y_i = b_0 + b_1 x_{1i} + b_2 x_{2i} + \cdots + b_k x_{ki} + e_i.$$

The least squares estimates are obtained by solving the set of normal equations developed in Section 10.2.

To solve these normal equations using matrix theory, we need to introduce a vector **y** of the observations, a vector **b** of the least squares estimates, and the **design matrix X**:

$$\mathbf{y} = \begin{bmatrix} y_1 \\ y_2 \\ \vdots \\ y_n \end{bmatrix}, \qquad \mathbf{b} = \begin{bmatrix} b_0 \\ b_1 \\ \vdots \\ b_k \end{bmatrix}, \qquad \mathbf{X} = \begin{bmatrix} 1 & x_{11} & x_{21} & \cdots & x_{k1} \\ 1 & x_{12} & x_{22} & \cdots & x_{k2} \\ \vdots & \vdots & \vdots & & \vdots \\ 1 & x_{1n} & x_{2n} & \cdots & x_{kn} \end{bmatrix}.$$

Apart from the initial element, the ith row of $\mathbf{X}$ represents those x values that give rise to the response y_i. Writing

$$
\mathbf{A} = \mathbf{X'X} = \begin{bmatrix}
n & \sum_{i=1}^{n} x_{1i} & \sum_{i=1}^{n} x_{2i} & \cdots & \sum_{i=1}^{n} x_{ki} \\
\sum_{i=1}^{n} x_{1i} & \sum_{i=1}^{n} x_{1i}^2 & \sum_{i=1}^{n} x_{1i} x_{2i} & \cdots & \sum_{i=1}^{n} x_{1i} x_{ki} \\
\vdots & \vdots & \vdots & & \vdots \\
\sum_{i=1}^{n} x_{ki} & \sum_{i=1}^{n} x_{ki} x_{1i} & \sum_{i=1}^{n} x_{ki} x_{2i} & \cdots & \sum_{i=1}^{n} x_{ki}^2
\end{bmatrix},
$$

and

$$
\mathbf{g} = \mathbf{X'y} = \begin{bmatrix}
g_0 = \sum_{i=1}^{n} y_i \\
g_1 = \sum_{i=1}^{n} x_{1i} y_i \\
\vdots \\
g_k = \sum_{i=1}^{n} x_{ki} y_i
\end{bmatrix},
$$

the normal equations can be put in the matrix form

$$
\mathbf{Ab} = \mathbf{g}.
$$

If the matrix $\mathbf{A}$ is nonsingular, we can write the solution for the regression coefficients as

$$
\mathbf{b} = \mathbf{A}^{-1}\mathbf{g} = (\mathbf{X'X})^{-1}\mathbf{X'y}.
$$

Thus one can obtain the prediction equation or regression equation by solving a set of $k + 1$ equations in a like number of unknowns. This involves the inversion of the $k + 1$ by $k + 1$ matrix $\mathbf{A}$. Techniques for inverting this matrix are explained in most textbooks on elementary determinants and matrices. Of course, there are many high-speed computer packages available for multiple regression problems, packages that not only print out estimates of the regression coefficients, but also provide other information relevant to making inferences concerning the regression equation.

Example 10.3 The percent survival of a certain type of animal semen after storage was measured at various combinations of concentrations of three materials used to increase chance of survival. The data are as follows:

y (% survival)	x_1 (weight %)	x_2 (weight %)	x_3 (weight %)
25.5	1.74	5.30	10.80
31.2	6.32	5.42	9.40
25.9	6.22	8.41	7.20
38.4	10.52	4.63	8.50
18.4	1.19	11.60	9.40
26.7	1.22	5.85	9.90
26.4	4.10	6.62	8.00
25.9	6.32	8.72	9.10
32.0	4.08	4.42	8.70
25.2	4.15	7.60	9.20
39.7	10.15	4.83	9.40
35.7	1.72	3.12	7.60
26.5	1.70	5.30	8.20

Estimate the multiple linear regression model for the given data.

Solution. From the experiment data we list the following sums of squares and products:

$$\sum_{i=1}^{13} y_i = 377.5 \qquad \sum_{i=1}^{13} y_i^2 = 11{,}400.15 \qquad \sum_{i=1}^{13} x_{1i} = 59.43$$

$$\sum_{i=1}^{13} x_{2i} = 81.82 \qquad \sum_{i=1}^{13} x_{3i} = 115.40 \qquad \sum_{i=1}^{13} x_{1i}^2 = 394.7255$$

$$\sum_{i=1}^{13} x_{2i}^2 = 576.7264 \qquad \sum_{i=1}^{13} x_{3i}^2 = 1035.9600 \qquad \sum_{i=1}^{13} x_{1i} y_i = 1877.567$$

$$\sum_{i=1}^{13} x_{2i} y_i = 2246.661 \qquad \sum_{i=1}^{13} x_{3i} y_i = 3337.780 \qquad \sum_{i=1}^{13} x_{1i} x_{2i} = 360.6621$$

$$\sum_{i=1}^{13} x_{1i} x_{3i} = 522.0780 \qquad \sum_{i=1}^{13} x_{2i} x_{3i} = 728.3100 \qquad n = 13.$$

The least squares estimating equations, $\mathbf{Ab} = \mathbf{g}$, are given by

$$\begin{bmatrix} 13 & 59.43 & 81.82 & 115.40 \\ 59.43 & 394.7255 & 360.6621 & 522.0780 \\ 81.82 & 360.6621 & 576.7264 & 728.3100 \\ 15.40 & 522.0780 & 728.3100 & 1035.9600 \end{bmatrix} \begin{bmatrix} b_0 \\ b_1 \\ b_2 \\ b_3 \end{bmatrix} = \begin{bmatrix} 377.5 \\ 1877.567 \\ 2246.661 \\ 3337.780 \end{bmatrix} .$$

From a computer readout we obtain the elements of the inverse matrix

$$\mathbf{A}^{-1} = \begin{bmatrix} 8.0648 & -0.0826 & -0.0942 & -0.7905 \\ -0.0826 & 0.0085 & 0.0017 & 0.0037 \\ -0.0942 & 0.0017 & 0.0166 & -0.0021 \\ -0.7905 & 0.0037 & -0.0021 & 0.0886 \end{bmatrix},$$

and then using the relation $\mathbf{b} = \mathbf{A}^{-1}\mathbf{g}$ the computer printout gives

$$b_0 = 39.1574, \quad b_1 = 1.0161, \quad b_2 = -1.8616, \quad b_3 = -0.3433.$$

Hence our estimated regression equation is

$$\hat{y} = 39.1574 + 1.0161x_1 - 1.8616x_2 - 0.3433x_3.$$

For the case of a single independent variable, the *degree* of the best fitting polynomial can often be determined by plotting a scatter diagram of the data obtained from an experiment that yields n pairs of observations of the form $\{(x_i, y_i); i = 1, 2, \ldots, n\}$. The set of normal equations derived in Section 10.2 for estimating the coefficients of the general polynomial of degree r assumes the form

$$\begin{bmatrix} n & \sum_{i=1}^{n} x_i & \sum_{i=1}^{n} x_i^2 & \cdots & \sum_{i=1}^{n} x_i^r \\ \sum_{i=1}^{n} x_i & \sum_{i=1}^{n} x_i^2 & \sum_{i=1}^{n} x_i^3 & \cdots & \sum_{i=1}^{n} x_i^{r+1} \\ \sum_{i=1}^{n} x_i^2 & \sum_{i=1}^{n} x_i^3 & \sum_{i=1}^{n} x_i^4 & \cdots & \sum_{i=1}^{n} x_i^{r+2} \\ \vdots & \vdots & \vdots & & \vdots \\ \sum_{i=1}^{n} x_i^r & \sum_{i=1}^{n} x_i^{r+1} & \sum_{i=1}^{n} x_i^{r+2} & \cdots & \sum_{i=1}^{n} x_i^{2r} \end{bmatrix} \begin{bmatrix} b_0 \\ b_1 \\ b_2 \\ \vdots \\ b_r \end{bmatrix} = \begin{bmatrix} \sum_{i=1}^{n} y_i \\ \sum_{i=1}^{n} x_i y_i \\ \vdots \\ \sum_{i=1}^{n} x_i^r y_i \end{bmatrix}.$$

Solving these $r + 1$ equations, we obtain the estimates $b_0, b_1, \ldots, b_r$ and thereby generate the polynomial regression prediction equation

$$\hat{y} = b_0 + b_1 x + b_2 x^2 + \cdots + b_r x^r.$$

The procedure for fitting a polynomial regression model can be generalized to the case of more than one independent variable. In fact, the student of regression analysis should at this stage have the facility for fitting any linear model in, say, k independent variables. Suppose, for example, that we have a response Y with $k = 2$ independent variables and a quadratic model is postulated of the type

$$y_i = \beta_0 + \beta_1 x_{1i} + \beta_2 x_{2i} + \beta_{11} x_{1i}^2 + \beta_{22} x_{2i}^2 + \beta_{12} x_{1i} x_{2i} + \varepsilon_i,$$

where y_i, $i = 1, 2, \ldots, n$, is the response to the combination (x_{1i}, x_{2i}) of the independent variables in the experiment. In this situation n must be at least 6, since there are six parameters to estimate by the least squares procedure. In addition, since the model contains quadratic terms in both variables, at least three levels of each variable must be used. The reader should easily verify that the least squares normal equations $(X'X)b = g$ are given by

$$
\begin{bmatrix}
n & \sum_{i=1}^{n} x_{1i} & \sum_{i=1}^{n} x_{2i} & \sum_{i=1}^{n} x_{1i}^2 & \sum_{i=1}^{n} x_{2i}^2 & \sum_{i=1}^{n} x_{1i}x_{2i} \\[2mm]
\sum_{i=1}^{n} x_{1i} & \sum_{i=1}^{n} x_{1i}^2 & \sum_{i=1}^{n} x_{1i}x_{2i} & \sum_{i=1}^{n} x_{1i}^3 & \sum_{i=1}^{n} x_{1i}x_{2i}^2 & \sum_{i=1}^{n} x_{1i}^2 x_{2i} \\[2mm]
\sum_{i=1}^{n} x_{2i} & \sum_{i=1}^{n} x_{1i}x_{2i} & \sum_{i=1}^{n} x_{2i}^2 & \sum_{i=1}^{n} x_{1i}^2 x_{2i} & \sum_{i=1}^{n} x_{2i}^3 & \sum_{i=1}^{n} x_{1i}x_{2i}^2 \\[2mm]
\sum_{i=1}^{n} x_{1i}^2 & \sum_{i=1}^{n} x_{1i}^3 & \sum_{i=1}^{n} x_{1i}^2 x_{2i} & \sum_{i=1}^{n} x_{1i}^4 & \sum_{i=1}^{n} x_{1i}^2 x_{2i}^2 & \sum_{i=1}^{n} x_{1i}^3 x_{2i} \\[2mm]
\sum_{i=1}^{n} x_{2i}^2 & \sum_{i=1}^{n} x_{1i}x_{2i}^2 & \sum_{i=1}^{n} x_{2i}^3 & \sum_{i=1}^{n} x_{1i}^2 x_{2i}^2 & \sum_{i=1}^{n} x_{2i}^4 & \sum_{i=1}^{n} x_{1i}x_{2i}^3 \\[2mm]
\sum_{i=1}^{n} x_{1i}x_{2i} & \sum_{i=1}^{n} x_{1i}^2 x_{2i} & \sum_{i=1}^{n} x_{1i}x_{2i}^2 & \sum_{i=1}^{n} x_{1i}^3 x_{2i} & \sum_{i=1}^{n} x_{1i}x_{2i}^3 & \sum_{i=1}^{n} x_{1i}^2 x_{2i}^2
\end{bmatrix}
\begin{bmatrix}
b_0 \\ b_1 \\ b_2 \\ b_{11} \\ b_{22} \\ b_{12}
\end{bmatrix}
=
\begin{bmatrix}
\sum_{i=1}^{n} y_i \\[2mm]
\sum_{i=1}^{n} x_{1i} y_i \\[2mm]
\sum_{i=1}^{n} x_{2i} y_i \\[2mm]
\sum_{i=1}^{n} x_{1i}^2 y_i \\[2mm]
\sum_{i=1}^{n} x_{2i}^2 y_i \\[2mm]
\sum_{i=1}^{n} x_{1i}x_{2i} y_i
\end{bmatrix}
$$

Example 10.4   The following data represent the percent of impurities that occurs at various temperatures and sterilizing times in a reaction associated with the manufacturing of a certain beverage.

Sterilizing Time, x_2 (minutes)	Temperature, x_1 (°C)		
	75	100	125
15	14.05	10.55	7.55
	14.93	9.48	6.59
20	16.56	13.63	9.23
	15.85	11.75	8.78
25	22.41	18.55	15.93
	21.66	17.98	16.44

Estimate the regression coefficients in the model

$$\mu_{Y|x_1, x_2} = \beta_0 + \beta_1 x_1 + \beta_2 x_2 + \beta_{11} x_1^2 + \beta_{22} x_2^2 + \beta_{12} x_1 x_2.$$

Solution. If we define $x_3 = x_1^2$, $x_4 = x_2^2$, and $x_5 = x_1 x_2$, then the computer program that was used in Example 10.3 can once again be used to invert the

6×6 **A** matrix. From the equation $\mathbf{b} = \mathbf{A}^{-1}\mathbf{g}$ we obtain

$$b_0 = 56.4668 \qquad b_{11} = 0.00081$$
$$b_1 = -0.36235 \qquad b_{22} = 0.08171$$
$$b_2 = -2.75299 \qquad b_{12} = 0.00314,$$

and our estimated regression equation is

$$\hat{y} = 56.4668 - 0.36235x_1 - 2.75299x_2 + 0.00081x_1^2$$
$$+ 0.08171x_2^2 + 0.00314x_1x_2.$$

Many of the principles and procedures associated with the estimation of polynomial regression functions fall into the category of **response surface methodology**, a collection of techniques that have been used quite successfully by scientists and engineers in many fields. Such problems as selecting a proper experimental design, particularly in cases where a large number of variables are in the model, and choosing "optimum" operating conditions on x_1, x_2, ..., x_k are often approached through the use of these methods. For an extensive exposure the reader is referred to *Response Surface Methodology* by Myers (see the Bibliography).

Exercises

1. Suppose in Exercise 7 on page 352 that we are also given the number of class periods missed by the 12 students taking the chemistry course. The complete data are recorded as follows:

Student	Chemistry Grade, y	Test Score, x_1	Classes Missed, x_2
1	85	65	1
2	74	50	7
3	76	55	5
4	90	65	2
5	85	55	6
6	87	70	3
7	94	65	2
8	98	70	5
9	81	55	4
10	91	70	3
11	76	50	1
12	74	55	4

(a) Fit a multiple linear regression equation of the form $\mu_{Y|x_1,x_2} = \beta_0 + \beta_1 x_1 + \beta_2 x_2$.
(b) Estimate the chemistry grade for a student who has an intelligence test score of 60 and missed 4 classes.

2. The following data represent a set of 10 experimental runs in which two independent variables x_1 and x_2 are controlled and values of a response, y, are observed:

y	x_1	x_2
61.5	2400	54.5
61.2	2450	56.4
32.0	2500	43.2
52.5	2700	65.2
31.5	2750	45.5
22.5	2800	47.5
53.0	2900	65.0
56.8	3000	66.5
34.8	3100	57.3
52.7	3200	68.0

Estimate the multiple linear regression equation $\mu_{Y|x_1,\,x_2} = \beta_0 + \beta_1 x_1 + \beta_2 x_2$.

3. A set of experimental runs was made to determine a way of predicting coking time y at various levels of oven width x_1 and flue temperature x_2. The coded data were recorded as follows:

y	x_1	x_2
6.40	1.32	1.15
15.05	2.69	3.40
18.75	3.56	4.10
30.25	4.41	8.75
44.85	5.35	14.82
48.94	6.20	15.15
51.55	7.12	15.32
61.50	8.87	18.18
100.44	9.80	35.19
111.42	10.65	40.40

Estimate the multiple linear regression equation $\mu_{Y|x_1,\,x_2} = \beta_0 + \beta_1 x_1 + \beta_2 x_2$.

4. An experiment was conducted to determine if the weight of an animal can be predicted after a given period of time on the basis of the initial weight of the animal and the amount of feed that was eaten. The following data, measured in kilograms, were recorded:

Final Weight, y	Initial Weight, x_1	Feed Eaten, x_2
95	42	272
77	33	226
80	33	259
100	45	292
97	39	311
70	36	183
50	32	173
80	41	236
92	40	230
84	38	235

(a) Fit a multiple regression equation of the form $\mu_{Y|x_1,\,x_2} = \beta_0 + \beta_1 x_1 + \beta_2 x_2$.

(b) Predict the final weight of an animal having an initial weight of 35 kilograms that is fed 250 kilograms of feed.

5. (a) Fit a multiple regression equation of the form $\mu_{Y|x} = \beta_0 + \beta_1 x + \beta_2 x^2$ to the data of Example 9.8 on page 342.

(b) Estimate the yield of the chemical reaction for a temperature of 225°C.

6. An experiment was conducted on a new model of a particular make of an automobile to determine the stopping distance at various speeds. The following data were recorded:

Speed, v (kilometers/hr)	35	50	65	80	95	110
Stopping distance, d (meters)	16	26	41	62	88	119

(a) Fit a multiple regression curve of the form $\mu_{D|v} = \beta_0 + \beta_1 v + \beta_2 v^2$.

(b) Estimate the stopping distance when the car is traveling at 70 kilometers per hour.

7. An experiment was conducted in order to determine if cerebral blood flow in human beings can be predicted from arterial oxygen tension (millimeters of mercury). Fifteen patients were used in the study and the following data were observed:

Blood flow, y	Arterial Oxygen Tension, x
84.33	603.40
87.80	582.50
82.20	556.20
78.21	594.60
78.44	558.90
80.01	575.20
83.53	580.10
79.46	451.20
75.22	404.00
76.58	484.00
77.90	452.40
78.80	448.40
80.67	334.80
86.60	320.30
78.20	350.30

Estimate the quadratic regression equation

$$\mu_{Y|x} = \beta_0 + \beta_1 x + \beta_2 x^2.$$

8. The following is a set of coded experimental data on the compressive strength of a particular alloy at various values of the concentration of some additive:

Concentration, x	Compressive Strength, y		
10.0	25.2	27.3	28.7
15.0	29.8	31.1	27.8
20.0	31.2	32.6	29.7
25.0	31.7	30.1	32.3
30.0	29.4	30.8	32.8

(a) Estimate the quadratic regression equation $\mu_{Y|x} = \beta_0 + \beta_1 x + \beta_2 x^2$.

(b) Test for lack of fit of the model.

9. The following data resulted from 15 experimental runs made on four independent variables and a single response y.

y	x_1	x_2	x_3	x_4
14.8	7.8	4.3	11.5	6.3
12.1	6.9	3.9	14.3	7.4
19.0	9.3	8.4	9.4	5.9
14.5	6.8	10.3	15.2	8.7
16.6	11.7	6.4	8.8	9.1
17.2	8.5	5.7	9.8	5.6
17.5	12.6	6.8	11.2	6.8
14.1	7.5	4.2	10.9	7.4
13.8	8.4	7.3	14.7	8.2
14.7	11.3	8.8	15.1	9.2
17.7	10.7	3.6	8.7	4.7
17.0	7.3	4.9	8.6	5.5
17.6	8.4	7.3	9.3	6.6
16.3	6.7	9.7	10.8	8.7
18.2	9.6	8.4	11.9	5.4

Estimate the multiple linear regression model relating y to $x_1, x_2, x_3,$ and x_4.

10. Given the data

x	0	1	2	3	4	5	6
y	1	4	5	3	2	3	4

(a) Fit the cubic model $\mu_{Y|x} = \beta_0 + \beta_1 x + \beta_2 x^2 + \beta_3 x^3$.

(b) Predict Y when $x = 2$.

11. The personnel department of a certain industrial firm used 12 subjects in a study to determine the relationship between job performance rating (y) and scores of four tests. The data are as follows:

y	x_1	x_2	x_3	x_4
11.2	56.5	71.0	38.5	43.0
14.5	59.5	72.5	38.2	44.8
17.2	69.2	76.0	42.5	49.0
17.8	74.5	79.5	43.4	56.3
19.3	81.2	84.0	47.5	60.2
24.5	88.0	86.2	47.4	62.0
21.2	78.2	80.5	44.5	58.1
16.9	69.0	72.0	41.8	48.1
14.8	58.1	68.0	42.1	46.0
20.0	80.5	85.0	48.1	60.3
13.2	58.3	71.0	37.5	47.1
22.5	84.0	87.2	51.0	65.2

Estimate the regression coeffficients in the model

$$\mu_{Y|x_1, x_2, x_3, x_4} = \beta_0 + \beta_1 x_1 + \beta_2 x_2 + \beta_3 x_3 + \beta_4 x_4.$$

12. Consider the following two-way table of experimental data:

x_1	x_2		
	0.5	1.0	1.5
1.0	18.0	20.9	19.8
	18.5	21.3	18.9
2.0	17.5	18.8	18.2
	16.8	18.5	17.9

Estimate the regression coefficients in the model

$$\mu_{Y|x_1, x_2} = \beta_0 + \beta_1 x_1 + \beta_2 x_2 + \beta_{22} x_2^2 + \beta_{12} x_1 x_2.$$

10.4 *Properties of the Least Squares Estimators*

The means and variances of the estimators B_0, B_1, ..., B_k are easily obtained under certain assumptions on the random errors E_1, E_2, ..., E_k that are identical to those made in the case of simple linear regression. Suppose that we assume these errors to be independent, each with zero mean and variance σ^2. It can then be shown that B_0, B_1, ..., B_k are, respectively, unbiased estimators of the regression coefficients β_0, β_1, ..., β_k. In addition, the variances of the B's are obtained through the elements of the inverse of the $\mathbf{A}$ matrix. One will note that the off-diagonal elements of $\mathbf{A} = \mathbf{X}'\mathbf{X}$ represent sums of products of elements in the columns of $\mathbf{X}$, while the diagonal elements of $\mathbf{A}$ represent sums of squares of elements in the columns of $\mathbf{X}$. The inverse matrix, $\mathbf{A}^{-1}$, apart from the multiplier σ^2, represents the **variance-covariance matrix** of the estimated regression coefficients. That is, the elements of the matrix $\mathbf{A}^{-1}\sigma^2$ display the variances of B_0, B_1, ..., B_k on the main diagonal and covariances on the off-diagonal. For example, in a $k = 2$ multiple linear regression problem, we might write

$$\mathbf{A}^{-1} = \begin{bmatrix} c_{00} & c_{01} & c_{02} \\ c_{10} & c_{11} & c_{12} \\ c_{20} & c_{21} & c_{22} \end{bmatrix}$$

with the elements below the main diagonal determined through the symmetry of the matrix. Then we can write

$$\sigma_{B_i}^2 = c_{ii}\sigma^2, \qquad\qquad i = 0, 1, 2$$
$$\sigma_{B_i B_j} = \operatorname{cov}(B_i, B_j) = c_{ij}\sigma^2, \qquad i \neq j.$$

Of course, the estimates of the variances and hence the standard errors of these estimators are obtained by replacing σ^2 with the appropriate estimate obtained through experimental data. An unbiased estimate of σ^2 is once again defined in terms of the error sum of squares, which is computed using the formula established in Theorem 10.1.

Theorem 10.1   *For the linear regression equation*

$$y_i = b_0 + b_1 x_{1i} + b_2 x_{2i} + \cdots + b_k x_{ki} + e_i,$$

$i = 1, 2, \ldots,\ n$, an unbiased estimate of σ^2 is given by

$$s^2 = \frac{SSE}{n - k - 1},$$

where

$$SSE = \sum_{i=1}^{n} e_i^2 = \sum_{i=1}^{n} (y_i - \hat{y}_i)^2.$$

One can easily see that Theorem 10.1 represents a generalization of Theorem 9.1 for the simple linear regression case. The proof is left for the reader. As in the simple linear regression case, the estimate s^2 is a measure of the variation in the prediction errors or residuals. Other important inferences regarding the fitted regression equation, based on the values of the individual residuals $e_i = y_i - \hat{y}_i$, $i = 1, 2, \ldots, n$, will be discussed in Sections 10.9 and 10.10.

The error and regression sum of squares take on the same form and play the same role as in the simple linear regression case. In fact, the sum-of-squares identity

$$\sum_{i=1}^{n} (y_i - \bar{y})^2 = \sum_{i=1}^{n} (\hat{y}_i - \bar{y})^2 + \sum_{i=1}^{n} (y - \hat{y}_i)^2$$

continues to hold and we retain our previous notation, namely,

$$SST = SSR + SSE$$

with

$$SST = \sum_{i=1}^{n} (y_i - \bar{y})^2 = \text{total sum of squares}$$

and

$$SSR = \sum_{i=1}^{n} (\hat{y}_i - \bar{y})^2 = \text{regression sum of squares}.$$

There are k degrees of freedom associated with SSR and, as always, SST has $n - 1$ degrees of freedom. Therefore, after subtraction, SSE has $n - k - 1$ degrees of freedom. Thus our estimate of σ^2 is again given by the error sum of squares divided by its degrees of freedom. Although all three of these sum of squares will appear on the printout of most multiple regression computer packages, they can also be easily calculated for small sets of data by using a calculator and the equivalent computing formulas of Theorem 10.2.

Theorem 10.2 *For the linear regression equation*

$$y_i = b_0 + b_1 x_{1i} + \cdots + b_k x_{ki} + e_i,$$

$i = 1, 2, \ldots, n$, *the error of squares may be written*

$$SSE = SST - SSR,$$

where

$$SST = S_{yy}$$

and

$$SSR = \sum_{j=0}^{k} b_j g_j - \frac{\left(\sum_{i=1}^{n} y_i\right)^2}{n}.$$

Proof. Let us first consider the case of $k = 2$ independent variables and then generalize to the general linear model containing k independent variables. Following the basic procedure outlined in Section 9.3, we may write

$$SSE = \sum_{i=1}^{n} (y_i - b_0 - b_1 x_{1i} - b_2 x_{2i})^2$$

$$= \sum_{i=1}^{n} [y_i - (b_0 + b_1 x_{1i} + b_2 x_{2i})]^2$$

$$= \sum_{i=1}^{n} [y_i^2 - y_i(b_0 + b_1 x_{1i} + b_2 x_{2i}) + (b_0 + b_1 x_{1i} + b_2 x_{2i})^2$$

$$- y_i(b_0 + b_1 x_{1i} + b_2 x_{2i})]$$

$$= \sum_{i=1}^{n} y_i^2 - b_0 \sum_{i=1}^{n} y_i - b_1 \sum_{i=1}^{n} x_{1i} y_i - b_2 \sum_{i=1}^{n} x_{2i} y_i$$

$$+ b_0 \sum_{i=1}^{n} (b_0 + b_1 x_{1i} + b_2 x_{2i} - y_i)$$

$$+ b_1 \sum_{i=1}^{n} (b_0 x_{1i} + b_1 x_{1i}^2 + b_2 x_{1i} x_{2i} - x_{1i} y_i)$$

$$+ b_2 \sum_{i=1}^{n} (b_0 x_{2i} + b_1 x_{1i} x_{2i} + b_2 x_{2i}^2 - x_{2i} y_i).$$

Using the normal equations of Section 10.2, we see that the last three sums are each equal to zero. Therefore,

$$SSE = \sum_{i=1}^{n} y_i^2 - b_0 \sum_{i=1}^{n} y_i - b_1 \sum_{i=1}^{n} x_{1i} y_i - b_2 \sum_{i=1}^{n} x_{2i} y_i$$

$$= S_{yy} - \left[b_0 g_0 + b_1 g_1 + b_2 g_2 - \frac{\left(\sum_{i=1}^{n} y_i\right)^2}{n} \right].$$

Generalizing to k independent variables, we have

$$SSE = S_{yy} - \left[\sum_{j=0}^{k} b_j g_j - \frac{\left(\sum_{i=1}^{n} y_i\right)^2}{n} \right]$$

$$= SST - SSR.$$

Example 10.5   For the data of Example 10.3 estimate σ^2.

Solution. The error sum of squares with $n - k - 1 = 9$ degrees of freedom can be found by writing

$$SSE = SST - SSR,$$

where

$$SST = S_{yy} = 11,400.15 - \frac{(377.5)^2}{13} = 438.13$$

and

$$SSR = \sum_{j=0}^{3} b_j g_j - \frac{\left(\sum_{i=1}^{13} y_i \right)^2}{13}$$

$$= (39.1574)(377.5) + (1.0161)(1877.567)$$

$$+ (-1.8616)(2246.6610) + (-0.3433)(3337.780) - \frac{(377.5)^2}{13}$$

$$= 399.45,$$

with 3 degrees of freedom. Our estimate of σ^2 is then given by

$$s^2 = \frac{438.13 - 399.45}{9} = 4.298.$$

The regression and error sum of squares give an intuitive indication of model adequacy (see Section 10.6). However, no form of hypothesis testing or confidence interval estimation can be accomplished without making the assumption that the errors E_i are normally distributed. In Section 10.5 we outline methods for finding confidence intervals and testing hypotheses in multiple linear regression analysis.

10.5 Inferences in Multiple Linear Regression

One of the most useful inferences that can be made regarding the ability of the regression equation to predict the response $\hat{y}_0$ corresponding to the values $x_{10}, x_{20}, \ldots, x_{k0}$ is the confidence interval on the mean response $\mu_{Y|x_{10}, x_{20}, \ldots, x_{k0}}$. In vector notation we are then interested in constructing a confidence interval on the mean response for the set of conditions given by $\mathbf{x}_0' = [1, x_{10}, x_{20}, \ldots, x_{k0}]$. We augment the conditions on the x's by the number 1 in order to facilitate using matrix notation. As in the $k = 1$ case, if we make the additional assumption that the errors are independent and normally distributed, then the B_j's are normal, with mean, variances, and covariances as indicated in Section 10.4, and hence

$$\hat{Y}_0 = B_0 + \sum_{j=1}^{k} B_j x_{j0}$$

is likewise normally distributed and is, in fact, an unbiased estimator for the **mean response** on which we are attempting to attach confidence intervals. The variance of $\hat{Y}_0$ written in matrix notation simply as a function of σ^2, $\mathbf{A}^{-1}$, and the condition vector $\mathbf{x}_0'$, is

$$\sigma_{\hat{Y}_0}^2 = \sigma^2 \mathbf{x}_0' \mathbf{A}^{-1} \mathbf{x}_0.$$

If this expression is expanded for a given case, say $k = 2$, it is easily seen that it appropriately accounts for the variances and covariances of the B_j's. After replacing σ^2 by s^2 as given in Section 9.4 the $100(1 - \alpha)\%$ confidence interval on $\mu_{Y|x_{10}, x_{20}, \ldots, x_{k0}}$ can be constructed from the statistic

$$T = \frac{\hat{Y}_0 - \mu_{Y|x_{10}, x_{20}, \ldots, x_{k0}}}{S\sqrt{\mathbf{x}_0' \mathbf{A}^{-1} \mathbf{x}_0}},$$

which has a t distribution with $n - k - 1$ degrees of freedom.

Confidence Interval for $\mu_{Y|x_{10}, x_{20}, \ldots, x_{k0}}$ *A* $(1 - \alpha)100\%$ *confidence interval for the* **mean response** $\mu_{Y|x_{10}, x_{20}, \ldots, x_{k0}}$ *is given by*

$$\hat{y}_0 - t_{\alpha/2} s \sqrt{\mathbf{x}_0' \mathbf{A}^{-1} \mathbf{x}_0} < \mu_{Y|x_{10}, x_{20}, \ldots, x_{k0}} < \hat{y}_0 + t_{\alpha/2} s \sqrt{\mathbf{x}_0' \mathbf{A}^{-1} \mathbf{x}_0},$$

where $t_{\alpha/2}$ is a value of the t distribution with $n - k - 1$ degrees of freedom.

The quantity $s\sqrt{\mathbf{x}_0' \mathbf{A}^{-1} \mathbf{x}_0}$ is often called the **standard error of prediction** and usually appears on the printout of many regression computer packages.

Example 10.6   Using the data of Example 10.3, construct a 95% confidence interval for the mean response when $x_1 = 3\%$, $x_2 = 8\%$, and $x_3 = 9\%$.

Solution. From the regression equation of Example 10.3 the estimated percent survival when $x_1 = 3\%$, $x_2 = 8\%$, and $x_3 = 9\%$ is

$$\hat{y} = 39.1574 + (1.0161)(3) - (1.8616)(8) - (0.3433)(9) = 24.2232.$$

Next we find

$$\mathbf{x}_0' \mathbf{A}^{-1} \mathbf{x}_0 = [1, 3, 8, 9] \begin{bmatrix} 8.064 & -0.0826 & -0.0942 & -0.7905 \\ -0.0826 & 0.0085 & 0.0017 & 0.0037 \\ -0.0942 & 0.0017 & 0.0166 & -0.0021 \\ -0.7905 & 0.0037 & -0.0021 & 0.0886 \end{bmatrix} \begin{bmatrix} 1 \\ 3 \\ 8 \\ 9 \end{bmatrix}$$

$$= 0.1267.$$

Previously, we found $s^2 = 4.298$ or $s = 2.073$, and using Table A.4 we see that $t_{0.025} = 2.262$ for 9 degrees of freedom. Therefore, a 95% confidence interval for the mean percent survival for $x_1 = 3\%$, $x_2 = 8\%$, and $x_3 = 9\%$ is given by

$$24.2232 - (2.262)(2.073)\sqrt{0.1267} < \mu_{Y|3, 8, 9} < 24.2232 \\ + (2.262)(2.073)\sqrt{0.1267}$$

or simply

$$22.5541 < \mu_{Y|3, 8, 9} < 25.8923.$$

A prediction interval for a single predicted response $\hat{y}_0$ is once again established by considering the differences $\hat{y}_0 - y_0$ of the random variable $\hat{Y}_0 - Y_0$. The sampling distribution can be shown to be normal with mean

$$\mu_{\hat{Y}_0 - Y_0} = 0$$

and variance

$$\sigma_{\hat{Y}_0 - Y_0}^2 = \sigma^2[1 + \mathbf{x}_0' \mathbf{A}^{-1}\mathbf{x}_0].$$

Thus the $(1 - \alpha)100\%$ prediction interval for a single predicted value y_0 can be constructed from the statistic

$$T = \frac{\hat{Y}_0 - Y_0}{S\sqrt{1 + \mathbf{x}_0' \mathbf{A}^{-1}\mathbf{x}_0}},$$

which has a t distribution with $n - k - 1$ degrees of freedom.

Prediction Interval for y_0　　*A $(1 - \alpha)100\%$ prediction interval for a **single response** y_0 is given by*

$$\hat{y}_0 - t_{\alpha/2} s \sqrt{1 + \mathbf{x}_0' \mathbf{A}^{-1}\mathbf{x}_0} < y_0 < \hat{y}_0 + t_{\alpha/2} s \sqrt{1 + \mathbf{x}_0' \mathbf{A}^{-1}\mathbf{x}_0},$$

where $t_{\alpha/2}$ is a value of the t distribution with $n - k - 1$ degrees of freedom.

Example 10.7　　Using the data of Example 10.3, construct a 95% prediction interval for an individual percent survival response when $x_1 = 3\%$, $x_2 = 8\%$, and $x_3 = 9\%$.

Solution. Referring to the results of Example 10.6, we find that the 95% prediction interval for the response y_0 when $x_1 = 3\%$, $x_2 = 8\%$, and $x_3 = 9\%$ is

$$24.2232 - (2.262)(2.073)\sqrt{1.1267} < y_0 < 24.2232 + (2.262)(2.073)\sqrt{1.1267},$$

which reduces to

$$19.2459 < y_0 < 29.2005.$$

A knowledge of the distributions of the individual coefficient estimators enables the experimenter to construct confidence intervals for the coefficients and to test hypotheses about them. One recalls that the B_j's ($j = 0, 1, 2, \ldots, k$)

are normally distributed with mean β_j and variance $c_{jj}\sigma^2$. Thus we can use the statistic

$$T = \frac{B_j - \beta_j}{S\sqrt{c_{jj}}}$$

with $n - k - 1$ degrees of freedom to test hypotheses and construct confidence intervals on β_j. For example, if we wish to test

$$H_0: \quad \beta_j = \beta_{j0},$$
$$H_1: \quad \beta_j \neq \beta_{j0},$$

we compute the statistic

$$t = \frac{b_j - \beta_{j0}}{s\sqrt{c_{jj}}}$$

and accept H_0 if

$$-t_{\alpha/2} < t < t_{\alpha/2},$$

where $t_{\alpha/2}$ has $n - k - 1$ degrees of freedom.

Example 10.8   For the model of Example 10.3, test the hypothesis that $\beta_2 = -2.5$ at the 0.05 level of significance against the alternative that $\beta_2 > -2.5$.

Solution

1. H_0: $\beta_2 = -2.5$.
2. H_1: $\beta_2 > -2.5$.
3. Choose a 0.05 level of significance.
4. Critical region: $t > 1.833$.
5. Computations:

$$t = \frac{b_2 - \beta_{20}}{s\sqrt{c_{22}}} = \frac{-1.8616 + 2.5}{2.073\sqrt{0.0166}} = 2.391.$$

6. Decision: Reject H_0 and conclude that $\beta_2 > -2.5$.

Exercises

1. For the data of Exercise 2 on page 363, estimate σ^2.

2. For the data of Exercise 3 on page 364, estimate σ^2.

3. For the data of Exercise 9 on page 365, estimate σ^2.

4. Obtain estimates of the variances and the covariance of the estimators B_1 and B_2 of Exercise 2 on page 363.

5. Referring to Exercise 9 on page 365, find the estimate of
   (a) $\sigma_{B_2}^2$.   (b) cov (B_1, B_4).

6. Using the data of Exercise 2 on page 363 and the estimate of σ^2 from Exercise 1, construct 95% confidence intervals for the predicted response and the mean response when $x_1 = 2500$ and $x_2 = 48.0$.

7. For Exercise 8 on page 365, construct a 90% confidence interval for the mean compressive strength when the concentration is $x = 19.5$ and a quadratic model is used.

8. Using the data of Exercise 9 on page 365 and the estimate of σ^2 from Exercise 3, construct 95% confidence intervals for the predicted response and

the mean response when $x_1 = 8.2$, $x_2 = 6.0$, $x_3 = 10.3$, and $x_4 = 5.8$.

9. For the model of Exercise 7 on page 364, test the hypothesis that $\beta_2 = 0$ at the 0.05 level of significance against the alternative that $\beta_2 \neq 0$.

10. For the model of Exercise 2 on page 363, test the hypothesis that $\beta_1 = 0$ at the 0.05 level of significance against the alternative that $\beta_1 \neq 0$.

11. For the model of Exercise 3 on page 364, test the hypothesis that $\beta_1 = 2$ at the 0.05 level of significance against the alternative $\beta_1 \neq 2$.

10.6 Adequacy of the Model

In many regression situations, individual coefficients are of importance to the experimenter. For example, in an economics application, β_1, β_2, ..., might have some particular significance, and thus confidence intervals and tests of hypotheses on these parameters are of interest to the economist. However, consider an industrial chemical situation in which the postulated model assumes that reaction yield is dependent linearly on reaction temperature and concentration of a certain catalyst. It is probably known that this is not the true model but an adequate representation, so the interest is likely to be not in the individual parameters but rather in the ability of the entire function to predict the true response in the range of the variables considered. Therefore, in this situation, one would put more emphasis on $\sigma_{\hat{Y}}^2$, confidence intervals on the mean response, and so forth, and likely deemphasize inferences on individual parameters.

The experimenter using regression analysis is also interested in deletion of variables when the situation dictates that, in addition to arriving at a workable prediction equation, he or she must find the "best regression" involving only variables that are useful predictors. There are a number of computer programs available for the practitioner that sequentially arrive at the so-called best regression equation depending on certain criteria. We shall discuss this further in Section 10.8.

One criterion that is commonly used to illustrate the adequacy of a fitted regression model is the **coefficient of multiple determination**:

$$R^2 = \frac{SSR}{SST} = \frac{\sum\limits_{i=1}^{n}(\hat{y}_i - \bar{y})^2}{\sum\limits_{i=1}^{n}(y_i - \bar{y})^2}.$$

This quantity merely indicates what proportion of the total variation in the response Y is explained by the fitted model. Often an experimenter will report

$R^2 \times 100\%$ and interpret the result as percentage variation explained by the postulated model. The square root of R^2 is called the **multiple correlation coefficient** between Y and the set $x_1, x_2, \ldots, x_k$. In Example 10.3, the value of R^2 indicating the proportion of variation explained by the three independent variables x_1, x_2, and x_3 is found to be

$$R^2 = \frac{SSR}{SST} = \frac{399.45}{438.13} = 0.9117,$$

which means that 91.17% of the variation in percent survival has been explained by the linear regression model.

The regression sum of squares can be used to give some indication concerning whether or not the model is an adequate explanation of the true situation. One can test the hypothesis H_0 that the regression is not significant by merely forming the ratio

$$f = \frac{SSR/k}{SSE/(n - k - 1)} = \frac{SSR/k}{s^2}$$

and rejecting H_0 at the α level of significance when $f > f_\alpha(k, n - k - 1)$. For the data of Example 10.3 we obtain

$$f = \frac{399.45/3}{4.298} = 30.98.$$

This value of F exceeds the tabulated critical point 6.99 of the F distribution for 3 and 9 degrees of freedom at the $\alpha = 0.01$ level. The result here should not be misinterpreted. Although it does indicate that the regression explained by the model is significant, this does not rule out the possibility that

1. The linear regression model in this set of x's is not the only model that can be used to explain the data; indeed, there might be other models with transformations on the x's that might give a larger value of the F statistic.
2. The model might have been more effective with the inclusion of other variables in addition to x_1, x_2, and x_3 or perhaps with the deletion of one or more of the variables in the model.

The addition of any single variable to a regression system will increase the regression sum of squares and thus reduce the error sum of squares. Consequently, we must decide whether the increase in regression is sufficient to warrant using it in the model. As one might expect, the use of unimportant variables can reduce the effectiveness of the prediction equation by increasing the variance of the estimated response. We shall pursue this point further by considering the importance of x_3 in Example 10.3. Initially, we can test

$$H_0: \quad \beta_3 = 0,$$
$$H_1: \quad \beta_3 \neq 0,$$

by using the t distribution with 9 degrees of freedom. We have

$$t = \frac{b_3 - 0}{s\sqrt{c_{33}}} = \frac{-0.3433}{2.073\sqrt{0.0886}} = -0.556,$$

which indicates that β_3 does not differ significantly from zero, and hence one may very well feel justified in removing x_3 from the model. Suppose that we consider the regression of Y on the set (x_1, x_2), the least squares normal equations now reducing to

$$\begin{bmatrix} 13 & 59.43 & 81.82 \\ 59.43 & 394.7255 & 360.6621 \\ 81.82 & 360.6621 & 576.7264 \end{bmatrix} \begin{bmatrix} b_0 \\ b_1 \\ b_2 \end{bmatrix} = \begin{bmatrix} 377.75 \\ 1877.5670 \\ 2246.6610 \end{bmatrix}.$$

The estimated regression coefficients for this reduced model are given by

$$b_0 = 36.094, \qquad b_1 = 1.031, \qquad b_2 = -1.870,$$

and the resulting regression sum of squares with 2 degreees of freedom is as follows:

$$R(\beta_1, \beta_2) = \sum_{j=0}^{2} b_j g_j - \frac{\left(\sum_{i=1}^{13} y_i\right)^2}{13}$$

$$= 398.12.$$

Here we use the notation $R(\beta_1, \beta_2)$ to indicate the regression sum of squares of the restricted model and it is not to be confused with SSR, the regression sum of squares of the original model with 3 degrees of freedom. The new error sum of squares is then given by

$$SST - R(\beta_1, \beta_2) = 438.13 - 398.12$$
$$= 40.01,$$

and the resulting error mean square with 10 degrees of freedom becomes

$$s^2 = \frac{40.01}{10} = 4.001.$$

The amount of variation in the response, the percent survival, which is attributed to x_3, the weight percent of the third additive, in the presence of the variables x_1 and x_2, is given by

$$R(\beta_3 \mid \beta_1, \beta_2) = SSR - R(\beta_1, \beta_2)$$
$$= 399.45 - 398.12$$
$$= 1.33,$$

which represents a small proportion of the entire regression variation. This amount of added regression is statistically insignificant as indicated by our previous test on β_3. An equivalent test involves the formation of the ratio

$$f = \frac{R(\beta_3 \mid \beta_1, \beta_2)}{s^2}$$

$$= \frac{1.33}{4.298} = 0.309,$$

which is a value of the F distribution with 1 and 9 degrees of freedom. Recall that the basic relationship between the t distribution with v degrees of freedom and the F distribution with 1 and v degrees of freedom is given by

$$t_{\alpha/2}^2 = f_\alpha(1, v)$$

and we note that the f value of 0.309 is indeed the square of the t value of 0.556.

We can provide additional support for deleting x_3 from the model by considering $\sigma_{\hat{Y}}^2$ under both the full and reduced regression equation. We first note that the estimate of σ^2 was reduced from 4.298 to 4.001 by deleting x_3 from the model. In the case of the full model an estimate of the variance of the estimated response $\hat{Y}_0$ when $x_1 = 4.00$, $x_2 = 5.5$, and $x_3 = 8.90$, is

$$\hat{\sigma}_{\hat{Y}_0}^2 = s^2 \mathbf{x}_0' \mathbf{A}^{-1} \mathbf{x}_0$$

$$= 4.298[1, 4.00, 5.5, 8.90]$$

$$\times \begin{bmatrix} 8.064 & -0.0826 & -0.0942 & -0.7905 \\ -0.0826 & 0.0085 & 0.0017 & 0.0037 \\ -0.0942 & 0.0017 & 0.0166 & -0.0021 \\ -0.7905 & 0.0037 & -0.0021 & 0.0886 \end{bmatrix} \begin{bmatrix} 1 \\ 4.00 \\ 5.5 \\ 8.90 \end{bmatrix}$$

$$= 0.3936.$$

For the reduced model our point of interest becomes $\mathbf{x}_0' = [1, 4.00, 5.5]$, our $\mathbf{A}$ matrix has been reduced to a 3×3 matrix with inverse given by

$$\mathbf{A}^{-1} = \begin{bmatrix} 1.0114 & -0.0494 & -0.1126 \\ -0.0494 & 0.0083 & 0.0018 \\ -0.1126 & 0.0018 & 0.0166 \end{bmatrix},$$

and the variance of the estimated response is now estimated to be

$$\hat{\sigma}_{\hat{Y}_0}^2 = 4.001[1, 4.00, 5.5] \begin{bmatrix} 1.0114 & -0.0494 & -0.1126 \\ -0.0494 & 0.0083 & 0.0018 \\ -0.1126 & 0.0018 & 0.0166 \end{bmatrix} \begin{bmatrix} 1 \\ 4.00 \\ 5.5 \end{bmatrix}$$

$$= 0.3668,$$

which represents a reduction over that found for the complete model.

To generalize the above concepts, one can assess the work of an independent variable x_i in the general multiple linear regression model

$$\mu_{Y|x_1, x_2, \ldots, x_k} = \beta_0 + \beta_1 x_1 + \cdots + \beta_k x_k$$

by observing the amount of regression attributed to x_i over that attributed to the other variables, that is, the regression on x_i *adjusted* for the other variables. This is computed by subtracting the regression sum of squares for a model with x_i removed, from SSR. For example, we say that x_1 is assessed by calculating

$$R(\beta_1 | \beta_2, \beta_3, \ldots, \beta_k) = SSR - R(\beta_2, \beta_3, \ldots, \beta_k),$$

where $R(\beta_2, \beta_3, \ldots, \beta_k)$ is the regression sum of squares with $\beta_1 x_1$ removed from the model. To test the hypothesis

$$H_0: \quad \beta_1 = 0,$$
$$H_1: \quad \beta_1 \neq 0,$$

compute

$$f = \frac{R(\beta_1 | \beta_2, \beta_3, \ldots, \beta_k)}{s^2}$$

and compare with $f_\alpha(1, n - k - 1)$.

In a similar manner we can test for the significance of a *set* of the variables. For example, to investigate simultaneously the importance of including x_1 and x_2 in the model, we test the hypothesis

$$H_0: \quad \beta_1 = \beta_2 = 0,$$
$$H_1: \quad \beta_1 \text{ and } \beta_2 \text{ are not both zero},$$

by computing

$$f = \frac{[R(\beta_1, \beta_2 | \beta_3, \beta_4, \ldots, \beta_k)]/2}{s^2}$$

$$= \frac{[SSR - R(\beta_3, \beta_4, \ldots, \beta_k)]/2}{s^2}$$

and comparing with $f_\alpha(2, n - k - 1)$. The number of degrees of freedom associated with the numerator, in this case 2, equals the number of variables in the set.

10.7 *Special Case of Orthogonality*

Prior to our original development of the general linear regression problem, the assumption was made that the independent variables were measured without error and are often controlled by the experimenter. Quite often they occur as a result of an elaborately *designed experiment*. In fact, one can increase the effectiveness of the resulting prediction equation with the use of a suitable experimental plan.

Suppose that we once again consider the $\mathbf{X}$ matrix as defined in Section 10.3. We can rewrite it to read

$$\mathbf{X} = [\mathbf{1}, \mathbf{x}_1, \mathbf{x}_2, \ldots, \mathbf{x}_k],$$

where $\mathbf{1}$ represents a column of ones and $\mathbf{x}_j$ is a column vector representing the levels of x_j. If

$$\mathbf{x}_p' \mathbf{x}_q = 0, \qquad p \neq q,$$

the variables x_p and x_q are said to be *orthogonal* to each other. There are certain obvious advantages to having a completely orthogonal situation whereby $\mathbf{x}_p' \mathbf{x}_q = 0$ for all possible p and q, $p \neq q$, and, in addition,

$$\sum_{i=1}^{n} x_{ji} = 0, \qquad j = 1, 2, \ldots, k.$$

The resulting $\mathbf{X'X}$ is a diagonal matrix and the normal equations in Section 10.3 reduce to

$$nb_0 = \sum_{i=1}^{n} y_i$$

$$b_1 \sum_{i=1}^{n} x_{1i}^2 = \sum_{i=1}^{n} x_{1i} y_i$$

$$\vdots \qquad \qquad \vdots$$

$$b_k \sum_{i=1}^{n} x_{ki}^2 = \sum_{i=1}^{n} x_{ki} y_i.$$

The most important advantage is that one is easily able to partition *SSR* into **single-degree-of-freedom components**, each of which corresponds to the amount of variation in Y accounted for by a given controlled variable. In the orthogonal situation we can write

$$SSR = \sum_{i=1}^{n} (\hat{y}_i - \bar{y})^2$$

$$= \sum_{i=1}^{n} (b_0 + b_1 x_{1i} + \cdots + b_k x_{ki} - b_0)^2$$

$$= b_1^2 \sum_{i=1}^{n} x_{1i}^2 + b_2^2 \sum_{i=1}^{n} x_{2i}^2 + \cdots + b_k^2 \sum_{i=1}^{n} x_{ki}^2$$

$$= R(\beta_1) + R(\beta_2) + \cdots + R(\beta_k).$$

The quantity $R(\beta_i)$ is the amount of regression sum of squares associated with a model involving a single independent variable x_i.

Table 10.1 Analysis of Variance for Orthogonal Variables

Source of Variation	Sum of Squares	Degrees of Freedom	Mean Square	Computed f
β_1	$R(\beta_1) = b_1^2 \sum_{i=1}^{n} x_{1i}^2$	1	$R(\beta_1)$	$\dfrac{R(\beta_1)}{s^2}$
β_2	$R(\beta_2) = b_2^2 \sum_{i=1}^{n} x_{2i}^2$	1	$R(\beta_2)$	$\dfrac{R(\beta_2)}{s^2}$
$\vdots$	$\vdots$	$\vdots$	$\vdots$	$\vdots$
β_k	$R(\beta_k) = b_k^2 \sum_{i=1}^{n} x_{ki}^2$	1	$R(\beta_k)$	$\dfrac{R(\beta_k)}{s^2}$
Error	SSE	$n - k - 1$	$s^2 = \dfrac{SSE}{n - k - 1}$	
Total	$SST = S_{yy}$	$n - 1$		

To test simultaneously for the significance of a set of m variables in an orthogonal situation, the regression sum of squares becomes

$$R(\beta_1, \beta_2, \ldots, \beta_m \mid \beta_{m+1}, \beta_{m+2}, \ldots, \beta_k) = R(\beta_1) + R(\beta_2) + \cdots + R(\beta_m)$$

and simplifies to

$$R(\beta_1 \mid \beta_2, \beta_3, \ldots, \beta_k) = R(\beta_1)$$

when evaluating a single independent variable. Therefore, the contribution of a given variable or set of variables is essentially found by *ignoring* the other variables in the model. Independent evaluations of the worth of the individual variables are accomplished using analysis-of-variance techniques as given in Table 10.1. The total variation in the response is partitioned into single-degree-of-freedom components plus the error term with $n - k - 1$ degrees of freedom. Each computed f value is used to test one of the hypotheses

$$\left. \begin{array}{l} H_0\colon \ \beta_i = 0 \\ H_1\colon \ \beta_i \neq 0 \end{array} \right\} \quad i = 1, 2, \ldots, k$$

by comparing with the critical point $f_\alpha(1, n - k - 1)$.

Example 10.9 Suppose that a scientist takes experimental data on the radius of a propellant grain Y as a function of powder temperature x_1, extrusion rate x_2, and die temperature x_3. Fit a linear regression model for predicting grain radius and determine the effectiveness of each variable in the model. The data are given as follows:

Grain Radius	Powder Temperature	Extrusion Rate	Die Temperature
82	150 (-1)	12 (-1)	220 (-1)
93	190 (1)	12 (-1)	220 (-1)
114	150 (-1)	24 (1)	220 (-1)
124	150 (-1)	12 (-1)	250 (1)
111	190 (1)	24 (1)	220 (-1)
129	190 (1)	12 (-1)	250 (1)
157	150 (-1)	24 (1)	250 (1)
164	190 (1)	24 (1)	250 (1)

Solution. Note that each variable is controlled at two levels and the experiment represents each of the eight possible combinations. The data on the independent variables are coded for convenience by means of the following formulas:

$$x_1 = \frac{\text{powder temperature} - 170}{20}$$

$$x_2 = \frac{\text{extrusion rate} - 18}{6}$$

$$x_3 = \frac{\text{die temperature} - 235}{15}.$$

The resulting levels of x_1, x_2, and x_3 take on the values -1 and $+1$ as indicated in the table of data. This particular *experimental design* affords the orthogonality that we are illustrating here. A more thorough treatment of this type of experimental layout will be given in Chapter 13. The **X** matrix is given by

$$\mathbf{X} = \begin{bmatrix} 1 & -1 & -1 & -1 \\ 1 & 1 & -1 & -1 \\ 1 & -1 & 1 & -1 \\ 1 & -1 & -1 & 1 \\ 1 & 1 & 1 & -1 \\ 1 & 1 & -1 & 1 \\ 1 & -1 & 1 & 1 \\ 1 & 1 & 1 & 1 \end{bmatrix}$$

and it is easy to verify the orthogonality conditions. One can now compute the coefficients

$$b_0 = \frac{\sum_{i=1}^{8} y_i}{8} = 121.75$$

$$b_1 = \frac{\sum_{i=1}^{8} x_{1i} y_i}{\sum_{i=1}^{8} x_{1i}^2} = \frac{20}{8} = 2.5$$

$$b_2 = \frac{\sum_{i=1}^{8} x_{2i} y_i}{\sum_{i=1}^{8} x_{2i}^2} = \frac{118}{8} = 14.75$$

$$b_3 = \frac{\sum_{i=1}^{8} x_{3i} y_i}{\sum_{i=1}^{8} x_{3i}^2} = \frac{174}{8} = 21.75,$$

so in terms of the *coded* variables, the prediction equation is given by

$$\hat{y} = 121.75 + 2.5x_1 + 14.75x_2 + 21.75x_3.$$

The analysis-of-variance table showing independent contributions to *SSR* for each variable is given in Table 10.2. The results, when compared to the $f_{0.05}(1, 4)$ critical point of 7.71, indicate that x_1 does not contribute significantly at the 0.05 level, while variables x_2 and x_3 are significant. In this example the estimate for σ^2 is 23.1250. As in the single-independent-variable case, it should be pointed out that this estimate does not solely contain experimental error variation unless the postulated model is correct. Otherwise, the estimate is

Table 10.2 Analysis of Variance on Grain Radius Data

Source of Variation	Sum of Squares	Degrees of Freedom	Mean Square	Computed f
β_1	$(2.5)^2(8) = 50$	1	50	2.16
β_2	$(14.75)^2(8) = 1740.50$	1	1740.50	75.26
β_3	$(21.75)^2(8) = 3784.50$	1	3784.50	163.65
Error	92.5	4	23.1250	
Total	5667.50	7		

"contaminated" by lack of fit in addition to pure error, and the lack of fit can be separated only if one obtains multiple experimental observations at the various (x_1, x_2, x_3) combinations.

Exercises

1. Compute and interpret the coefficient of multiple determination for the variables of Exercise 3 on page 364.

2. Test whether the regression explained by the model in Exercise 3 on page 364 is significant at the 0.01 level of signficance.

3. Test whetehr the regression explained by the model in Exercise 9 on page 365 is significant at the 0.01 level of significance.

4. For the model of Exercise 9 on page 365, test the hypothesis

$$H_0: \quad \beta_1 = \beta_2 = 0$$
$$H_1: \quad \beta_1 \text{ and } \beta_2 \text{ are not both zero.}$$

5. Repeat Exercise 10 on page 373 using an F statistic.

6. A small experiment is conducted to fit a multiple regression equation relating the yield y to temperature x_1, reaction time x_2, and concentration of one of the reactants x_3. Two levels of each variable were chosen and measurements corresponding to the coded independent variables were recorded as

follows:

y	x_1	x_2	x_3
7.6	-1	-1	-1
8.4	1	-1	-1
9.2	-1	1	-1
10.3	-1	-1	1
9.8	1	1	-1
11.1	1	-1	1
10.2	-1	1	1
12.6	1	1	1

(a) Using the coded variables, estimate the multiple linear regression equation

$$\mu_{Y|x_1, x_2, x_3} = \beta_0 + \beta_1 x_1 + \beta_2 x_2 + \beta_3 x_3.$$

(b) Partition SSR, the regression sum of squares, into three single-degree-of-freedom components attributable to x_1, x_2, and x_3, respectively. Show an analysis-of-variance table, indicating significance tests on each variable.

10.8 Sequential Methods for Model Selection

At times the significance tests outlined in Section 10.6 are quite adequate in determining which variables should be used in the final regression model. These tests are certainly effective if the experiment can be planned and the variables are orthogonal to each other. Even if the variables are not orthogonal, the individual t tests can be of some use in many problems where the number of variables under investigation is small. However, there are many problems in which it is necessary to use more elaborate techniques for screening variables, particularly when the experiment exhibits a substantial deviation from orthogonality. Useful measures of **multicollinearity** (linear dependency) among the independent variables are provided by the sample correlation coefficients $r_{x_i x_j}$. Since we are concerned only with linear dependency among independent variables, no confusion will result if we drop the x's from our notation and simply write $r_{x_i x_j} = r_{ij}$, where

$$r_{ij} = \frac{S_{ij}}{\sqrt{S_{ii} S_{jj}}}.$$

It should be noted that the r_{ij}'s do not give true estimates of population correlation coefficients in the strict sense, since the x's are actually not random variables in the context discussed here. Thus the term "correlation," albeit standard, is perhaps a misnomer.

When one or more of these sample correlation coefficients deviate substantially from zero, it can be quite difficult to find the most effective subset of variables for inclusion in our prediction equation. In fact, for some problems the multicollinearity will be so extreme that a suitable predictor cannot be found unless all possible subsets of the variables are investigated. Of course, the latter approach is prohibitive for a large number of variables. Informative discussions of model selection in regression by Thompson and Cady and Hocking are cited in the Bibliography. Other procedures for detection of multicollinearity are discussed in Montgomery and Peck, also cited in the Bibliography.

The user of multiple linear regression attempts to accomplish one of three objectives:

1. Obtain estimates of individual coefficients in a complete model.
2. Screen variables to determine which have a significant effect on the response.
3. Arrive at the most effective prediction equation.

In (1) it is known a priori that all variables are to be included in the model. In (2) prediction is secondary, while in (3) individual regression coefficients are not as important as the quality of the estimated response $\hat{y}$. In each of the situations above, multicollinearity in the experiment can have a profound effect on the success of the regression.

In this section some standard sequential procedures for selecting variables are discussed. They are based on the notion that a single variable or a collection of variables should not appear in the estimating equation unless they result in a significant increase in the regression sum of squares or, equivalently, a significant increase in R^2, the coefficient of multiple determination.

Consider the data in Table 10.3 in which measurements were taken on 9 infants. The purpose of the experiment was to arrive at a suitable estimating equation relating the length of an infant to all or a subset of the independent

Table 10.3 Data Relating to Infant Length[a]

Infant Length, y (cm)	Age, x_1 (days)	Length at Birth, x_2 (cm)	Weight at Birth, x_3 (kg)	Chest Size at Birth, x_4 (cm)
57.5	78	48.2	2.75	29.5
52.8	69	45.5	2.15	26.3
61.3	77	46.3	4.41	32.2
67.0	88	49.0	5.52	36.5
53.5	67	43.0	3.21	27.2
62.7	80	48.0	4.32	27.7
56.2	74	48.0	2.31	28.3
68.5	94	53.0	4.30	30.3
69.2	102	58.0	3.71	28.7

[a] Data analyzed by the Statistical Consulting Center, Virginia Polytechnic Institute and State University, Blacksburg, Virginia, 1976.

variables. The sample correlation coefficients, indicating the linear dependency among the independent variables, are displayed in the symmetric matrix

$$\begin{array}{cccc} x_1 & x_2 & x_3 & x_4 \end{array}$$
$$\begin{bmatrix} 1.0000 & 0.9523 & 0.5340 & 0.3900 \\ 0.9523 & 1.0000 & 0.2626 & 0.1549 \\ 0.5340 & 0.2626 & 1.0000 & 0.7847 \\ 0.3900 & 0.1549 & 0.7847 & 1.0000 \end{bmatrix}.$$

Note that there appears to be an appreciable amount of multicollinearity. Using the least squares technique outlined in Section 10.3, the estimated regression equation using the complete model was fitted and is given by

$$\hat{y} = 7.1475 + 0.1000x_1 + 0.7264x_2 + 3.0758x_3 + 0.0300x_4.$$

The value of s^2 with 4 degrees of freedom is 0.7414, and the value for the coefficient of determination for this model is found to be 0.9907. Regression sum of squares measuring the variation attributed to each individual variable in the presence of the others, and the corresponding t values, are given in Table 10.4.

A two-tailed critical region with 4 degrees of freedom at the 0.05 level of significance is given by $|t| > 2.776$. Of the four computed t values, only variable x_3 appears to be significant. However, it should be recalled that although the t statistic described in Section 10.6 measures the worth of a variable adjusted for all other variables, it does not detect the potential importance of a variable in combination with a subset of the variables. For example, consider

Table 10.4 t Values for the Regression Data of Table 10.3

Variable x_1	Variable x_2	Variable x_3	Variable x_4
$R(\beta_1 \mid \beta_2, \beta_3, \beta_4)$ $= 0.0644$ $t = 0.2947$	$R(\beta_2 \mid \beta_1, \beta_3, \beta_4)$ $= 0.6334$ $t = 0.9243$	$R(\beta_3 \mid \beta_1, \beta_2, \beta_4)$ $= 6.2523$ $t = 2.9040$	$R(\beta_4 \mid \beta_1, \beta_2, \beta_3)$ $= 0.0241$ $t = -0.1805$

the model with only the variables x_2 and x_3 in the equation. The data analysis gives the regression function

$$\hat{y} = 2.1833 + 0.9576x_2 + 3.3253x_3,$$

with $R^2 = 0.9905$, certainly not a substantial reduction from $R^2 = 0.9907$ for the complete model. However, unless the performance characteristics of this particular combination had been observed, one would not be aware of its predictive potential. This, of course, lends support for a methodology that observes all possible regressions or a systematic sequential procedure designed to test several subsets.

One standard procedure for searching for the "optimum subset" of variables in the absence of orthogonality is a technique called **stepwise regression**. It is based on the procedure of sequentially introducing the variables into the model one at a time. The description of the stepwise routine will be better understood by the reader if the methods of **forward selection** and **backward elimination** are described first.

Forward selection is based on the notion that variables should be inserted one at a time until a satisfactory regression equation is found. The procedure is as follows:

STEP 1 Choose the variable that gives the largest regression sum of squares when performing a simple linear regression with y, or equivalently, that which gives the largest value of R^2. We shall call this initial variable x_1.

STEP 2 Choose the variable that when inserted in the model gives the largest increase in R^2, in the presence of x_1, over the R^2 found in step 1. This, of course, is the variable x_j, for which

$$R(\beta_j \mid \beta_1) = R(\beta_1, \beta_j) - R(\beta_1)$$

is largest. Let us call this variable x_2. The regression model with x_1 and x_2 is then fitted and R^2 observed.

STEP 3 Choose the variable x_j that gives the largest value of

$$R(\beta_j \mid \beta_1, \beta_2) = R(\beta_1, \beta_2, \beta_j) - R(\beta_1, \beta_2),$$

again resulting in the largest increase of R^2 over that given in step 2. Calling this variable x_3, we now have a regression model involving $x_1, x_2,$ and x_3.

This process is continued until the most recent variable inserted fails to induce a significant increase in the explained regression. Such an increase can be determined at each step by using the appropriate F test or t test. For example, in step 2 the value

$$f = \frac{R(\beta_2 \mid \beta_1)}{s^2}$$

can be determined to test the appropriateness of x_2 in the model. Here the value of s^2 is the error mean square for the model containing the variables x_1 and x_2. Similarly, in step 3, the ratio

$$f = \frac{R(\beta_3 \mid \beta_1, \beta_2)}{s^2}$$

tests the appropriateness of x_3 in the model. Now, however, the value for s^2 is the error mean square for the model that contains the three variables x_1, x_2, and x_3. If $f < f_\alpha(1, n - 3)$ at step 2, for a prechosen significant level, x_2 is not included and the process is terminated, resulting in a simple linear equation relating y and x_1. However, if $f > f_\alpha(1, n - 3)$, we proceed to step 3. Again, if $f < f_\alpha(1, n - 4)$ at step 3, x_3 is not included and the process is terminated with the appropriate regression equation containing the variables x_1 and x_2.

Backward elimination involves the same concepts as forward selection except that one begins with all the variables in the model. Suppose, for example that there are five variables under consideration. The steps are as follows:

STEP 1 Fit a regression equation with all five variables included in the model. Choose the variable that gives the smallest value of the regression sum of squares **adjusted for the others**. Suppose that this variable is x_2. Remove x_2 from the model if

$$f = \frac{R(\beta_2 \mid \beta_1, \beta_3, \beta_4, \beta_5)}{s^2}$$

is insignificant.

STEP 2 Fit a regression equation using the remaining variables x_1, x_3, x_4, and x_5 and repeat step 1. Suppose that variable x_5 is chosen this time. Once again if

$$f = \frac{R(\beta_5 \mid \beta_1, \beta_3, \beta_4)}{s^2}$$

is insignificant, the variable x_5 is removed from the model. At each step the s^2 used in the F test is the error mean square for the regression model at that stage.

This process is repeated until at some step the variable with the smallest adjusted regression sum of squares results in a significant f value for some predetermined significance level.

Stepwise regression is accomplished with a slight but important modification of the forward selection procedure. The modification involves further testing at each stage to ensure the continued effectiveness of variables that had been inserted into the model at an earlier stage. This represents an improvement over forward selection, since it is quite possible that a variable entering the regression equation at an early stage might have been rendered unimportant or redundant because of relationships that exist between it and other variables entering at later stages. Therefore, at a stage in which a new variable has been entered into the regression equation through a significant increase in R^2 as determined by the F test, all the variables already in the model are subjected to F tests (or, equivalently to t tests) in light of this new variable, and are deleted if they do not display a significant f value. The procedure is continued until a stage is reached in which no additional variables can be inserted or deleted. We illustrate the stepwise procedure in the following example.

Example 10.10 Using the techniques of stepwise regression, find an appropriate linear regression model for predicting the length of infants for the data of Table 10.3.

Solution

STEP 1 In considering each variable separately, four individual simple linear regression equations are fitted. The following pertinent regression sums of squares are computed:

$$R(\beta_1) = 288.1468 \qquad R(\beta_2) = 215.3013$$
$$R(\beta_3) = 186.1065 \qquad R(\beta_4) = 100.8594.$$

Variable x_1 very clearly gives the largest regression sum of squares. The error mean square for the equation involving only x_1 is $s^2 = 4.7276$, and since

$$f = \frac{R(\beta_1)}{s^2} = \frac{288.1468}{4.7276} = 60.9500,$$

which exceeds $f_{0.05}(1, 7) = 5.59$, the variable x_1 is entered into the model.

STEP 2 Three regression equations are fitted at this stage, all containing x_1. The important results for the combinations (x_1, x_2), (x_1, x_3), and (x_1, x_4) are

$$R(\beta_2 \mid \beta_1) = 23.8703, \qquad R(\beta_3 \mid \beta_1) = 29.3086, \qquad R(\beta_4 \mid \beta_1) = 13.8178.$$

Variable x_3 displays the largest regression sum of squares in the presence of x_1. The regression involving x_1 and x_3 gives a new value of $s^2 = 0.6307$, and since

$$f = \frac{R(\beta_3 \mid \beta_1)}{s^2} = \frac{29.3086}{0.6307} = 46.47,$$

which exceeds $f_{0.05}(1, 6) = 5.99$, the variable x_3 is included along with x_1 in the model. Now we must subject x_1 in the presence of x_3 to a significance test. We find that $R(\beta_1 \mid \beta_3) = 131.349$, and hence

$$f = \frac{R(\beta_1 \mid \beta_3)}{s^2} = \frac{131.349}{0.6307} = 208.26,$$

which is highly significant. Therefore, x_1 is retained along with x_3.

STEP 3 With x_1 and x_3 already in the model, we now require $R(\beta_2 \mid \beta_1, \beta_3)$ and $R(\beta_4 \mid \beta_1, \beta_3)$ in order to determine which, if any, of the remaining two variables is entered at this stage. From the regression analysis using x_2 along with x_1 and x_3, we find $R(\beta_2 \mid \beta_1, \beta_3) = 0.7948$, and when x_4 is used along with x_1 and x_3, we obtain $R(\beta_4 \mid \beta_1, \beta_3) = 0.1855$. The value of s^2 is 0.5979 for the (x_1, x_2, x_3) combination and 0.7198 for the (x_1, x_2, x_4) combination. Since neither f value is significant at the $\alpha = 0.05$ level, the final regression model includes only the variables x_1 and x_3. The estimating equation is found to be

$$\hat{y} = 20.1084 + 0.4136 x_1 + 2.0253 x_3$$

and the coefficient of determination for this model is $R^2 = 0.9882$.

Although (x_1, x_3) is the combination chosen by stepwise regression, it is not necessarily the combination of two variables that gives the largest value of R^2. In fact, we have already observed that the combination (x_2, x_3) gives an $R^2 = 0.9905$. Of course, the stepwise procedure never actually observed this combination. A rational argument could be made that there is actually a negligible difference in performance between these two estimating equations, at least in terms of percent variation explained. It is interesting to observe, however, that the backward elimination procedure gives the combination (x_2, x_3) in the final equation (see Exercise 3 on page 396).

The main function of each of the procedures outlined in this section is to expose the variables to a systematic methodology designed to ensure the eventual inclusion of the best combinations of the variables. Obviously, there is no assurance that this will happen in all problems, and, of course, it is possible that the multicollinearity is so extensive that one has no alternative but to resort to estimation procedures other than least squares. These estimation procedures are discussed in Section 10.11.

The sequential procedures discussed here represent three of many such methods that have been put forth in the literature and appear in various regression computer packages that are available. These methods are designed to be computationally efficient but, of course, do not give results for all possible subsets of the variables. As a result, the procedures are most effective in data sets that involve a **large number of variables**. In regression problems involving a relatively small number of variables, certainly when $k \leq 5$, modern regression computer packages allow for the computation and summarization of quantitative information on **all models** for every possible subset of the variables. Illustrations will be given in Section 10.10.

10.9 *Study of Residuals*

It was suggested earlier in this chapter that the residuals, or errors in the regression fit, often carry information that can be very informative to the data analyst. The $e_i = y_i - \hat{y}_i$, $i = 1, 2, \ldots, n$, which are the numerical counterpart to the ε_i's, the model errors, often shed light on the possible violation of assumptions or the presence of "suspect" data points. Suppose that we let the vector $\mathbf{x}_i$ denote the values of the regressor variables corresponding to the ith data point, supplemented by a 1 in the initial position. That is,

$$\mathbf{x}_i' = [1, x_{1i}, x_{2i}, \ldots, x_{ki}].$$

Consider the quantity

$$h_{ii} = \mathbf{x}_i' \mathbf{A}^{-1} \mathbf{x}_i, \qquad i = 1, 2, \ldots, n.$$

The reader should recognize that h_{ii} was used in the computation of the confidence intervals on the mean response in Section 10.5. Apart from σ^2, h_{ii} represents the variance of the fitted value $\hat{Y}_i$. The h_{ii} values are the diagonal elements of the **HAT matrix** given by

$$\mathbf{H} = \mathbf{X}(\mathbf{X}'\mathbf{X})^{-1}\mathbf{X}'$$
$$= \mathbf{X}\mathbf{A}^{-1}\mathbf{X}'$$

which plays an important role in any study of residuals and in other modern aspects of regression analysis (see the reference to Montgomery and Peck listed in the Bibliography). The term *HAT matrix* is derived from the fact that $\mathbf{H}$ generates the "y-hats" or the fitted values when multiplied by the vector $\mathbf{y}$ of observed responses. That is,

$$\hat{\mathbf{y}} = \mathbf{X}(\mathbf{X}'\mathbf{X})^{-1}\mathbf{X}'\mathbf{y} = \mathbf{H}\mathbf{y},$$

where $\hat{\mathbf{y}}$ is the vector whose ith element is $\hat{y}_i$.

If we make the usual assumptions that the E_i are independent and normally distributed with zero mean and variance σ^2, the statistical properties of the residuals are easily characterized. Then

$$E(E_i) = E(Y_i - \hat{Y}_i) = 0$$

and

$$\sigma_{E_i}^2 = (1 - h_{ii})\sigma^2$$

for $i = 1, 2, \ldots, n$. See the Montgomery and Peck reference for details. It can be shown that the HAT diagonal values are bounded according to the inequality

$$0 < h_{ii} < 1.$$

In addition, $\sum\limits_{i=1}^{n} h_{ii} = k + 1$, the number of regression parameters. As a result, any data point whose HAT diagonal element is large, that is, well above the average value of $(k + 1)/n$, is in a position in the data set where the variance of $\hat{Y}_i$ is relatively large, and the variance of a residual is relatively small. As a result, the data analyst can easily gain some insight on how large a residual may become before its deviation from zero can be attributed to something other than mere chance. Many of the commercial regression computer packages produce the set of **studentized residuals**

$$r_i = \frac{e_i}{s\sqrt{1 - h_{ii}}}, \qquad i = 1, 2, \ldots, n.$$

Here each residual has been divided by an estimate of its standard deviation, creating a *t-like* statistic that is designed to give the analyst a scale-free quantity that provides information regarding the *size* of the residual. In addition, standard computer packages often give values of another set of studentized-type residuals, called the **R-Student values** and defined by

$$t_i = \frac{e_i}{s_{-i}\sqrt{1 - h_{ii}}}, \qquad i = 1, 2, \ldots, n,$$

where s_{-i} is an estimate of the error standard deviation, calculated with the ith data point deleted.

The two studentized residuals can be viewed as diagnostic procedures for detecting any deviation from the usual assumption that $E(E_i) = 0$ for a specific value of i. If there is reason to believe, for experimental or other reasons, that the ith observation is an **outlier** exerting a large influence on the fitted model, r_i or t_i may be informative. The R-Student values can be expected to be more sensitive to outliers than the r_i values. In fact, under the condition that $E(E_i) = 0$, t_i is a value of a random variable following a t distribution with $n - 1 - (k + 1) = n - k - 2$ degrees of freedom. Thus a two sided t-test can be used to provide information for detecting whether or not the ith point is an outlier.

Though the R-Student statistic, t_i, produces an exact t-test for detection of an outlier at a specific data location, the t distribution would not apply for simultaneously testing for outliers at all locations. As a result, the studentized residuals or R-Student values should be used strictly as diagnostic tools *without* formal hypothesis testing as the mechanism. The implication is that these statistics highlight data points where the error of fit is larger than what is expected by chance. Large R-Student values in magnitude suggest a need for "checking" the data with whatever resources possible. The practice of eliminating observations from regression data sets should not be done indiscriminantly. Further information regarding the use of outlier diagnostics is given in the publication by Cook and Weisberg, listed in the Bibliography.

Example 10.11 In a biological experiment conducted at the Virginia Polytechnic Institute and State University by the Department of Entomology, n experimental runs were made with two different methods for capturing grasshoppers. The methods are: Drop net catch and sweep net catch. The average number of grasshoppers caught in a set of field quadrants on a given date is recorded for each of the two methods. An additional regressor variable, the average plant height in the quadrants, was also recorded. The experimental data are as follows:

Observation	Drop Net Catch, y	Sweep Net Catch, x_1	Plant Height (cm), x_2
1	18.0000	4.15476	52.705
2	8.8750	2.02381	42.069
3	2.0000	0.15909	34.766
4	20.0000	2.32812	27.622
5	2.3750	0.25521	45.879
6	2.7500	0.57292	97.472
7	3.3333	0.70139	102.062
8	1.0000	0.13542	97.790
9	1.3333	0.12121	88.265
10	1.7500	0.10937	58.737
11	4.1250	0.56250	42.386
12	12.8750	2.45312	31.274
13	5.3750	0.45312	31.750
14	28.0000	6.68750	35.401
15	4.7500	0.86979	64.516
16	1.7500	0.14583	25.241
17	0.1333	0.01562	36.354

The goal is to be able to estimate grasshopper catch by using only the sweep net method, which is less costly. There was some concern about the validity of the fourth data point. The observed catch that was reported using the drop net method seemed unusually high, given the other conditions and, indeed, it was felt that the figure might be erroneous. Fit a model of the type

$$\mu_{Y|x_1, x_2} = \beta_0 + \beta_1 x_1 + \beta_2 x_2$$

to the 17 data points and study the residuals to determine if data point 4 is an outlier.

Solution. A computer program generated the fitted regression model

$$\hat{y} = 3.6870 + 4.1050x_1 - 0.0367x_2$$

along with the statistics $R^2 = 0.9244$ and $s^2 = 5.580$. The residuals and other diagnostic information were also generated and recorded as follows:

Obs.	y_i	$\hat{y}_i$	$y_i - \hat{y}_i$	h_{ii}	$s\sqrt{1 - h_{ii}}$	r_i	t_i
1	18.000	18.809	-0.809	0.2291	2.074	-0.390	-0.3780
2	8.875	10.452	-1.577	0.0766	2.270	-0.695	-0.6812
3	2.000	3.065	-1.065	0.1364	2.195	-0.485	-0.4715
4	20.000	12.231	7.769	0.1256	2.209	3.517	9.9315
5	2.375	3.052	-0.677	0.0931	2.250	-0.301	-0.2909
6	2.750	2.464	0.286	0.2276	2.076	0.138	0.1329
7	3.333	2.823	0.510	0.2669	2.023	0.252	0.2437
8	1.000	0.656	0.344	0.2318	2.071	0.166	0.1601
9	1.333	0.947	0.386	0.1691	2.153	0.179	0.1729
10	1.750	1.982	-0.232	0.0852	2.260	-0.103	-0.0989
11	4.125	4.442	-0.317	0.0884	2.255	-0.140	-0.1353
12	12.875	12.610	0.265	0.1152	2.222	0.119	0.1149
13	5.375	4.383	0.992	0.1339	2.199	0.451	0.4382
14	28.000	29.841	-1.841	0.6233	1.450	-1.270	-1.3005
15	4.750	4.891	-0.141	0.0699	2.278	-0.062	-0.0598
16	1.750	3.360	-1.610	0.1891	2.127	-0.757	-0.7447
17	0.133	2.418	-2.285	0.1386	2.193	-1.042	-1.0454

As expected, the residual at the fourth location appears to be unusually high, namely 7.769. The vital issue here is whether or not this residual is larger than one would expect by chance. The residual standard error for point 4 is 2.209. The R-Student value t_4 is found to be 9.9315. Viewing this as a value of a random variable having a t distribution with 13 degrees of freedom, one would certainly conclude that the residual of the fourth observation is estimating something greater than 0 and that the suspected measurement error is supported by the study of residuals. Notice that no other residual results in an R-Student value that produces any cause for alarm.

10.10 Cross Validation and PRESS Residuals

In many regression problems the experimenter must choose between various alternative models or model forms that are developed from the same data set. Quite often, in fact, the model that best predicts or estimates mean response is required. The experimenter should take into account the relative sizes of the s^2

values for the candidate models, and certainly the general nature of the confidence intervals on the mean response. One must also consider how well the model predicts response values that were **not used in building the candidate models**. The models should be subjected to **cross validation**. What is required, then, are cross-validation errors rather than fitting errors. Such errors in prediction are the **PRESS residuals**

$$\delta_i = y_i - \hat{y}_{i, -i}, \qquad i = 1, 2, \ldots, n,$$

where $\hat{y}_{i, -i}$ is the prediction of the ith data point by a model that did not make use of the ith point in the calculation of the coefficients. These PRESS residuals are easily calculated from the formula

$$\delta_i = \frac{e_i}{1 - h_{ii}}, \qquad i = 1, 2, \ldots, n.$$

The derivation can be found in the Montgomery and Peck reference.

Criteria that make use of the PRESS residuals are given by

$$\sum_{i=1}^{n} |\delta_i| \qquad \text{and} \qquad \text{PRESS} = \sum_{i=1}^{n} \delta_i^2.$$

The term *PRESS* is an acronym for the **Prediction Sum of Squares**. We suggest that both of these criteria be used. It is possible for PRESS to be dominated by one or only a few large PRESS residuals. Clearly, the criteria $\sum_{i=1}^{n} |\delta_i|$ is less sensitive to a small number of large values.

In the following example a "case-study" illustration is provided in which many candidate models are fit to a set of data and the best model is chosen. The sequential procedures described in Section 10.8 are not used. Rather, the role of the PRESS residuals and other statistical values in selecting the best regression equation is illustrated.

Example 10.12 Leg strength is a necessary ingredient of a successful punter in American football. One measure of the quality of a good punt is the "hang time." This is the time that the ball hangs in the air before being caught by the punt returner. To determine what leg strength factors influence hang time and to develop an empirical model for predicting this response, a study on "*The Relationship Between Selected Physical Performance Variables and Football Punting Ability*" was conducted by the Department of Health, Physical Education, and Recreation at the Virginia Polytechnic Institute and State University in 1983. Thirteen punters were chosen for the experiment and each punted a football 10 times. The average hang time, along with the strength measures used in the analysis, were recorded as follows:

Punter	Hang Time (sec), y	RLS, x_1	LLS, x_2	RHF, x_3	LHF, x_4	Power, x_5
1	4.75	170	170	106	106	240.57
2	4.07	140	130	92	93	195.49
3	4.04	180	170	93	78	152.99
4	4.18	160	160	103	93	197.09
5	4.35	170	150	104	93	266.56
6	4.16	150	150	101	87	260.56
7	4.43	170	180	108	106	219.25
8	3.20	110	110	86	92	132.68
9	3.02	120	110	90	86	130.24
10	3.64	130	120	85	80	205.88
11	3.68	120	140	89	83	153.92
12	3.60	140	130	92	94	154.64
13	3.85	160	150	95	95	240.57

Each regressor variable is defined as follows:

1. *RLS*—right leg strength, pounds.
2. *LLS*—left leg strength, pounds.
3. *RHF*—right hamstring muscle flexibility, degrees.
4. *LHF*—left hamstring muscle flexibility, degrees.
5. *Power*—overall leg strength, foot pounds.

Determine the most appropriate model for predicting hang time.

Solution. In the search for the "best" of the candidate models for predicting hang time, the information of Table 10.5 was obtained from a regression computer package. The models are ranked in ascending order of the values of the PRESS statistic. This display provides enough information on all possible models to enable the user to eliminate from consideration all but a few models. The model $x_2 x_5$ (*LLS* and *Power*) appears to be superior for predicting punter hang time. Also note that all models with low PRESS, low s^2, low $\sum_{i=1}^{n} |\delta_i|$ and high R^2 values contain these two variables.

In order to gain some insight from the residuals of the fitted regression

$$\hat{y}_i = b_0 + b_2 x_{2i} + b_5 x_{5i}$$

the residuals and PRESS residuals were generated. The actual prediction model (see Exercise 1 on page 396) is given by

$$\hat{y} = 1.10765 + 0.01370x_2 + 0.00429x_5.$$

Table 10.5 Comparing Different Regression Models

Model	s^2	$\sum \lvert \delta_i \rvert$	PRESS	R^2
$x_2 x_5$	0.036907	1.93583	0.54683	0.871300
$x_1 x_2 x_5$	0.041001	2.06489	0.58998	0.871321
$x_2 x_4 x_5$	0.037708	2.18797	0.59915	0.881658
$x_2 x_3 x_5$	0.039636	2.09553	0.66182	0.875606
$x_1 x_2 x_4 x_5$	0.042265	2.42194	0.67840	0.882093
$x_1 x_2 x_3 x_5$	0.044578	2.26283	0.70958	0.875642
$x_2 x_3 x_4 x_5$	0.042421	2.55789	0.86236	0.881658
$x_1 x_3 x_5$	0.053664	2.65276	0.87325	0.831580
$x_1 x_4 x_5$	0.056279	2.75390	0.89551	0.823375
$x_1 x_5$	0.059621	2.99434	0.97483	0.792094
$x_2 x_3$	0.056153	2.95310	0.98815	0.804187
$x_1 x_3$	0.059400	3.01436	0.99697	0.792864
$x_1 x_2 x_3 x_4 x_5$	0.048302	2.87302	1.00920	0.882096
x_2	0.066894	3.22319	1.04564	0.743404
$x_3 x_5$	0.065678	3.09474	1.05708	0.770971
$x_1 x_2$	0.068402	3.09047	1.09726	0.761474
x_3	0.074518	3.06754	1.13555	0.714161
$x_1 x_3 x_4$	0.065414	3.36304	1.15043	0.794705
$x_2 x_3 x_4$	0.062082	3.32392	1.17491	0.805163
$x_2 x_4$	0.063744	3.59101	1.18531	0.777716
$x_1 x_2 x_3$	0.059670	3.41287	1.26558	0.812730
$x_3 x_4$	0.080605	3.28004	1.28314	0.718921
$x_1 x_4$	0.069965	3.64415	1.30194	0.756023
x_1	0.080208	3.31562	1.30275	0.692334
$x_1 x_3 x_4 x_5$	0.059169	3.37362	1.36867	0.834936
$x_1 x_2 x_4$	0.064143	3.89402	1.39834	0.798692
$x_3 x_4 x_5$	0.072505	3.49695	1.42036	0.772450
$x_1 x_2 x_3 x_4$	0.066088	3.95854	1.52344	0.815633
x_5	0.111779	4.17839	1.72511	0.571234
$x_4 x_5$	0.105648	4.12729	1.87734	0.631593
x_4	0.186708	4.88870	2.82207	0.283819

Residuals, hat diagonal values, and PRESS residuals are listed in Table 10.6. Note the relatively good fit of the two variable regression model to the data. The PRESS residuals reflect the capability of the regression equation to predict hang time if independent predictions were to be made. For example, for punter number 4, the hang time of 4.180 would encounter a prediction error of 0.039 if the model constructed by using the remaining 12 punters were used. For this model, the average prediction error or cross validation

Table 10.6 PRESS Residuals

Punter	y_i	$\hat{y}_i$	$e_i = y_i - \hat{y}_i$	h_{ii}	δ_i
1	4.750	4.470	0.280	0.198	0.349
2	4.070	3.728	0.342	0.118	0.388
3	4.040	4.094	−0.054	0.444	−0.097
4	4.180	4.146	0.034	0.132	0.039
5	4.350	4.307	0.043	0.286	0.060
6	4.160	4.281	−0.121	0.250	−0.161
7	4.430	4.515	−0.085	0.298	−0.121
8	3.200	3.184	0.016	0.294	0.023
9	3.020	3.174	−0.154	0.301	−0.220
10	3.640	3.636	0.004	0.231	0.005
11	3.680	3.687	−0.007	0.152	−0.008
12	3.600	3.553	0.047	0.142	0.055
13	3.850	4.196	−0.346	0.154	−0.409

error is given by

$$\frac{\sum_{i=1}^{n} |\delta_i|}{13} = 0.1489 \text{ seconds},$$

which is quite small compared to the average hang time for the 13 punters.

Exercises

1. Consider the "hang time" punting data given in Example 10.12 on page 393. Using only the variables x_2 and x_5,
   (a) verify the regression equation shown on page 394;
   (b) predict punter hang time for a punter with $LLS = 180$ pounds and $Power = 260$ foot pounds;
   (c) construct a 95% confidence interval for the mean hang time of a punter with $LLS = 180$ pounds and $Power = 260$ foot pounds.

2. For the data of Exercise 11 on page 365, use the techniques of

   (a) *forward selection* with a 0.05 level of significance to choose a linear regression model;
   (b) *backward elimination* with a 0.05 level of significance to choose a linear regression model;
   (c) *stepwise regression* with a 0.05 level of significance to choose a linear regression model.

3. Use the techniques of *backward elimination* with $\alpha = 0.05$ to choose a prediction equation for the data of Table 10.3 on page 384.

4. For the punter data given in Example 10.12 on page 393, an additional response "punting distance" was also recorded. The following are average distance values for each of the 13 punters:

Punter	Distance, y (ft)
1	162.50
2	144.00
3	147.50
4	163.50
5	192.00
6	171.75
7	162.00
8	104.93
9	105.67
10	117.59
11	140.25
12	150.17
13	165.17

y	x
812.52	1
822.50	2
1211.50	3
1348.00	4
1301.00	8
2567.50	9
2526.50	10
2755.00	11
4390.50	12
5581.50	13
5548.00	14
6086.00	15
5764.00	16
8903.00	17

(a) Using the distance data rather than the hang times, estimate a multiple linear regression model of the type

$$\mu_{Y|x_1, x_2, x_3, x_4, x_5}$$
$$= \beta_0 + \beta_1 x_1 + \beta_2 x_2 + \beta_3 x_3 + \beta_4 x_4 + \beta_5 x_5$$

for predicting punting distance.

(b) Use stepwise regression with a significance level of 0.10 to select a combination of variables.

(c) Generate values for s^2, R^2, PRESS, and $\sum_{i=1}^{13} |\delta_i|$ for the entire set of 31 models. Use this information to determine the best combination of variables for predicting punting distance.

5. The following is a set of data for y, the amount of money (thousands of dollars) contributed to the alumni association at Virginia Tech by the Class of 1960, and x, the number of years following graduation:

(a) Fit a regression model of the type

$$\mu_{Y|x} = \beta_0 + \beta_1 x.$$

(b) Fit a quadratic model of the type

$$\mu_{Y|x} = \beta_0 + \beta_1 x + \beta_{11} x^2.$$

(c) Determine which of the models in (a) or (b) is preferable. Use s^2, R^2, and the PRESS residuals to support your decision.

6. For the model of Exercise 4(a), test the hypothesis

$$H_0: \quad \beta_4 = 0,$$
$$H_1: \quad \beta_4 \neq 0,$$

at the 0.05 level of significance.

7. For the quadratic model of Exercise 5(b), give estimates of the variances and covariance of the estimates of β_1 and β_{11}.

10.11 Ridge Regression

Sequential methods for arriving at a suitable regression equation cannot be used if one requires that all variables in the experiment contribute to the predicted response $\hat{y}$. If there is an excessive amount of multicollinearity among the regressor variables, the **A** matrix approaches a near-singular condition, resulting in extremely large values along the diagonal of $\mathbf{A}^{-1}$. In other words, the least squares procedure yields unbiased estimators for the regression coefficients but estimators that can have large variances. These large variances result in two practical difficulties with least squares estimators in the presence of severe multicollinearity: (1) The estimators can be very unstable, that is, sensitive to small, seemingly unimportant perturbations in the data; (2) the estimators tend to give results for the coefficients that are too large in magnitude, either positive or negative. This results from the fact that B_j^2 can have a large positive bias induced by multicollinearity even though B_j is unbiased. Since the correlation among the independent variables is often a natural phenomenon, one cannot always alleviate the difficulty brought about by the multicollinearity simply by changing the experimental plan or by obtaining additional data.

A solution to the problem of multicollinearity is to abandon the usual least squares procedure and resort to biased estimation techniques. In using a biased estimation procedure, one is essentially willing to allow for a certain amount of bias in the estimates in order to reduce their variances.

The biased estimates obtained here for the regression coefficients β_0, β_1, ..., β_k in the model

$$y = \beta_0 + \beta_1 x_1 + \beta_2 x_2 + \cdots + \beta_k x_k + \varepsilon,$$

are denoted by b_0^*, b_1^*, ..., b_k^* and are called **ridge regression estimates**. Suppose that we write the multiple linear regression model as

$$\mathbf{y} = \mathbf{X}\boldsymbol{\beta} + \boldsymbol{\varepsilon},$$

where **X** is defined as in Section 10.3 and $\boldsymbol{\beta}$ is given by

$$\boldsymbol{\beta} = \begin{bmatrix} \beta_0 \\ \beta_1 \\ \vdots \\ \beta_k \end{bmatrix}.$$

In what follows it is convenient to make a simple location and scale change on each regressor variable in the data (see Example 10.13) so that $\mathbf{1}'\mathbf{x}_j = 0$ and $\mathbf{x}_j'\mathbf{x}_j = 1$. This results in $\mathbf{X}'\mathbf{X}$ being a matrix of correlations among the regressor variables. Since **A** is a real $k + 1$ square symmetric matrix with **eigenvalues** or **characteristic roots** λ_1, λ_2, ..., λ_{k+1}, then from the theory of matrix algebra there exists an orthogonal matrix **P** such that

$$\mathbf{P}'\mathbf{A}\mathbf{P} = \mathbf{P}\mathbf{A}\mathbf{P}' = \text{diag}(\lambda_1, \lambda_2, ..., \lambda_{k+1}).$$

The rows of **P** are the **normalized eigenvectors** of the **A** matrix. Methods for obtaining the matrix **P** are demonstrated in most matrix algebra textbooks; however, for our purposes **P** is more easily obtained by using one of the computer programs that are readily available for high-speed computers.

Since $\mathbf{P}$ is orthogonal, the multiple linear regression model can be written in **canonical form** as

$$\mathbf{y} = \mathbf{X}^*\boldsymbol{\alpha} + \boldsymbol{\varepsilon},$$

where

$$\mathbf{X}^* = \mathbf{XP'}$$

and

$$\boldsymbol{\alpha} = \begin{bmatrix} \alpha_0 \\ \alpha_1 \\ \vdots \\ \alpha_k \end{bmatrix} = \mathbf{P}\boldsymbol{\beta}.$$

The estimates $\hat{\alpha}_0^*, \hat{\alpha}_1^*, \ldots, \hat{\alpha}_k^*$ of $\alpha_1, \alpha_2, \ldots, \alpha_k$, respectively, are related to the ridge regression estimates by the equation

$$\hat{\boldsymbol{\alpha}}^* = \mathbf{Pb}^*.$$

Once the $\hat{\alpha}_j^*$'s are determined, the ridge regression estimates are found by writing

$$\mathbf{b}^* = \mathbf{P'}\hat{\boldsymbol{\alpha}}^*.$$

Our problem therefore reduces to finding estimates of the $\hat{\alpha}_j^*$'s.

The reader can view the biased estimation that we discuss here as falling into two classes, **ridge** and **generalized ridge regression**, the former being a special case of the latter. For generalized ridge regression, consider regression performed on the model in canonical form. Also note that since the $\mathbf{P}$ matrix is orthogonal,

$$\hat{\mathbf{a}}^{*\prime}\hat{\mathbf{a}}^* = \mathbf{b}^{*\prime}\mathbf{b}^*.$$

Generalized ridge regression essentially reduces the length of the coefficient vector (which is often too large with least squares under multicollinearity) from that of least squares. The error sum of squares for the canonical model

$$\mathbf{y} = \mathbf{X}^*\hat{\boldsymbol{\alpha}}^* + \mathbf{e},$$

is minimized with constraints applied of the type $\alpha_j^{*2} = \rho_j$, $j = 0, 1, 2, \ldots, k$, where the ρ_j are positive finite constants. In other words, least squares is applied but the squared coefficients, and hence the coefficients themselves, are not allowed to grow arbitrarily large. The minimization process requires the use of $k + 1$ **Lagrangian multipliers**, which we shall designate as $d_0, d_1, d_2, \ldots, d_k$. We shall omit the details. However, when the derivatives with respect to the estimates are equated to zero, we obtain the system of equations

$$(\mathbf{A}^* + \mathbf{D})\hat{\boldsymbol{\alpha}}^* = \mathbf{g}^*,$$

in which

$$\mathbf{A}^* = \mathbf{X}^{*\prime}\mathbf{X}^*$$

and

$$\mathbf{g}^* = \mathbf{X}^{*\prime}\mathbf{y}.$$

The solution of this system of equations leads to the estimates of the ridge regression coefficients as stated in Definition 10.1.

Definition 10.1 *The **generalized ridge regression estimates** are given by*

$$\mathbf{b}^* = \mathbf{P}'\hat{\boldsymbol{\alpha}}^*,$$

where

$$\hat{\boldsymbol{\alpha}}^* = (\mathbf{A}^* + \mathbf{D})^{-1}\mathbf{g}^*$$

and

$$\mathbf{D} = \mathrm{diag}\,(d_0, d_1, \ldots, d_k)$$

with $d_j > 0$ for $j = 0, 1, 2, \ldots, k$.

By restricting the magnitudes of our coefficients in the minimization procedure, we have, in effect, added constants to the diagonal elements of $\mathbf{A}^*$ and thereby introduced the bias into our estimates. However, the addition of these constants to the diagonal elements of $\mathbf{A}^*$ causes the matrix to behave as if the variables are orthogonal to each other. Consequently, by requiring the d_j's to be positive in Definition 10.1, the elements on the diagonal of $(\mathbf{A}^* + \mathbf{D})^{-1}$ will be smaller, indicating greater stability in the estimates of the coefficients.

Definition 10.1 outlines the method for finding generalized ridge regression estimates but gives no insight into the specific numerical values of the d_j's that one should add to the main diagonal elements of the $\mathbf{A}^*$ matrix. It would seem reasonable that the optimal values assigned to the d_j's should yield $\hat{\alpha}_j^*$'s and hence b_j^*'s that minimize the quantity

$$\sum_{j=0}^{k} E(B_j^* - \beta_j)^2.$$

This is accomplished when the d_j's are given by

$$d_j = \frac{\sigma^2}{\alpha_j^2}, \qquad j = 0, 1, 2, \ldots, k.$$

Unfortunately, σ^2 and the α_j's are unknown and estimates must be used. In practice one estimates σ^2 by s^2 using ordinary least squares procedures. The estimates of the α_j's are of course given in Definition 10.1 once the d_j's are known. To circumvent this seemingly hopeless situation, one must resort to the following iterative procedure.

Generalized Ridge Regression

STEP 1 Using ordinary least squares procedures on the canonical model, estimate the α_j's by computing

$$\hat{\boldsymbol{\alpha}} = \mathbf{A}^{*-1}\mathbf{g}^*$$

and estimate σ^2 by s^2.

STEP 2 Use the value of s^2 and the $\hat{\alpha}_j$'s from step 1 to compute

$$\hat{d}_j = \frac{s^2}{\hat{\alpha}_j^2}, \qquad j = 0, 1, 2, \ldots, k.$$

STEP 3 Use the $\hat{d}_j$'s to solve the expression

$$\boldsymbol{\alpha}^* = (\mathbf{A}^* + \mathbf{D})^{-1}\mathbf{g}^*$$

and thus obtain initial estimates of the $\hat{\alpha}_j^*$'s. Next compute

$$\hat{\boldsymbol{\alpha}}^{*'}\hat{\boldsymbol{\alpha}}^* = \sum_{j=0}^{k} \hat{\alpha}_j^{*2}.$$

STEP 4 Repeat steps 2 and 3 using the $\hat{\alpha}_j^*$'s from step 3 and again compute $\hat{\boldsymbol{\alpha}}^{*'}\hat{\boldsymbol{\alpha}}^*$.

STEP 5 Continue this iterative procedure and terminate only when stability is achieved in $\hat{\boldsymbol{\alpha}}^{*'}\hat{\boldsymbol{\alpha}}^*$.

STEP 6 The generalized ridge regression coefficients are now computed from the formula $\mathbf{b}^* = \mathbf{P}'\hat{\boldsymbol{\alpha}}^*$.

Example 10.13 Compute the generalized regression estimates for the parameters of the multiple linear regression model

$$\mu_{Y|x_1, x_2, x_3, x_4} = \beta_0 + \beta_1 x_1 + \beta_2 x_2 + \beta_3 x_3 + \beta_4 x_4$$

for the data of Table 10.3.

Solution. Suppose that we code the data by letting

$$x_j' = \frac{x_j - \bar{x}_j}{\sqrt{n-1}\, s_j}, \qquad j = 1, 2, 3, 4,$$

where

$$s_j^2 = \frac{\sum_{i=1}^{n}(x_{ji} - \bar{x}_j)^2}{n-1},$$

and find the generalized ridge regression estimates for the equation

$$\mu_{Y|x_1', x_2', x_3', x_4'} = \beta_0' + \beta_1' x_1' + \beta_2' x_2' + \beta_3' x_3' + \beta_4' x_4'.$$

In this form we find that the first column of $\mathbf{X}$ (column of ones) is orthogonal to the other columns. Consequently, the ridge regression estimate b_0^* of β_0' is simply $\bar{y} = 60.9667$, the average of the y values, while b_1^*, b_2^*, b_3^*, and b_4^* can be found from the iterative procedure described above but with the first column of $\mathbf{X}$ deleted. Now the resulting 4×4 $\mathbf{A}$ matrix is just the correlation matrix given in Example 10.10.

To find the x' values of the $\mathbf{X}$ matrix and the corresponding correlation matrix $\mathbf{A}$ for the data of Table 10.3, we must first compute

$$\begin{array}{ll} \bar{x}_1 = 81.0000 & s_1 = 11.5866 \\ \bar{x}_2 = 48.7778 & s_2 = 4.3979 \\ \bar{x}_3 = 3.1311 & s_3 = 1.1178 \\ \bar{x}_4 = 29.6333 & s_4 = 3.1149. \end{array}$$

Then the $\mathbf{P}$ matrix, whose rows are the normalized eigenvectors of the correlation matrix, is given by

$$\mathbf{P} = \begin{bmatrix} 0.5729 & 0.4776 & 0.7987 & 0.4416 \\ -0.3586 & -0.5778 & 0.4592 & 0.5716 \\ 0.0381 & -0.1453 & 0.7071 & -0.6910 \\ -0.7361 & 0.6457 & 0.2011 & 0.0294 \end{bmatrix}.$$

Next one can determine $\mathbf{X}^* = \mathbf{X}'\mathbf{P}$ and $\mathbf{A}^* = \mathbf{X}^{*\prime}\mathbf{X}^*$, and then proceed to step 1 of the iterative procedure. Solving $\hat{\boldsymbol{\alpha}} = \mathbf{A}^{*-1}\mathbf{g}^*$ gives

$$\hat{\alpha}_1 = 10.9281$$
$$\hat{\alpha}_2 = -2.0833$$
$$\hat{\alpha}_3 = 5.8716$$
$$\hat{\alpha}_4 = 5.3672.$$

In Section 10.8 we found the error mean square for the complete model to be $s^2 = 0.7414$. Therefore, from step 2,

$$\hat{d}_1 = 0.0062$$
$$\hat{d}_2 = 0.1708$$
$$\hat{d}_3 = 0.0215$$
$$\hat{d}_4 = 0.0257.$$

Steps 3, 4, and 5 are now carried out and the results displayed in Table 10.6.

As a result of the convergence in $\hat{\boldsymbol{\alpha}}^{*\prime}\hat{\boldsymbol{\alpha}}^*$, the process was terminated after the third iteration. From step 6 the generalized ridge regression coefficients are found to be

$$\hat{b}_1^* = 7.0687$$
$$\hat{b}_2^* = 5.4613$$
$$\hat{b}_3^* = 8.3069$$
$$\hat{b}_4^* = 0.2284$$

and the estimated regression equation, in terms of the coded variables, is given by

$$\hat{y} = 60.9667 + 7.0687x_1' + 5.4613x_2' + 8.3069x_3' + 0.2284x_4',$$

or, converting back to the original variables, assumes the form

$$\hat{y} = 13.0851 + 0.2157x_1 + 0.4390x_2 + 2.6274x_3 + 0.0259x_4.$$

Table 10.6 Iterations in Generalized Ridge Regression

Iteration	$\hat{d}_1$	$\hat{d}_2$	$\hat{d}_3$	$\hat{d}_4$	$\hat{\alpha}_1^*$	$\hat{\alpha}_2^*$	$\hat{\alpha}_3^*$	$\hat{\alpha}_4^*$	$\hat{\alpha}^{*\prime}\hat{\alpha}^*$
1	0.0062	0.1708	0.0215	0.0257	10.9017	−1.8291	5.3221	0.6011	150.8787
2	0.0062	0.2216	0.0262	2.0521	10.9015	−1.7650	5.2161	0.0085	149.1657
3	0.0062	0.2380	0.0272	10,320	10.9015	−1.7453	5.1923	0.0000	148.8488

The generalized ridge regression described here is one of several formal procedures for biased estimation, the purpose of which is to reduce the variance of the estimators of the regression coefficients, even though the resulting estimators are biased. The procedure is conceptually very important. However, since it requires the computation of k of the d_j parameters, many researchers do not find it practical. Ordinary ridge regression obtains regression estimates by minimizing the error sum of squares for the model

$$\mathbf{y} = \mathbf{Xb^*} + \mathbf{e}$$

subject to the *single* constraint that $\sum_{j=0}^{k} b_j^{*2} = \rho$, where ρ is again a finite positive constant. The method of Lagrange multipliers requires the differentiation of

$$L = \sum_{i=1}^{n}(y_i - b_0^* - b_1^* x_{1i} - b_2^* x_{2i} - \cdots - b_k^* x_{ki})^2 + d\left(\sum_{j=0}^{k} b_j^{*2} - \rho\right)$$

with respect to $b_0, b_1, \ldots, b_k$. When these derivatives are equated to zero, we obtain the system of equations

$$(\mathbf{A} + d\mathbf{I})\mathbf{b^*} = \mathbf{g},$$

which can be solved for the ridge regression estimates of the coefficients in terms of the Lagrange multiplier d.

Definition 10.2   **Ridge regression estimates** *are given by*

$$\mathbf{b^*} = (\mathbf{A} + d\mathbf{I})^{-1}\mathbf{g},$$

where $d > 0$.

In Definition 10.2 the ridge regression estimates are computed for various increasing values of d, beginning with $d = 0$, until we determine a value of d for which all the regression coefficients appear to have stabilized. Several calculations may be required before the estimates of the coefficients reach stability. By plotting the values of the coefficients against the successive values of d, we obtain a curve referred to as the **ridge trace**.

The purpose of the ridge trace is to indicate, for a given set of data, a set of estimates that are reasonable. Sometimes, unfortunately, it is difficult to determine an appropriate value of d for which the estimates of all the coefficients have stabilized. The method for obtaining the ridge regression estimates as stated in Definition 10.2 is illustrated in Example 10.14.

Example 10.14 Use Definition 10.2 to estimate the ridge regression coefficients of the parameters β_1', β_2', β_3', and β_4' in Example 10.13 for various values of d and then plot the ridge trace.

Solution. The computations were carried out on a computer and are summarized in Table 10.7.

The ridge trace is illustrated in Figure 10.1. It was decided that all the coefficients had sufficiently stabilized at $d = 0.1$. Therefore, the ridge regression estimates of the coefficients of the coded variables are given by

$$b_1^* = 6.7426$$
$$b_2^* = 5.6694$$
$$b_3^* = 7.1994$$
$$b_4^* = 0.8067.$$

As one would expect, the estimates are not the same as those computed by the generalization ridge regression procedure. Observe from the ridge trace how relatively unstable all the estimates seem to be in the vicinity of $d = 0$, where the b_j^*'s are the unbiased least squares estimates.

Table 10.7 Ridge Regression Estimates

d	b_1^*	b_2^*	b_3^*	b_4^*
0.000	3.2803	9.0361	9.7245	-0.2647
0.004	5.4449	7.1270	9.0451	-0.2790
0.012	6.3413	6.3189	8.6327	-0.1789
0.020	6.5993	6.0695	8.4053	-0.0634
0.100	6.7426	5.6694	7.1994	0.8067
0.200	6.5000	5.4970	6.3675	1.3844
0.300	6.2533	5.3273	5.8221	1.7013
0.400	6.0225	5.1583	5.4220	1.8870

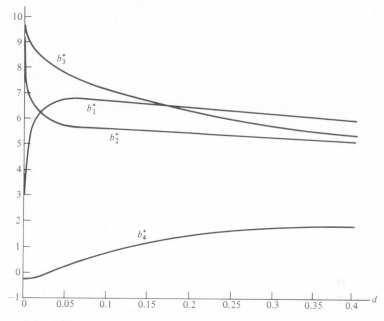

Figure 10.1   Ridge trace.

Exercises

1. Compute the generalized ridge regression estimates for the parameters of the multiple linear regression equation

$$\mu_{Y|x_1, x_2, x_3, x_4} = \beta_0 + \beta_1 x_1 + \beta_2 x_2 + \beta_3 x_3 + \beta_4 x_4$$

for the data of Exercise 11 on page 365.

2. Use Definition 10.2 to compute ridge regression estimates for the model in Exercise 1. Plot the ridge trace.

11

Analysis of Variance

11.1 Analysis-of-Variance Technique

In the estimation and hypotheses testing material covered in Chapters 7 and 8, we were restricted in each case to considering no more than two population parameters. Such was the case, for example, in testing for the equality of two population means using independent samples from normal populations with common but unknown variance, where it was necessary to obtain a pooled estimate of σ^2. It would seem obvious and desirable that the reader be able to extend the techniques developed so far to cover tests of hypotheses in which there are, say, k population means. One very common procedure that has been developed to test such a hypothesis involves the analysis of variance.

The analysis of variance is certainly not a new technique if the reader has followed the material on regression theory. We used the analysis-of-variance approach to partition the total sum of squares into a portion due to regression and a portion due to error. Further, we were able, in some cases, to conveniently partition SSR into meaningful components for the purpose of testing relevant hypotheses on the parameters in the model. The term *analysis of variance* describes a technique whereby the total variation is being analyzed or divided into meaningful components.

Regression problems in which the model contains *quantitative variables* (like those we have discussed to this point) are not the only type in which the analysis of variance plays an important role. In this chapter we shall present

Table 11.1 Absorption of Moisture in Concrete Aggregates

	Aggregate (weight %)					
	1	2	3	4	5	
	551	595	639	417	563	
	457	580	615	449	631	
	450	508	511	517	522	
	731	583	573	438	613	
	499	633	648	415	656	
	632	517	677	555	679	
Total	3320	3416	3663	2791	3664	16,854
Mean	553.33	569.33	610.50	465.17	610.67	561.80

and study other types of models in which this technique is used. The degree of difficulty of the analysis depends on the complexity of the problem.

Suppose in an industrial experiment that an engineer is interested in how the mean absorption of moisture in concrete varies among five different concrete aggregates. The samples are exposed to moisture for 48 hours. It is decided that 6 samples are to be tested for each aggregate, requiring a total of 30 samples to be tested. The data are recorded in Table 11.1.

The model for this situation may be considered as follows. There are 6 observations taken from each of 5 populations with means μ_1, μ_2, ..., μ_5, respectively. We might wish to test

$$H_0: \quad \mu_1 = \mu_2 = \cdots = \mu_5,$$
$$H_1: \quad \text{At least two of the means are not equal.}$$

In addition, we might be interested in making individual comparisons among these 5 population means.

In the analysis-of-variance procedure, it is assumed that whatever variation exists between the aggregate averages is attributed to (1) variation in absorption among observations within aggregate types, and (2) variation due to aggregate types, that is, due to differences in the chemical composition of the aggregates. The **within-aggregate variation** is, of course, brought about by various causes. Perhaps humidity and temperature conditions were not kept entirely constant throughout the experiment. It is possible that there was a certain amount of heterogeneity in the batches of raw materials that were used. At any rate, we shall consider the within-sample variation to be **chance** or **random variation**, and part of the goal of the analysis of variance is to determine if the differences between the 5 sample means are what one would expect due to random variation alone or if indeed there is also a contribution from the systematic variation attributed to the aggregate types. The procedure essentially, then, separates the total variability into the following important two components:

1. Variability between aggregates, measuring systematic and random variation.
2. Variability within aggregates, measuring only random variation.

There remains then the task of determining if component 1 is significantly larger than component 2.

Many pointed questions appear at this stage concerning the preceding problem. For example, how many samples must be tested in each aggregate? This is a question that continually haunts the practitioner. In addition, what if the within-sample variation is so large that it is difficult for a statistical procedure to detect the systematic differences? Can we systematically control extraneous sources of variation and thus remove them from the portion we call random variation? We shall attempt to answer these and other questions in the following sections.

11.2 One-Way Classification

Random samples of size n are selected from each of k populations. The k different populations are classified on the basis of a single criterion such as different treatments or groups. Today the term **treatment** is used very generally to refer to the various classifications, whether they be different aggregates, different analysts, different fertilizers, or different regions of the country. It will be assumed that the k populations are independent and normally distributed with means $\mu_1, \mu_2, \ldots, \mu_k$ and common variance σ^2. We wish to derive appropriate methods for testing the hypothesis

$$H_0: \quad \mu_1 = \mu_2 = \cdots = \mu_k,$$
$$H_1: \quad \text{At least two of the means are not equal.}$$

Let y_{ij} denote the jth observation from the ith treatment and arrange the data as in Table 11.2. Here, $T_i.$ is the total of all observations in the sample from the ith treatment, $\bar{y}_i.$ is the mean of all observations in the sample from the ith treatment, $T..$ is the total of all nk observations, and $\bar{y}..$ is the mean of all nk observations. Each observation may be written in the form

$$y_{ij} = \mu_i + \varepsilon_{ij},$$

where ε_{ij} measures the deviation of the jth observation of the ith sample from the corresponding treatment mean. The ε_{ij} term represents random error and

Table 11.2 *k* Random Samples

	Treatment						
	1	2	$\cdots$	i	$\cdots$	k	
	y_{11}	y_{21}	$\cdots$	y_{i1}	$\cdots$	y_{k1}	
	y_{12}	y_{22}	$\cdots$	y_{i2}	$\cdots$	y_{k2}	
	$\vdots$	$\vdots$		$\vdots$		$\vdots$	
	y_{1n}	y_{2n}	$\cdots$	y_{in}	$\cdots$	y_{kn}	
Total	$T_1.$	$T_2.$	$\cdots$	$T_i.$	$\cdots$	$T_k.$	$T..$
Mean	$\bar{y}_1.$	$\bar{y}_2.$	$\cdots$	$\bar{y}_i.$	$\cdots$	$\bar{y}_k.$	$\bar{y}..$

plays the same role as the error terms in the regression models. An alternative and preferred form of this equation is obtained by substituting $\mu_i = \mu + \alpha_i$, subject to the constraint $\sum_{i=1}^{k} \alpha_i = 0$. Hence we may write

$$y_{ij} = \mu + \alpha_i + \varepsilon_{ij},$$

where μ is just the **grand mean** of all the μ_i's; that is,

$$\mu = \frac{\sum_{i=1}^{k} \mu_i}{k},$$

and α_i is called the **effect** of the *i*th treatment.

The null hypothesis that the *k* population means are equal against the alternative that at least two of the means are unequal may now be replaced by the equivalent hypothesis,

$$H_0: \quad \alpha_1 = \alpha_2 = \cdots = \alpha_k = 0,$$
$$H_1: \quad \text{At least one of the } \alpha_i\text{'s is not equal to zero.}$$

Our test will be based on a comparison of two independent estimates of the common population variance σ^2. These estimates will be obtained by splitting the total variability of our data, designated by the double summation $\sum_{i=1}^{k} \sum_{j=1}^{n} (y_{ij} - \bar{y}..)^2$, into two components.

Theorem 11.1   Sum-of-Squares Identity

$$\sum_{i=1}^{k} \sum_{j=1}^{n} (y_{ij} - \bar{y}..)^2 = n \sum_{i=1}^{k} (\bar{y}_i. - \bar{y}..)^2 + \sum_{i=1}^{k} \sum_{j=1}^{n} (y_{ij} - \bar{y}_i.)^2.$$

Proof

$$\sum_{i=1}^{k} \sum_{j=1}^{n} (y_{ij} - \bar{y}..)^2 = \sum_{i=1}^{k} \sum_{j=1}^{n} [(\bar{y}_i. - \bar{y}..) + (y_{ij} - \bar{y}_i.)]^2$$

$$= \sum_{i=1}^{k} \sum_{j=1}^{n} [(\bar{y}_i. - \bar{y}..)^2 + 2(\bar{y}_i. - \bar{y}..)(y_{ij} - \bar{y}_i.)$$

$$+ (y_{ij} - \bar{y}_i.)^2]$$

$$= \sum_{i=1}^{k} \sum_{j=1}^{n} (\bar{y}_i. - \bar{y}..)^2$$

$$+ 2 \sum_{i=1}^{k} \sum_{j=1}^{n} (\bar{y}_i. - \bar{y}..)(y_{ij} - \bar{y}_i.)$$

$$+ \sum_{i=1}^{k} \sum_{j=1}^{n} (y_{ij} - \bar{y}_i.)^2.$$

The middle term is zero, since

$$\sum_{j=1}^{n} (y_{ij} - \bar{y}_i.) = \sum_{j=1}^{n} y_{ij} - n\bar{y}_i. = \sum_{j=1}^{n} y_{ij} - n \frac{\sum_{j=1}^{n} y_{ij}}{n} = 0.$$

The first sum does not have j as a subscript and therefore may be written as

$$\sum_{i=1}^{k} \sum_{j=1}^{n} (\bar{y}_i. - \bar{y}..)^2 = n \sum_{i=1}^{k} (\bar{y}_i. - \bar{y}..)^2.$$

Hence

$$\sum_{i=1}^{k} \sum_{j=1}^{n} (y_{ij} - \bar{y}..)^2 = n \sum_{i=1}^{k} (\bar{y}_i. - \bar{y}..)^2 + \sum_{i=1}^{k} \sum_{j=1}^{n} (y_{ij} - \bar{y}_i.)^2.$$

It will be convenient in what follows to identify the terms of the sum-of-squares identity by the following notation:

$$SST = \sum_{i=1}^{k} \sum_{j=1}^{n} (y_{ij} - \bar{y}..)^2 = \text{total sum of squares}$$

$$SSA = n \sum_{i=1}^{k} (\bar{y}_i. - \bar{y}..)^2 = \text{treatment sum of squares}$$

$$SSE = \sum_{i=1}^{k} \sum_{j=1}^{n} (y_{ij} - \bar{y}_i.)^2 = \text{error sum of squares}.$$

The sum-of-squares identity can then be represented symbolically by the equation

$$SST = SSA + SSE.$$

As implied earlier, we need to compare the appropriate measure of the between-treatment variation with the within-treatment variation in order to detect significant differences in the observations due to the treatment effects. Suppose that we look at the expected value of the treatment sum of squares.

Theorem 11.2

$$E(SSA) = (k - 1)\sigma^2 + n \sum_{i=1}^{k} \alpha_i^2.$$

Proof. Looking upon *SSA* as a random variable whose values would undoubtedly vary if the experiment were repeated several times, we now write

$$SSA = n \sum_{i=1}^{k} (\bar{Y}_{i.} - \bar{Y}_{..})^2.$$

From the model

$$Y_{ij} = \mu + \alpha_i + E_{ij}$$

we obtain

$$\bar{Y}_{i.} = \mu + \alpha_i + \bar{E}_{i.}.$$
$$\bar{Y}_{..} = \mu + \bar{E}_{..},$$

since $\sum_{i=1}^{k} \alpha_i = 0$. Hence

$$SSA = n \sum_{i=1}^{k} (\alpha_i + \bar{E}_{i.} - \bar{E}_{..})^2$$

and

$$E(SSA) = n \sum_{i=1}^{k} \alpha_i^2 + n \sum_{i=1}^{k} E(\bar{E}_{i.}^2) - nkE(\bar{E}_{..}^2) + 2n \sum_{i=1}^{k} \alpha_i E(\bar{E}_{i.}).$$

Recalling that the E_{ij}'s are independent variables with mean zero and variance σ^2, we find that

$$E(\bar{E}_{i.}^2) = \frac{\sigma^2}{n}, \qquad E(\bar{E}_{..}^2) = \frac{\sigma^2}{nk}, \qquad E(\bar{E}_{i.}) = 0.$$

Therefore,

$$E(SSA) = n \sum_{i=1}^{k} \alpha_i^2 + k\sigma^2 - \sigma^2$$

$$= (k - 1)\sigma^2 + n \sum_{i=1}^{k} \alpha_i^2.$$

One estimate of σ^2, based on $k - 1$ degrees of freedom, is given by the treatment mean square

$$s_1^2 = \frac{SSA}{k - 1}.$$

If H_0 is true and thus each α_i in Theorem 11.2 is equal to zero, we see that

$$E\left(\frac{SSA}{k - 1}\right) = \sigma^2$$

and s_1^2 is an unbiased estimate of σ^2. However, if H_1 is true, we have

$$E\left(\frac{SSA}{k-1}\right) = \sigma^2 + \frac{n \sum_{i=1}^{k} \alpha_i^2}{k-1}$$

and s_1^2 estimates σ^2 plus an additional term, which measures variation due to the systematic effects.

A second and independent estimate of σ^2, based on $k(n-1)$ degrees of freedom, is the familiar formula

$$s^2 = \frac{SSE}{k(n-1)}.$$

The estimate s^2 is unbiased regardless of the truth or falsity of the null hypothesis (see Exercise 1 on page 420). It is important to note that the sum-of-squares identity has not only partitioned the total variability of the data, but also the total number of degrees of freedom. That is,

$$nk - 1 = k - 1 + k(n-1).$$

When H_0 is true, the ratio

$$f = \frac{s_1^2}{s^2}$$

is a value of the random variable F having the F distribution with $k-1$ and $k(n-1)$ degrees of freedom. Since s_1^2 overestimates σ^2 when H_0 is false, we have a one-tailed test with the critical region entirely in the right tail of the distribution. The null hypothesis H_0 is rejected at the α level of significance when

$$f > f_\alpha[k-1, k(n-1)].$$

The previously defined formulas for SST, SSA, and SSE are not in the best form for doing computations. In actual practice, one computes SST and SSA by the following equivalent formulas and then, making use of the sum-of-squares identity, obtains SSE by subtraction.

Sum-of-Squares Computing Formulas; Equal Sample Sizes

$$SST = \sum_{i=1}^{k} \sum_{j=1}^{n} y_{ij}^2 - \frac{T_{..}^2}{nk}$$

$$SSA = \frac{\sum_{i=1}^{k} T_{i.}^2}{n} - \frac{T_{..}^2}{nk}.$$

$$SSE = SST - SSA.$$

Table 11.3 Analysis of Variance for the One-Way Classification

Source of Variation	Sum of Squares	Degrees of Freedom	Mean Square	Computed f
Treatments	SSA	$k - 1$	$s_1^2 = \dfrac{SSA}{k - 1}$	$\dfrac{s_1^2}{s^2}$
Error	SSE	$k(n - 1)$	$s^2 = \dfrac{SSE}{k(n - 1)}$	
Total	SST	$nk - 1$		

The computations in an analysis-of-variance problem are usually summarized in tabular form as shown in Table 11.3.

Example 11.1 Test the hypothesis $\mu_1 = \mu_2 = \cdots = \mu_5$ at the 0.05 level of significance for the data of Table 11.1 on absorption of moisture by various types of cement aggregates.

Solution

1. H_0: $\mu_1 = \mu_2 = \cdots = \mu_5$.
2. H_1: At least two of the means are not equal.
3. $\alpha = 0.05$.
4. Critical region: $f > 2.76$ with $v_1 = 4$ and $v_2 = 25$ degrees of freedom.
5. Computations:

$$SST = 551^2 + 457^2 + \cdots + 679^2 - \frac{16{,}854^2}{30}$$

$$= 9{,}677{,}954 - 9{,}468{,}577$$

$$= 209{,}377$$

$$SSA = \frac{3320^2 + 3416^2 + \cdots + 3664^2}{6} - 9{,}468{,}577$$

$$= 85{,}356$$

$$SSE = 209{,}377 - 85{,}356 = 124{,}021.$$

These results and the remaining computations are exhibited in Table 11.4.
6. Decision: Reject H_0 and conclude that the aggregates do not have the same mean absorption.

Table 11.4 Analysis of Variance for the Data of Table 11.1

Source of Variation	Sum of Squares	Degrees of Freedom	Mean Square	Computed f
Aggregates	85,356	4	21,339	4.30
Error	124,021	25	4,961	
Total	209,377	29		

In experimental work one often loses some of the desired observations. For example, an experiment might be conducted to determine if college students obtain different grades on the average for classes meeting at different times of the day. Because of dropouts during the semester it is entirely possible to conclude the experiment with unequal numbers of students in the various sections. The previous analysis for equal sample size will still be valid by slightly modifying the sum of squares formulas. We now assume the k random samples to be of size $n_1, n_2, \ldots, n_k$, respectively, with $N = \sum_{i=1}^{k} n_i$. The computing formulas for SST, SSA, and SSE are now given by

Sum-of-Squares Computing Formulas; Unequal Sample Sizes

$$SST = \sum_{i=1}^{k} \sum_{j=1}^{n_i} y_{ij}^2 - \frac{T_{..}^2}{N},$$

$$SSA = \sum_{i=1}^{k} \frac{T_{i.}^2}{n_i} - \frac{T_{..}^2}{N},$$

$$SSE = SST - SSA.$$

The degrees of freedom are then partitioned as before: $N - 1$ for SST, $k - 1$ for SSA, and $N - 1 - (k - 1) = N - k$ for SSE.

Example 11.2 Part of the study *"Serum Inorganic Phosphorus Levels in Children with Seizure Disorders Taking Anticonvulsant Drugs,"* conducted at the Virginia Polytechnic Institute and State University in 1982, was designed to measure serum alkaline phosphatase activity levels (Bessey–Lowry Units) in children with seizure disorders who were receiving anticonvulsant therapy under the care of a private physician. Forty-five subjects were found for the study and categorized into four drug groups:

> G-1: control (not receiving anticonvulsants and having
>   no history of seizure disorders),

Table 11.5 Serum Alkaline Phosphatase Activity
Level

Drug Group				
G-1		G-2	G-3	G-4
49.20	97.50	97.07	62.10	110.60
44.54	105.00	73.40	94.95	57.10
45.80	58.05	68.50	142.50	117.60
95.84	86.60	91.85	53.00	77.71
30.10	58.35	106.60	175.00	150.00
36.50	72.80	0.57	79.50	82.90
82.30	116.70	0.79	29.50	111.50
87.85	45.15	0.77	78.40	
105.00	70.35	0.81	127.50	
95.22	77.40			

G-2: phenobarbital,

G-3: carbamazepine,

G-4: other anticonvulsants.

From blood samples collected on each subject the serum alkaline phosphatase
activity level was determined and recorded in Table 11.5.
Test the hypothesis at the 0.05 level of significance that the average serum
alkaline phosphatase activity level is the same for the four drug groups.

Solution

1. H_0: $\mu_1 = \mu_2 = \mu_3 = \mu_4$.
2. H_1: At least two of the means are not equal.
3. $\alpha = 0.05$.
4. Critical region: $f > 2.836$, by interpolating in Table A.6.
5. Computations: $T_1. = 1460.25$, $T_2. = 440.36$, $T_3. = 842.45$, $T_4. = 707.41$,
   and $T.. = 3450.47$. Therefore,

$$SST = 49.20^2 + 44.54^2 + \cdots + 111.50^2 - \frac{3450.47^2}{45}$$

$$= 331{,}886.9701 - 264{,}572.0716 = 67{,}314.8985,$$

$$SSA = \frac{1460.25^2}{20} + \frac{440.36^2}{9} + \frac{842.45^2}{9} + \frac{707.41^2}{7} - \frac{3450.47^2}{45}$$

$$= 278{,}510.6729 - 264{,}572.0716 = 13{,}938.6013,$$

$$SSE = 67{,}314.8985 - 13{,}938.6013 = 53{,}376.2972.$$

5. Computations: First compute

$$s_1^2 = 662.862, \qquad s_2^2 = 2219.781, \qquad s_3^2 = 2168.434, \qquad s_4^2 = 946.032,$$

and then

$$s_p^2 = \frac{(19)(662.862) + (8)(2219.781) + (8)(2168.434) + (6)(946.032)}{41}$$

$$= 1301.861.$$

Now,

$$b = \frac{[(662.862)^{19}(2219.781)^8(2168.434)^8(946.032)^6]^{1/41}}{1301.861}$$

$$= 0.8557.$$

6. Decision: Accept the hypothesis and conclude that the population variances of the four drug groups are equal.

Although Bartlett's test is most often used in testing for homogeneity of variances, there are other methods available. A method due to Cochran provides a computationally simple procedure, but it is restricted to situations in which the sample sizes are equal. **Cochran's test** is particularly useful in detecting if one variance is much larger than the others. The statistic that is used is given by

$$G = \frac{\text{largest } S_i^2}{\displaystyle\sum_{i=1}^{k} S_i^2}$$

and the hypothesis of equality of variances is rejected if $g > g_\alpha$, where the value of g_α is obtained from Table A.11.

To illustrate Cochran's test, let us refer again to the data of Table 11.1 on the absorption of moisture in concrete aggregates. Were we justified in assuming equal variances when we performed the analysis of variance in Example 11.1? We find that

$$s_1^2 = 12{,}134, \qquad s_2^2 = 2303, \qquad s_3^2 = 3594, \qquad s_4^2 = 3319, \qquad s_5^2 = 3455.$$

Therefore,

$$g = \frac{12{,}134}{24{,}805} = 0.4892,$$

which does not exceed the tabled value $g_{0.05} = 0.5065$. Hence we conclude that the assumption of equal variances is reasonable.

Exercises

1. Show that the error mean square

$$s^2 = \frac{SSE}{k(n-1)}$$

for the analysis of variance in a one-way classification is an unbiased estimate of σ^2.

2. Show that the computing formula for SSA, in the analysis of variance of the one-way classification, is equivalent to the corresponding term in the identity of Theorem 11.1 on page 410.

3. Six different machines are being considered for use in manufacturing rubber seals. The machines are being compared with respect to tensile strength of the product. A random sample of 4 seals from each machine is used to determine whether or not the mean tensile strength varies from machine to machine. The following are the tensile-strength measurements in kilograms per square centimeter $\times 10^{-1}$:

Machine					
1	2	3	4	5	6
17.5	16.4	20.3	14.6	17.5	18.3
16.9	19.2	15.7	16.7	19.2	16.2
15.8	17.7	17.8	20.8	16.5	17.5
18.6	15.4	18.9	18.9	20.5	20.1

Perform the analysis of variance at the 0.05 level of significance and indicate whether or not the mean tensile strengths differ significantly for the 6 machines.

4. The data in the following table represent the number of hours of relief provided by 5 different brands of headache tablets administered to 25 subjects experiencing fevers of 38°C or more. Perform the analysis of variance, and test the hypothesis at the 0.05 level of significance that the mean number of hours of relief provided by the tablets is the same for all 5 brands.

Tablet				
A	B	C	D	E
5	9	3	2	7
4	7	5	3	6
8	8	2	4	9
6	6	3	1	4
3	9	7	4	7

5. In the article *"Shelf-Space Strategy in Retailing,"* published in the *Proceedings: Southern Marketing Association* (1975), the effect of shelf height on the supermarket sales of canned dog food is investigated. An experiment was conducted at a small supermarket for a period of 8 days on the sales of a single brand of dog food, referred to as Arf dog food, involving three levels of shelf height: knee level, waist level, and eye level. During each day the shelf height of the canned dog food was randomly changed on three different occasions. The remaining sections of the gondola that housed the given brand were filled with a mixture of dog food brands that were both familar and unfamiliar to customers in this particular geographic area. Sales, in hundreds of dollars, of Arf dog food per day for the three shelf heights are as follows:

Shelf Height		
Knee Level	Waist Level	Eye Level
77	88	85
82	94	85
86	93	87
78	90	81
81	91	80
86	94	79
77	90	87
81	87	93

Is there a significant difference in the average daily sales of this dog food based on shelf height? Use a 0.01 level of significance.

6. Immobilization of free-ranging white-tailed deer by drugs allows researchers the opportunity to closely examine deer and gather valuable physiological information. In the study *"Influence of Physical Restraint and Restraint-Facilitating Drugs on Blood Measurements of White-Tailed Deer and Other Selected Mammals"* conducted at the Virginia Polytechnic Institute and State University in 1976, wildlife biologists tested the "knock down" time (time from injection to immobilization) of three different immobilizing drugs. Immobilization, in this case, is defined as the point where the animal no longer has enough muscle control to remain standing. Thirty male white-tailed deer were randomly assigned to each of three treatments. Group A received 5 milligrams of liquid succinylcholine chloride (SCC); group B received 8 milligrams of powdered SCC; and group C received 200 miligrams of phencyclidine hydrochloride. Knock down times, in minutes, were recorded as follows:

Group		
A	*B*	*C*
11	10	4
5	7	4
14	16	6
7	7	3
10	7	5
7	5	6
23	10	8
4	10	3
11	6	7
11	12	3

Perform an analysis of variance at the 0.01 level of significance and determine whether or not the average knock down time for the 3 drugs is the same.

7. It has been shown that the fertilizer magnesium ammonium phosphate, $MgNH_4Po_4$ is an effective supplier of the nutrients necessary for plant growth. The compounds supplied by this fertilizer are highly soluble in water, allowing the fertilizer to be applied directly on the soil surface or mixed with the growth substrate during the potting process. A study on the *"Effect of Magnesium Ammonium Phosphate on Height of Chrysanthemums"* was conducted at George Mason University in 1980 to determine a possible optimum level of fertilization, based on the enhanced vertical growth response of the chrysanthemums. Forty chrysanthemum seedlings were divided into 4 groups each containing 10 plants. Each was planted in a similar pot containing a uniform growth medium. To each group of plants an increasing concentration of $MgNH_4Po_4$, measured in grams per bushel, was added. The 4 groups of plants were grown under uniform conditions in a greenhouse for a period of four weeks. The treatments and the respective changes in heights, measured in centimeters, are shown in the following table:

Treatment			
50 gm/bu	100 gm/bu	200 gm/bu	400 gm/bu
13.2	16.0	7.8	21.0
12.4	12.6	14.4	14.8
12.8	14.8	20.0	19.1
17.2	13.0	15.8	15.8
13.0	14.0	17.0	18.0
14.0	23.6	27.0	26.0
14.2	14.0	19.6	21.1
21.6	17.0	18.0	22.0
15.0	22.2	20.2	25.0
20.0	24.4	23.2	18.2

Can we conclude at the 0.05 level of significance that different concentrations of $MgNH_4Po_4$ affect the average attained height of chrysanthemums?

8. Three sections of the same elementary mathematics course are taught by three teachers. The final grades were recorded as follows:

Teacher		
A	B	C
73	88	68
89	78	79
82	48	56
43	91	91
80	51	71
73	85	71
66	74	87
60	77	41
45	31	59
93	78	68
36	62	53
77	76	79
	96	15
	80	
	56	

Is there a significant difference in the average grades given by the three teachers? Use a 0.05 level of significance.

9. The mitochondrial enzyme NAPH : NAD transhydrogenase of the common rat tapeworm (*Hymenolegsis diminuta*) catalyzes a hydrogen in transfer from NADPH to NAD, producing NADH. This enzyme is known to serve a vital role in the tapeworm's anaerobic metabolism, and it has recently been hypothesized that it may serve as a proton exhange pump, transferring protons across the mitochondrial membrane. A study on the *"Effect of Various Substrate Concentrations on the Conformational Variation of the NADPH:NAD Transhydrogenase of Hymenolepsis Diminuta"* conducted in 1983 at Bowling Green State University, was designed to assess the ability of this enzyme to undergo conformation or shape changes. Changes in the specific activity of the enzyme caused by variations in the concentration of NADP could be interpreted as supporting the theory of conformational change. The enzyme in question is located in the inner membrane of the tapeworm's mitochondria. These tapeworms were homogenized, and through a series of centrifugations, the enzyme was isolated. Various concentrations of NADP were then added to the isolated enzyme solution, and the mixture was then incubated in a water bath at 56°C for 3 minutes. The enzyme was then analyzed on a dual-beam spectrophotometer, and the following results were calculated in terms of the specific activity of the enzyme in nanomoles per minute, per milligram of protein:

NADP Concentration (nm)				
0	80	160	360	
11.01	11.38	11.02	6.04	10.31
12.09	10.67	10.67	8.65	8.30
10.55	12.33	11.50	7.76	9.48
11.26	10.08	10.31	10.13	8.89
			9.36	

Test the hypothesis at the 0.01 level that the average specific activity is the same for the four concentrations.

10. Four laboratories are being used to perform chemical analyses. Samples of the same material are sent to the laboratories for analysis as part of the study to determine whether or not they give, on the average, the same results. The analytical results for the four laboratories are as follows:

Laboratory			
A	B	C	D
58.7	62.7	55.9	60.7
61.4	64.5	56.1	60.3
60.9	63.1	57.3	60.9
59.1	59.2	55.2	61.4
58.2	60.3	58.1	62.3

(a) Use Bartlett's test to show that the within-laboratory variances are not significantly different at the $\alpha = 0.05$ level of significance.

(b) Perform the analysis of variance and give conclusions concerning the laboratories.

11. Use Bartlett's test at the 0.05 level of significance to test for homogeneity of variances in Exercise 8.

12. Use Bartlett's test at the 0.01 level of significance to test for homogeneity of variances in Exercise 9.

13. The number of bacteria in six containers of milk were recorded by each of four observers. The bacteria counts are as follows:

Observer			
A	B	C	D
230	184	205	196
241	72	156	210
336	214	308	284

Observer			
A	B	C	D
128	348	118	312
253	68	247	125
124	330	104	99

Use Cochran's test at the 0.05 level of significance to test for homogeneity of variances.

14. Use Cochran's test at the 0.01 level of significance to test for homogeneity of variances in Exercise 6.

11.4 Single-Degree-of-Freedom Comparisons

The analysis of variance in a one-way classification or the one-factor experiment, as it is often called, merely indicates whether or not the hypothesis of equal treatment means can be rejected. Usually, an experimenter would prefer his analysis to probe deeper than this. For instance, in Example 11.1, by rejecting the null hypothesis we concluded that the means are not all equal, but we still do not know where the differences exist among the aggregates. The engineer might have the feeling from the outset that aggregates 1 and 2 should have similar absorption properties due to similar composition and that the same is true for aggregates 3 and 5 but that the two groups possibly differ a great deal. It would seem, then, appropriate to test the hypothesis

$$H_0: \quad \mu_1 + \mu_2 - \mu_3 - \mu_5 = 0,$$
$$H_1: \quad \mu_1 + \mu_2 - \mu_3 - \mu_5 \neq 0.$$

We notice that the hypothesis is a linear function of the population means in which the coefficients sum to zero.

Definition 11.1   *Any linear function of the form*

$$\omega = \sum_{i=1}^{k} c_i \mu_i, \qquad where \quad \sum_{i=1}^{k} c_i = 0,$$

*is called a **comparison** or **contrast** in the treatment means.*

The experimenter can often make multiple comparisons by testing the significance of contrasts in the treatment means, that is, by testing a hypothesis of the type

$$H_0: \quad \sum_{i=1}^{k} c_i \mu_i = 0,$$

$$H_1: \quad \sum_{i=1}^{k} c_i \mu_i \neq 0,$$

where $\sum_{i=1}^{k} c_i = 0$. The test is conducted by first computing a similar contrast in the sample means,

$$w = \sum_{i=1}^{k} c_i \bar{y}_{i\,.}\,.$$

Since $\bar{Y}_{1\,.}, \bar{Y}_{2\,.}, \ldots, \bar{Y}_{k\,.}$ are independent random variables having normal distributions with means $\mu_1, \mu_2, \ldots, \mu_k$ and variances $\sigma^2/n_1, \sigma^2/n_2, \ldots, \sigma^2/n_k$, respectively, Theorem 6.11 assures us that w is a value of the normal random variable W with mean

$$\mu_W = \sum_{i=1}^{k} c_i \mu_i$$

and variance

$$\sigma_W^2 = \sigma^2 \sum_{i=1}^{k} \frac{c_i^2}{n_i}.$$

Therefore, when H_0 is true, $\mu_W = 0$ and, by Example 6.5 on page 171, the statistic

$$\frac{W^2}{\sigma_W^2} = \frac{\left(\sum_{i=1}^{k} c_i \bar{Y}_{i\,.}\right)^2}{\sigma^2 \sum_{i=1}^{k} (c_i^2/n_i)}$$

is distributed as a chi-square random variable with 1 degree of freedom. Our hypothesis is tested at the α level of significance by computing

$$f = \frac{\left(\sum_{i=1}^{k} c_i \bar{y}_{i\,.}\right)^2}{s^2 \sum_{i=1}^{k} (c_i^2/n_i)} = \frac{\left[\sum_{i=1}^{k} (c_i T_{i\,.}/n_i)\right]^2}{s^2 \sum_{i=1}^{k} (c_i^2/n_i)} = \frac{SSw}{s^2},$$

where f is a value of the random variable F having the F distribution with 1 and $N - k$ degrees of freedom and

$$SSw = \frac{\left[\sum_{i=1}^{k} (c_i T_{i\,.}/n_i)\right]^2}{\sum_{i=1}^{k} (c_i^2/n_i)}.$$

When the sample sizes are all equal to n,

$$SSw = \frac{\left(\sum\limits_{i=1}^{k} c_i T_{i\cdot}\right)^2}{n \sum\limits_{i=1}^{k} c_i^2}.$$

The quantity SSw, called the **contrast sum of squares**, indicates the portion of SSA that is explained by the contrast in question.

Definition 11.2   *The two contrasts*

$$\omega_1 = \sum_{i=1}^{k} b_i \mu_i \qquad and \qquad \omega_2 = \sum_{i=1}^{k} c_i \mu_i$$

are said to be **orthogonal** *if* $\sum\limits_{i=1}^{k} b_i c_i / n_i = 0$ *or, when the n_i's are all equal to n, if*

$$\sum_{i=1}^{k} b_i c_i = 0.$$

If ω_1 and ω_2 are orthogonal, then the quantities SSw_1 and SSw_2 are components of SSA each with a single degree of freedom. The treatment sum of squares with $k - 1$ degrees of freedom can be partitioned into at most $k - 1$ independent single-degree-of-freedom contrast sum of squares satisfying the identity

$$SSA = SSw_1 + SSw_2 + \cdots + SSw_{k-1}$$

if the contrasts are orthogonal to each other.

Example 11.4   Referring to Example 11.1, find the contrast sum of squares corresponding to the orthogonal contrasts

$$\omega_1 = \mu_1 + \mu_2 - \mu_3 - \mu_5$$
$$\omega_2 = \mu_1 + \mu_2 + \mu_3 + \mu_5 - 4\mu_4$$

and carry out appropriate tests of significance.

Solution. It is obvious that the two contrasts are orthogonal, since $(1)(1) + (1)(1) + (-1)(1) + (0)(-4) + (-1)(1) = 0$. The second contrast indicates a comparison between aggregates 1, 2, 3, and 5, and aggregate 4. One can write down two additional contrasts orthogonal to the first two such as

$$\omega_3 = \mu_1 - \mu_2 \qquad \text{(aggregate 1 versus aggregate 2)}$$
$$\omega_4 = \mu_3 - \mu_5 \qquad \text{(aggregate 3 versus aggregate 5)}.$$

From the data of Table 11.1, we have

$$SSw_1 = \frac{(3320 + 3416 - 3663 - 3664)^2}{6[(1)^2 + (1)^2 + (-1)^2 + (-1)^2]} = 14,553$$

$$SSw_2 = \frac{[3320 + 3416 + 3663 + 3664 - 4(2791)]^2}{6[(1)^2 + (1)^2 + (1)^2 + (1)^2 + (-4)^2]} = 70,035.$$

A more extensive analysis-of-variance table is then given by Table 11.7. We note that the two contrast sum of squares account for nearly all the aggregate sum of squares. Although there is a significant difference between aggregates in their absorption properties, the contrast w_1 is not significant when compared to the critical value $f_{0.05}(1, 25) = 4.24$. However, the f value of 14.12 for w_2 is significant and the hypothesis

$$H_0: \quad \mu_1 + \mu_2 + \mu_3 + \mu_5 = 4\mu_4$$

is rejected.

Orthogonal contrasts are used when the experimenter is interested in partitioning the treatment variation into independent components. There are several choices available in selecting the orthogonal contrasts except for the last one. Normally, the experimenter would have certain contrasts that are of interest to him. Such was the case in our example, where chemical composition suggested that aggregates (1, 2) and (3, 5) constitute distinct groups with different absorption properties, a postulation that was not supported by the significance test. However, the second comparison supports the conclusion that aggregate 4 seems to "stand out" from the rest. In this case the complete partitioning of SSA was not necessary, since two of the four possible independent comparisons accounted for a majority of the variation in treatments.

Table 11.7 Analysis of Variance Using Orthogonal Contrasts

Source of Variation	Sum of Squares	Degrees of Freedom	Mean Square	Computed f
Aggregates	85,356	4	21,339	4.30
(1, 2) vs. (3, 5)	$\{14{,}553$	$\{1$	$\{14{,}553$	2.93
(1, 2, 3, 5) vs. 4	$\{70{,}035$	$\{1$	$\{70{,}035$	14.12
Error	124,021	25	4,961	
Total	209,377	29		

11.5 Multiple-Range Test

The analysis of variance is a powerful procedure for testing the homogeneity of a set of means. However, if we reject the null hypothesis and accept the stated alternative—that the means are not all equal—we still do not know which of the population means are equal and which are different. Several tests are available that separate a set of significantly different means into subsets of homogeneous means. The test that we shall study in this section is called **Duncan's multiple-range test**.

Let us assume that the analysis-of-variance procedure has led to a rejection of the null hypothesis of equal population means. It is also assumed that the k random samples are all of equal size n. The range of any subset of p sample means must exceed a certain value before we consider any of the p population means to be different. This value is called the **least significant range** for the p means, and is denoted by R_p, where

$$R_p = r_p \cdot s_{\bar{x}} = r_p \sqrt{\frac{s^2}{n}}.$$

The sample variance s^2, which is an estimate of the common variance σ^2, is obtained from the error mean square in the analysis of variance. The values of the quantity r_p, called the **least significant studentized-range**, depend on the desired level of significance and the number of degrees of freedom of the error mean square. These values may be obtained from Table A.12 for $p = 2, 3, \ldots,$ 10 means.

To illustrate the multiple-range test procedure, let us consider a hypothetical example in which 6 treatments are compared in a one-way classification with 5 observations per treatment. The error mean square, obtained from the analysis-of-variance table, is $s^2 = 2.45$ with 24 degrees of freedom. First, we arrange the sample means in increasing order of magnitude:

$\bar{y}_2.$	$\bar{y}_5.$	$\bar{y}_1.$	$\bar{y}_3.$	$\bar{y}_6.$	$\bar{y}_4.$
14.50	16.75	19.84	21.12	22.90	23.20

Let $\alpha = 0.05$. Then the values of r_p are obtained from Table A.12, with $v = 24$ degrees of freedom, for $p = 2, 3, 4, 5,$ and 6. Finally, we obtain R_p by multiplying each r_p by $\sqrt{s^2/n} = \sqrt{2.45/5} = 0.7$. The results of these computations are summarized as follows:

p	2	3	4	5	6
r_p	2.919	3.066	3.160	3.226	3.276
R_p	2.043	2.146	2.212	2.258	2.293

Comparing these least significant ranges with the differences in ordered means, we arrive at the following conclusions:

1. Since $\bar{y}_{4.} - \bar{y}_{2.} = 8.70 > R_6 = 2.293$, we conclude that $\bar{y}_{4.}$ and $\bar{y}_{2.}$ are significantly different.
2. Comparing $\bar{y}_{4.} - \bar{y}_{5.}$ and $\bar{y}_{6.} - \bar{y}_{2.}$ with R_5, we conclude that $\bar{y}_{4.}$ is significantly greater than $\bar{y}_{5.}$ and $\bar{y}_{6.}$ is significantly greater than $\bar{y}_{2.}$.
3. Comparing $\bar{y}_{4.} - \bar{y}_{1.}$, $\bar{y}_6 - \bar{y}_{5.}$, and $\bar{y}_{3.} - \bar{y}_{2.}$ with R_4, we conclude that each difference is significant.
4. Comparing $\bar{y}_{4.} - \bar{y}_{3.}$, $\bar{y}_{6.} - \bar{y}_{1.}$, $\bar{y}_{3.} - \bar{y}_{5.}$, and $\bar{y}_{1.} - \bar{y}_{2.}$ with R_3, we find all differences significant except for $\bar{y}_{4.} - \bar{y}_{3.}$. Therefore, $\bar{y}_{4.}$, $\bar{y}_{3.}$, and $\bar{y}_{6.}$ constitute a subset of homogeneous means.
5. Comparing $\bar{y}_{3.} - \bar{y}_{1.}$, $\bar{y}_{1.} - \bar{y}_{5.}$, and $\bar{y}_{5.} - \bar{y}_{2.}$ with R_2, we conclude that only $\bar{y}_{3.}$ and $\bar{y}_{1.}$ are not significantly different.

It is customary to summarize the above conclusions by drawing a line under any subset of adjacent means that are not significantly different. Thus we have

$\bar{y}_{2.}$	$\bar{y}_{5.}$	$\bar{y}_{1.}$	$\bar{y}_{3.}$	$\bar{y}_{6.}$	$\bar{y}_{4.}$
14.50	16.75	19.84	21.12	22.90	23.20

One can immediately observe from this manner of presentation that $\mu_4 > \mu_1$, $\mu_4 > \mu_5$, $\mu_4 > \mu_2$, $\mu_6 > \mu_1$, $\mu_6 > \mu_5$, $\mu_6 > \mu_2$, $\mu_3 > \mu_5$, $\mu_3 > \mu_2$, $\mu_1 > \mu_5$, $\mu_1 > \mu_2$, and $\mu_5 > \mu_2$, while all other pairs of population means are not considered significantly different.

11.6 Comparing Treatments with a Control

In many scientific and engineering problems one is not interested in drawing inferences regarding all possible comparisons among the treatment means of the type $\mu_i - \mu_j$. Rather, the experiment often dictates the need for comparing simultaneously each *treatment* with a *control*. A test procedure for determining significant differences between each treatment mean and the control, at a single joint significance level α, has been developed by C. W. Dunnett. To illustrate Dunnett's procedure, let us consider the experiment data of Table 11.8 for the one-way classification in which the effect of three catalysts on the yield of a reaction is being studied. A fourth treatment, no catalyst, was used as a control.

In general we wish to test the k hypotheses

$$\left. \begin{array}{l} H_0: \quad \mu_0 = \mu_i \\ H_1: \quad \mu_0 \neq \mu_i \end{array} \right\} i = 1, 2, \ldots, k,$$

Table 11.8 Yield of Reaction

Control	Catalyst 1	Catalyst 2	Catalyst 3
50.7	54.1	52.7	51.2
51.5	53.8	53.9	50.8
49.2	53.1	57.0	49.7
53.1	52.5	54.1	48.0
52.7	54.0	52.5	47.2
$\bar{y}_{0.} = 51.44$	$\bar{y}_{1.} = 53.50$	$\bar{y}_{2.} = 54.04$	$\bar{y}_{3.} = 49.38$

where μ_0 represents the mean yield for the population of measurements in which the control is used. The usual analysis-of-variance assumptions, as outlined in Section 11.2 are expected to remain valid. To test the null hypotheses specified by H_0 against two-sided alternatives for an experimental situation in which there are k treatments, excluding the control, and n observations per treatment, we first calculate the values

$$d_i = \frac{\bar{y}_{i.} - \bar{y}_{0.}}{\sqrt{2s^2/n}}, \qquad i = 1, 2, \ldots, k.$$

The sample variance s^2 is obtained, as before, from the error mean square in the analysis of variance. Now, the critical region for rejecting H_0, at the α level of significance, is established by the inequality

$$|d_i| > d_{\alpha/2}(k, v),$$

where v is the number of degrees of freedom for the error mean square. The values of the quantity $d_{\alpha/2}(k, v)$ for a two-tailed test are given in Table A.13 (see Statistical Tables) for $\alpha = 0.05$ and $\alpha = 0.01$ for various values of k and v.

Example 11.5 For the data of Table 11.8, test hypotheses comparing each catalyst with the control, using two-sided alternatives. Choose $\alpha = 0.05$ as the joint significance level.

Solution. The error sum of squares with 16 degrees of freedom is obtained from the analysis-of-variance table using all $k + 1$ treatments or by direct computation from the formula

$$SSE = \sum_{i=0}^{k} \sum_{j=1}^{n} y_{ij}^2 - \frac{\sum_{i=0}^{k} T_{i.}^2}{n}$$

$$= 54{,}371.960 - 54{,}335.148$$

$$= 36.812.$$

Then the error mean square is given by

$$s^2 = \frac{36.812}{16} = 2.30075$$

and

$$\sqrt{\frac{2s^2}{n}} = \sqrt{\frac{(2)(2.30075)}{5}} = 0.9593.$$

Hence

$$d_1 = \frac{53.50 - 51.44}{0.9593} = 2.147,$$

$$d_2 = \frac{54.04 - 51.44}{0.9593} = 2.710,$$

$$d_3 = \frac{49.38 - 51.44}{0.9593} = -2.147.$$

From Table A.13 of the Appendix the critical value for $\alpha = 0.05$ is found to be $d_{0.025}$ (3, 16) = 2.59. Since $|d_1| < 2.59$, and $|d_3| < 2.59$, we conclude that only the mean yield for catalyst 2 is significantly different from the mean yield of the reaction using the control.

Many practical applications dictate the need for a one-tailed test in comparing treatments with a control. Certainly, when a pharmacologist is concerned with the comparison of various dosages of a drug on the effect of reducing cholesterol level and his control is zero dosage, it is of interest to determine if each dosage produces a significantly larger reduction than that of the control. Table A.14 gives the critical values $d_\alpha(k, v)$ for one-sided alternatives.

Exercises

1. Extend the analysis of variance in Exercise 10 on page 422 to make significance tests on the following contrasts:
   (a) *B* versus *A, C*, and *D*;
   (b) *C* versus *A* and *D*;
   (c) *A* versus *D*.

2. The study *"Loss of Nitrogen Through Sweat by Preadolescent Boys Consuming Three Levels of Dietary Protein"* was conducted by the Department of Human Nutrition and Foods at the Vir- ginia Polytechnic Institute and State University in 1975 to determine perspiration nitrogen loss at various dietary protein levels. Twelve pre- adolescent boys ranging in age from 7 years, 8 months to 9 years, 8 months, and judged to be clinically healthy, were used in the experiment. Each boy was subjected to one of three controlled diets in which 29, 54, or 84 grams of protein per day were consumed. The following data represent the body perspiration nitrogen loss, in milligrams,

collected during the last two days of the experimental period:

Protein Level		
29 Grams	54 Grams	84 Grams
190	318	390
266	295	321
270	271	396
	438	399
	402	

(a) Perform an analysis of variance at the 0.05 level of significance to show that the mean perspiration nitrogen loss at the three protein levels are different.

(b) Use a single-degree-of-freedom contrast with $\alpha = 0.05$ to compare the mean perspiration nitrogen loss for boys who consume 29 grams of protein per day versus boys who consume 54 and 84 grams of protein.

3. The purpose of the study *"The Incorporate of a Chelating Agent into a Flame Retardant Finish of a Cotton Flannelette and the Evaluation of Selected Fabric Properties"* conducted at the Virginia Polytechnic Institute and State University in 1974 was to evaluate the use of a chelating agent as part of the flame retardant finish of cotton flannelette by determining its effects upon flammability after the fabric is laundered under specific conditions. Two baths were prepared, one with carboxymethyl cellulose and one without. Twelve pieces of fabric were laundered 5 times in Bath I, and 12 other pieces of fabric were laundered 10 times in Bath I. This was repeated using 24 additional pieces of cloth in Bath II. After the washings, the lengths of fabric that burned and the burn times were measured. For convenience, let us define the following treatments:

> Treatment 1:   5 launderings in Bath I,
> Treatment 2:   5 launderings in Bath II,
> Treatment 3:   10 launderings in Bath I,
> Treatment 4:   10 launderings in Bath II.

Burn times, in seconds, were recorded as follows:

Treatment			
1	2	3	4
13.7	6.2	27.2	18.2
23.0	5.4	16.8	8.8
15.7	5.0	12.9	14.5
25.5	4.4	14.9	14.7
15.8	5.0	17.1	17.1
14.8	3.3	13.0	13.9
14.0	16.0	10.8	10.6
29.4	2.5	13.5	5.8
9.7	1.6	25.5	7.3
14.0	3.9	14.2	17.7
12.3	2.5	27.4	18.3
12.3	7.1	11.5	9.9

(a) Perform an analysis of variance using a 0.01 level of significance and determine whether there are any significant differences among the treatment means.

(b) Use single-degree-of-freedom contrasts with $\alpha = 0.01$ to compare the mean burn time of treatment 1 versus treatment 2 and also treatment 3 versus treatment 4.

4. Use Duncan's multiple-range test, with a 0.05 level of significance, to analyze the means of the 5 different brands of headache tablets in Exercise 4 on page 420.

5. Use Duncan's multiple-range test in Exercise 10 on page 422, with a 0.01 level of significance, to determine which laboratories differ, on the average, in their analysis.

6. An investigation was conducted to determine the source of reduction in yield of a certain chemical product. It was known that the loss in yield occurred in the mother liquor, that is, the material removed at the filtration stage. It was felt that different blends of the original material may result in different yield reductions at the mother liquor stage. The following are results of the percent reduction for 3 batches at each of 4 preselected blends:

Blend			
1	2	3	4
25.6	25.2	20.8	31.6
24.3	28.6	26.7	29.8
27.9	24.7	22.2	34.3

(a) Perform the analysis of variance at the $\alpha = 0.05$ level of significance.

(b) Use Duncan's multiple-range test to determine which blends differ.

7. In the study *"An Evaluation of the Removal Method for Estimating Benthic Populations and Diversity"* conducted by the Virginia Polytechnic Institute and State University on the Jackson River in 1983, 5 different sampling procedures were used to determine the species count. Twenty samples were selected at random and each of the 5 sampling procedures were repeated 4 times. The species counts were recorded as follows:

		Sampling Procedure		
Deple-tion	Modified Hess	Surber	Substrate Removal Kicknet	Kick-net
85	75	31	43	17
55	45	20	21	10
40	35	9	15	8
77	67	37	27	15

(a) Is there a significant difference in the average species count for the different sampling procedures? Use a 0.01 level of significance.

(b) Use Duncan's multiple-range test with $\alpha = 0.05$ to find which sampling procedures differ.

8. The financial structure of a firm refers to the way the firm's assets are divided by equity and debt, and the financial leverage refers to the percentage of assets financed by debt. In the paper *"The Effect of Financial Leverage on Return"* (1980), Tai Ma of the Virginia Polytechnic Institute and State University claims that financial leverage can

be used to increase the rate of return on equity. To say it in another way, stockholders can receive higher returns on equity with the same amount of investment by the use of financial leverage. The following data show the rates of return on equity using 3 different levels of financial leverage and a control level (zero debt) for 24 randomly selected firms:

Financial Leverage			
Control	Low	Medium	High
2.1	6.2	9.6	10.3
5.6	4.0	8.0	6.9
3.0	8.4	5.5	7.8
7.8	2.8	12.6	5.8
5.2	4.2	7.0	7.2
2.6	5.0	7.8	12.0

Source: S & P's Machinery Industry Survey, 1975.

(a) Perform the analysis of variance at the 0.05 level of significance.

(b) Use Dunnett's test at the 0.01 level of significance to determine whether the mean rates of return on equity at the low, medium, and high levels of financial leverage are higher than at the control level.

9. In the following biological experiment 4 concentrations of a certain chemical are used to enhance the growth in centimeters of a certain type of plant over time. Five plants are used at each concentration and the growth in each plant is measured. The following growth data are taken. A control (no chemical) is also applied.

	Concentration			
Control	1	2	3	4
6.8	8.2	7.7	6.9	5.9
7.3	8.7	8.4	5.8	6.1
6.3	9.4	8.6	7.2	6.9
6.9	9.2	8.1	6.8	5.7
7.1	8.6	8.0	7.4	6.1

Use Dunnett's two-sided test at the 0.05 level of significance to simultaneously compare the concentrations with the control.

10. Three catalysts are used in a chemical process with a control (no catalyst) being included. The following are yield data from the process:

Control	Catalyst		
	1	2	3
74.5	77.5	81.5	78.1
76.1	82.0	82.3	80.2
75.9	80.6	81.4	81.5
78.1	84.9	79.5	83.0
76.2	81.0	83.0	82.1

Use Dunnett's test at the $\alpha = 0.01$ level of significance to determine if a significantly higher yield is obtained with the catalysts than with no catalyst.

11.7 Comparing a Set of Treatments in Blocks

It often becomes necessary in analysis-of-variance problems to design the experiment in such a way that the experimental error variation due to extraneous sources can be systematically controlled. In the preceding development of the one-way analysis-of-variance problem, it was assumed that conditions remained relatively homogeneous for the **experimental units** used in the various test runs. For example, in a chemical experiment designed to determine if there is a difference in mean reaction yield among four catalysts, samples of materials to be tested are drawn from the same batches of raw materials, while other conditions, such as temperature and concentration of reactants, are held constant. In this case the time of day for the experimental runs might represent the experimental units, and if the experimenter feels that there could possibly be a slight time effect, he or she would *randomize* the assignment of the catalysts to the runs to counteract the possible trend. This type of experimental strategy, whereby the treatments (catalysts) are assigned randomly to the experimental units, is called a **completely randomized design**. As a second example of such a design, consider an experiment to compare four methods of measuring a particular physical property of a fluid substance. Suppose the sampling process is destructive; that is, once a sample of the substance has been measured by one method, it cannot be measured again by one of the other methods. If it is decided that five measurements are to be taken for each method, then 20 samples of the material are selected from a large batch *at random* and are used in the experiment to compare the four measuring devices. The experimental units are the randomly selected samples. Any variation from sample to sample will appear in the error variation, as measured by s^2 in the analysis.

If the variation due to heterogeneity in experimental units is so large that the sensitivity of detecting treatment differences is reduced due to an inflated value of s^2, a better plan might be to "block off" variation due to these units and thus reduce the extraneous variation to that accounted for by smaller or more homogeneous blocks. The simplest design calling for this strategy is a **randomized block design**. For example, suppose that in the previous catalyst illustration it is known a priori that there definitely is a significant day-to-day effect on the yield and that we can measure the yield for four catalysts on a given day. Rather than assign the four catalysts to the 20 test runs completely at random, we choose, say, five days and run each of the four catalysts on each day, randomly assigning the catalysts to the runs within days. In this way the day-to-day variation is removed in the analysis and consequently the experimental error, which still includes any time trend *within days*, more accurately represents chance variation. Each day is referred to as a **block**.

The classical example, using a randomized block design, is an agricultural experiment in which different fertilizers are being compared for their ability to increase the yield of a particular crop. Rather than assign fertilizers at random to many plots over a large area of variable soil composition, one should assign the fertilizers to smaller blocks comprised of homogeneous plots. The variation between these blocks, which is most likely significant compared to the uniformity of the plots within a block, is then removed from the experimental error in the analysis of variance.

The most straightforward of the randomized block designs is one in which we randomly assign each treatment once to every block. Such an experimental layout is called a **randomized complete block design,** each block constituting a single **replication** of the treatments.

11.8 Randomized Complete Block Designs

A typical layout for the randomized complete block design using 3 measurements in 4 blocks is as follows:

Block 1	Block 2	Block 3	Block 4
t_2	t_1	t_3	t_2
t_1	t_3	t_2	t_1
t_3	t_2	t_1	t_3

The t's denote the assignment to blocks of each of the 3 treatments. Of course, the true allocation of treatments to units within blocks is done at random.

Once the experiment has been completed, the data can be recorded as in the following 3×4 array

Treatment	Block			
	1	2	3	4
1	y_{11}	y_{12}	y_{13}	y_{14}
2	y_{21}	y_{22}	y_{23}	y_{24}
3	y_{31}	y_{32}	y_{33}	y_{34}

where y_{11} represents the response obtained by using treatment 1 in block 1, y_{12} represents the response obtained by using treatment 1 in block 2, ..., and y_{34} represents the response obtained by using treatment 3 in block 4.

Let us now generalize and consider the case of k treatments assigned to b blocks. The data may be summarized as shown in the $k \times b$ rectangular array of Table 11.9. It will be assumed that the y_{ij}, $i = 1, 2, \ldots, k$ and $j = 1, 2, \ldots, b$, are values of independent random variables having normal distributions with means μ_{ij} and common variance σ^2. In Table 11.9 we define

$T_{i.} =$ sum of the observations for the ith treatment

$T_{.j} =$ sum of the observations in the jth block

$T_{..} =$ sum of all bk observations

$\bar{y}_{i.} =$ mean of the observations for the ith treatment

$\bar{y}_{.j} =$ mean of the observations in the jth block

$\bar{y}_{..} =$ mean of all bk observations.

Let $\mu_{i.}$ represent the average (rather than the total) of the b population means for the ith treatment. That is,

$$\mu_{i.} = \frac{\sum\limits_{j=1}^{b} \mu_{ij}}{b}.$$

Similarly, the average of the population means for the jth block, $\mu_{.j}$, is defined by

$$\mu_{.j} = \frac{\sum\limits_{i=1}^{k} \mu_{ij}}{k},$$

and the average of the bk population means, μ, is defined by

$$\mu = \frac{\sum\limits_{i=1}^{k} \sum\limits_{j=1}^{b} \mu_{ij}}{bk}.$$

Table 11.9 $k \times b$ Array for a Randomized Complete Block Design

Treatment	Block 1	2	$\cdots$	j	$\cdots$	b	Total	Mean
1	y_{11}	y_{12}	$\cdots$	y_{1j}	$\cdots$	y_{1b}	$T_{1.}$	$\bar{y}_{1.}$
2	y_{21}	y_{22}	$\cdots$	y_{2j}	$\cdots$	y_{2b}	$T_{2.}$	$\bar{y}_{2.}$
$\vdots$	$\vdots$	$\vdots$		$\vdots$		$\vdots$	$\vdots$	$\vdots$
i	y_{i1}	y_{i2}	$\cdots$	y_{ij}	$\cdots$	y_{ib}	$T_{i.}$	$\bar{y}_{i.}$
$\vdots$	$\vdots$	$\vdots$		$\vdots$		$\vdots$	$\vdots$	$\vdots$
k	y_{k1}	y_{k2}	$\cdots$	y_{kj}	$\cdots$	y_{kb}	$T_{k.}$	$\bar{y}_{k.}$
Total	$T_{.1}$	$T_{.2}$	$\cdots$	$T_{.j}$	$\cdots$	$T_{.b}$	$T_{..}$	
Mean	$\bar{y}_{.1}$	$\bar{y}_{.2}$	$\cdots$	$\bar{y}_{.j}$	$\cdots$	$\bar{y}_{.b}$		$\bar{y}_{..}$

To determine if part of the variation in our observations is due to differences among the treatments, we consider the test

$$H_0': \quad \mu_{1.} = \mu_{2.} = \cdots = \mu_{k.} = \mu,$$
$$H_1': \quad \text{the } \mu_i.\text{'s are not all equal.}$$

Similarly, to determine if part of the variation is due to differences among the blocks, we consider the test

$$H_0'': \quad \mu_{.1} = \mu_{.2} = \cdots = \mu_{.b} = \mu,$$
$$H_1'': \quad \text{the } \mu_{.j}\text{'s are not equal.}$$

Each observation may be written in the form

$$y_{ij} = \mu_{ij} + \varepsilon_{ij},$$

where ε_{ij} measures the deviation of the observed value y_{ij} from the population mean μ_{ij}. The preferred form of this equation is obtained by substituting

$$\mu_{ij} = \mu + \alpha_i + \beta_j,$$

where α_i is, as before, the effect of the ith treatment and β_j is the effect of the jth block. It is assumed that the treatment and block effects are additive. Hence we may write

$$y_{ij} = \mu + \alpha_i + \beta_j + \varepsilon_{ij}.$$

Notice that the model resembles that of the one-way classification, the essential difference being the introduction of the block effect β_j. The basic concept is much like that of the one-way classification except that we must account in the analysis for the additional effect due to blocks, since we are now systematically controlling variation in *two directions*. If we now impose the restrictions that

$$\sum_{i=1}^{k} \alpha_i = 0 \qquad \text{and} \qquad \sum_{j=1}^{b} \beta_j = 0,$$

then

$$\mu_{i\cdot} = \frac{\sum_{j=1}^{b} (\mu + \alpha_i + \beta_j)}{b} = \mu + \alpha_i$$

and

$$\mu_{\cdot j} = \frac{\sum_{i=1}^{k} (\mu + \alpha_i + \beta_j)}{k} = \mu + \beta_j.$$

The null hypothesis that the k treatment means $\mu_{i\cdot}$ are equal, and therefore equal to μ, is now equivalent to testing the hypothesis

$$H_0': \quad \alpha_1 = \alpha_2 = \cdots = \alpha_k = 0,$$
$$H_1': \quad \text{at least one of the } \alpha_i\text{'s is not equal to zero.}$$

Similarly, the null hypothesis that the b block means $\mu_{\cdot j}$ are equal is equivalent to testing the hypothesis

$$H_0'': \quad \beta_1 = \beta_2 = \cdots = \beta_b = 0,$$
$$H_1'': \quad \text{at least one of the } \beta_j\text{'s is not equal to zero.}$$

Each of these tests will be based on a comparison of independent estimates of the common population variance σ^2. These estimates will be obtained by splitting the total sum of squares of our data into three components by means of the following identity.

Theorem 11.3   Sum-of-Squares Identity

$$\sum_{i=1}^{k} \sum_{j=1}^{b} (y_{ij} - \bar{y}_{\cdot\cdot})^2 = b \sum_{i=1}^{k} (\bar{y}_{i\cdot} - \bar{y}_{\cdot\cdot})^2 + k \sum_{j=1}^{b} (\bar{y}_{\cdot j} - \bar{y}_{\cdot\cdot})^2$$

$$+ \sum_{i=1}^{k} \sum_{j=1}^{b} (y_{ij} - \bar{y}_{i\cdot} - \bar{y}_{\cdot j} + \bar{y}_{\cdot\cdot})^2.$$

Proof

$$\sum_{i=1}^{k}\sum_{j=1}^{b}(y_{ij}-\bar{y}..)^2 = \sum_{i=1}^{k}\sum_{j=1}^{b}[(\bar{y}_i.-\bar{y}..)+(\bar{y}._j-\bar{y}..)+(y_{ij}-\bar{y}_i.-\bar{y}._j+\bar{y}..)]^2$$

$$= \sum_{i=1}^{k}\sum_{j=1}^{b}(\bar{y}_i.-\bar{y}..)^2 + \sum_{i=1}^{k}\sum_{j=1}^{b}(\bar{y}._j-\bar{y}..)^2$$

$$+ \sum_{i=1}^{k}\sum_{j=1}^{b}(y_{ij}-\bar{y}_i.-\bar{y}._j+\bar{y}..)^2$$

$$+ 2\sum_{i=1}^{k}\sum_{j=1}^{b}(\bar{y}_i.-\bar{y}..)(\bar{y}._j-\bar{y}..)$$

$$+ 2\sum_{i=1}^{k}\sum_{j=1}^{b}(\bar{y}_i.-\bar{y}..)(y_{ij}-\bar{y}_i.-\bar{y}._j+\bar{y}..)$$

$$+ 2\sum_{i=1}^{k}\sum_{j=1}^{b}(\bar{y}._j-\bar{y}..)(y_{ij}-\bar{y}_i.-\bar{y}._j+\bar{y}..).$$

The cross-product terms are all equal to zero. Hence

$$\sum_{i=1}^{k}\sum_{j=1}^{b}(y_{ij}-\bar{y}..)^2 = b\sum_{i=1}^{k}(\bar{y}_i.-\bar{y}..)^2 + k\sum_{j=1}^{b}(\bar{y}._j-\bar{y}..)^2$$

$$+ \sum_{i=1}^{k}\sum_{j=1}^{b}(y_{ij}-\bar{y}_i.-\bar{y}._j+\bar{y}..)^2.$$

The sum-of-squares identity may be presented symbolically by the equation

$$SST = SSA + SSB + SSE,$$

where

$$SST = \sum_{i=1}^{k}\sum_{j=1}^{b}(y_{ij}-\bar{y}..)^2 = \text{total sum of squares}$$

$$SSA = b\sum_{i=1}^{k}(\bar{y}_i.-\bar{y}..)^2 = \text{treatment sum of squares}$$

$$SSB = k\sum_{j=1}^{b}(\bar{y}._j-\bar{y}..)^2 = \text{block sum of squares}$$

$$SSE = \sum_{i=1}^{k}\sum_{j=1}^{b}(y_{ij}-\bar{y}_i.-\bar{y}._j+\bar{y}..)^2 = \text{error sum of squares}.$$

Following the procedure outlined in Theorem 11.2, where we interpret the sum of squares as functions of the independent random variables $Y_{11}, Y_{12}, \ldots, Y_{kb}$, we can show that the expected values of the treatment, block, and error sum of squares are given by

$$E(SSA) = (k - 1)\sigma^2 + b \sum_{i=1}^{k} \alpha_i^2$$

$$E(SSB) = (b - 1)\sigma^2 + k \sum_{j=1}^{b} \beta_j^2$$

$$E(SSE) = (b - 1)(k - 1)\sigma^2.$$

One estimate of σ^2, based on $k - 1$ degrees of freedom, is given by

$$s_1^2 = \frac{SSA}{k - 1}.$$

If the treatment effects $\alpha_1 = \alpha_2 = \cdots = \alpha_k = 0$, s_1^2 is an unbiased estimate of σ^2. However, if the treatment effects are not all zero, we have

$$E\left(\frac{SSA}{k - 1}\right) = \sigma^2 + \frac{b \sum_{i=1}^{k} \alpha_i^2}{k - 1}$$

and s_1^2 overestimates σ^2. A second estimate of σ^2, based on $b - 1$ degrees of freedom, is given by

$$s_2^2 = \frac{SSB}{b - 1}.$$

The estimate s_2^2 is an unbiased estimate of σ^2 when the block effects $\beta_1 = \beta_2 = \cdots = \beta_b = 0$. If the block effects are not all zero, then

$$E\left(\frac{SSB}{b - 1}\right) = \sigma^2 + \frac{k \sum_{j=1}^{b} \beta_j^2}{b - 1}$$

and s_2^2 will overestimate σ^2. A third estimate of σ^2, based on $(k - 1)(b - 1)$ degrees of freedom and independent of s_1^2 and s_2^2, is given by

$$s^2 = \frac{SSE}{(k - 1)(b - 1)},$$

which is unbiased regardless of the truth or falsity of either null hypothesis.

To test the null hypothesis that the treatment effects are all equal to zero, we compute the ratio

$$f_1 = \frac{s_1^2}{s^2},$$

which is a value of the random variable F_1 having the F distribution with $k - 1$ and $(k - 1)(b - 1)$ degrees of freedom when the null hypothesis is true. The null hypothesis is rejected at the α level of significance when $f_1 > f_\alpha[k - 1, (k - 1)(b - 1)]$.

Similarly, to test the null hypothesis that the block effects are all equal to zero, we compute the ratio

$$f_2 = \frac{s_2^2}{s^2},$$

which is a value of the random variable F_2 having the F distribution with $b - 1$ and $(k - 1)(b - 1)$ degrees of freedom when the null hypothesis is true. In this case the null hypothesis is rejected at the α level of significance when $f_2 > f_\alpha[b - 1, (k - 1)(b - 1)]$.

In practice we first compute SST, SSA, and SSB, and then, using the sum-of-squares identity, obtain SSE by subtraction. The degrees of freedom associated with SSE are also usually obtained by subtraction; that is,

$$(k - 1)(b - 1) = (bk - 1) - (k - 1) - (b - 1).$$

Preferred computing formulas for the sums of squares are given as follows:

Sum-of-Squares Computing Formulas

$$SST = \sum_{i=1}^{k} \sum_{j=1}^{b} y_{ij}^2 - \frac{T_{..}^2}{bk}$$

$$SSA = \frac{\sum_{i=1}^{k} T_{i.}^2}{b} - \frac{T_{..}^2}{bk}$$

$$SSB = \frac{\sum_{j=1}^{b} T_{.j}^2}{k} - \frac{T_{..}^2}{bk}.$$

$$SSE = SST - SSA - SSB.$$

Table 11.10 Analysis of Variance for the Randomized Complete Block Design

Source of Variation	Sum of Squares	Degrees of Freedom	Mean Square	Computed f
Treatments	SSA	$k - 1$	$s_1^2 = \dfrac{SSA}{k - 1}$	$f_1 = \dfrac{s_1^2}{s^2}$
Blocks	SSB	$b - 1$	$s_2^2 = \dfrac{SSB}{b - 1}$	$f_2 = \dfrac{s_2^2}{s^2}$
Error	SSE	$(k - 1)(b - 1)$	$s^2 = \dfrac{SSE}{(b - 1)(k - 1)}$	
Total	SST	$bk - 1$		

The computations in an analysis-of-variance problem for a randomized complete block design may be summarized as shown in Table 11.10 on page 440.

Example 11.6 Four different machines, M_1, M_2, M_3, and M_4 are to be considered in the assembling of a particular product. It is decided that six different operators are to be used in a randomized block experiment to compare the machines. The machines are assigned in a random order to each operator. The operation of the machines requires a certain amount of physical dexterity, and it is anticipated that there will be a difference among the operators in the speed with which they operate the machines. The following times, in seconds, were recorded for assembling the given product:

| Operator 1 | M_2 39.8 | M_4 41.3 | M_3 40.2 | M_1 42.5 |

| Operator 2 | M_3 40.5 | M_1 39.3 | M_2 40.1 | M_4 42.2 |

| Operator 3 | M_2 40.5 | M_1 39.6 | M_4 43.5 | M_3 41.3 |

| Operator 4 | M_4 44.2 | M_2 42.3 | M_1 39.9 | M_3 43.4 |

| Operator 5 | M_1 42.9 | M_3 44.9 | M_2 42.5 | M_4 45.9 |

| Operator 6 | M_2 43.1 | M_4 42.3 | M_3 45.1 | M_1 43.6 |

(a) Test the hypothesis H'_0, at the 0.05 level of significance, that the machines perform at the same mean rate of speed.
(b) Test the hypothesis H''_0, at the 0.05 level of significance, that the operators perform at the same mean rate of speed.

Solution

1. (a) H'_0: $\alpha_1 = \alpha_2 = \alpha_3 = \alpha_4 = 0$ (machine effects are zero).
   (b) H''_0: $\beta_1 = \beta_2 = \cdots = \beta_6 = 0$ (operator effects are zero).

2. (a) H_1':   At least one of the α_i's is not equal to zero.
   (b) H_1'':   At least one of the β_j's is not equal to zero.
3. $\alpha = 0.05$.
4. Critical regions: (a) $f_1 > 3.29$. (b) $f_2 > 2.90$.
5. Computations: To aid in the computations, the data are displayed in the 4×6 rectangular array of Table 11.11.

Table 11.11 Time in Seconds to Assemble Product

Machine	Operator						Total
	1	2	3	4	5	6	
1	42.5	39.3	39.6	39.9	42.9	43.6	247.8
2	39.8	40.1	40.5	42.3	42.5	43.1	248.3
3	40.2	40.5	41.3	43.4	44.9	45.1	255.4
4	41.3	42.2	43.5	44.2	45.9	42.3	259.4
Total	163.8	162.1	164.9	169.8	176.2	174.1	1010.9

Now,

$$SST = 42.5^2 + 39.8^2 + \cdots + 42.3^2 - \frac{1010.9^2}{24} = 81.86$$

$$SSA = \frac{247.8^2 + 248.3^2 + 255.4^2 + 259.4^2}{6} - \frac{1010.9^2}{24} = 15.93$$

$$SSB = \frac{163.8^2 + 162.1^2 + \cdots + 174.1^2}{4} - \frac{1010.9^2}{24} = 42.09$$

$$SSE = 81.86 - 15.93 - 42.09 = 23.84.$$

These and the remaining computations are exhibited in Table 11.12.

Table 11.12 Analysis of Variance for the Data of Table 11.11

Source of Variation	Sum of Squares	Degrees of Freedom	Mean Square	Computed f
Machines	15.93	3	5.31	3.34
Operators	42.09	5	8.42	5.30
Error	23.84	15	1.59	
Total	81.86	23		

6. Decision:
   (a) Reject H_0' and conclude that the machines do not perform at the same mean rate of speed.
   (b) Reject H_0'', as expected, and conclude that the operators do not perform at the same mean rate of speed.

In Chapter 8 we presented a procedure for comparing means when the observations were *paired*. The procedure involved "subtracing out" the effect due to the homogeneous pair and thus working with differences. This is a special case of a randomized complete block design with $k = 2$ treatments. The n homogeneous units to which the treatments were assigned take on the role of blocks.

If there is heterogeneity in the experimental units, the experimenter should not be misled into believing that it is always advantageous to reduce the experimental error through the use of small homogeneous blocks. Indeed, there may be instances where it would not be desirable to block. The purpose in reducing the error variance is to increase the *sensitivity* of the test for detecting differences in the treatment means. This is reflected in the power of the test procedure. (The power of the analysis-of-variance test procedure is discussed more extensively in Section 11.12.) The power for detecting certain differences among the treatment means increases with a decrease in the error variance. However, the power is also affected by the degrees of freedom with which this variance is estimated, and blocking reduces the degrees of freedom that are available from $k(b - 1)$ for the one-way classification to $(k - 1)(b - 1)$. So one could lose power by blocking if there is not a significant reduction in the error variance.

Another important assumption that is implicit in writing the model for a randomized complete block design is that the treatment and block effects were additive. This is equivalent to stating that $\mu_{ij} - \mu_{ij'} = \mu_{i'j} - \mu_{i'j'}$ or $\mu_{ij} - \mu_{i'j} = \mu_{ij'} - \mu_{i'j'}$ for every value of i, i', j, and j'. That is, the difference between the population means for blocks j and j' is the same for every treatment and the difference between the population means for treatments i and i' is the same for every block. The parallel lines of Figure 11.1(a) illustrate a set of mean responses for which the treatment and block effects are additive, whereas the intersecting lines of Figure 11.1(b) show a situation in which treatment and block effects are said to **interact**. Referring to Example 11.6, we see that if operator 3 is 0.5 second faster on the average than operator 2 when machine 1 is used, then operator 3 will still be 0.5 second faster on the average than operator 2 when machine 2, 3, or 4 is used. Similarly, if operator 1 is 1.2 seconds faster on the average using machine 2 than on machine 4, then operator 2, 3, ..., 6 will also be 1.2 seconds faster on the average using machine 2 than on machine 4.

In many experiments the assumption of additivity does not hold and the analysis of Section 11.8 leads to erroneous conclusions. Suppose, for instance, that operator 3 is 0.5 second faster on the average than operator 2 when

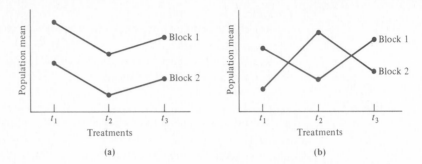

Figure 11.1 Population means for (a) additive effects, and (b) interacting effects.

machine 1 is used but is 0.2 second slower on the average than operator 2 when machine 2 is used. The operators and machines are now interacting.

An inspection of Table 11.11 suggests the presence of interaction. This apparent interaction may be real or it may be due to experimental error. The analysis of Example 11.6 was based on the assumption that the apparent interaction was due entirely to experimental error. If the total variability of our data was in part due to an interaction effect, this source of variation remained a part of the error sum of squares, causing the error mean square to overestimate σ^2, and thereby increased the probability of committing a type II error. We have, in effect, assumed an incorrect model. If we let $(\alpha\beta)_{ij}$ denote the interaction effect of the ith treatment and the jth block, we can write a more appropriate model in the form

$$y_{ij} = \mu + \alpha_i + \beta_j + (\alpha\beta)_{ij} + \varepsilon_{ij},$$

on which we impose the additional restrictions $\displaystyle\sum_{i=1}^{k} (\alpha\beta)_{ij} = \sum_{j=1}^{b} (\alpha\beta)_{ij} = 0$. One can now very easily verify that

$$E\left[\frac{SSE}{(b-1)(k-1)}\right] = \sigma^2 + \frac{\displaystyle\sum_{i=1}^{k}\sum_{j=1}^{b} (\alpha\beta)_{ij}^2}{(b-1)(k-1)}.$$

Thus the error mean square is seen to be a biased estimate of σ^2 when existing interaction has been ignored. It would seem necessary at this point to arrive at a procedure for the detection of interaction for cases where there is suspicion that it exists. Such a procedure requires the availability of an unbiased and independent estimate of σ^2. Unfortunately, the randomized block design does not lend itself to such a test unless the experimental setup is altered. This is discussed extensively in Chapter 12.

11.9 Latin Squares

The randomized block design is very effective in reducing the experimental error by removing one source of variation. Another design that is particularly useful in controlling two sources of variation, and at the same time reduces the required number of treatment combinations, is called the **Latin square**. Suppose that we are interested in the yields of 4 varieties of wheat using 4 different fertilizers over a period of 4 years. The total number of treatment combinations for a completely randomized design would be 64. By selecting the same number of categories for all three criteria of classification, we may select a Latin square design and perform the analysis of variance using the results of only 16 treatment combinations. A typical Latin square, selected at random from all possible 4 × 4 squares, might be the following:

		Column		
Row	1	2	3	4
1	A	B	C	D
2	D	A	B	C
3	C	D	A	B
4	B	C	D	A

The four letters *A*, *B*, *C*, and *D* represent the 4 varieties of wheat that are referred to as the treatments. The rows and columns, represented by the 4 fertilizers and the 4 years, respectively, are the two sources of variation that we wish to control. We now see that each treatment occurs exactly once in each row and each column. With such a balanced arrangement the analysis of variance enables one to separate the variation due to the different fertilizers and different years from the error sum of squares and thereby obtain a more accurate test for differences in the yielding capabilities of the 4 varieties of wheat. When there is interaction present between any of the sources of variation, the *f* values in the analysis of variance are no longer valid. In this case, the Latin square design would be inappropriate.

We shall now generalize and consider an $r \times r$ Latin square where y_{ijk} denotes an observation in the *i*th row and *j*th column corresponding to the *k*th letter. Note that once *i* and *j* are specified for a particular Latin square, we automatically know the letter given by *k*. For example, when $i = 2$ and $j = 3$ in the 4 × 4 Latin square above, we have $k = $ B. Hence *k* is a function of *i* and *j*. If α_i and β_j are the effects of the *i*th row and *j*th column, τ_k the effect of the *k*th treatment, μ the grand mean, and ε_{ijk} the random error, then we can write

$$y_{ijk} = \mu + \alpha_i + \beta_j + \tau_k + \varepsilon_{ijk}$$

on which we impose the restrictions

$$\sum_i \alpha_i = \sum_j \beta_j = \sum_k \tau_k = 0.$$

As before, the y_{ijk} are assumed to be values of independent random variables having normal distributions with means

$$\mu_{ijk} = \mu + \alpha_i + \beta_j + \tau_k$$

and common variance σ^2.

The three hypotheses to be tested are as follows:

1. H_0': $\alpha_1 = \alpha_2 = \cdots = \alpha_r = 0,$
   H_1': At least one of the α_i's is not equal to zero.

2. H_0'': $\beta_1 = \beta_2 = \cdots = \beta_r = 0,$
   H_1'': At least one of the β_j's is not equal to zero.

3. H_0''': $\tau_1 = \tau_2 = \cdots = \tau_r = 0,$
   H_1''': At least one of the τ_k's is not equal to zero.

Each of these tests will be based on a comparison of independent estimates of σ^2 provided by splitting the total sum of squares of our data into four components by means of the following identity. The reader is asked to provide the proof in Exercise 13 on page 453.

Theorem 11.4   Sum-of-Squares Identity

$$\sum_i \sum_j \sum_k (y_{ijk} - \bar{y}\ldots)^2 = r \sum_i (\bar{y}_i.. - \bar{y}\ldots)^2 + r \sum_j (\bar{y}._j. - \bar{y}\ldots)^2$$

$$+ r \sum_k (\bar{y}.._k - \bar{y}\ldots)^2 + \sum_i \sum_j \sum_k (y_{ijk} - \bar{y}_i.. - \bar{y}._j. - \bar{y}.._k + 2\bar{y}\ldots)^2$$

Symbolically, we write the sum-of-squares identity as

$$SST = SSR + SSC + SSTr + SSE,$$

where *SSR* and *SSC* are called the row sum of squares and column sum of squares, respectively; *SSTr* is called the treatment sum of squares; and *SSE* is the error sum of squares. The degrees of freedom are partitioned according to the identity

$$r^2 - 1 = (r - 1) + (r - 1) + (r - 1) + (r - 1)(r - 2).$$

Dividing each of the sum of squares on the right side of the sum-of-squares identity by their corresponding number of degrees of freedom, we obtain the four independent estimates

$$s_1^2 = \frac{SSR}{r-1}, \quad s_2^2 = \frac{SSC}{r-1}, \quad s_3^2 = \frac{SSTr}{r-1}, \quad s^2 = \frac{SSE}{(r-1)(r-2)}$$

of σ^2. Interpreting the sums of squares as functions of independent random variables, it is not difficult to verify that

$$E(S_1^2) = E\left[\frac{SSR}{r-1}\right] = \sigma^2 + \frac{r\sum_i \alpha_i^2}{r-1},$$

$$E(S_2^2) = E\left[\frac{SSC}{r-1}\right] = \sigma^2 + \frac{r\sum_j \beta_j^2}{r-1},$$

$$E(S_3^2) = E\left[\frac{SSTr}{r-1}\right] = \sigma^2 + \frac{r\sum_k \tau_k^2}{r-1},$$

$$E(S^2) = E\left[\frac{SSE}{(r-1)(r-2)}\right] = \sigma^2,$$

from which we immediately conclude that all four estimates of σ^2 are unbiased when H_0', H_0'', and H_0''' are true. All three tests of hypotheses are now carried out by computing the appropriate f values as indicated in Table 11.13.

Table 11.13 Analysis of Variance for an $r \times r$ Latin Square

Source of Variation	Sum of Squares	Degrees of Freedom	Mean Square	Computed f
Rows	SSR	$r-1$	$s_1^2 = \dfrac{SSR}{r-1}$	$f_1 = \dfrac{s_1^2}{s^2}$
Columns	SSC	$r-1$	$s_2^2 = \dfrac{SSC}{r-1}$	$f_2 = \dfrac{s_2^2}{s^2}$
Treatments	SSTr	$r-1$	$s_3^2 = \dfrac{SSTr}{r-1}$	$f_3 = \dfrac{s_3^2}{s^2}$
Error	SSE	$(r-1)(r-2)$	$s^2 = \dfrac{SSE}{(r-1)(r-2)}$	
Total	SST	$r^2 - 1$		

Since the subscript k of the observation y_{ijk} is a function of i and j, it will be to our advantage in writing the computing formulas for the sums of squares to introduce the following notation:

$$T_{i..} = \text{sum of the observations in the } i\text{th row,}$$
$$T_{.j.} = \text{sum of the observations in the } j\text{th column,}$$
$$T_{..k} = \text{sum of the observations for treatment } k,$$
$$T_{...} = \text{sum of all } r^2 \text{ observations.}$$

The sums of squares are now easily computed for the following formulas:

Sum-of-Squares Computing Formulas

$$SST = \sum_i \sum_j \sum_k y_{ijk}^2 - \frac{T_{...}^2}{r^2},$$

$$SSR = \frac{\sum_i T_{i..}^2}{r} - \frac{T_{...}^2}{r^2},$$

$$SSC = \frac{\sum_j T_{.j.}^2}{r} - \frac{T_{...}^2}{r^2},$$

$$SSTr = \frac{\sum_k T_{..k}^2}{r} - \frac{T_{...}^2}{r^2},$$

$$SSE = SST - SSR - SSC - SSTr.$$

Example 11.7   To illustrate the analysis of a Latin square design, let us return to the experiment in which the letters A, B, C, and D represent 4 varieties of wheat; the rows represent 4 different fertilizers; and the columns account for 4 different years. The data in Table 11.14 are the yields for the 4 varieties of wheat measured in kilograms per plot. It is assumed that the various sources of variation do not interact. Using a 0.05 level of significance, test the hypotheses that (a) H_0': there is no difference in the average yields of wheat when different kinds of fertilizer are used; (b) H_0'': there is no difference in the average yield of wheat due to different years; and (c) H_0''': there is no difference in the average yields of the 4 varieties of wheat.

Solution

1. (a) H_0': $\alpha_1 = \alpha_2 = \alpha_3 = \alpha_4 = 0$,
   (b) H_0'': $\beta_1 = \beta_2 = \beta_3 = \beta_4 = 0$,
   (c) H_0''': $\tau_A = \tau_B = \tau_C = \tau_D = 0$.

Table 11.14 Yields of Wheat in Kilograms per Plot

Fertilizer Treatment	Year			
	1981	1982	1983	1984
t_1	A 70	B 75	C 68	D 81
t_2	D 66	A 59	B 55	C 63
t_3	C 59	D 66	A 39	B 42
t_4	B 41	C 57	D 39	A 55

2. (a) H_1': at least one of the α_i's is not equal to zero,
   (b) H_1'': at least one of the β_j's is not equal to zero,
   (c) H_1''': at least one of the τ_k's is not equal to zero.
3. $\alpha = 0.05$.
4. Critical regions: (a) $f_1 > 4.76$, (b) $f_2 > 4.76$, (c) $f_3 > 4.76$.
5. Computations: From Table 11.14, we find the row, column, and treatment totals to be

$$T_{1..} = 294, \quad T_{2..} = 243, \quad T_{3..} = 206, \quad T_{4..} = 192,$$
$$T_{.1.} = 236, \quad T_{.2.} = 257, \quad T_{.3.} = 201, \quad T_{.4.} = 241,$$
$$T_{..A} = 223, \quad T_{..B} = 213, \quad T_{..C} = 247, \quad T_{..D} = 252.$$

Hence

$$SST = 70^2 + 75^2 + \cdots + 55^2 - \frac{935^2}{16} = 2500,$$

$$SSR = \frac{294^2 + 243^2 + 206^2 + 192^2}{4} - \frac{935^2}{16} = 1557,$$

$$SSC = \frac{236^2 + 257^2 + 201^2 + 241^2}{4} - \frac{935^2}{16} = 418,$$

$$SSTr = \frac{223^2 + 213^2 + 247^2 + 252^2}{4} - \frac{935^2}{16} = 264,$$

$$SSE = 2500 - 1557 - 418 - 264 = 261.$$

These results, along with the remaining computations, are given in Table 11.15.

Table 11.15 Analysis of Variance for the Data of Table 11.14

Source of Variation	Sum of Squares	Degrees of Freedom	Mean Square	Computed f
Rows	1557	3	519.000	11.93
Columns	418	3	139.333	3.20
Treatments	264	3	88.000	2.02
Error	261	6	43.500	
Total	2500	15		

6. Decisions:
   (a) Reject H_0' and conclude that a difference in the average yields of wheat exists when different kinds of fertilizer are used.
   (b) Accept H_0'' and conclude that there is no difference in the average yields of wheat due to different years.
   (c) Accept H_0''' and conclude that there is no difference in the average yields of the four varieties of wheat.

Exercises

1. Show that the computing formula for SSB, in the analysis of variance of the randomized complete block design, is equivalent to the corresponding term in the identity of Theorem 11.3.

2. For the randomized block design with k treatments and b blocks, show that

$$E(SSB) = (b-1)\sigma^2 + k \sum_{j=1}^{b} \beta_j^2.$$

3. Four kinds of fertilizer f_1, f_2, f_3, and f_4 are used to study the yield of beans. The soil is divided into 3 blocks each containing 4 homogeneous plots. The yields in kilograms per plot and the corresponding treatments are as follows:

Block 1	Block 2	Block 3
$f_1 = 42.7$	$f_3 = 50.9$	$f_4 = 51.1$
$f_3 = 48.5$	$f_1 = 50.0$	$f_2 = 46.3$
$f_4 = 32.8$	$f_2 = 38.0$	$f_1 = 51.9$
$f_2 = 39.3$	$f_4 = 40.2$	$f_3 = 53.5$

(a) Conduct an analysis of variance at the 0.05 level of significance using the randomized complete block model.
(b) Use single-degree-of-freedom contrasts and a 0.01 level of significance to compare the fertilizers (f_1, f_3) versus (f_2, f_4) and f_1 versus f_3.

4. Three varieties of potatoes are being compared for yield. The experiment was conducted by assigning each variety at random to 3 equal-size plots at each of 4 different locations. The following yields for varieties A, B, and C, in 100 kilograms per plot, were recorded:

Location 1	Location 2	Location 3	Location 4
B 13	C 21	C 9	A 11
A 18	A 20	B 12	C 10
C 12	B 23	A 14	B 17

Perform a two-way analysis of variance to test the hypothesis that there is no difference in the yielding capabilities of the 3 varieties of potatoes. Use a 0.05 level of significance. Was it necessary to plant each variety of potato at each location to reach a valid conclusion concerning the 3 varieties?

5. The following data are the percents of foreign additives measured by 5 analysts for 3 similar brands of strawberry jam, *A*, *B*, and *C*:

Analyst 1 Analyst 2 Analyst 3 Analyst 4 Analyst 5

B	C	B	A	C
2.7	7.5	2.8	1.7	8.1

C	A	A	B	A
3.6	1.6	2.7	1.9	2.0

A	B	C	C	B
3.8	5.2	6.4	2.6	4.8

Perform the analysis of variance and test the hypothesis, at the 0.05 level of significance, that
(a) there is no difference in the percents of foreign additives due to different analysts;
(b) the percent of foreign additives is the same for all 3 brands of jam.

6. The following data represent the final grades obtained by 5 students in mathematics, English, French, and biology:

	Subject			
Student	Mathematics	English	French	Biology
1	68	57	73	61
2	83	94	91	86
3	72	81	63	59
4	55	73	77	66
5	92	68	75	87

Use a 0.05 level of significance to test the hypothesis that

(a) the courses are of equal difficulty;
(b) the students have equal ability.

7. In a study on *"The Periphyton of the South River, Virginia: Mercury Concentration, Productivity, and Autotropic Index Studies"* conducted by the Department of Environmental Sciences and Engineering at the Virginia Polytechnic Institute and State University in 1979, the total mercury concentration in periphyton total solids is being measured at 6 different stations on 6 different dates. The following data was recorded:

	Station					
Date	*CA*	*CB*	*E1*	*E2*	*E3*	*E4*
April 8	0.45	3.24	1.33	2.04	3.93	5.93
June 23	0.10	0.10	0.99	4.31	9.92	6.49
July 1	0.25	0.25	1.65	3.13	7.39	4.43
July 8	0.09	0.06	0.92	3.66	7.88	6.24
July 15	0.15	0.16	2.17	3.50	8.82	5.39
July 23	0.17	0.39	4.30	2.91	5.50	4.29

Use a 0.01 level of significance to test the hypothesis that
(a) the mean mercury concentration is the same for the 6 stations;
(b) the mean mercury concentration is the same for the 6 dates on which measurements were collected.

8. A nuclear power facility produces a vast amount of heat which is usually discharged into aquatic systems. This heat raises the temperature of the aquatic system resulting in a greater concentration of *chlorophyll a*, which in turn extends the growing season. To study this effect, water samples were collected monthly at 3 stations for a period of 12 months. Station *A* is located closest to a potential heated water discharge, station *C* is located farthest away from the discharge, while station *B* is located halfway between stations *A* and *C*. The following concentrations of *chlorophyll a* were recorded:

Month	Station		
	A	B	C
January	9.867	3.723	4.410
February	14.035	8.416	11.100
March	10.700	20.723	4.470
April	13.853	9.168	8.010
May	7.067	4.778	34.080
June	11.670	9.145	8.990
July	7.357	8.463	3.350
August	3.358	4.086	4.500
September	4.210	4.233	6.830
October	3.630	2.320	5.800
November	2.953	3.843	3.480
December	2.640	3.610	3.020

Perform an analysis of variance and test the hypothesis, at the 0.05 level of significance, that

(a) there is no difference in the mean concentrations of *chlorophyll a* at the 3 stations;

(b) the mean concentrations of *chlorophyll a* are the same for all 12 months.

9. In a study conducted by the Department of Health and Physical Education at the Virginia Polytechnic Institute and State University in 1983, 3 diets were assigned for a period of 3 days to each of 6 subjects in a randomized block design. The subjects, playing the role of blocks, were assigned the following 3 diets in a random order:

Diet 1: mixed fat and carbohydrates,

Diet 2: high fat,

Diet 3: high carbohydrates.

At the end of the 3-day period each subject was put on a treadmill and the time to exhaustion, in seconds, was measured. The following data were recorded:

Diet	Subject					
	1	2	3	4	5	6
1	84	35	91	57	56	45
2	91	48	71	45	61	61
3	122	53	110	71	91	122

Perform the analysis of variance, separating out the diet, subject, and error sum of squares. Use a 0.01 level of significance to determine if there are significant differences among the diets.

10. Organic arsenicals are used by forestry personnel as silvicides. The amount of arsenic that is taken into the body when exposed to these silvicides is a major health problem. It is important that the amount of exposure be determined quickly so that a field worker with a high level of arsenic can be removed from the job. In an experiment reported in the paper *"A Rapid Method for the Determination of Arsenic Concentrations in Urine at Field Locations,"* published in the *Amer. Ind. Hyg. Assoc. J.* (Vol. 37, 1976), urine specimens from 4 forest service personnel were divided equally into 3 samples so that each individual could be analyzed for arsenic by a university laboratory, by a chemist using a portable system, and by a forest employee after a brief orientation. The following arsenic levels, in parts per million, were recorded:

Individual	Analyst		
	Employee	Chemist	Laboratory
1	0.05	0.05	0.04
2	0.05	0.05	0.04
3	0.04	0.04	0.03
4	0.15	0.17	0.10

Perform an analysis of variance and test the hypothesis, at the 0.05 level of significance, that

(a) there is no difference in the arsenic levels for the 3 methods of analysis;

(b) the arsenic level is the same for all 4 individuals.

11. Scientists in the Department of Plant Pathology at Virginia Tech devised an experiment in which 5 different treatments were applied to 6 different locations in an apple orchard to determine if there were significant differences in growth among the treatments. Treatments 1 through 4 represent different herbicides and treatment 5 represents a control. The growth period was from May to November in 1982, and the new growth, measured in centimeters, for samples selected from the 6 locations in the orchard were recorded as follows:

Treatment	Locations					
	1	2	3	4	5	6
1	455	72	61	215	695	501
2	622	82	444	170	437	134
3	695	56	50	443	701	373
4	607	650	493	257	490	262
5	388	263	185	103	518	622

Perform an analysis of variance, separating out the treatment, location, and error sum of squares. Use a 0.01 level of significance to test the hypothesis that there are no differences among the treatment means.

12. In the paper *"Self-Control and Therapist Control in the Behavioral Treatment of Overweight Women"* published in *Behavioral Research and Therapy* (Vol. 10, 1972), two reduction treatments and a control treatment were studied for their effects on the weight change of obese women. The two reduction treatments involved were, respectively, a self-induced weight reduction program and a therapist controlled reduction program. Each of 10 subjects were assigned to the 3 treatment programs in a random order and measured for weight loss. The following weight changes were recorded:

Subject	Treatment		
	Control	Self-Induced	Therapist
1	1.00	−2.25	−10.50
2	3.75	−6.00	−13.50
3	0.00	−2.00	0.75
4	−0.25	−1.50	−4.50
5	−2.25	−3.25	−6.00
6	−1.00	−1.50	4.00
7	−1.00	−10.75	−12.25
8	3.75	−0.75	−2.75
9	1.50	0.00	−6.75
10	−0.50	−3.75	−7.00

Perform an analysis of variance and test the hypothesis, at the 0.01 level of significance, that

(a) there is no difference in the mean weight losses for the 3 treatments;

(b) different subjects have no affect on the mean weight losses.

13. Verify the sum-of-squares indentity of Theorem 11.4 on page 446.

14. Show that the computing formula for $SSTr$, in the analysis of variance of the Latin square design, is equivalent to the corresponding term in the sum-of-squares identity of Theorem 11.4.

15. For the $r \times r$ Latin square design, show that

$$E(SSTr) = (r-1)\sigma^2 + r \sum_k \tau_k^2.$$

16. The mathematics department of a large university wishes to evaluate the teaching capabilities of 4 professors. In order to eliminate any effects due to different mathematics courses and different times of the day, it was decided to conduct an experiment using a Latin square design in which the letters A, B, C, and D represent the 4 different professors. Each professor taught one section of each of 4 different courses scheduled at each of 4 different times during the day. The data in the following table show the grades assigned by these professors to 16 students of approximately equal ability:

Time Period	Course			
	Algebra	Geometry	Statistics	Calculus
1	A 84	B 79	C 63	D 97
2	B 91	C 82	D 80	A 93
3	C 59	D 70	A 77	B 80
4	D 75	A 91	B 75	C 68

Use a 0.05 level of significance to test the hypothesis that
(a) there is no difference in the grades due to different time periods;
(b) the courses are of equal difficulty;
(c) different professors have no effect on the grades.

17. A manufacturing firm wants to investigate the effects of 5 color additives on the setting time of a new concrete mix. Variations in the setting times can be expected from day to day changes in temperature and humidity and also from the different workers who prepare the test molds. To eliminate these extraneous sources of variation, a 5 × 5 Latin square design was used in which the letters A, B, C, D, and E represent the 5 additives. The setting times, in hours, for the 25 molds are shown in the following table:

Worker	Day 1	2	3	4	5
1	D 10.7	E 10.3	B 11.2	A 10.9	C 10.5
2	E 11.3	C 10.5	D 12.0	B 11.5	A 10.3
3	A 11.8	B 10.9	C 10.5	D 11.3	E 7.5
4	B 14.1	A 11.6	E 11.0	C 11.7	D 11.5
5	C 14.5	D 11.5	A 11.5	E 12.7	B 10.9

At the 0.05 level of significance, can we say that the color additives have any effect on the setting time of the concrete mix?

18. In 1983 the Department of Dairy Science at the Virginia Polytechnic Institute and State University conducted an experiment to study the effect of feed rations, differing by source of protein, on the average daily milk production of cows. There were 5 rations used in the experiment. A 5 × 5 Latin square was used in which the rows represented different cows and the columns were different lactation periods. The following data, recorded in kilograms, was analyzed by the Statistical Consulting Center at Virginia Tech:

Lactation Periods	Cows 1	2	3	4	5
1	A 33.1	B 34.4	C 26.4	D 34.6	E 33.9
2	C 30.7	D 28.7	E 24.9	A 28.8	B 28.0
3	D 28.7	E 28.8	A 20.0	B 31.9	C 22.7
4	B 31.4	A 22.3	E 18.7	C 31.0	D 21.3
5	B 28.9	C 22.3	D 15.8	E 30.9	A 19.0

At the 0.01 level of significance can we conclude that rations with different sources of protein have an effect on the average daily milk production of cows?

11.10 Random Effects Models

Throughout this chapter we have dealt with analysis-of-variance procedures in which the primary goal was to study the effect on some response of certain fixed or predetermined treatments. Experiments in which the treatments or treatment levels are preselected by the experimenter as opposed to being chosen randomly are called **fixed effects experiments** or **model I experiments**. For the fixed effects model, inferences were made only on those particular treatments used in the experiment.

It is often important that the experimenter be able to draw inferences about a population of treatments by means of an experiment in which the treatments used are chosen randomly from the population. For example, a biologist may be interested in whether or not there is a significant variance is some physiological characteristic due to animal type. The animal types actually used in the experiment are then chosen randomly and represent the treatment effects. A chemist may be interested in studying the effect of analytical laboratories on the chemical analysis of a substance. He is not concerned with particular laboratories but rather with a large population of laboratories. He might then select a group of laboratories at random and allocate samples to each for analysis. The statistical inference would then involve (1) testing whether or not the laboratories contribute a nonzero variance to the analytical results, and (2) estimating the variance due to laboratories and the variance within laboratories.

The one-way **random effects model**, often referred to as **model II**, is written like the fixed effects model but with the terms taking on different meanings. The response

$$y_{ij} = \mu + \alpha_i + \varepsilon_{ij}$$

is now a value of the random variable

$$Y_{ij} = \mu + A_i + E_{ij}$$

with $i = 1, 2, \ldots, k$ and $j = 1, 2, \ldots, n$, where the A_i's are normally and independently distributed with mean zero and variance σ_α^2 and are independent of the E_{ij}'s. As for the fixed effects model, the E_{ij}'s are also normally and independently distributed with mean zero and variance σ^2. Note that for a model II experiment the random variable $\sum_{i=1}^{k} A_i$ assumes the value

$$\sum_{i=1}^{k} \alpha_i \neq 0.$$

Theorem 11.5   *For the random effects one-way analysis-of-variance model*

$$E(SSA) = (k - 1)\sigma^2 + n(k - 1)\sigma_\alpha^2$$

and

$$E(SSE) = k(n - 1)\sigma^2.$$

Proof. From the model

$$Y_{ij} = \mu + A_i + E_{ij}$$

we obtain

$$\bar{Y}_{i\cdot} = \mu + A_i + \bar{E}_{i\cdot},$$
$$\bar{Y}_{\cdot\cdot} = \mu = \bar{A}_{\cdot} + \bar{E}_{\cdot\cdot\cdot}$$

Hence

$$SSA = n \sum_{i=1}^{k} (\bar{Y}_{i\cdot} - \bar{Y}_{\cdot\cdot})^2$$

$$= n \sum_{i=1}^{k} [(A_i - \bar{A}\cdot) + (\bar{E}_{i\cdot} - \bar{E}_{\cdot\cdot})]^2$$

and

$$E(SSA) = n \sum_{i=1}^{k} E(A_i^2) - nkE(\bar{A}^2\cdot) + n \sum_{i=1}^{k} E(\bar{E}_{i\cdot}^2) - nkE(\bar{E}^2_{\cdot\cdot})$$

$$= nk\sigma_\alpha^2 - n\sigma_\alpha^2 + k\sigma^2 - \sigma^2$$

$$= (k-1)\sigma^2 + n(k-1)\sigma_\alpha^2.$$

Following the same steps as above, we also find that

$$SSE = \sum_{i=1}^{k} \sum_{j=1}^{n} (Y_{ij} - \bar{Y}_{i\cdot})^2$$

$$= \sum_{i=1}^{k} \sum_{j=1}^{n} (E_{ij} - \bar{E}_{i\cdot})^2$$

and therefore

$$E(SSE) = \sum_{i=1}^{k} \sum_{j=1}^{n} E(E_{ij}^2) - n \sum_{i=1}^{k} E(\bar{E}_{i\cdot}^2)$$

$$= nk\sigma^2 - k\sigma^2$$

$$= k(n-1)\sigma^2.$$

Table 11.16 shows the expected mean squares for both a model I and a model II experiment. The computations for a model II experiment are carried out in exactly the same way as for a model I experiment. That is, the sum-of-squares, degrees-of-freedom, and mean-square columns in an analysis-of-variance table are the same for both models.

Table 11.16 Expected Mean Squares for the One-Way Classification

Source of Variation	Degrees of Freedom	Mean Squares	Expected Mean Squares	
			Model I	Model II
Treatments	$k-1$	s_1^2	$\sigma^2 + \dfrac{n \sum_{i=1}^{k} \alpha_i^2}{k-1}$	$\sigma^2 + n\sigma_\alpha^2$
Error	$k(n-1)$	s^2	σ^2	σ^2
Total	$nk-1$			

For the random effects model, the hypothesis that the treatment effects are all zero is written

$$H_0: \quad \sigma_\alpha^2 = 0,$$
$$H_1: \quad \sigma_\alpha^2 \neq 0,$$

which says that the different treatments contribute nothing to the variability of the response. It is obvious from Table 11.16 that s_1^2 and s^2 are both estimates of σ^2 when H_0 is true and that the ratio

$$f = \frac{s_1^2}{s^2}$$

is a value of the random variable F having the F distribution with $k - 1$ and $k(n - 1)$ degrees of freedom. The null hypothesis is rejected at the α level of significance when

$$f > f_\alpha[k - 1, k(n - 1)].$$

Table 11.16 can also be used to estimate the **variance components** σ^2 and σ_α^2. Since s_1^2 estimates $\sigma^2 + n\sigma_\alpha^2$ and s^2 estimates σ^2,

$$\hat{\sigma}^2 = s^2$$

$$\hat{\sigma}_\alpha^2 = \frac{s_1^2 - s^2}{n}.$$

Example 11.8 The following data are coded observations on the yield of a chemical process using 5 batches of raw material selected randomly:

	\multicolumn{5}{c}{Batch}				
	1	2	3	4	5	
	9.7	10.4	15.9	8.6	9.7	
	5.6	9.6	14.4	11.1	12.8	
	8.4	7.3	8.3	10.7	8.7	
	7.9	6.8	12.8	7.6	13.4	
	8.2	8.8	7.9	6.4	8.3	
	7.7	9.2	11.6	5.9	11.7	
	8.1	7.6	9.8	8.1	10.7	
Total	55.6	59.7	80.7	58.4	75.3	329.7

Show that the batch variance component is significantly greater than zero and obtain its estimate.

Solution. The total, batch, and error sum of squares are given by

$$SST = 9.7^2 + 5.6^2 + \cdots + 10.7^2 - \frac{329.7^2}{35}$$

$$= 194.64$$

$$SSA = \frac{55.6^2 + 59.7^2 + \cdots + 75.3^2}{7} - \frac{329.7^2}{35}$$

$$= 72.60$$

$$SSE = 194.64 - 72.60 = 122.04.$$

These results, with the remaining computations, are given in Table 11.17. The f ratio is significant at the $\alpha = 0.05$ level, indicating that the hypothesis of a zero batch component is rejected. An estimate of the batch variance component is given by

$$\hat{\sigma}_\alpha^2 = \frac{18.15 - 4.07}{7} = 2.01.$$

In a randomized complete block experiment in which the blocks represent days it is conceivable that the experimenter would like his results to apply not only to the actual days used in the analysis but to every day in the year. He would then select the days on which he runs his experiment as well as the treatments at random and use the random effects model

$$Y_{ij} = \mu + A_i + B_j + E_{ij},$$

$i = 1, 2, \ldots, k$ and $j = 1, 2, \ldots, b$, with the A_i, B_j, and E_{ij} being independent random variables with means zero and variances σ_α^2, σ_β^2, and σ^2, respectively. The expected mean squares for a model II randomized complete block design are obtained using the same procedure as for the one-way classification and are presented along with those for a model I experiment in Table 11.18.

Again the computations for the individual sum of squares and degrees of freedom are identical to those of the fixed effects model. The hypothesis

$$H_0: \quad \sigma_\alpha^2 = 0,$$
$$H_1: \quad \sigma_\alpha^2 \neq 0,$$

Table 11.17 Analysis of Variance for Example 11.8

Source of Variation	Sum of Squares	Degrees of Freedom	Mean Square	Computed f
Batches	72.60	4	18.15	4.46
Error	122.04	30	4.07	
Total	194.64	34		

Table 11.18 Expected Mean Squares for the Randomized Complete Block Design

Source of Variation	Degrees of Freedom	Mean Square	Expected Mean Squares	
			Model I	Model II
Treatments	$k-1$	s_1^2	$\sigma^2 + \dfrac{b\sum\limits_{i=1}^{k}\alpha_i^2}{k-1}$	$\sigma^2 + b\sigma_\alpha^2$
Blocks	$b-1$	s_2^2	$\sigma^2 + \dfrac{k\sum\limits_{j=1}^{b}\beta_j^2}{b-1}$	$\sigma^2 + k\sigma_\beta^2$
Error$(k-1)$		$s^2\sigma^2$		
Total	$bk-1$			

is carried out by computing

$$f = \frac{s_1^2}{s^2}$$

and rejecting H_0 when $f > f_\alpha[k-1, (b-1)(k-1)]$. Similarly, we can test

$$H_0: \quad \sigma_\beta^2 = 0,$$
$$H_1: \quad \sigma_\beta^2 \neq 0,$$

by comparing

$$f = \frac{s_2^2}{s^2}$$

with the critical point $f_\alpha[b-1, (b-1)(k-1)]$.

The unbiased estimates of the variance components are given by

$$\hat{\sigma}^2 = s^2$$

$$\hat{\sigma}_\alpha^2 = \frac{s_1^2 - s^2}{b}$$

$$\hat{\sigma}_\beta^2 = \frac{s_2^2 - s^2}{k}.$$

For the Latin square design, the random effects model is written

$$Y_{ijk} = \mu + A_i + B_j + T_k + E_{ijk},$$

$i = 1, 2, \ldots, r; j = 1, 2, \ldots, r;$ and $k = A, B, C, \ldots,$ with $A_i, B_j, T_k,$ and E_{ijk} being independent random variables with means zero and variances $\sigma_\alpha^2, \sigma_\beta^2, \sigma_\tau^2,$ and $\sigma^2,$ respectively. The derivation of the expected mean squares for a model II Latin square design is straightforward and, for comparison, are presented along with those for a model I experiment in Table 11.19.

Table 11.19 Expected Mean Squares for a Latin Square Design

Source of Variation	Degrees of Freedom	Mean Square	Expected Mean Squares	
			Model I	Model II
Rows	$r - 1$	s_1^2	$\sigma^2 + \dfrac{r \sum_i \alpha_i^2}{r - 1}$	$\sigma^2 + r\sigma_\alpha^2$
Columns	$r - 1$	s_2^2	$\sigma^2 + \dfrac{r \sum_j \beta_j^2}{r - 1}$	$\sigma^2 + r\sigma_\beta^2$
Treatments	$r - 1$	s_3^2	$\sigma^2 + \dfrac{r \sum_k \tau_k^2}{r - 1}$	$\sigma^2 + r\sigma_\tau^2$
Error	$(r - 1)(r - 2)$	s^2	σ^2	σ^2
Total	$r^2 - 1$			

Tests of hypotheses concerning the various variance components are made by computing the ratios of appropriate mean squares as indicated in Table 11.19, and comparing with corresponding f values from Table A.6.

11.11 Regression Approach to Analysis of Variance

So far, we have treated the regression models and the analysis-of-variance models as two separate and unrelated topics. Although this has become the accepted approach in dealing with these procedures on an elementary level, one can treat an analysis-of-variance model as a special case of a multiple linear regression model. In this section we shall show the relationship between the two models and indicate how the analysis of variance techniques can be developed through a regression approach.

Suppose that we consider two models, the multiple linear regression model

$$y_i = \beta_0 + \beta_1 x_{1i} + \beta_2 x_{2i} + \cdots + \beta_k x_{ki} + \varepsilon_i$$

and the one-way classification analysis-of-variance model

$$y_{ij} = \mu + \alpha_i + \varepsilon_{ij}.$$

Traditionally, the two are presented as methods for handling different practical problems, the regression model being a means of arriving at a procedure for predicting some response as a function of one or more quantitative independent variables, and the analysis-of-variance model for arriving at significance tests on multiple population means. However, any mathematical model that is linear in the parameters, such as the analysis-of-variance model, can be considered a special case of the multiple linear regression model. We can use conventional matrix notation to describe how each observation is expressed as a function of the parameters for the two models. For the regression model

$$\mathbf{y} = \mathbf{X}\boldsymbol{\beta} + \boldsymbol{\varepsilon},$$

or, more explicitly,

$$\begin{bmatrix} y_1 \\ y_2 \\ \vdots \\ y_n \end{bmatrix} = \begin{bmatrix} 1 & x_{11} & x_{21} & \cdots & x_{k1} \\ 1 & x_{12} & x_{22} & \cdots & x_{k2} \\ \vdots & \vdots & \vdots & & \vdots \\ 1 & x_{1n} & x_{2n} & \cdots & x_{kn} \end{bmatrix} \begin{bmatrix} \beta_0 \\ \beta_1 \\ \vdots \\ \beta_k \end{bmatrix} + \begin{bmatrix} \varepsilon_1 \\ \varepsilon_2 \\ \vdots \\ \varepsilon_n \end{bmatrix},$$

where the $\mathbf{y}$ vector on the left of the equality sign is the array of responses in the experiment. The $\mathbf{X}$ matrix has already been described and used in Section 10.3. The $\boldsymbol{\beta}$ vector is the vector of parameters appearing in the model, and the $\boldsymbol{\varepsilon}$ vector completes the model by the addition of the random error. The reader will recall that the least squares estimates $b_0, b_1, \ldots, b_k$ of the parameters $\beta_0, \beta_1, \ldots, \beta_k$ are obtained by solving the equation

$$\mathbf{A}\mathbf{b} = \mathbf{g},$$

where $\mathbf{A} = \mathbf{X}'\mathbf{X}$ is a nonsingular matrix and $\mathbf{g} = \mathbf{X}'\mathbf{y}$ is a vector whose elements are sums of products of elements in the columns of $\mathbf{X}$ and the elements in the vector $\mathbf{y}$. Thus the estimates are given by

$$\mathbf{b} = \mathbf{A}^{-1}\mathbf{g}.$$

Consider now the analysis-of-variance model in matrix form:

$$\begin{bmatrix} y_{11} \\ y_{12} \\ \vdots \\ y_{1n} \\ \cdots \\ y_{21} \\ y_{22} \\ \vdots \\ y_{2n} \\ \cdots \\ \vdots \\ \cdots \\ y_{k1} \\ y_{k2} \\ \vdots \\ y_{kn} \end{bmatrix} = \begin{bmatrix} 1 & 1 & 0 & \cdots & 0 \\ 1 & 1 & 0 & \cdots & 0 \\ \vdots & \vdots & \vdots & & \vdots \\ 1 & 1 & 0 & \cdots & 0 \\ \cdots & & & & \\ 1 & 0 & 1 & \cdots & 0 \\ 1 & 0 & 1 & \cdots & 0 \\ \vdots & \vdots & \vdots & & \vdots \\ 1 & 0 & 1 & \cdots & 0 \\ \cdots & & & & \\ \vdots & \vdots & \vdots & & \vdots \\ \cdots & & & & \\ 1 & 0 & 0 & \cdots & 1 \\ 1 & 0 & 0 & \cdots & 1 \\ \vdots & \vdots & \vdots & & \vdots \\ 1 & 0 & 0 & \cdots & 1 \end{bmatrix} \begin{bmatrix} \mu \\ \alpha_1 \\ \alpha_2 \\ \vdots \\ \alpha_k \end{bmatrix} + \begin{bmatrix} \varepsilon_{11} \\ \varepsilon_{12} \\ \vdots \\ \varepsilon_{1n} \\ \cdots \\ \varepsilon_{21} \\ \varepsilon_{22} \\ \vdots \\ \varepsilon_{2n} \\ \cdots \\ \vdots \\ \cdots \\ \varepsilon_{k1} \\ \varepsilon_{k2} \\ \vdots \\ \varepsilon_{kn} \end{bmatrix}.$$

Again each observation is expressed as a function of the parameters. Here the very important $\mathbf{X}$ matrix, the matrix of experimental conditions, consists of ones and zeros. Similar formulations can be written for the randomized complete block model.

Let us apply the least squares approach to the one-way analysis-of-variance model. The normal equations are given by

$$
\begin{bmatrix}
nk & n & n & \cdots & n \\
n & n & 0 & \cdots & 0 \\
n & 0 & n & \cdots & 0 \\
\vdots & \vdots & \vdots & & \vdots \\
n & 0 & 0 & \cdots & n
\end{bmatrix}
\begin{bmatrix}
\hat{\mu} \\
\hat{\alpha}_1 \\
\hat{\alpha}_2 \\
\vdots \\
\hat{\alpha}_k
\end{bmatrix}
=
\begin{bmatrix}
T_{..} \\
T_1 \\
T_2 \\
\vdots \\
T_k
\end{bmatrix}
$$

At this stage it is simple to illustrate why a distinction is made in the presentation of the two models. The last k columns of the $\mathbf{A}$ matrix for the analysis-of-variance model add to the first column and thus the matrix is **singular**, implying that there is no unique solution to the estimating equations. This initially seems like a serious drawback as far as the model is concerned. In fact, we say that the parameters in the model are not **estimable**. The reader will recall that significance tests were performed on the population means, $\mu_1 = \mu + \alpha_1$, $\mu_2 = \mu + \alpha_2$, ..., $\mu_k = \mu + \alpha_k$, and, in formulating the test procedure, the linear constraint $\sum_{i=1}^{k} \alpha_i = 0$ was applied. Thus the α_i's take on the role of deviations (plus or minus) of the treatment or population means from the overall mean μ. Testing equality of population means then becomes equivalent to testing that the α_i's $(i = 1, 2, \ldots, k)$ are all zero.

With the constraint that the α_i's sum to zero, the estimating equations can be solved to yield

$$
\hat{\mu} = \frac{T_{..}}{nk} = \bar{y}_{..}
$$

$$
\hat{\alpha}_i = \frac{T_{i.}}{n} - \frac{T_{..}}{nk} = \bar{y}_{i.} - \bar{y}_{..}, \qquad i = 1, 2, \ldots, k.
$$

While these estimates are not unique, since they are dependent on the constraint that was applied to the α_i's, they do give us a basis for using the general regression procedure outlined in Section 10.6 to determine if the deletion of the α_i's from the model significantly increases the error sum of squares, thereby providing us, in the regression context, with a test of hypothesis of no significant treatment effects.

If we approach the hypothesis testing problem for the one-way analysis-of-variance model following the multiple regression procedures in Chapter 10, we might begin by computing the regression sum of squares for the parameters α_1, α_2, ..., α_k. These parameters take on the same role as the coefficients β_1, β_2,

..., β_k in the multiple linear regression model. We would then compute the regression sum of squares

$$R(\alpha_1, \alpha_2, \ldots, \alpha_k) = SSR$$

$$= b_0 g_0 + b_1 g_1 + \cdots + b_k g_k - \frac{\left(\sum_{i=1}^{k} \sum_{j=1}^{n} y_{ij}^2\right)}{nk}$$

$$= b_1 g_1 + \cdots + b_k g_k$$

$$= \hat{\alpha}_1 g_1 + \cdots + \hat{\alpha}_k g_k.$$

The right side of the estimating equations gives $g_1 = T_{1.}, g_2 = T_{2.}, \ldots, g_k = T_{k.}$. Hence

$$R(\alpha_1, \alpha_2, \ldots, \alpha_k) = \sum_{i=1}^{k} \left(\frac{T_{i.}}{n} - \frac{T_{..}}{nk}\right) T_{i.}$$

$$= \sum_{i=1}^{k} \frac{T_{i.}^2}{n} - \frac{T_{..}^2}{nk}$$

$$= SSA,$$

with $k - 1$ degrees of freedom rather than k. One degree of freedom is lost on account of the single linear restraint imposed on the treatment effects. The error sum of squares with $(nk - 1) - (k - 1) = k(n - 1)$ degrees of freedom is given by

$$SSE = SST - R(\alpha_1, \alpha_2, \ldots, \alpha_k)$$

$$= SST - SSA,$$

which is identical to the expression developed earlier in this chapter.

The hypothesis that the regression on the α_i's is insignificant; that is, $\alpha_i = 0$ for all i's, is tested by forming the ratio

$$f = \frac{R(\alpha_1, \alpha_2, \ldots, \alpha_k)/(k - 1)}{SSE/k(n - 1)} = \frac{SSA/(k - 1)}{s^2}.$$

A value of $f > f_\alpha[(k - 1), \ k(n - 1)]$ implies that regression is significantly increased and consequently the error sum of squares is significantly decreased by including the treatment effects in the model.

The regression approach to analysis-of-variance-type models can be extended to the randomized complete block and Latin square designs discussed in Sections 11.8 and 11.9, and also to the factorial models of Chapter 12.

11.12 Power of Analysis-of-Variance Tests

As we indicated earlier, the research worker is often plagued by the problem of not knowing how large a sample to choose. In conducting a one-way fixed effects analysis of variance with n observations per treatment, the main objective is to test the hypothesis of equality of treatment means.

$$H_0: \quad \alpha_1 = \alpha_2 = \cdots = \alpha_k = 0,$$
$$H_1: \quad \text{at least one of the } \alpha_i\text{'s is not equal to zero.}$$

Quite often, however, the experimental error variance, σ^2, is so large that the test procedure will be insentive to actual differences among the k treatment means. In Section 11.2 the expected values of the mean squares for the one-way model were given by

$$E(S_1^2) = E\left(\frac{SSA}{k-1}\right) = \sigma^2 + \frac{n\sum_{i=1}^{k}\alpha_i^2}{k-1}$$

$$E(S^2) = E\left(\frac{SSE}{k(n-1)}\right) = \sigma^2.$$

Thus for a given deviation from the null hypothesis H_0, as measured by $n\sum_{i=1}^{k}\alpha_i^2/(k-1)$, large values of σ^2 decrease the chance of obtaining a value $f = s_1^2/s^2$ that is in the critical region for the test. The sensitivity of the test describes the ability of the procedure to detect differences in the population means and is measured by the power of the test (see Section 8.5), which is merely $1 - \beta$, where β is the probability of accepting a false hypothesis. We can interpret the power for our analysis-of-variance tests, then, as the probability that the F statistic is in the critical region when, in fact, the null hypothesis is false and the treatment means do differ. For the one-way analysis-of-variance test, the power, $1 - \beta$, is given by

$$1 - \beta = P\left[\frac{S_1^2}{S^2} > f_\alpha(v_1, v_2) \text{ when } H_1 \text{ is true}\right]$$

$$= P\left[\frac{S_1^2}{S^2} > f_\alpha(v_1, v_2) \text{ when } \sum_{i=1}^{k}\alpha_i^2 \neq 0\right].$$

The term $f_\alpha(v_1, v_2)$ is, of course, the upper tail critical point of the F distribution with v_1 and v_2 degrees of freedom. For given values of $\sum_{i=1}^{k}\alpha_i^2/(k-1)$ and σ^2, the power can be increased by using a larger sample size n. The problem becomes one of designing the experiment with a value of n so that the power requirements are met. For example, we might require that for specific values of $\sum_{i=1}^{k}\alpha_i^2 \neq 0$ and σ^2, the hypothesis be rejected with probability 0.9. When the power of the test is low, it severely limits the scope of the inferences that can be drawn from the experimental data.

Fixed Effects Case

In the analysis of variance the power depends on the distribution of the F ratio under the alternative hypothesis that the treatment means differ. Therefore, in the case of the one-way fixed effects model, we require the distribution of S_1^2/S^2 when, in fact, $\sum_{i=1}^{k} \alpha_i^2 \neq 0$. Of course, when the hypothesis is true, $\alpha_i = 0$ for $i = 1, 2, \ldots, k$, and the statistic follows the F distribution with $k - 1$ and $N - k$ degrees of freedom. If $\sum_{i=1}^{k} \alpha_i^2 \neq 0$, the ratio follows a **noncentral F distribution**.

The basic random variable of the noncentral F is denoted by F'. Let $f'_\alpha(v_1, v_2, \lambda)$ be a value of F' with parameters v_1, v_2, and λ. The parameters v_1 and v_2 of the distribution are the degrees of freedom associated with S_1^2 and S^2, respectively, and λ is called the **noncentrality parameter**. When $\lambda = 0$, the noncentral F simply reduces to the ordinary F distribution with v_1 and v_2 degrees of freedom.

For the fixed effects, one-way analysis of variance with sample sizes $n_1, n_2, \ldots, n_k$, we define

$$\lambda = \frac{\sum_{i=1}^{k} n_i \alpha_i^2}{2\sigma^2}.$$

If we have tables of the noncentral F at our disposal, the power for detecting a particular alternative is obtained by evaluating the following probability:

$$1 - \beta = P\left[\frac{S_1^2}{S^2} > f_\alpha(k - 1, N - k) \text{ when } \lambda = \frac{\sum_{i=1}^{k} n_i \alpha_i^2}{2\sigma^2}\right]$$

$$= P[F' > f_\alpha(k - 1, N - k)].$$

Although the noncentral F is normally defined in terms of λ, it is more convenient, for purposes of tabulation, to work with

$$\phi^2 = \frac{2\lambda}{v_1 + 1}.$$

Table A.15 gives graphs of the power of the analysis of variance as a function of ϕ for various values of v_1, v_2, and the significance level α. These **power charts** can be used not only for the fixed effects models discussed in this chapter, but also for the multifactor models of Chapter 12. It remains now to give a procedure whereby the noncentrality parameter λ, and thus ϕ, can be found for these fixed effects cases.

Table 11.20 Noncentrality Parameter λ and ϕ^2
for Fixed Effects Models

	One-way Classification	Randomized Complete Block	Latin Square
λ	$\dfrac{\sum\limits_i n_i \alpha_i^2}{2\sigma^2}$	$\dfrac{b\sum\limits_i \alpha_i^2}{2\sigma^2}$	$\dfrac{r\sum\limits_k \tau_k^2}{2\sigma^2}$
ϕ^2	$\dfrac{\sum\limits_i n_i \alpha_i^2}{k\sigma^2}$	$\dfrac{b\sum\limits_i \alpha_i^2}{k\sigma^2}$	$\dfrac{\sum\limits_k \tau_k^2}{\sigma^2}$

The noncentrality parameter λ can be written in terms of the **expected values of the numerator mean square** of the F ratio in the analysis of variance. We have

$$\lambda = \frac{v_1[E(S_i^2)]}{2\sigma^2} - \frac{v_1}{2}$$

and thus

$$\phi^2 = \frac{[E(S_i^2) - \sigma^2]}{\sigma^2} \cdot \frac{v_1}{v_1 + 1}.$$

Expressions for λ and ϕ^2 for the one-way classification, the randomized complete block design, and the Latin square design are given in Table 11.20.

Note from Table A.15 that for given values of v_1 and v_2, the power of the test increases with increasing values of ϕ. The value of λ depends, of course, on σ^2, and in a practical problem one may often need to substitute the error mean square as an estimate in determining ϕ^2.

Example 11.9 In a randomized block experiment 4 treatments are to be compared in 6 blocks, resulting in 15 degrees of freedom for error. Are 6 blocks sufficient if the power of our test for detecting differences among the treatment means, at the 0.05 level of significance, is to be at least 0.8 when the true means are $\mu_1. = 5.0$, $\mu_2. = 7.0$, $\mu_3. = 4.0$, and $\mu_4. = 4.0$? An estimate of σ^2 to be used in the computation of the power is given by $\hat{\sigma}^2 = 2.0$.

Solution. Recall that the treatment means are given by $\mu_i. = \mu + \alpha_i$. If we invoke the restriction that $\sum\limits_{i=1}^{4} \alpha_i = 0$, we have

$$\mu = \frac{\sum\limits_{i=1}^{4} \mu_i.}{4} = 5.0,$$

and then $\alpha_1 = 0$, $\alpha_2 = 2.0$, $\alpha_3 = -1.0$, and $\alpha_4 = -1.0$. Therefore,

$$\phi^2 = \frac{b \sum_{i=1}^{k} \alpha_i^2}{k\sigma^2} = \frac{(6)(6)}{(4)(2)} = 4.5,$$

from which we obtain $\phi = 2.121$. Using Table A.15, the power is found to be approximately 0.89 and thus the power requirements are met. This means that if the value of $\sum_{i=1}^{4} \alpha_i^2 = 6$ and $\sigma^2 = 2.0$, the use of 6 blocks will result in rejecting the hypothesis of equal treatment means with probability 0.89.

Random Effects Case

In the fixed effects case, the computation of power requires the use of the noncentral F distribution. Such is not the case in the random effects model. In fact, the power is computed very simply by the use of the standard F tables. Consider, for example, the one-way random effects model, n observations per treatment, with the hypothesis

$$H_0: \quad \sigma_\alpha^2 = 0,$$
$$H_1: \quad \sigma_\alpha^2 \neq 0.$$

When H_1 is true, the ratio

$$f = \frac{SSA/[(k-1)(\sigma^2 + n\sigma_\alpha^2)]}{SSE/k(n-1)\sigma^2} = \frac{s_1^2}{s^2(1 + n\sigma_\alpha^2/\sigma^2)}$$

is a value of the random variable F having the F distribution with $k - 1$ and $k(n - 1)$ degrees of freedom. The problem becomes one, then, of determining the probability of rejecting H_0 under the condition that the true treatment variance component $\sigma_\alpha^2 \neq 0$. We have then

$$1 - \beta = P\left\{\frac{S_1^2}{S^2} > f_\alpha[(k-1), k(n-1)] \text{ when } H_1 \text{ is true}\right\}$$

$$= P\left\{\frac{S_1^2}{S^2(1 + n\sigma_\alpha^2/\sigma^2)} > \frac{f_\alpha[(k-1), k(n-1)]}{1 + n\sigma_\alpha^2/\sigma^2}\right\}$$

$$= P\left\{F > \frac{f_\alpha[(k-1), k(n-1)]}{1 + n\sigma_\alpha^2/\sigma^2}\right\}.$$

Note that as n increases, the value $f_\alpha[(k-1), k(n-1)]/(1 + n\sigma_\alpha^2/\sigma^2)$ approaches zero, resulting in an increase in the power of the test. An illustration of the power for this kind of situation is given in Figure 11.2. The lighter shaded area is the significance level α, while the entire shaded area is the power of the test.

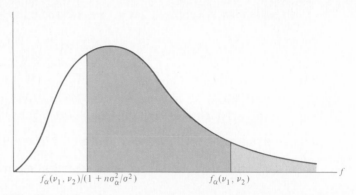

$$f_\alpha(\nu_1, \nu_2)/(1 + n\sigma_\alpha^2/\sigma^2) \qquad\qquad f_\alpha(\nu_1, \nu_2)$$

Figure 11.2 Power for the random effects one-way analysis of variance.

Example 11.10 Suppose in a one-way classification it is of interest to test for the significance of the variance component σ_α^2. Four treatments are to be used in the experiment with 5 observations per treatment. What will be the probability of rejecting the hypothesis $\sigma_\alpha^2 = 0$, when in fact the treatment variance component is $(3/4)\sigma^2$?

Solution. Using an $\alpha = 0.05$ significance level, we have

$$1 - \beta = P\left\{F > \frac{f_{0.05}(3, 16)}{1 + (5)(3)/4}\right\}$$

$$= P\left[F > \frac{f_{0.05}(3, 16)}{4.75}\right]$$

$$= P\left(F > \frac{3.24}{4.75}\right)$$

$$= P(F > 0.864).$$

Using Theorem 6.19 on page 209 and then Table A-7c of *Introduction to Statistics*, 3rd ed., by Dixon and Massey, we see that

$$1 - \beta \simeq 0.46.$$

Therefore, only about 46% of the time will the test procedure detect a variance component that is $(3/4)\sigma^2$.

Exercises

1. The following data show the effect of 4 operators, chosen randomly, on the output of a particular machine:

Operator			
1	2	3	4
175.4	168.5	170.1	175.2
171.7	162.7	173.4	175.7
173.0	165.0	175.7	180.1
170.5	164.1	170.7	183.7

(a) Perform a model II analysis of variance at the 0.05 level of significance.
(b) Compute an estimate of the operator variance component and the experimental error variance component.

2. Assuming a random effects model, show that

$$E(SSB) = (b - 1)\sigma^2 + k(b - 1)\sigma_\beta^2$$

for the randomized complete block design.

3. An experiment is conducted in which 4 treatments are to be compared in 5 blocks. The following data are generated:

	Block				
Treatment	1	2	3	4	5
1	12.8	10.6	11.7	10.7	11.0
2	11.7	14.2	11.8	9.9	13.8
3	11.5	14.7	13.6	10.7	15.9
4	12.6	16.5	15.4	9.6	17.1

(a) Assuming a random effects model, test the hypothesis at the 0.05 level of significance that there is no difference between treatment means.
(b) Compute estimates of the treatment and block variance components.

4. Assuming a random effects model, show that

$$E(SSTr) = (r - 1)(\sigma^2 + r\sigma_\tau^2)$$

for the Latin square design.

5. (a) Using a regression approach for the randomized complete block design, obtain the normal equations $\mathbf{Ab} = \mathbf{g}$ in matrix form.
(b) Show that $R(\beta_1, \beta_2, \ldots, \beta_b | \alpha_1, \alpha_2, \ldots, \alpha_k) = SSB$.

6. In Exercise 1, if we are interested in testing for the significance of the operator variance component, do we have large enough samples to ensure with a probability as large as 0.95 a significant variance component if the true σ_α^2 is $1.5\sigma^2$? If not, how many runs are necessary for each operator? Use a 0.05 level of significance.

7. If one assumes a fixed effects model in Exercise 3 and uses an $\alpha = 0.05$ level test, how many blocks are needed in order that we accept the hypothesis of equality of treatment means with probability 0.1 when, in fact,

$$\frac{\sum_{i=1}^{4} \alpha_i^2}{\sigma^2} = 2.0?$$

8. Verify the values given for λ and ϕ^2 in Table 11.20 on page 466 for the randomized complete block design.

12
Factorial Experiments

12.1 Two-Factor Experiments

Consider a situation in which it is of interest to study the effect of **two factors** A and B on some response. For example, in a chemical experiment we would like to simultaneously vary the reaction pressure and reaction time and study the effect of each on the yield. In a biological experiment, it is of interest to study the effect of drying time and temperature on the amount of solids (percent by weight) left in samples of yeast. The term **factor** is used in a general sense to denote any feature of the experiment such as temperature, time, or pressure that may be varied from trial to trial. We define the **levels** of a factor to be the actual values used in the experiment.

In each of these cases it is important not only to determine if the two factors have an influence on the response, but also if there is a significant interaction between the two factors. As far as terminology is concerned, the experiment described here is a two-way classification or a two-factor experiment and the experimental design may be either a completely randomized design, in which the various treatment combinations are arranged randomly to all the experimental units, or a randomized complete block design, where each level of one of the factors determines a block to which all levels of the other factor are randomly assigned. In the case of the yeast example, the various treatment combinations of temperature and drying time would be assigned randomly to the samples of yeast if we are using a completely randomized design. On the other hand, if the various temperatures represent different blocks, and the drying times are selected in a random order for each temperature, then we have a randomized block experiment.

12.2 *Interaction in Two-Factor Experiments*

In the randomized block model discussed previously it was assumed that one observation on each treatment is taken in each block. If the model assumption is correct, that is, if blocks and treatments are the only real effects and interaction does not exist, the expected value of the error mean square is the experimental error variance σ^2. Suppose, however, that there is interaction occurring between treatments and blocks as indicated by the model

$$y_{ij} = \mu + \alpha_i + \beta_j + (\alpha\beta)_{ij} + \varepsilon_{ij}$$

of Section 11.8. The expected value of the error mean square was then given as

$$E\left[\frac{SSE}{(b-1)(k-1)}\right] = \sigma^2 + \frac{\sum\limits_{i=1}^{k}\sum\limits_{j=1}^{b}(\alpha\beta)_{ij}^2}{(b-1)(k-1)}.$$

The treatment and block effects do not appear in the expected error mean square but the interaction effects do. Thus, if there is interaction in the model, the error mean square reflects variation due to experimental error *plus* an interaction contribution and, for this experimental plan, there is no way of separating them.

From an experimenter's point of view it should seem necessary to arrive at a significance test on the existence of interaction by separating true error variation from that due to interaction. The effects of factors *A* and *B*, often called the **main effects**, take on a different meaning in the presence of interaction. In the previous biological example the effect that drying time has on the amount of solids left in the yeast might very well depend on the temperature that the samples are exposed to. In general, there could very well be experimental situations in which factor *A* has a positive effect on the response at one level of factor *B*, while at a different level of factor *B* the effect of *A* is negative. We use the term **positive effect** here to indicate that the yield or response increases as the levels of a given factor increase according to some defined order. In the same sense a **negative effect** corresponds to a decrease in yield for increasing levels of the factor. Consider, for example, the following hypothetical data taken on two factors each at three levels:

	B			
A	b_1	b_2	b_3	Total
a_1	4.4	8.8	5.2	18.4
a_2	7.5	8.5	2.4	18.4
a_3	9.7	7.9	0.8	18.4
Total	21.6	25.2	8.4	55.2

Clearly, the effect of A is positive at b_1 and negative at b_3. These differences in the levels of A at different levels of B are of interest to the experimenter but an ordinary significance test on factor A would yield a value of zero for SSA, since the totals for each level of A are all of the same magnitude. We say, then, that the presence of interaction is *masking* the effect of factor A. Therefore, if we consider the average influence of A, over all three levels of B, there is no effect. However, this is most likely not what is pertinent to the experimenter.

Before drawing any final conclusions resulting from tests of significance on the main effects and interaction effects, the experimenter should first observe whether or not the test for interaction is significant. If interaction is not significant, then the results of the tests on the main effects are meaningful. However, if interaction should be significant, then only those tests on the main effects that turn out to be significant are meaningful. Nonsignificant main effects in the presence of interaction might well be a result of masking and dictate the need to observe the influence of each factor at fixed levels of the other.

Interaction and experimental error are separated in the two-factor experiment only if multiple observations are taken at the various treatment combinations. To ease the computations involved, there should be the same number, n, of observations at each combination. These should be true replications, not just repeated measurements. For example, in the yeast illustration, if we take $n = 2$ observations at each combination of temperature and drying time, there should be 2 separate samples and not merely repeated measurements on the same sample. This is important because now the measure of experimental error comes from variation between readings *within* treatment combinations and thus indicates true or pure experimental error.

12.3 *Two-Factor Analysis of Variance*

To present general formulas for the analysis of variance of a two-factor experiment using repeated observations in a completely randomized design, we shall consider the case of n replications of the treatment combinations determined

Table 12.1 Two-Factor Experiment with n Replications

A	B				Total	Mean
	1	2	$\cdots$	b		
1	y_{111}	y_{121}	$\cdots$	y_{1b1}	$T_1..$	$\bar{y}_1..$
	y_{112}	y_{122}	$\cdots$	y_{1b2}		
	$\vdots$	$\vdots$		$\vdots$		
	y_{11n}	y_{12n}	$\cdots$	y_{1bn}		
2	y_{211}	y_{221}	$\cdots$	y_{2b1}	$T_2..$	$\bar{y}_2..$
	y_{212}	y_{222}	$\cdots$	y_{2b2}		
	$\vdots$	$\vdots$		$\vdots$		
	y_{21n}	y_{22n}	$\cdots$	y_{2bn}		
$\vdots$	$\vdots$	$\vdots$		$\vdots$	$\vdots$	$\vdots$
a	y_{a11}	y_{a21}	$\cdots$	y_{ab1}	$T_a..$	$\bar{y}_a..$
	y_{a12}	y_{a22}	$\cdots$	y_{ab2}		
	$\vdots$	$\vdots$		$\vdots$		
	y_{a1n}	y_{a2n}	$\cdots$	y_{abn}		
Total	$T._{1}.$	$T._{2}.$	$\cdots$	$T._{b}.$	$T...$	
Mean	$\bar{y}._{1}.$	$\bar{y}._{2}.$	$\cdots$	$\bar{y}._{b}.$		$\bar{y}...$

by a levels of factor A and b levels of factor B. The observations may be classified by means of a rectangular array in which the rows represent the levels of factor A and the columns represent the levels of factor B. Each treatment combination defines a cell in our array. Thus we have ab cells, each cell containing n observations. Denoting the kth observation taken at the ith level of factor A and the jth level of factor B by y_{ijk}, the abn observations are shown in Table 12.1.

The observations in the (ij)th cell constitute a random sample of size n from a population that is assumed to be normally distributed with mean μ_{ij} and variance σ^2. All ab populations are assumed to have the same variance σ^2. Let us define the following useful symbols, some of which are used in Table 12.1:

$T_{ij}.$ = sum of the observations in the (ij)th cell

$T_i..$ = sum of the observations for the ith level of factor A

$T._{j}.$ = sum of the observations for the jth level of factor B

$T...$ = sum of all abn observations

$\bar{y}_{ij}.$ = mean of the observations in the (ij)th cell

$\bar{y}_i..$ = mean of the observations for the ith level of factor A

$\bar{y}._{j}.$ = mean of the observations for the jth level of factor B

$\bar{y}...$ = mean of all abn observations.

Each observation in Table 12.1 may be written in the form

$$y_{ijk} = \mu_{ij} + \varepsilon_{ijk},$$

where ε_{ijk} measures the deviations of the observed y_{ijk} values in the (ij)th cell from the population mean μ_{ij}. If we let $(\alpha\beta)_{ij}$ denote the interaction effect of the ith level of factor A and the jth level of factor B, α_i the effect of the ith level of factor A, β_j the effect of the jth level of factor B, and μ the overall mean, we can write

$$\mu_{ij} = \mu + \alpha_i + \beta_j + (\alpha\beta)_{ij},$$

and then

$$y_{ijk} = \mu + \alpha_i + \beta_j + (\alpha\beta)_{ij} + \varepsilon_{ijk},$$

on which we impose the restrictions

$$\sum_{i=1}^{a} \alpha_i = 0, \qquad \sum_{j=1}^{b} \beta_j = 0, \qquad \sum_{i=1}^{a} (\alpha\beta)_{ij} = 0, \qquad \sum_{j=1}^{b} (\alpha\beta)_{ij} = 0.$$

The three hypotheses to be tested are as follows:

1. H_0': $\quad \alpha_1 = \alpha_2 = \cdots = \alpha_a = 0,$
   H_1': $\quad$ at least one of the α_i's is not equal to zero.

2. H_0'': $\quad \beta_1 = \beta_2 = \cdots = \beta_b = 0,$
   H_1'': $\quad$ at least one of the β_j's is not equal to zero.

3. H_0''': $\quad (\alpha\beta)_{11} = (\alpha\beta)_{12} = \cdots = (\alpha\beta)_{ab} = 0,$
   H_1''': $\quad$ at least one of the $(\alpha\beta)_{ij}$'s is not equal to zero.

Each of these tests will be based on a comparison of independent estimates of σ^2 provided by the splitting of the total sum of squares of our data into 4 components by means of the following identity.

Theorem 12.1   Sum-of-Squares Identity

$$\sum_{i=1}^{a} \sum_{j=1}^{b} \sum_{k=1}^{n} (y_{ijk} - \bar{y}...)^2 = bn \sum_{i=1}^{a} (\bar{y}_{i}.. - \bar{y}...)^2$$

$$+ an \sum_{j=1}^{b} (\bar{y}._{j}. - \bar{y}...)^2$$

$$+ n \sum_{i=1}^{a} \sum_{j=1}^{b} (\bar{y}_{ij}. - \bar{y}_{i}.. - \bar{y}._{j}. + \bar{y}...)^2$$

$$+ \sum_{i=1}^{a} \sum_{j=1}^{b} \sum_{k=1}^{n} (y_{ijk} - \bar{y}_{ij}.)^2.$$

Proof

$$\sum_{i=1}^{a} \sum_{j=1}^{b} \sum_{k=1}^{n} (y_{ijk} - \bar{y}...)^2 = \sum_{i=1}^{a} \sum_{j=1}^{b} \sum_{k=1}^{n} [(\bar{y}_i.. - \bar{y}...) + (\bar{y}._j. - \bar{y}...)$$

$$+ (\bar{y}_{ij}. - \bar{y}_i.. - \bar{y}._j. + \bar{y}...) + (y_{ijk} - \bar{y}_{ij}.)]^2$$

$$= \sum_{i=1}^{a} \sum_{j=1}^{b} \sum_{k=1}^{n} (\bar{y}_i.. - \bar{y}...)^2$$

$$+ \sum_{i=1}^{a} \sum_{j=1}^{b} \sum_{k=1}^{n} (\bar{y}._j. - \bar{y}...)^2$$

$$+ \sum_{i=1}^{a} \sum_{j=1}^{b} \sum_{k=1}^{n} (\bar{y}_{ij}. - \bar{y}_i.. - \bar{y}._j. + \bar{y}...)^2$$

$$+ \sum_{i=1}^{a} \sum_{j=1}^{b} \sum_{k=1}^{n} (y_{ijk} - \bar{y}_{ij}.)^2$$

$$+ \; 6 \text{ cross-product terms.}$$

The cross-product terms are all equal to zero. Hence

$$\sum_{i=1}^{a} \sum_{j=1}^{b} \sum_{k=1}^{n} (y_{ijk} - \bar{y}...)^2 = bn \sum_{i=1}^{a} (\bar{y}_i.. - \bar{y}...)^2 + an \sum_{j=1}^{b} (\bar{y}._j. - \bar{y}...)^2$$

$$+ n \sum_{i=1}^{a} \sum_{j=1}^{b} (\bar{y}_{ij}. - \bar{y}_i.. - \bar{y}._j. + \bar{y}...)^2$$

$$+ \sum_{i=1}^{a} \sum_{j=1}^{b} \sum_{k=1}^{n} (y_{ijk} - \bar{y}_{ij}.)^2.$$

Symbolically, we write the sum of squares identity as

$$SST = SSA + SSB + SS(AB) + SSE,$$

where SSA and SSB are called the sum of squares for the main effects A and B, respectively, $SS(AB)$ is called the interaction sum of squares for A and B, and SSE is the error sum of squares. The degrees of freedom are partitioned according to the identity

$$abn - 1 = (a - 1) + (b - 1) + (a - 1)(b - 1) + ab(n - 1).$$

Dividing each of the sum of squares on the right side of the sum of squares identity by their corresponding number of degrees of freedom, we obtain the 4 independent estimates

$$s_1^2 = \frac{SSA}{a - 1}, \qquad s_2^2 = \frac{SSB}{b - 1}, \qquad s_3^2 = \frac{SS(AB)}{(a - 1)(b - 1)}, \qquad s^2 = \frac{SSE}{ab(n - 1)}$$

of σ^2. If we interpret the sum of squares as functions of the independent random variables Y_{111}, Y_{112}, ..., Y_{abn}, it is not difficult to verify that

$$E(S_1^2) = E\left[\frac{SSA}{a-1}\right] = \sigma^2 + \frac{nb \sum_{i=1}^{a} \alpha_i^2}{a-1}$$

$$E(S_2^2) = E\left[\frac{SSB}{b-1}\right] = \sigma^2 + \frac{na \sum_{j=1}^{b} \beta_j^2}{b-1}$$

$$E(S_3^2) = E\left[\frac{SS(AB)}{(a-1)(b-1)}\right] = \sigma^2 + \frac{n \sum_{i=1}^{a} \sum_{j=1}^{b} (\alpha\beta)_{ij}^2}{(a-1)(b-1)}$$

$$E(S^2) = E\left[\frac{SSE}{ab(n-1)}\right] = \sigma^2,$$

from which we immediately conclude that all four estimates of σ^2 are unbiased when H_0', H_0'', and H_0''' are true.

To test the hypothesis H_0', that the effects of factors A are all equal to zero, we compute the ratio

$$f_1 = \frac{s_1^2}{s^2},$$

which is a value of the random variable F_1 having the F distribution with $a-1$ and $ab(n-1)$ degrees of freedom when H_0' is true. The null hypothesis is rejected at the α level of significance when $f_1 > f_\alpha[a-1, ab(n-1)]$. Similarly, to test the hypothesis H_0'', that the effects of factor B are all equal to zero, we compute the ratio

$$f_2 = \frac{s_2^2}{s^2},$$

which is a value of the random variable F_2 having the F distribution with $b-1$ and $ab(n-1)$ degrees of freedom when H_0'' is true. This hypothesis is rejected at the α level of significance when $f_2 > f_\alpha[b-1, ab(n-1)]$. Finally, to test the hypothesis H_0''', that the interaction effects are all equal to zero, we compute the ratio

$$f_3 = \frac{s_3^2}{s^2},$$

which is a value of the random variable F_3 having the F distribution with $(a-1)(b-1)$ and $ab(n-1)$ degrees of freedom when H_0''' is true. We conclude that interaction is present when $f_3 > f_\alpha[(a-1)(b-1), ab(n-1)]$.

As indicated in Section 12.2, it is advisable to conduct the test for interaction before attempting to draw inferences on the main effects. If interaction is not significant, the experimenter can proceed to test the main effects.

Table 12.2 Analysis of Variance for the Two-Factor Experiment with n Replications

Source of Variation	Sum of Squares	Degrees of Freedom	Mean Square	Computed f
Main effect				
A	SSA	$a - 1$	$s_1^2 = \dfrac{SSA}{a - 1}$	$f_1 = \dfrac{s_1^2}{s^2}$
B	SSB	$b - 1$	$s_2^2 = \dfrac{SSB}{b - 1}$	$f_2 = \dfrac{s_2^2}{s^2}$
Two-factor interaction				
AB	$SS(AB)$	$(a - 1)(b - 1)$	$s_3^2 = \dfrac{SS(AB)}{(a - 1)(b - 1)}$	$f_3 = \dfrac{s_3^2}{s^2}$
Error	SSE	$ab(n - 1)$	$s^2 = \dfrac{SSE}{ab(n - 1)}$	
Total	SST	$abn - 1$		

However, a significant interaction could very well imply that the data should be analyzed in a somewhat different manner—perhaps observing the effect of factor A at fixed levels of factor B, and so forth.

The computations in an analysis-of-variance problem, for a two-factor experiment with n replications, are usually summarized as in Table 12.2.

The sums of squares are usually obtained by constructing the following table of totals:

A	B				Total
	1	2	$\cdots$	b	
1	$T_{11\cdot}$	$T_{12\cdot}$	$\cdots$	$T_{1b\cdot}$	$T_{1\cdot\cdot}$
2	$T_{21\cdot}$	$T_{22\cdot}$	$\cdots$	$T_{2b\cdot}$	$T_{2\cdot\cdot}$
$\vdots$	$\vdots$	$\vdots$		$\vdots$	$\vdots$
a	$T_{a1\cdot}$	$T_{a2\cdot}$	$\cdots$	$T_{ab\cdot}$	$T_{a\cdot\cdot}$
Total	$T_{\cdot1\cdot}$	$T_{\cdot2\cdot}$	$\cdots$	$T_{\cdot b\cdot}$	$T_{\cdot\cdot\cdot}$

and using the following computing formulas:

Sum-of-Squares Computing Formulas

$$SST = \sum_{i=1}^{a} \sum_{j=1}^{b} \sum_{k=1}^{n} y_{ijk}^2 - \frac{T_{...}^2}{abn}$$

$$SSA = \frac{\sum_{i=1}^{a} T_{i..}^2}{bn} - \frac{T_{...}^2}{abn}$$

$$SSB = \frac{\sum_{j=1}^{b} T_{.j.}^2}{an} - \frac{T_{...}^2}{abn}$$

$$SS(AB) = \frac{\sum_{i=1}^{a} \sum_{j=1}^{b} T_{ij.}^2}{n} - \frac{\sum_{i=1}^{a} T_{i..}^2}{bn} - \frac{\sum_{j=1}^{b} T_{.j.}^2}{an} + \frac{T_{...}^2}{abn}$$

$$SSE = SST - SSA - SSB - SS(AB).$$

Example 12.1 In an experiment conducted to determine which of 3 different missile systems is preferable, the propellant burning rate for 24 static firings was measured. Four different propellant types were used. The experiment yielded duplicate observations of burning rates at each combination of the treatments. The data, after coding, are given in Table 12.3.

Use a 0.05 level of significance to test the following hypotheses: (a) H_0': There is no difference in the mean propellant burning rates when different missile systems are used. (b) H_0'': There is no difference in the mean propellant burning rates of the 4 propellant types. (c) H_0''': There is no interaction between the different missile systems and the different propellant types.

Table 12.3 Propellant Burning Rates

Missile System	Propellant Type			
	b_1	b_2	b_3	b_4
a_1	34.0 32.7	30.1 32.8	29.8 26.7	29.0 28.9
a_2	32.0 33.2	30.2 29.8	28.7 28.1	27.6 27.8
a_3	28.4 29.3	27.3 28.9	29.7 27.3	28.8 29.1

Solution

1. (a) H_0': $\alpha_1 = \alpha_2 = \alpha_3 = 0$.
   (b) H_0'': $\beta_1 = \beta_2 = \beta_3 = \beta_4 = 0$.
   (c) H_0''': $(\alpha\beta)_{11} = (\alpha\beta)_{12} = \cdots = (\alpha\beta)_{34} = 0$.
2. (a) H_1':   at least one of the α_i's is not equal to zero.
   (b) H_1'':   at least one of the β_j's is not equal to zero.
   (c) H_1''': at least one of the $(\alpha\beta)_{ij}$'s is not equal to zero.
3. $\alpha = 0.05$.
4. Critical regions: (a) $f_1 > 3.89$, (b) $f_2 > 3.49$, (c) $f_3 > 3.00$.
5. Computations: From Table 12.3 we first construct the following table of totals:

	b_1	b_2	b_3	b_4	Total
a_1	66.7	62.9	56.5	57.9	244.0
a_2	65.2	60.0	56.8	55.4	237.4
a_3	57.7	56.2	57.0	57.9	228.8
Total	189.6	179.1	170.3	171.2	710.2

Now

$$SST = 34.0^2 + 32.7^2 + \cdots + 29.1^2 - \frac{710.2^2}{24}$$

$$= 21{,}107.68 - 21{,}016.00 = 91.68$$

$$SSA = \frac{244.0^2 + 237.4^2 + 228.8^2}{8} - \frac{710.2^2}{24}$$

$$= 21{,}030.52 - 21{,}016.00 = 14.52$$

$$SSB = \frac{189.6^2 + 179.1^2 + 170.3^2 + 171.2^2}{6} - \frac{710.2^2}{24}$$

$$= 21{,}056.08 - 21{,}016.00 = 40.08$$

$$SS(AB) = \frac{66.7^2 + 65.2^2 + \cdots + 57.9^2}{2} - 21{,}030.52$$

$$- 21{,}056.08 + 21{,}016.00$$

$$= 22.17$$

$$SSE = 91.68 - 14.52 - 40.08 - 22.17 = 14.91.$$

These results, with the remaining computations, are given in Table 12.4.

Table 12.4 Analysis of Variance for the Data of Table 12.3

Source of Variation	Sum of Squares	Degrees of Freedom	Mean Square	Computed f
Missile system	14.52	2	7.26	5.85
Propellant type	40.08	3	13.36	10.77
Interaction	22.17	6	3.70	2.98
Error	14.91	12	1.24	
Total	91.68	23		

6. Decision:
   (a) Reject H_0' and conclude that different missile systems result in different mean propellant burning rates.
   (b) Reject H_0'' and conclude that the mean propellant burning rates are not the same for the four propellant types.
   (c) Accept H_0''' and conclude that there is no interaction between the different missile systems and the different propellant types.
   Since there is no significant interaction, the tests on the missile systems and propellant burning rates are meaningful.

Example 12.2 Referring to Example 12.1, choose two orthogonal contrasts to partition the sum of squares for the missile systems into single-degree-of-freedom components to be used in comparing systems 1 and 2 with 3 and system 1 versus system 2.

Solution. The contrast for comparing systems 1 and 2 with 3 is given by

$$\omega_1 = \mu_1. + \mu_2. - 2\mu_3..$$

A second contrast, orthogonal to ω_1, for comparing system 1 with system 2, is given by $\omega_2 = \mu_1. - \mu_2..$ The single-degree-of-freedom sum of squares are

$$SSw_1 = \frac{[244.0 + 237.4 - (2)(228.8)]^2}{(8)[(1)^2 + (1)^2 + (-2)^2]} = 11.80$$

and

$$SSw_2 = \frac{(244.0 - 237.4)^2}{(8)[(1)^2 + (-1)^2]} = 2.72.$$

Notice that $SSw_1 + SSw_2 = SSA$, as expected. The computed f values corresponding to w_1 and w_2 are, respectively,

$$f_1 = \frac{11.80}{1.24} = 9.5$$

and

$$f_2 = \frac{2.72}{1.24} = 2.2.$$

Compared to the critical value $f_{0.05}(1, 12) = 4.75$, we find f_1 to be significant. Thus the first contrast indicates that the hypothesis

$$H_0: \quad \frac{\mu_{1.} + \mu_{2.}}{2} = \mu_{3.}.$$

is rejected. Since $f_2 < 4.75$, the mean burning rates of the first and second systems are not significantly different.

If the hypothesis of no interaction in Example 12.1 is true, as indicated by the f ratio in the analysis-of-variance table, we are able to make the *general* comparisons of Example 12.2 regarding our missile systems rather than separate comparisons for each propellant. Similarly, we might make general comparisons among the propellants rather than separate comparisons for each missile system. For example, we could compare propellants 1 and 2 with 3 and 4 and also propellant 1 versus propellant 2. The resulting f ratios, each with 1 and 12 degrees of freedom, turn out to be 24.86 and 7.41, respectively, and both are significant at the 0.05 level of significance. The indication is then that propellant 1 gives the highest mean burning rate. A prudent experimenter might be somewhat cautious in making overall conclusions in a problem such as this one, where the f ratio for interaction is barely below the 0.05 critical value. This is far from overwhelming evidence that interaction between the factors does not exist. In fact, a quick inspection of the cell totals points out possible evidence of interaction. For example, the overall evidence, 189.6 versus 171.2, certainly indicates that propellant 1 is superior, in terms of a higher burning rate, to propellant 4. However, if we restrict ourselves to system 3, where we have a total of 57.7 for propellant 1 as opposed to 57.9 for propellant 4, there appears to be little or no difference between propellants 1 and 4. In fact, there appears to be a stabilization of burning rates for the different propellants if we operate with system 3. There is certainly overall evidence, 244.0 versus 228.8, that indicates that system 1 gives a higher burning rate than system 3, but if we restrict ourselves to propellant 4, this conclusion does not appear to hold.

Exercises

---

1. An experiment was conducted to study the effect of temperature and type of oven on the life of a particular component being tested. Four types of ovens and 3 temperature levels were used in the experiment. Twenty-four pieces were assigned randomly, 2 to each combination of treatments, and the following results recorded:

Temperature (degrees)	Oven			
	O_1	O_2	O_3	O_4
500	227 221	214 259	225 236	260 229
550	187 208	181 179	232 198	246 273
600	174 202	198 194	178 213	206 219

Using a 0.05 level of significance test the hypothesis that

(a) different temperatures have no effect on the life of the component;

(b) different ovens have no effect on the life of the component;

(c) the type of oven and temperature do not interact.

2. To ascertain the stability of vitamin C in reconstituted frozen orange juice concentrate stored in a refrigerator for a period of up to one week, the study "*Vitamin C Retention in Reconstituted Frozen Orange Juice*" was conducted by the Department of Human Nutrition and Foods at the Virginia Polytechnic Institute and State University in 1975. Three types of frozen orange juice concentrate were tested using 3 different time periods. The time periods refer to the number of days from when the orange juice was blended until it was tested. The results, in milligrams of ascorbic acid per liter, were recorded as follows:

Brand	Time (days)					
	0		3		7	
Richfood	52.6 49.8	54.2 46.5	49.4 42.8	49.2 53.2	42.7 40.4	48.8 47.6
Sealed-Sweet	56.0 49.6	48.0 48.4	48.8 44.0	44.0 42.4	49.2 42.0	44.0 43.2
Minute Maid	52.5 51.8	52.0 53.6	48.0 48.2	47.0 49.6	48.5 45.2	43.4 47.6

Use a 0.05 level of significance to test the hypothesis that

(a) there is no difference in ascorbic acid contents among the different brands of orange juice concentrate;

(b) there is no difference in ascorbic acid contents for the different time periods;

(c) the brands of orange juice concentrate and the number of days from the time the juice was blended until it is tested do not interact.

3. Three strains of rats were studied under 2 environmental conditions for their performance in a maze test. The error scores for the 48 rats were recorded as follows:

Environment	Strain					
	Bright		Mixed		Dull	
Free	28 22 25 36	12 23 10 86	33 36 41 22	83 14 76 58	101 33 122 35	94 56 83 23
Restricted	72 48 25 91	32 93 31 19	60 35 83 99	89 126 110 118	136 38 64 87	120 153 128 140

Use a 0.01 level of significance to test the hypothesis that

(a) there is no difference in error scores for different environments;

(b) there is no difference in error scores for different strains;

(c) the environments and strains of rats do not interact.

4. Corrosion fatigue in metals has been defined as the simultaneous action of cyclic stress and chemical attack on a metal structure. A widely used technique for minimizing corrosion-fatigue damage in aluminum involves the application of a protective coating. In the study "*Effect of Humidity and Several Surface Coatings on the Fatigue Life of 2024-T351 Aluminum Alloy*" conducted by the Department of Mechanical Engineering at the Virginia Polytechnic Institute and State University in

1979, 3 different levels of humidity

Low: 20–25% relative humidity,
Medium: 55–60% relative humidity,
High: 86–91% relative humidity,

and 3 types of surface coatings

Uncoated: no coating,
Anodized: sulfuric acid
                      anodic oxide coating,
Conversion: chromate chemical
                      conversion coating,

were used. The corrosion fatigue data, expressed in thousands of cycles to failure, were recorded as follows:

Coating	Relative Humidity					
	Low		Medium		High	
Uncoated	361	469	314	522	1344	1216
	466	937	244	739	1027	1097
	1069	1357	261	134	663	1011
Anodized	114	1032	322	471	78	466
	1236	92	306	130	387	107
	533	211	68	398	130	327
Conversion	130	1482	252	874	586	524
	841	529	105	755	402	751
	1595	754	847	573	846	529

(a) Perform an analysis of variance with $\alpha = 0.05$ to test for significant main and interaction effects.

(b) Use Duncan's multiple-range test at the 0.05 level of significance to determine which humidity levels result in different corrosion fatigue damage.

5. To determine which muscles need to be subjected to a conditioning program in order to improve one's performance on the flat serve used in tennis, the study "*An Electromyographic-Cinematrographic Analysis of the Tennis Serve*" was conducted by the Department of Health, Physical Education and Recreation at the Virginia Polytechnic Institute and State University in 1978. Five different muscles

1:   anterior deltoid,
2:   pectorial major,
3:   posterior deltoid,
4:   middle deltoid,
5:   triceps,

were tested on each of 3 subjects, and the experiment was carried out 3 times for each treatment combination. The electromyographic data, recorded during the serve, are as follows:

Subject	Muscle				
	1	2	3	4	5
1	32	5	58	10	19
	59	1.5	61	10	20
	38	2	66	14	23
2	63	10	64	45	43
	60	9	78	61	61
	50	7	78	71	42
3	43	41	26	63	61
	54	43	29	46	85
	47	42	23	55	95

Use a 0.01 level of significance to test the hypothesis that

(a) different subjects have equal electromyographic measurements;

(b) different muscles have no effect on electromyographic measurements;

(c) subjects and types of muscle do not interact.

6. A study was made to determine if humidity conditions have an effect on the force required to pull apart pieces of glued plastic. Three types of plastic were tested using 4 different levels of humidity. The results, in kilograms, are given as follows:

Plastic Type	Humidity			
	30%	50%	70%	90%
A	39.0	33.1	33.8	33.0
	42.8	37.8	30.7	32.9

	Humidity			
Plastic Type	30%	50%	70%	90%
B	36.9 41.0	27.2 26.8	29.7 29.1	28.5 27.9
C	27.4 30.3	29.2 29.9	26.7 32.0	30.9 31.5

(a) Assuming a model I experiment, perform an analysis of variance and test the hypothesis of no interaction between humidity and plastic type at the 0.05 level of significance.

(b) Using only plastics *A* and *B* and the value of s^2 from part (a), once again test for the presence of interaction at the 0.05 level of significance.

(c) Use a single-degree-of-freedom comparison and the value of s^2 from part (a) to compare, at the 0.05 level of significance, the force required at 30% humidity versus 50%, 70%, and 90% humidity.

(d) Using only plastic *C* and the value of s^2 from part (a), repeat part (c).

12.4 *Three-Factor Experiments*

In this section we consider an experiment with three factors *A*, *B*, and *C* at *a*, *b*, and *c* levels, respectively, in a completely randomized experimental design. Assume again that we have *n* observations for each of the *abc* treatment combinations. We shall proceed to outline significance tests for the three main effects and interactions involved. It is hoped that the reader can then use the description given here to generalize the analysis to $k > 3$ factors.

The model for the three-factor experiment is given by

$$y_{ijkl} = \mu + \alpha_i + \beta_j + \gamma_k + (\alpha\beta)_{ij} + (\alpha\gamma)_{ik} + (\beta\gamma)_{jk} + (\alpha\beta\gamma)_{ijk} + \varepsilon_{ijkl},$$

$i = 1, 2, \ldots, a; j = 1, 2, \ldots, b; k = 1, 2, \ldots, c;$ and $l = 1, 2, \ldots, n$—where α_i, β_j, and γ_k are the main effects; $(\alpha\beta)_{ij}$, $(\alpha\gamma)_{ik}$, and $(\beta\gamma)_{jl}$ are the two-factor interaction effects that have the same interpretation as in the two-factor experiment. The term $(\alpha\beta\gamma)_{ijk}$ is called the **three-factor interaction effect**, a term that represents a nonadditivity of the $(\alpha\beta)_{ij}$ over the different levels of the factor *C*. As before, the sum of all main effects is zero and the sum over any subscript of the two- and three-factor interaction effects is zero. In many experimental situations these higher-order interactions are insignificant and their mean squares reflect only random variation, but we shall outline the analysis in its most general detail.

Again, in order that valid significance tests can be made, we must assume that the errors are values of independent and normally distributed random variables, each with zero mean and common variance σ^2.

The general philosophy concerning the analysis is the same as that discussed for the one- and two-factor experiments. The sum of squares is partitioned into eight terms, each representing a source of variation from which we obtain independent estimates of σ^2 when all the main effects and interaction effects are zero. If the effects of any given factor or interaction are not all zero, then the mean square will estimate the error variance plus a component due to the systematic effect in question.

Let us proceed directly to the computational procedure for obtaining the sums of squares in the three-factor analysis of variance. We require the following notation:

$T_{....}$ = sum of all $abcn$ observations

$T_{i...}$ = sum of the observations for the ith level of factor A

$T_{.j..}$ = sum of the observations for the jth level of factor B

$T_{..k.}$ = sum of the observations for the kth level of factor C

$T_{ij..}$ = sum of the observations for the ith level of A and the jth level of B

$T_{i.k.}$ = sum of the observations for the ith level of A and the kth level of C

$T_{.jk.}$ = sum of the observations for the jth level of B and kth level of C

$T_{ijk.}$ = sum of the observations for the (ijk)th treatment combination.

In practice it is advantageous to construct the following two-way tables of totals and subtotals:

A	B				Total
	1	2	$\cdots$	b	
1	$T_{11k.}$	$T_{12k.}$	$\cdots$	$T_{1bk.}$	$T_{1.k.}$
2	$T_{21k.}$	$T_{22k.}$	$\cdots$	$T_{2bk.}$	$T_{2.k.}$
$\vdots$	$\vdots$	$\vdots$		$\vdots$	$\vdots$
a	$T_{a1k.}$	$T_{a2k.}$	$\cdots$	$T_{abk.}$	$T_{a.k.}$
Total	$T_{.1k.}$	$T_{.2k.}$	$\cdots$	$T_{.bk.}$	$T_{..k.}$

$k = 1, 2, \ldots, c$

A	B				Total
	1	2	$\cdots$	b	
1	$T_{11..}$	$T_{12..}$	$\cdots$	$T_{1b..}$	$T_{1...}$
2	$T_{21..}$	$T_{22..}$	$\cdots$	$T_{2b..}$	$T_{2...}$
$\vdots$	$\vdots$	$\vdots$		$\vdots$	$\vdots$
a	$T_{a1..}$	$T_{a2..}$	$\cdots$	$T_{ab..}$	$T_{a...}$
Total	$T_{.1..}$	$T_{.2..}$	$\cdots$	$T_{.b..}$	$T_{....}$

A	C				Total
	1	2	$\cdots$	c	
1	$T_{1\cdot1\cdot}$	$T_{1\cdot2\cdot}$	$\cdots$	$T_{1\cdot c\cdot}$	$T_{1\cdots}$
2	$T_{2\cdot1\cdot}$	$T_{2\cdot2\cdot}$	$\cdots$	$T_{2\cdot c\cdot}$	$T_{2\cdots}$
$\vdots$	$\vdots$	$\vdots$		$\vdots$	$\vdots$
a	$T_{a\cdot1\cdot}$	$T_{a\cdot2\cdot}$	$\cdots$	$T_{a\cdot c\cdot}$	$T_{a\cdots}$
Total	$T_{\cdot\cdot1\cdot}$	$T_{\cdot\cdot2\cdot}$	$\cdots$	$T_{\cdot\cdot c\cdot}$	$T_{\cdots\cdot}$

B	C				Total
	1	2	$\cdots$	c	
1	$T_{\cdot11\cdot}$	$T_{\cdot12\cdot}$	$\cdots$	$T_{\cdot1c\cdot}$	$T_{\cdot1\cdot\cdot}$
2	$T_{\cdot21\cdot}$	$T_{\cdot22\cdot}$	$\cdots$	$T_{\cdot2c\cdot}$	$T_{\cdot2\cdot\cdot}$
$\vdots$	$\vdots$	$\vdots$		$\vdots$	$\vdots$
b	$T_{\cdot b1\cdot}$	$T_{\cdot b2\cdot}$	$\cdots$	$T_{\cdot bc\cdot}$	$T_{\cdot b\cdot\cdot}$
Total	$T_{\cdot\cdot1\cdot}$	$T_{\cdot\cdot2\cdot}$	$\cdots$	$T_{\cdot\cdot c\cdot}$	$T_{\cdots\cdot}$

The sums of squares are computed by substituting the appropriate totals into the following computational formulas:

$$SST = \sum_{i=1}^{a} \sum_{j=1}^{b} \sum_{k=1}^{c} \sum_{l=1}^{n} y_{ijkl}^2 - \frac{T_{\cdots\cdot}^2}{abcn}$$

$$SSA = \frac{\sum_{i=1}^{a} T_{i\cdots}^2}{bcn} - \frac{T_{\cdots\cdot}^2}{abcn}$$

$$SSB = \frac{\sum_{j=1}^{b} T_{\cdot j\cdot\cdot}^2}{acn} - \frac{T_{\cdots\cdot}^2}{abcn}$$

$$SSC = \frac{\sum_{k=1}^{c} T_{\cdot\cdot k\cdot}^2}{abn} - \frac{T_{\cdots\cdot}^2}{abcn}$$

$$SS(AB) = \frac{\sum_{i=1}^{a} \sum_{j=1}^{b} T_{ij\cdot\cdot}^2}{cn} - \frac{\sum_{i=1}^{a} T_{i\cdots}^2}{bcn} - \frac{\sum_{j=1}^{b} T_{\cdot j\cdot\cdot}^2}{acn} + \frac{T_{\cdots\cdot}^2}{abcn}$$

$$SS(AC) = \frac{\sum\limits_{i=1}^{a}\sum\limits_{k=1}^{c} T_{i\cdot k\cdot}^2}{bn} - \frac{\sum\limits_{i=1}^{a} T_{i\cdots}^2}{bcn} - \frac{\sum\limits_{k=1}^{c} T_{\cdot\cdot k\cdot}^2}{abn} + \frac{T_{\cdots}^2}{abcn}$$

$$SS(BC) = \frac{\sum\limits_{j=1}^{b}\sum\limits_{k=1}^{c} T_{\cdot jk\cdot}^2}{an} - \frac{\sum\limits_{j=1}^{b} T_{\cdot j\cdots}^2}{acn} - \frac{\sum\limits_{k=1}^{c} T_{\cdot\cdot k\cdot}^2}{abn} + \frac{T_{\cdots}^2}{abcn}$$

$$SS(ABC) = \frac{\sum\limits_{i=1}^{a}\sum\limits_{j=1}^{b}\sum\limits_{k=1}^{c} T_{ijk\cdot}^2}{n} - \frac{\sum\limits_{i=1}^{a}\sum\limits_{j=1}^{b} T_{ij\cdots}^2}{cn} - \frac{\sum\limits_{i=1}^{a}\sum\limits_{k=1}^{c} T_{i\cdot k\cdot}^2}{bn}$$

$$- \frac{\sum\limits_{j=1}^{b}\sum\limits_{k=1}^{c} T_{\cdot jk\cdot}^2}{an} + \frac{\sum\limits_{i=1}^{a} T_{i\cdots}^2}{bcn} + \frac{\sum\limits_{j=1}^{b} T_{\cdot j\cdots}^2}{acn} + \frac{\sum\limits_{k=1}^{c} T_{\cdot\cdot k\cdot}^2}{abn} - \frac{T_{\cdots}^2}{abcn},$$

and SSE, as usual, is obtained by subtraction. The computations in an analysis-of-variance problem, for a three-factor experiment with n replications, are summarized in Table 12.5.

Table 12.5 Analysis of Variance for the Three-Factor Experiment with n Replications

Source of Variation	Sum of Squares	Degrees of Freedom	Mean Square	Computed f
Main effect				
$\quad A$	SSA	$a-1$	s_1^2	$f_1 = \dfrac{s_1^2}{s^2}$
$\quad B$	SSB	$b-1$	s_2^2	$f_2 = \dfrac{s_2^2}{s^2}$
$\quad C$	SSC	$c-1$	s_3^2	$f_3 = \dfrac{s_3^2}{s^2}$
Two-factor interaction				
$\quad AB$	$SS(AB)$	$(a-1)(b-1)$	s_4^2	$f_4 = \dfrac{s_4^2}{s^2}$
$\quad AC$	$SS(AC)$	$(a-1)(c-1)$	s_5^2	$f_5 = \dfrac{s_5^2}{s^2}$
$\quad BC$	$SS(BC)$	$(b-1)(c-1)$	s_6^2	$f_6 = \dfrac{s_6^2}{s^2}$
Three-factor interaction				
$\quad ABC$	$SS(ABC)$	$(a-1)(b-1)(c-1)$	s_7^2	$f_7 = \dfrac{s_7^2}{s^2}$
Error	SSE	$abc(n-1)$	s^2	
Total	SST	$abcn-1$		

For the three-factor experiment with a single replicate we may use the analysis of Table 12.5 by setting $n = 1$ and using the ABC interaction sum of squares for SSE. In this case we are assuming that the $(\alpha\beta\gamma)_{ijk}$ interaction effects are all equal to zero so that

$$E\left[\frac{SS(ABC)}{(a-1)(b-1)(c-1)}\right] = \sigma^2 + \frac{n \sum_{i=1}^{a} \sum_{j=1}^{b} \sum_{k=1}^{c} (\alpha\beta\gamma)_{ijk}^2}{(a-1)(b-1)(c-1)}$$

$$= \sigma^2.$$

That is, $SS(ABC)$ represents variation due only to experimental error. Its mean square thereby provides an unbiased estimate of the error variance. With $n = 1$ and $SSE = SS(ABC)$, the error sum of squares is found by subtracting the sums of squares of the main effects and two-factor interactions from the total sum of squares.

Example 12.3 In the production of a particular material three variables are of interest: A the operator effect (three operators), B the catalyst used in the experiment (three catalysts), and C the washing time of the product following the cooling process (15 minutes and 20 minutes). Three runs were made at each combination of factors. It was felt that all interactions among the factors should be studied. The coded yields are as follows:

				Washing Time, C		
		15 minutes			20 minutes	
		B			B	
A	1	2	3	1	2	3
1	10.7	10.3	11.2	10.9	10.5	12.2
	10.8	10.2	11.6	12.1	11.1	11.7
	11.3	10.5	12.0	11.5	10.3	11.0
2	11.4	10.2	10.7	9.8	12.6	10.8
	11.8	10.9	10.5	11.3	7.5	10.2
	11.5	10.5	10.2	10.9	9.9	11.5
3	13.6	12.0	11.1	10.7	10.2	11.9
	14.1	11.6	11.0	11.7	11.5	11.6
	14.5	11.5	11.5	12.7	10.9	12.2

Perform an analysis of variance to test for significant effects.

Solution. First we construct the following two-way tables:

C (15 minutes)	B			
	1	2	3	Total
A				
1	32.8	31.0	34.8	98.6
2	34.7	31.6	31.4	97.7
3	42.2	35.1	33.6	110.9
Total	109.7	97.7	99.8	307.2

C (20 minutes)	B			
	1	2	3	Total
A				
1	34.5	31.9	34.9	101.3
2	32.0	30.0	32.5	94.5
3	35.1	32.6	35.7	103.4
Total	101.6	94.5	103.1	299.2

A	B			
	1	2	3	Total
1	67.3	62.9	69.7	199.9
2	66.7	61.6	63.9	192.2
3	77.3	67.7	69.3	214.3
Total	211.3	192.2	202.9	606.4

		C	
A	1	2	Total
1	98.6	101.3	199.9
2	97.7	94.5	192.2
3	110.9	103.4	214.3
Total	307.2	299.2	606.4

		C	
B	1	2	Total
1	109.7	101.6	211.3
2	97.7	94.5	192.2
3	99.8	103.1	202.9
Total	307.2	299.2	606.4

Now

$$SST = 10.7^2 + 10.8^2 + \cdots + 12.2^2 - \frac{606.4^2}{54}$$
$$= 6872.84 - 6809.65 = 63.19$$

$$SSA = \frac{199.9^2 + 192.2^2 + 214.3^2}{18} - \frac{606.4^2}{54}$$
$$= 6823.63 - 6809.65 = 13.98$$

$$SSB = \frac{211.3^2 + 192.2^2 + 202.9^2}{18} - \frac{606.4^2}{54}$$
$$= 6819.83 - 6809.65 = 10.18$$

$$SSC = \frac{307.2^2 + 299.2^2}{27} - \frac{606.4^2}{54}$$
$$= 6810.83 - 6809.65 = 1.18$$

$$SS(AB) = \frac{67.3^2 + 66.7^2 + \cdots + 69.3^2}{6} - 6823.63 - 6819.83 + 6809.65$$
$$= 4.78$$

$$SS(AC) = \frac{98.6^2 + 97.7^2 + \cdots + 103.4^2}{9} - 6823.63 - 6810.83 + 6809.65$$
$$= 2.92$$

$$SS(BC) = \frac{109.7^2 + 97.7^2 + \cdots + 103.1^2}{9} - 6819.83 - 6810.83 + 6809.65$$
$$= 3.64$$

$$SS(ABC) = \frac{32.8^2 + 34.7^2 + \cdots + 35.7^2}{3} - 6838.59 - 6827.73 - 6824.65$$
$$+ 6823.63 + 6819.83 + 6810.83 - 6809.65$$
$$= 4.89$$

$$SSE = 63.19 - 13.98 - 10.18 - 1.18 - 4.78 - 2.92 - 3.64 - 4.89$$
$$= 21.62.$$

Table 12.6 Analysis of Variance for Example 12.3

Source of Variation	Sum of Squares	Degrees of Freedom	Mean Square	Computed f
Main effects				
A	13.98	2	6.99	11.65
B	10.18	2	5.09	8.48
C	1.18	1	1.18	1.97
Two-factor interaction				
AB	4.78	4	1.20	2.00
AC	2.92	2	1.46	2.43
BC	3.64	2	1.82	3.03
Three-factor interaction				
ABC	4.89	4	1.22	2.03
Error	21.62	36	0.60	
Total	63.19	53		

These results, with the remaining computations, are given in Table 12.6. None of the interactions show a significant effect at the $\alpha = 0.05$ level. The operator and catalyst effects are significant, while the washing time has no significant effect on the yield for the range used in the experiment.

12.5 Specific Multifactor Models

We have described the three-factor model and its analysis in the most general form by including all possible interactions in the model. Of course, there are many situations in which it is known a priori that the model should not contain certain interactions. We can then take advantage of this knowledge by combining or pooling the sums of squares corresponding to negligible interactions with the error sum of squares to form a new estimator for σ^2 with a larger number of degrees of freedom. For example, in a metallurgy experiment designed to study the effect on film thickness of three important processing variables, suppose it is known that factor A, acid concentration, does not interact with factors B and C. The sums of squares SSA, SSB, SSC, and $SS(BC)$ are computed using the methods described in Section 12.4. The mean squares for the remaining effects will now all independently estimate the error variance σ^2. Therefore, we form our new error mean square by pooling $SS(AB)$, $SS(AC)$,

$SS(ABC)$, and SSE, along with the corresponding degrees of freedom. The resulting denominator for the significance tests is then the error mean square given by

$$s^2 = \frac{SS(AB) + SS(AC) + SS(ABC) + SSE}{(a-1)(b-1) + (a-1)(c-1) + (a-1)(b-1)(c-1) + abc(n-1)}.$$

Computationally, of course, one obtains the pooled sum of squares and the pooled degrees of freedom by subtraction once SST and the sums of squares for the existing effects are computed. The analysis-of-variance table would then take the form of Table 12.7.

In our analysis of the two-factor experiment in Section 12.3 a completely randomized design was used. By interpreting the levels of factor A in Table 12.7 as different blocks, we then have the analysis-of-variance procedure for a two-factor experiment in a randomized block design. For example, if we interpret the operators in Example 12.3 as blocks and assume no interaction between blocks and the other two factors, the analysis of variance takes the form of Table 12.8 rather than that given in Table 12.6. The reader can easily verify that the error mean square is also given by

$$s^2 = \frac{4.78 + 2.92 + 4.89 + 21.62}{4 + 2 + 4 + 36} = 0.74,$$

which demonstrates the pooling of the sums of squares for the nonexisting interaction effects.

Table 12.7 Analysis of Variance with Factor A Noninteracting

Source of Variation	Sum of Squares	Degrees of Freedom	Mean Square	Computed f
Main effect				
A	SSA	$a-1$	s_1^2	$f_1 = \dfrac{s_1^2}{s^2}$
B	SSB	$b-1$	s_2^2	$f_2 = \dfrac{s_2^2}{s^2}$
C	SSC	$c-1$	s_3^2	$f_3 = \dfrac{s_3^2}{s^2}$
Two-factor interaction				
BC	$SS(BC)$	$(b-1)(c-1)$	s_4^2	$f_4 = \dfrac{s_4^2}{s^2}$
Error	SSE	Subtraction	s^2	
Total	SST	$abcn-1$		

Table 12.8 Analysis of Variance for a Two-Factor Experiment in a Randomized Block Design

Source of Variation	Sum of Squares	Degrees of Freedom	Mean Square	Computed f
Blocks	13.98	2	6.99	9.45
Main effect				
B	10.18	2	5.09	6.88
C	1.18	1	1.18	1.59
Two-factor interaction				
BC	3.64	2	1.82	2.46
Error	34.21	46	0.74	
Total	63.19	53		

Exercises

1. The following data are taken in a study involving three factors A, B, and C, all fixed effects:

	C_1			C_2			C_3		
	B_1	B_2	B_3	B_1	B_2	B_3	B_1	B_2	B_3
A_1	15.0	14.8	15.9	16.8	14.2	13.2	15.8	15.5	19.2
	18.5	13.6	14.8	15.4	12.9	11.6	14.3	13.7	13.5
	22.1	12.2	13.6	14.3	13.0	10.1	13.0	12.6	11.1
A_2	11.3	17.2	16.1	18.9	15.4	12.4	12.7	17.3	7.8
	14.6	15.5	14.7	17.3	17.0	13.6	14.2	15.8	11.5
	18.2	14.2	13.4	16.1	18.6	15.2	15.9	14.6	12.2

(a) Perform tests of significance on all interactions at the $\alpha = 0.05$ level.
(b) Perform tests of significance on the main effects at the $\alpha = 0.05$ level.

(c) Give an explanation of how a significant interaction has masked the effect of factor C.

2. The method of X-ray fluorescence is an important analytical tool for determining the concentration of material in solid missile propellants. In the paper "*An X-ray Fluorescence Method for Analyzing Polybutadiene-Acrylic Acid (PBAA) Propellants,*" *Quarterly Report*, RK-TR-62-1, Army Ordinance Missile Command (1962), it is postulated that the propellant mixing process and analysis time have an influence on the homogeneity of the material and hence on the accuracy of X-ray intensity measurements. An experiment was conducted using 3 factors: A, the mixing conditions (4 levels); B, the analysis time (2 levels); and C, the method of loading propellant into sample holders (hot and room temperature). The following data, which represent the analysis in weight percent of ammonium perchlorate in a particular propellant, were recorded:

		Method of Loading, C		
		Hot		Room Temperature
		B		B
A	1	2	1	2
1	38.62 37.20 38.02	38.45 38.64 38.75	39.82 39.15 39.78	39.82 40.26 39.72
2	37.67 37.57 37.85	37.81 37.75 37.91	39.53 39.76 39.90	39.56 39.25 39.04
3	37.51 37.74 37.58	37.21 37.42 37.79	39.34 39.60 39.62	39.74 39.49 39.45
4	37.52 37.15 37.51	37.60 37.55 37.91	40.09 39.63 39.67	39.36 39.38 39.00

Coating	Humidity	Shear Stress		
		13,000 psi	17,000 psi	20,000 psi
Uncoated	Low (20–25% RH)	4580 10,126 1341 6414 3549	5252 897 1465 2694 1017	361 466 1069 469 937
	Medium (50–60% RH)	2858 8829 10,914 4067 2595	799 3471 685 810 3409	314 244 261 522 739
	High (86–91% RH)	6489 5248 6816 5860 5901	1862 2710 2632 2131 2470	1344 1027 663 1216 1097
Chromated	Low (20–25% RH)	5395 2768 1821 3604 4106	4035 2022 914 2036 3524	130 841 1595 1482 529
	Medium (50–60% RH)	4833 7414 10,022 7463 21,906	1847 1684 3042 4482 996	252 105 847 874 755
	High (86–91% RH)	3287 5200 5493 4145 3336	1319 929 1263 2236 1392	586 402 846 524 751

Perform an analysis of variance with $\alpha = 0.01$ to test for significant main and interaction effects.

3. Corrosion fatigue in metals has been defined as the simultaneous action of cyclic stress and chemical attack on a metal structure. In the study "*Effect of Humidity and Several Surface Coatings on the Fatigue Life of 2024-T351 Aluminum Alloy*" conducted by the Department of Mechanical Engineering at the Virginia Polytechnic Institute and State University in 1979, a technique involving the application of a protective chromate coating was used to minimize corrosion-fatigue damage in aluminum. Three factors were used in the investigation with 5 replicates for each treatment combination: coating, at 2 levels, and humidity and shear stress, both with 3 levels. The fatigue data, recorded in thousands of cycles to failure, are as follows:

Perform an analysis of variance with $\alpha = 0.01$ to test for significant main and interaction effects.

4. Consider an experimental situation involving factors A, B, and C, where we assume a three-way fixed effects model of the form

$$y_{ijkl} = \mu + \alpha_i + \beta_j + \gamma_k + (\beta\gamma)_{jk} + \varepsilon_{ijkl}.$$

All other interactions are considered to be non-existent or negligible. The data were recorded as follows:

	B_1			B_2		
	C_1	C_2	C_3	C_1	C_2	C_3
A_1	4.0	3.4	3.9	4.4	3.1	3.1
	4.9	4.1	4.3	3.4	3.5	3.7
A_2	3.6	2.8	3.1	2.7	2.9	3.7
	3.9	3.2	3.5	3.0	3.2	4.2

	B_1			B_2		
	C_1	C_2	C_3	C_1	C_2	C_3
A_3	4.8	3.3	3.6	3.6	2.9	2.9
	3.7	3.8	4.2	3.8	3.3	3.5
A_4	3.6	3.2	3.2	2.2	2.9	3.6
	3.9	2.8	3.4	3.5	3.2	4.3

(a) Perform a test of significance on the BC interaction at the $\alpha = 0.05$ level.
(b) Perform tests of significance on the main effects A, B, and C using a pooled error mean square at the $\alpha = 0.05$ level.

12.6 Model II Factorial Experiments

In a two-factor experiment with random effects we have the model

$$Y_{ijk} = \mu + A_i + B_j + (AB)_{ij} + E_{ijk},$$

$i = 1, 2, \ldots, a; j = 1, 2, \ldots, b;$ and $k = 1, 2, \ldots, n$, where the A_i, B_j, $(AB)_{ij}$, and E_{ijk} are independent random variables with zero means and variances σ_α^2, σ_β^2, $\sigma_{\alpha\beta}^2$, and σ^2, respectively. The sum of squares for the model II experiments are computed in exactly the same way as for the model I experiments. We are now interested in testing hypotheses of the form

$$H_0': \ \sigma_\alpha^2 = 0, \qquad H_0'': \ \sigma_\beta^2 = 0, \qquad H_0''': \ \sigma_{\alpha\beta}^2 = 0,$$
$$H_1': \ \sigma_\alpha^2 \neq 0, \qquad H_1'': \ \sigma_\beta^2 \neq 0, \qquad H_1''': \ \sigma_{\alpha\beta}^2 \neq 0,$$

where the denominator in the f ratio is not necessarily the error mean square. The appropriate denominator can be determined by examining the expected values of the various mean squares. These are shown in Table 12.9.

From Table 12.9 we see that H_0' and H_0'' are tested by using s_3^2 in the denominator of the f ratio, while H_0''' is tested using s^2 in the denominator. The unbiased estimates of the variance components are given by

$$\hat{\sigma}^2 = s^2$$

$$\hat{\sigma}_{\alpha\beta}^2 = \frac{s_3^2 - s^2}{n}$$

$$\hat{\sigma}_\alpha^2 = \frac{s_1^2 - s_3^2}{bn}$$

Table 12.9 Expected Mean Squares for a Model II Two-Factor Experiment

Source of Variation	Degrees of Freedom	Mean Square	Expected Mean Square
A	$a - 1$	s_1^2	$\sigma^2 + n\sigma_{\alpha\beta}^2 + bn\sigma_{\alpha}^2$
B	$b - 1$	s_2^2	$\sigma^2 + n\sigma_{\alpha\beta}^2 + an\sigma_{\beta}^2$
AB	$(a - 1)(b - 1)$	s_3^2	$\sigma^2 + n\sigma_{\alpha\beta}^2$
Error	$ab(n - 1)$	s^2	σ^2
Total	$abn - 1$		

$$\hat{\sigma}_{\beta}^2 = \frac{s_2^2 - s_3^2}{an}.$$

The expected mean squares for the three-factor experiment with random effects in a completely randomized design are shown in Table 12.10. It is evident from the expected mean squares of Table 12.10 that one can form appropriate f ratios for testing all two-factor and three-factor interaction variance components. However, to test a hypothesis of the form

$$H_0: \quad \sigma_{\alpha}^2 = 0,$$
$$H_1: \quad \sigma_{\alpha}^2 \neq 0,$$

there appears to be no appropriate f ratio unless we have found one or more of the two-factor interaction variance components not significant. Suppose, for example, that we have compared s_5^2 with s_7^2 and found $\sigma_{\alpha\gamma}^2$ to be negligible. We could then argue that the term $\sigma_{\alpha\gamma}^2$ should be dropped from all the expected mean squares of Table 12.10; then the ratio s_1^2/s_4^2 provides a test for

Table 12.10 Expected Mean Squares for a Model II Three-Factor Experiment

Source of Variation	Degrees of Freedom	Mean Square	Expected Mean Square
A	$a - 1$	s_1^2	$\sigma^2 + n\sigma_{\alpha\beta\gamma}^2 + cn\sigma_{\alpha\beta}^2 + bn\sigma_{\alpha\gamma}^2 + bcn\sigma_{\alpha}^2$
B	$b - 1$	s_2^2	$\sigma^2 + n\sigma_{\alpha\beta\gamma}^2 + cn\sigma_{\alpha\beta}^2 + an\sigma_{\beta\gamma}^2 + acn\sigma_{\beta}^2$
C	$c - 1$	s_3^2	$\sigma^2 + n\sigma_{\alpha\beta\gamma}^2 + bn\sigma_{\alpha\gamma}^2 + an\sigma_{\beta\gamma}^2 + abn\sigma_{\gamma}^2$
AB	$(a - 1)(b - 1)$	s_4^2	$\sigma^2 + n\sigma_{\alpha\beta\gamma}^2 + cn\sigma_{\alpha\beta}^2$
AC	$(a - 1)(c - 1)$	s_5^2	$\sigma^2 + n\sigma_{\alpha\beta\gamma}^2 + bn\sigma_{\alpha\gamma}^2$
BC	$(b - 1)(c - 1)$	s_6^2	$\sigma^2 + n\sigma_{\alpha\beta\gamma}^2 + an\sigma_{\beta\gamma}^2$
ABC	$(a - 1)(b - 1)(c - 1)$	s_7^2	$\sigma^2 + n\sigma_{\alpha\beta\gamma}^2$
Error	$abc(n - 1)$	s^2	σ^2
Total	$abcn - 1$		

the significance of the variance component σ_α^2. Therefore, if we are to test hypotheses concerning the variance components of the main effects, it is necessary first to investigate the significance of the two-factor interaction components. An approximate test derived by Satterthwaite may be used when certain two-factor interaction variance components are found to be significant and hence must remain a part of the expected mean square.

Example 12.4 In a study to determine which are the important sources of variation in an industrial process, 3 measurements are taken on yield for 3 operators chosen randomly and 4 batches of raw materials chosen randomly. It was decided that a significance test should be made at the 0.05 level of significance to determine if the variance components due to batches, operators, and interaction are significant. In addition, estimates of variance components are to be computed. The data are as follows, with the response being percent by weight:

Operators	Batches			
	1	2	3	4
1	66.9	68.3	69.0	69.3
	68.1	67.4	69.8	70.9
	67.2	67.7	67.5	71.4
2	66.3	68.1	69.7	69.4
	65.4	66.9	68.8	69.6
	65.8	67.6	69.2	70.0
3	65.6	66.0	67.1	67.9
	66.3	66.9	66.2	68.4
	65.2	67.3	67.4	68.7

Solution. The sums of squares are found in the usual way with the results given by

$$SST \text{ (total)} = 84.5564$$
$$SSA \text{ (operators)} = 18.2106$$
$$SSB \text{ (batches)} = 50.1564$$
$$SS(AB) \text{ (interaction)} = 5.5161$$
$$SSE \text{ (error)} = 10.6733.$$

All other computations are carried out and exhibited in Table 12.11. Since $f_{0.05}(2, 6) = 5.14$, $f_{0.05}(3, 6) = 4.76$, and $f_{0.05}(6, 24) = 2.51$, we find the operator and batch variance components to be significant, while the interaction

Table 12.11 Analysis of Variance for Example 12.4

Source of Variation	Sum of Squares	Degrees of Freedom	Mean Square	Computed f
Operators	18.2106	2	9.1053	9.90
Batches	50.1564	3	16.7188	18.18
Interaction	5.5161	6	0.9194	2.07
Error		24		
Total	84.5564	35		

variance is not significant at the $\alpha = 0.05$ level. Estimates of the main effect variance components are given by

$$\hat{\sigma}_\alpha^2 = \frac{9.1053 - 0.9194}{12} = 0.68$$

$$\hat{\sigma}_\beta^2 = \frac{16.7188 - 0.9144}{9} = 1.76.$$

12.7 Choice of Sample Size

Our study of factorial experiments throughout this chapter has been restricted to the use of a completely randomized design with the exception of Section 12.5, where we demonstrated the analysis of a two-factor experiment in a randomized block design. The completely randomized design is very easy to lay out and the analysis is simple to perform; however, it should be used only when the number of treatment combinations is small and the experimental material is homogeneous. Although the randomized block design is ideal for dividing a large group of heterogeneous units into subgroups of homogeneous units, it is generally difficult to obtain uniform blocks with enough units to which a large number of treatment combinations may be assigned. This disadvantage may be overcome by choosing a design from the catalog of **incomplete block designs**. These designs allow one to investigate differences among t treatments arranged in b blocks each containing k experimental units, where $k < t$.

Once the experimenter has selected a completely randomized design, he must decide if the number of replications is sufficient to yield tests in the analysis of variance with high power. If not, he must add additional replications, which in turn may force him into a randomized complete block design. Had he started with a randomized block design, it would still be

Table 12.12 Noncentrality Parameter λ and ϕ^2 for Two-Factor and Three-Factor Models

	Two-Factor Experiments		Three-Factor Experiments		
	A	B	A	B	C
λ	$\dfrac{bn\sum\limits_{i=1}^{a}\alpha_i^2}{2\sigma^2}$	$\dfrac{an\sum\limits_{j=1}^{b}\beta_j^2}{2\sigma^2}$	$\dfrac{bcn\sum\limits_{i=1}^{a}\alpha_i^2}{2\sigma^2}$	$\dfrac{acn\sum\limits_{j=1}^{b}\beta_j^2}{2\sigma^2}$	$\dfrac{abn\sum\limits_{k=1}^{c}\gamma_k^2}{2\sigma^2}$
ϕ^2	$\dfrac{bn\sum\limits_{i=1}^{a}\alpha_i^2}{a\sigma^2}$	$\dfrac{an\sum\limits_{j=1}^{b}\beta_j^2}{b\sigma^2}$	$\dfrac{bcn\sum\limits_{i=1}^{a}\alpha_i^2}{a\sigma^2}$	$\dfrac{acn\sum\limits_{j=1}^{b}\beta_j^2}{b\sigma^2}$	$\dfrac{abn\sum\limits_{k=1}^{c}\gamma_k^2}{c\sigma^2}$

necessary to determine if the number of blocks is sufficient to yield powerful tests. Basically, then, we are back to the question of sample size.

The power of a fixed effects test or the probability of rejecting H_0 when the alternative H_1 is true, for a given sample size, is found from Table A.15 by computing the noncentrality parameter λ and the function ϕ discussed in Section 11.12. Expressions for λ and ϕ^2 for the two-factor and three-factor fixed effects experiments are given in Table 12.12.

The results of Section 11.12 for the random effects model can be extended very easily to the two- and three-factor models. Once again the general procedure is based on the values of the expected mean squares. For example, if we are testing $\sigma_\alpha^2 = 0$ in a two-factor experiment by computing the ratio s_1^2/s_3^2 (see Table 12.9), then

$$f = \frac{s_1^2/(\sigma^2 + n\sigma_{\alpha\beta}^2 + bn\sigma_\alpha^2)}{s_3^2/(\sigma^2 + n\sigma_{\alpha\beta}^2)}$$

is a value of the random variable F having the F distribution with $a-1$ and $(a-1)(b-1)$ degrees of freedom, and the power of the test is given by

$$1 - \beta = P\left\{\frac{S_1^2}{S_3^2} > f_\alpha[(a-1),(a-1)(b-1)] \text{ when } \sigma_\alpha^2 \neq 0\right\}$$

$$= P\left\{F > \frac{f_\alpha[(a-1),(a-1)(b-1)](\sigma^2 + n\sigma_{\alpha\beta}^2)}{\sigma^2 + n\sigma_{\alpha\beta}^2 + bn\sigma_\alpha^2}\right\}.$$

Exercises

1. To estimate the various components of variability in a filtration process, the percent of material lost in the mother liquor was measured for 12 experimental conditions, 3 runs on each condition. Three filters and 4 operators were selected at random to use in the experiment resulting in the following measurements:

	Operator			
Filter	1	2	3	4
1	16.2	15.9	15.6	14.9
	16.8	15.1	15.9	15.2
	17.1	14.5	16.1	14.9
2	16.6	16.0	16.1	15.4
	16.9	16.3	16.0	14.6
	16.8	16.5	17.2	15.9
3	16.7	16.5	16.4	16.1
	16.9	16.9	17.4	15.4
	17.1	16.8	16.9	15.6

   (a) Test the hypothesis of no interaction variance component between filters and operators at the $\alpha = 0.05$ level of significance.
   (b) Test the hypotheses that the operators and the filters have no effect on the variability of the filtration process at the $\alpha = 0.05$ level of significance.
   (c) Estimate the components of variance due to filters, operators, and experimental error.

2. Assuming a model II experiment for Exercise 2 on page 483, estimate the variance components for brand of orange juice concentrate, for number of days from when orange juice was blended until it was tested, and for experimental error.

3. Consider the following analysis of variance for a model II experiment:

Source of Variation	Degrees of Freedom	Mean Square
A	3	140
B	1	480
C	2	325
AB	3	15
AC	6	24
BC	2	18
ABC	6	2
Error	24	5
Total	47	

   Test for significant variance components among all main effects and interaction effects at the 0.01 level of significance
   (a) by using a pooled estimate of error when appropriate;
   (b) by not pooling sums of squares of insignificant effects.

4. Are 2 observations for each treatment combination in Exercise 4 on page 495 sufficient if the power of our test for detecting differences among the levels of factor C at the 0.05 level of significance is to be at least 0.8 when $\gamma_1 = -0.2$, $\gamma_2 = 0.4$, and $\gamma_3 = -0.2$? Use the same pooled estimate of σ^2 that was used in the analysis of variance.

5. Using the estimates of the variance components in Exercise 1, evaluate the power when we test the variance component due to filters to be zero.

13

2^k *Factorial Experiments*

13.1 Introduction

In almost any experimental study in which statistical procedures are applied to a collection of scientific data, the methods involve performing certain operations or computations on the sample information, followed by the drawing of inferences about the population or populations studied. Often there are characteristics of the experiment that are subject to the control of the experimenter, quantities such as sample size, number of levels of the factors, treatment combinations to be used, and so forth. These *experimental parameters* can often have a great effect on the precision with which hypotheses are tested or estimation is accomplished.

We have already been exposed to certain experimental design concepts. The sampling plan for the simple t test on the mean of a normal population and also the analysis of variance involve randomly allocating pre-chosen treatments to experimental units. The randomized block design, where treatments are assigned to units within relatively homogeneous blocks, involves restricted randomization.

In this chapter we give special attention to experimental designs in which the experimental plan calls for the study of the effect on a response of k factors, each at two levels. These are commonly known as 2^k factorial experiments. We often denote the levels as "high" and "low," even though this notation may be arbitrary in the case of qualitative variables. The complete factorial design requires that each level of every factor occur with each level of every other factor, giving a total of 2^k treatment combinations. We shall denote the higher levels of the factors $A, B, C, \ldots$ by the letters $a, b, c, \ldots$ and the lower levels of each factor by the notation (1). In the presence of other letters we omit the symbol (1). For example, the treatment combination in a 2^4 experiment that contains the high levels of factors B and C and the low levels of factors A and D is written simply as bc. The treatment combination that consists of the low level of all factors in the experiment is denoted by the symbol (1). In the case of a 2^3 experiment, the eight possible treatment combinations are (1), a, b, c, ab, ac, bc, and abc.

The factorial experiment allows the effect of each and every factor to be estimated and tested independently through the usual analysis of variance. In addition, the interaction effects are easily assessed. The disadvantage, of course, with the factorial experiment is the excessive amount of experimentation that is required. For example, if it is desired to study the effect of eight variables, $2^8 = 256$ treatment combinations are required. In many instances we can obtain considerable information by using only a fraction of the experimental runs. This type of design is called a **fractional factorial design** and will be considered in Sections 13.6 and 13.7.

13.2 Analysis of Variance

Consider initially a 2^2 factorial plan in which there are n experimental observations per treatment combination. Extending our previous notation, we now interpret the symbols (1), a, b, and ab to be the total yields for each of the four treatment combinations. Table 13.1 gives a two-way table of these total yields.

Table 13.1 2^2 Factorial Experiment

		B		Mean
A		(1)	b	$\dfrac{b + (1)}{2n}$
		a	ab	$\dfrac{ab + a}{2n}$
Mean		$\dfrac{a + (1)}{2n}$	$\dfrac{ab + b}{2n}$	

Let us define the following contrasts among the treatment totals:

$$A \text{ contrast} = ab + a - b - (1)$$
$$B \text{ contrast} = ab - a + b - (1)$$
$$AB \text{ contrast} = ab - a - b + (1).$$

Clearly, there will be exactly one single-degree-of-freedom contrast for the means of each factor A and B, which we shall write as

$$w_A = \frac{ab + a - b - (1)}{2n} = \frac{A \text{ contrast}}{2n}$$

and

$$w_B = \frac{ab + b - a - (1)}{2n} = \frac{B \text{ contrast}}{2n}.$$

The contrast w_A is seen to be the difference between the mean response at the low and high levels of factor A. In fact, we call w_A the **main effect** of A. Similarly, w_B is the main effect of factor B. Apparent interaction in the data is observed by inspecting the difference between $ab - b$ and $a - (1)$ or between $ab - b$ and $b - (1)$ in Table 13.1. If, for example, $ab - a \simeq b - (1)$ or $ab - a - b + (1) \simeq 0$, a line connecting the responses for each level of factor A at the high level of factor B will be approximately parallel to a line connecting the response for each level of factor A at the low level of factor B. The nonparallel lines of Figure 13.1 suggest the presence of interaction. To test whether this apparent interaction is significant, a third contrast in the treatment totals orthogonal to the main effect contrasts, called the **interaction effect**, is constructed by evaluating

$$w_{AB} = \frac{ab - a - b + (1)}{2n} = \frac{AB \text{ contrast}}{2n}.$$

We take advantage of the fact that in the 2^2 factorial, or for that matter in the general 2^k factorial experiment, each main effect and interaction effect has

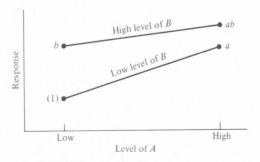

Figure 13.1 Response suggesting apparent interaction.

associated with it a single degree of freedom. Therefore, we can write $2^k - 1$ orthogonal single-degree-of-freedom contrasts in the treatment combinations, each representing variation due to some main or interaction effect. Thus, under the usual independence and normality assumptions in the experimental model, we can make tests to determine if the contrast reflects systematic variation or merely chance or random variation. The sums of squares for each contrast are found by following the procedures given in Section 11.4. Writing $T_1.. = b + (1)$, $T_2.. = ab + a$, $c_1 = -1$, and $c_2 = 1$, where $T_1..$ and $T_2..$ are the totals of $2n$ observations, we have

$$SSA = SSw_A = \frac{\left(\sum_{i=1}^{2} c_i T_i.. \right)^2}{2n \sum_{i=1}^{2} c_i^2}$$

$$= \frac{[ab + a - b - (1)]^2}{2^2 n} = \frac{(A \text{ contrast})^2}{2^2 n},$$

with 1 degree of freedom. Similarly, we find that

$$SSB = \frac{[ab + b - (1) - a]^2}{2^2 n} = \frac{(B \text{ contrast})^2}{2^2 n}$$

and

$$SS(AB) = \frac{[ab + (1) - a - b]^2}{2^2 n} = \frac{(AB \text{ contrast})^2}{2^2 n},$$

each with 1 degree of freedom, while the error sum of squares, with $2^2(n - 1)$ degrees of freedom, is obtained by subtraction from the formula

$$SSE = SST - SSA - SSB - SS(AB).$$

In computing the sums of squares for the main effects A and B and the interaction effect AB, it is convenient to present the total yields of the treatment combinations along with the appropriate algebraic signs for each contrast as in Table 13.2. The main effects are obtained as simple comparisons

Table 13.2 Signs for Contrasts in a 2^2 Factorial Experiment

Treatment Combination	Factorial Effect		
	A	B	AB
(1)	−	−	+
a	+	−	−
b	−	+	−
ab	+	+	+

Table 13.3 Signs for Contrasts in a 2^3 Factorial Experiment

Treatment Combination	Factorial Effect						
	A	B	C	AB	AC	BC	ABC
(1)	−	−	−	+	+	+	−
a	+	−	−	−	−	+	+
b	−	+	−	−	+	−	+
c	−	−	+	+	−	−	+
ab	+	+	−	+	−	−	−
ac	+	−	+	−	+	−	−
bc	−	+	+	−	−	+	−
abc	+	+	+	+	+	+	+

between the low and high levels. Therefore, we assign a positive sign to the treatment combination that is at the high level of a given factor and a negative sign to the treatment combination at the lower level. The positive and negative signs for the interaction effect are obtained by multiplying the corresponding signs of the contrasts of the interacting factors.

Let us now consider an experiment using three factors A, B, and C with levels (1), a; (1), b; and (1), c, respectively. This is a 2^3 factorial experiment giving the eight treatment combinations (1), a, b, c, ab, ac, bc, and abc. The treatment combinations and the appropriate algebraic signs for each contrast used in computing the sums of squares for the main effects and interaction effects are presented in Table 13.3.

An inspection of Table 13.3 reveals that for the 2^3 experiment any two contrasts among the seven are orthogonal and therefore the seven effects are assessed independently. The sum of squares for, say the ABC interaction with 1 degree of freedom, is given by

$$SS(ABC) = \frac{[abc + a + b + c - (1) - ab - ac - bc]^2}{2^3 n}.$$

For a 2^k factorial experiment the single-degree-of-freedom sums of squares for the main effects and interaction effects are obtained by squaring the appropriate contrasts in the treatment totals and dividing by $2^k n$, where n is the number of replications of the treatment combinations.

The orthogonality property has the same importance here as it did in the material on comparisons discussed in Chapter 11. Orthogonality of contrasts implies that the estimated effects and thus the sums of squares are independent. This independence is easily illustrated in a 2^3 factorial experiment if the yields, with factor A at its high level, are increased by an amount x in Table 13.3. Only the A contrast leads to a larger sum of squares, since the x

effect cancels out in the formation of the six remaining contrasts as a result of the two positive and two negative signs associated with treatment combinations in which A is at the high level.

13.3 Yates' Technique for Computing Contrasts

It is laborious to write out the table of positive and negative signs for large experiments. A systematic tabular technique for deriving the factorial effects has been developed by Yates. The treatment combinations and the observations must be written down in **standard form**. For one factor the standard form is (1), a. For two factors we add b and ab, derived by multiplying the first two treatment combinations by the additional letter b. For three factors we add c, ac, bc, and abc, derived by multiplying the first four treatment combinations by the additional letter c, and so on. In the case of three factors the standard order is then

$$(1), \quad a, \quad b, \quad ab, \quad c, \quad ac, \quad bc, \quad abc.$$

Yates' method is carried out in the following steps:

1. Place the treatment combinations and the corresponding total yields in a column in standard order.
2. Obtain the top half of a column marked (1) by adding the first two yields, then the next two, and so on. The bottom half is obtained by subtracting the first from the second of each of these same pairs.

Table 13.4 Yates' Technique for a 2^3 Factorial Experiment

Treatment Combination	(1)	(2)	(3)	Identification
(1)	$(1) + a$	$(1) + a + b + ab$	$(1) + a + b + ab + c + ac + bc + abc$	Total
a	$b + ab$	$c + ac + bc + abc$	$a - (1) + ab - b + ac - c + abc - bc$	A contrast
b	$c + ac$	$a - (1) + ab - b$	$b + ab - (1) - a + bc + abc - c - ac$	B contrast
ab	$bc + abc$	$ac - c + abc - bc$	$ab - b - a + (1) + abc - bc - ac + c$	AB contrast
c	$a - (1)$	$b + ab - (1) - a$	$c + ac + bc + abc - (1) - a - b - ab$	C contrast
ac	$ab - b$	$bc + abc - c - ac$	$ac - c + abc - bc - a + (1) - ab + b$	AC contrast
bc	$ac - c$	$ab - b - a + (1)$	$bc + abc - c - ac - b - ab + (1) + a$	BC contrast
abc	$abc - bc$	$abc - bc - ac + c$	$abc - bc - ac + c - ab + b + a - (1)$	ABC contrast

3. Repeat the operation using the results in column (1) to obtain column (2). This operation is continued until we have k columns for a 2^k experiment.

4. The first value of the kth column will be the grand total of the yields in the experiment. Each remaining number will be a contrast in the treatment totals. Finally, the sum of squares for the main effects and interaction effects are obtained by squaring the entries in column (k) and dividing by $2^k n$, where n is the number of replications and 2^k is the sum of the squares of the coefficients of the individual contrasts.

As an illustration we outline the procedure in Table 13.4 for the 2^3 factorial experiment.

Example 13.1   In a metallurgy experiment it is desired to test the effect of four factors and their interactions on the concentration (percent by weight) of a particular phosphorus compound in casting material. The variables are A, percent phosphorus in the refinement; B, percent remelted material; C, fluxing time; and D, holding time. The four factors are varied in a 2^4 factorial experiment with two castings taken and the content measured at each treatment combination. Using Yates' technique, perform the analysis of variance of the following data:

Treatment Combination	Weight % of Phosphorus Compound		
	Replication 1	Replication 2	Total
(1)	30.3	28.6	58.9
a	28.5	31.4	59.9
b	24.5	25.6	50.1
ab	25.9	27.2	53.1
c	24.8	23.4	48.2
ac	26.9	23.8	50.7
bc	24.8	27.8	52.6
abc	22.2	24.9	47.1
d	31.7	33.5	65.2
ad	24.6	26.2	50.8
bd	27.6	30.6	58.2
abd	26.3	27.8	54.1
cd	29.9	27.7	57.6
acd	26.8	24.2	51.0
bcd	26.4	24.9	51.3
$abcd$	26.9	29.3	56.2
Total	428.1	436.9	865.0

Table 13.5 Yates' Technique for a 2^4 Example

Treatment Combination	Treatment Total	(1)	(2)	(3)	(4)	Sum of Squares
(1)	58.9	118.8	222.0	420.6	865.0	
a	59.9	103.2	198.6	444.4	−19.2	11.52
b	50.1	98.9	228.3	1.0	−19.6	12.00
ab	53.1	99.7	216.1	−20.2	15.8	7.80
c	48.2	116.0	4.0	−14.8	−35.6	39.61
ac	50.7	112.3	−3.0	−4.8	9.8	3.00
bc	52.6	108.6	−18.5	−6.0	19.0	11.28
abc	47.1	107.5	−1.7	21.8	−8.8	2.42
d	65.2	1.0	−15.6	−23.4	23.8	17.70
ad	50.8	3.0	0.8	−12.2	−21.2	14.05
bd	58.2	2.5	−3.7	−7.0	10.0	3.13
abd	54.1	−5.5	−1.1	16.8	27.8	24.15
cd	57.6	−14.4	2.0	16.4	11.2	3.92
acd	51.0	−4.1	−8.0	2.6	23.8	17.70
bcd	51.3	−6.6	10.3	−10.0	−13.8	5.95
abcd	56.2	4.9	11.5	1.2	11.2	3.92

Solution. Table 13.5 gives the 16 treatment totals and outlines Yates' technique for computing the individual sums of squares. The total sum of squares is given by

$$SST = 30.3^2 + 28.5^2 + \cdots + 29.3^2 - \frac{865^2}{32} = 217.51.$$

We can now set up the analysis of variance as in Table 13.6. Note that the interactions BC, AD, ABD, and ACD are significant when compared with $f_{0.05}(1, 16) = 4.49$. The tests on the main effects (which in the presence of interactions may be regarded as the effects *averaged* over the levels of the other factors) indicate significance in each case.

   Very often the experimenter knows in advance that certain interactions in a 2^k factorial experiment are negligible and should not be included in the model. For example, in a 2^4 factorial experiment he may postulate a model that contains only two-factor interaction effects and then pool the sums of squares and corresponding degrees of freedom of the remaining higher-ordered interactions with the pure error. In fact, this is often done in lieu of taking several replications.

Table 13.6 Analysis of Variance for the Data of Table 13.5

Source of Variation	Sum of Squares	Degrees of Freedom	Mean Square	Computed f
Main effect				
A	11.52	1	11.52	4.68
B	12.00	1	12.00	4.90
C	39.61	1	39.61	16.10
D	17.70	1	17.70	7.20
Two-factor interaction				
AB	7.80	1	7.80	3.17
AC	3.00	1	3.00	1.22
AD	14.05	1	14.05	5.71
BC	11.28	1	11.28	4.59
BD	3.13	1	3.13	1.27
CD	3.92	1	3.92	1.59
Three-factor interaction				
ABC	2.42	1	2.42	0.98
ABD	24.15	1	24.15	9.82
ACD	17.70	1	17.70	7.20
BCD	5.95	1	5.95	2.42
Four-factor interaction				
ABCD	3.92	1	3.92	1.59
Error	39.36	16	2.46	
Total	217.51	31		

Exercises

1. The following data were obtained from a 2^3 factorial experiment replicated three times:

Treatment Combination	Replicate 1	Replicate 2	Replicate 3
(1)	12	19	10
a	15	20	16
b	24	16	17
ab	23	17	27

Treatment Combination	Replicate 1	Replicate 2	Replicate 3
c	17	25	21
ac	16	19	19
bc	24	23	29
abc	28	25	20

Evaluate the sums of squares for all factorial effects by the contrast method.

2. In an experiment conducted by the Mining Engineering Department at the Virginia Polytechnic Institute and State University in 1979 to study a particular filtering system for coal, a coagulant was added to a solution in a tank containing coal and sludge, which was then placed in a recirculation system in order that the coal could be washed. Three factors were varied in the experimental process:

Factor A: percent solids circulated initially in the overflow,

Factor B: flow rate of the polymer,

Factor C: pH of the tank.

The amount of solids in the underflow of the cleansing system determines how clean the coal has become. Two levels of each factor were used and two experimental runs were made for each of the $2^3 = 8$ combinations. The responses, percent solids by weight, in the underflow of the circulation system are as follows:

Treatment Combination	Response	
	Replicate 1	Replicate 2
(1)	4.65	5.81
a	21.42	21.35
b	12.66	12.56
ab	18.27	16.62
c	7.93	7.88
ac	13.18	12.87
bc	6.51	6.26
abc	18.23	17.83

Assuming that all interactions are potentially important, do a complete analysis of the data. Use a 0.01 level of significance.

3. The effects of four factors on some response are to be studied. Each factor is varied at two levels in a 2^4 factorial arrangement and the following data recorded:

Treatment Combination	Response
(1)	23.8
a	19.6

Treatment Combination	Response
b	29.9
ab	25.7
c	26.5
ac	22.6
bc	32.6
abc	28.6
d	21.6
ad	17.5
bd	27.5
abd	23.7
cd	24.6
acd	20.9
bcd	31.1
abcd	26.7

Assuming all three- and four-factor interactions to be negligible, analyze the given data by Yates' technique. Use a 0.01 level of significance.

4. A preliminary experiment is conducted to study the effects of four factors and their interactions on the output of a certain machining operation. Two runs are made at each of the treatment combinations in order to supply a measure of pure experimental error. Two levels of each factor are used, resulting in the following data:

Treatment Combination	Replicate 1	Replicate 2
(1)	7.9	9.6
a	9.1	10.2
b	8.6	5.8
c	10.4	12.0
d	7.1	8.3
ab	11.1	12.3
ac	16.4	15.5
ad	7.1	8.7
bc	12.6	15.2
bd	4.7	5.8
cd	7.4	10.9
abc	21.9	21.9
abd	9.8	7.8
acd	13.8	11.2
bcd	10.2	11.1
abcd	12.8	14.3

Use Yates' method to make tests on all main effects and interactions at the 0.05 level of significance.

5. In the study *"An X-Ray Fluorescence Method for Analyzing Polybutadiene-Acrylic Acid (PBAA) Propellants,"* Quarterly Reports, RK-TR-62-1, Army Ordinance Missile Command (1962), an experiment was conducted to determine whether or not there is a significant difference in the amount of aluminum achieved in the analysis between certain levels of the following processing variables:

  A: mixing time
        level 1—2 hours
        level 2—4 hours,

  B: blade speed
        level 1—36 rpm
        level 2—78 rpm,

  C: condition of nitrogen passed over
           propellant
        level 1—dry
        level 2—72% relative humidity,

  D: physical state of propellant
        level 1—uncured
        level 2—cured.

The following data were recorded:

Obser-vation	Physical State	Mixing Time	Blade Speed	Nitrogen Condition	Aluminum
1	1	1	2	2	16.3
2	1	2	2	2	16.0
3	1	1	1	1	16.2
4	1	2	1	2	16.1
5	1	1	1	2	16.0
6	1	2	1	1	16.0
7	1	2	2	1	15.5
8	1	1	2	1	15.9
9	2	1	2	2	16.7
10	2	2	2	2	16.1
11	2	1	1	1	16.3
12	2	2	1	2	15.8
13	2	1	1	2	15.9
14	2	2	1	1	15.9
15	2	2	2	1	15.6
16	2	1	2	1	15.8

Assuming all three- and four-factor interactions to be negligible, analyze the data by Yates' technique. Use a 0.05 level of significance.

13.4 Factorial Experiments in Incomplete Blocks

The 2^k factorial experiment lends itself to partitioning into *incomplete blocks*. For a k-factor experiment, it is often useful to use a design in 2^p blocks ($p < k$) when the entire 2^k treatment combinations cannot be applied under homogeneous conditions. The disadvantage with this experimental setup is that certain effects are completely sacrificed as a result of the blocking, the amount of sacrifice depending on the number of blocks required. For example, suppose that the eight treatment combinations in a 2^3 factorial experiment must be run in two blocks of size 4. One possible arrangement is given by

Block 1	Block 2
(1)	*a*
ab	*b*
ac	*c*
bc	*abc*

If one assumes the usual model with the additive block effect, this effect cancels out in the formation of the contrasts on all effects except *ABC*. To illustrate, let x denote the contribution to the yield due to the difference between blocks. Writing the yields as

Block 1	Block 2
(1)	$a + x$
ab	$b + x$
ac	$c + x$
bc	$abc + x$

we see that the *ABC* contrast and also the contrast comparing the 2 blocks are both given by

$$ABC \text{ contrast} = (abc + x) + (c + x) + (b + x) + (a + x) - (1) - ab$$
$$- ac - bc$$
$$= abc + a + b + c - (1) - ab - ac - bc + 4x.$$

Therefore, we are measuring the *ABC* effect plus the block effect and there is no way of assessing the *ABC* interaction effect independent of blocks. We say then that the *ABC* interaction is **completely confounded with blocks**. By necessity, information on *ABC* has been sacrificed. On the other hand, the block effect cancels out in the formation of all other contrasts. For example, the *A* contrast is given by

$$A \text{ contrast} = (abc + x) + (a + x) + ab + ac - (b + x) - (c + x) - bc - (1)$$
$$= abc + a + ab + ac - b - c - bc - (1),$$

as in the case of a completely randomized design. We say that the effects *A*, *B*, *C*, *AB*, *AC*, and *BC* are orthogonal to blocks. Generally, for a 2^k factorial experiment in 2^p blocks, the number of effects confounded with blocks is 2^{p-1}, which is equivalent to the degrees of freedom for blocks.

When two blocks are to be used with a 2^k factorial, one effect, usually a high-order interaction, is chosen as the **defining contrast**. This effect is to be confounded with blocks. The additional $2^k - 2$ effects are orthogonal with the defining contrast and thus with blocks.

Suppose that we represent the defining contrast as $A^{\gamma_1} B^{\gamma_2} C^{\gamma_3} \ldots$, where γ_i is either 0 or 1. This generates the expression

$$L = \gamma_1 + \gamma_2 + \cdots + \gamma_k,$$

which in turn is evaluated for each of the 2^k treatment combinations by setting γ_i equal to 0 or 1 according as to whether the treatment combination contains the ith factor at its high or low level. The L values are then reduced (modulo 2) to either 0 or 1 and thereby determine to which block the treatment combinations are assigned. In other words, the treatment combinations are divided into two blocks according to whether the L values leave a remainder of 0 or 1 when divided by 2.

Example 13.2 Determine the values of L (modulo 2) for a 2^3 factorial experiment when the defining contrast is ABC.

Solution. With ABC the defining contrast, we have

$$L = \gamma_1 + \gamma_2 + \gamma_3,$$

which is applied to each treatment combination as follows:

$$
\begin{aligned}
(1)\colon \quad & L = 0 + 0 + 0 = 0 = 0 && \text{(modulo 2)}\\
a\colon \quad & L = 1 + 0 + 0 = 1 = 1 && \text{(modulo 2)}\\
b\colon \quad & L = 0 + 1 + 0 = 1 = 1 && \text{(modulo 2)}\\
ab\colon \quad & L = 1 + 1 + 0 = 2 = 0 && \text{(modulo 2)}\\
c\colon \quad & L = 0 + 0 + 1 = 1 = 1 && \text{(modulo 2)}\\
ac\colon \quad & L = 1 + 0 + 1 = 2 = 0 && \text{(modulo 2)}\\
bc\colon \quad & L = 0 + 1 + 1 = 2 = 0 && \text{(modulo 2)}\\
abc\colon \quad & L = 1 + 1 + 1 = 3 = 1 && \text{(modulo 2).}
\end{aligned}
$$

The blocking arrangement, in which ABC is confounded, is given as before by

Block 1	Block 2
(1)	a
ab	b
ac	c
bc	abc

The A, B, C, AB, AC, and BC effects and sums of squares are computed in the usual way, ignoring blocks.

The block containing the treatment combination (1) in Example 13.2 is called the **principal block**. This block forms an algebraic group with respect to multiplication when the exponents are reduced to the modulo 2 base. For example, the property of closure holds, since $(ab)(bc) = ab^2c = ac$, $(ab)(ab) = a^2b^2 = (1)$, and so forth.

If the experimenter is required to allocate the treatment combinations to four blocks, two defining contrasts are chosen by the experimenter. A third effect, known as their **generalized interaction**, is automatically confounded with blocks, these three effects corresponding to the three degrees of freedom for blocks. The procedure for constructing the design is best explained through an example. Suppose it is decided that for a 2^4 factorial AB and CD are the defining contrasts. The third effect confounded, their generalized interaction, is formed by multiplying together the initial two modulo 2. Thus the effect

$$(AB)(CD) = ABCD$$

is also confounded with blocks. We construct the design by calculating the expressions

$$L_1 = \gamma_1 + \gamma_2 \qquad (AB)$$
$$L_2 = \gamma_3 + \gamma_4 \qquad (CD)$$

modulo 2 for each of the 16 treatment combinations to generate the following blocking scheme:

Block 1	Block 2	Block 3	Block 4
(1)	a	c	ac
ab	b	abc	bc
cd	acd	d	ad
abcd	bcd	abd	bd

$L_1 = 0$	$L_1 = 1$	$L_1 = 0$	$L_1 = 1$
$L_2 = 0$	$L_2 = 0$	$L_2 = 1$	$L_2 = 1$

A shortcut procedure can be used to construct the remaining blocks after the principal block has been generated. We begin by placing any treatment combination not in the principal block in a second block and build the block by multiplying (modulo 2) by the treatment combinations in the principal block. In the preceding example the second, third, and fourth blocks are generated as follows:

Block 2	Block 3	Block 4
$a(1) = a$	$c(1) = c$	$ac(1) = ac$
$a(ab) = b$	$c(ab) = abc$	$ac(ab) = bc$
$a(cd) = acd$	$c(cd) = d$	$ac(cd) = ad$
$a(abcd) = bcd$	$c(abcd) = abd$	$ac(abcd) = bd$

The analysis for the case of four blocks is quite simple. All effects that are orthogonal to blocks (those other than the defining contrasts) are computed in the usual fashion. In fact, Yates' technique can be used on the entire experiment, but the sums of squares for the three confounded effects are then added together to form the sum of squares due to blocks.

The general scheme for the 2^k factorial experiment in 2^p blocks is not difficult. We select p defining contrasts such that none is the generalized interaction of any two in the group. Since there are $2^p - 1$ degrees of freedom for blocks, we have $2^p - 1 - p$ additional effects confounded with blocks. For example, in a 2^6 factorial experiment in eight blocks, we might choose ACF, $BCDE$, and $ABDF$ as the defining contrasts. Then

$$(ACF)(BCDE) = ABDEF$$
$$(ACF)(ABDF) = BCD$$
$$(BCDE)(ABDF) = ACEF$$
$$(ACF)(BCDE)(ABDF) = E$$

are the additional four effects confounded with blocks. This is not a desirable blocking scheme, since one of the confounded effects is the main effect E. The design is constructed by evaluating

$$L_1 = \gamma_1 + \gamma_3 + \gamma_6$$
$$L_2 = \gamma_2 + \gamma_3 + \gamma_4 + \gamma_5$$
$$L_3 = \gamma_1 + \gamma_2 + \gamma_4 + \gamma_6$$

and assigning treatment combinations to blocks according to the following scheme:

$$\text{Block 1:} \quad L_1 = 0, \quad L_2 = 0, \quad L_3 = 0$$
$$\text{Block 2:} \quad L_1 = 0, \quad L_2 = 0, \quad L_3 = 1$$
$$\text{Block 3:} \quad L_1 = 0, \quad L_2 = 1, \quad L_3 = 0$$
$$\text{Block 4:} \quad L_1 = 0, \quad L_2 = 1, \quad L_3 = 1$$
$$\text{Block 5:} \quad L_1 = 1, \quad L_2 = 0, \quad L_3 = 0$$
$$\text{Block 6:} \quad L_1 = 1, \quad L_2 = 0, \quad L_3 = 1$$
$$\text{Block 7:} \quad L_1 = 1, \quad L_2 = 1, \quad L_3 = 0$$
$$\text{Block 8:} \quad L_1 = 1, \quad L_2 = 1, \quad L_3 = 1.$$

The shortcut procedure that was illustrated for the case of four blocks also applies here. Hence we can construct the remaining seven blocks from the principal block.

Example 13.3 It is of interest to study the effect of five factors on some response with the assumption that interactions involving three, four, and five of the factors are negligible. We shall divide the 32 treatment combinations into four blocks using the defining contrasts $BCDE$ and $ABCD$. Thus $(BCDE)(ABCD) = AE$ is also confounded with blocks. The experimental design and the observations are given in Table 13.7.

   The allocation of treatment combinations to experimental units within blocks is, of course, random. By pooling the unconfounded three, four, and five factor interactions to form the error term, perform the analysis of variance for the data of Table 13.7.

Table 13.7 Data for a 2^5 Experiment in Four Blocks

Block 1	Block 2	Block 3	Block 4
$(1) = 30.6$	$a = 32.4$	$b = 32.6$	$e = 30.7$
$bc = 31.5$	$abc = 32.4$	$c = 31.9$	$bce = 31.7$
$bd = 32.4$	$abd = 32.1$	$d = 33.3$	$bde = 32.2$
$cd = 31.5$	$acd = 35.3$	$bcd = 33.0$	$cde = 31.8$
$abe = 32.8$	$be = 31.5$	$ae = 32.0$	$ab = 32.0$
$ace = 32.1$	$ce = 32.7$	$abce = 33.1$	$ac = 33.1$
$ade = 32.4$	$de = 33.4$	$abde = 32.9$	$ad = 32.2$
$abcde = 31.8$	$bcde = 32.9$	$acde = 35.0$	$abcd = 32.3$

Solution. The sums of squares for each of the 31 contrasts are computed by Yates' method and the block sum of squares is found to be

$$SS(\text{blocks}) = SS(ABCD) + SS(BCDE) + SS(AE)$$
$$= 7.538.$$

Table 13.8 Analysis of Variance for the Data of Table 13.7

Source of Variation	Sum of Squares	Degrees of Freedom	Mean Square	Computed f
Main effect				
A	3.251	1	3.251	6.32
B	0.320	1	0.320	0.62
C	1.361	1	1.361	2.64
D	4.061	1	4.061	7.89
E	0.005	1	0.005	0.01
Two-factor interaction				
AB	1.531	1	1.531	2.97
AC	1.125	1	1.125	2.18
AD	0.320	1	0.320	0.62
BC	1.201	1	1.201	2.33
BD	1.711	1	1.711	3.32
BE	0.020	1	0.020	0.04
CD	0.045	1	0.045	0.09
CE	0.001	1	0.001	0.002
DE	0.001	1	0.001	0.002
Blocks ($ABCD$, $BCDE$, AE)	7.538	3	2.513	4.88
Error	7.208	14	0.515	

The analysis of variance is given in Table 13.8. None of the two-factor interactions are significant at the $\alpha = 0.05$ level when compared to $f_{0.05}(1, 14) = 4.60$. The main effects A and D are significant and both give positive effects on the response as we go from the low to the high level. Notice that the block effects are also significant when compared to $f_{0.05}(3, 14) = 3.34$. However, there is no way to determine whether the significant block effects are due to actual differences in the blocks or perhaps due to a significant interaction that has been confounded with blocks.

13.5 *Partial Confounding*

It is possible to confound any effect with blocks by the methods described in Section 13.4. Suppose that we consider a 2^3 factorial experiment in two blocks with three complete replications. If ABC is confounded with blocks in all three replicates, we can proceed as before and determine single-degree-of-freedom sums of squares for all main effects and two-factor interaction effects. The sum of squares for blocks has 5 degrees of freedom, leaving $23 - 5 - 6 = 12$ degrees of freedom for error.

Now let us confound ABC in one replicate, AC in the second, and BC in the third. The plan for this type of experiment would be as follows:

Block			Block			Block	
1	2		1	2		1	2
abc	*ab*		*abc*	*ab*		*abc*	*ab*
a	*ac*		*ac*	*bc*		*bc*	*ac*
b	*bc*		*b*	*a*		*a*	*b*
c	(1)		(1)	*c*		(1)	*c*

           Replicate 1   Replicate 2   Replicate 3
       ABC confounded AC confounded BC confounded

The effects ABC, AC, and BC are said to be **partially confounded with blocks**. These three effects can be estimated from two of the three replicates. The ratio 2/3 serves as a measure of the extent of the confounding. Yates calls this ratio the **relative information** on the confounded effects. This ratio gives the amount of information available on the partially confounded effect relative to that available on an unconfounded effect.

The analysis-of-variance layout is given in Table 13.9. The sums of squares for blocks and for the unconfounded effects A, B, C, and AB are found in the usual way. The sums of squares for AC, BC, and ABC are computed from the

Table 13.9 Analysis of Variance
with Partial Confounding

Source of Variation	Degrees of Freedom
Blocks	5
A	1
B	1
C	1
AB	1
AC	1′
BC	1′
ABC	1′
Error	11
Total	23

two replicates in which the particular effect is not confounded. One must be careful to divide by 16 instead of 24 when obtaining the sums of squares for the partially confounded effects, since we are only using 16 observations. In Table 13.9 the primes are inserted with the degrees of freedom as a reminder that these effects are partially confounded and require special calculations.

Exercises

1. In a 2^3 factorial experiment with 3 replications, show the block arrangement and indicate by means of an analysis-of-variance table the effects to be tested and their degrees of freedom, when the *AB* interaction is confounded with blocks.

2. The following coded data represent the strength of a certain type of bread-wrapper stock produced under 16 different conditions, the latter representing two levels of each of four process variables. An operator effect was introduced into the model, since it was necessary to obtain half the experimental runs under operator 1 and half under operator 2. It was felt that operators do have an effect on the quality of the product.

Operator 1	Operator 2
(1) = 18.8	*a* = 14.7
ab = 16.5	*b* = 15.1
ac = 17.8	*c* = 14.7
bc = 17.3	*abc* = 19.0
d = 13.5	*ad* = 16.9
abd = 17.6	*bd* = 17.5
acd = 18.5	*cd* = 18.2
bcd = 17.6	*abcd* = 20.1

(a) Assuming all interactions are negligible, make significance tests for the factors *A*, *B*, *C*, and *D*. Use a 0.05 level of significance.

(b) What interaction is confounded with operators?

3. Divide the treatment combinations of a 2^4 factorial experiment into four blocks by confounding *ABC* and *ABD*. What additional effect is also confounded with blocks?

4. An experiment was conducted to determine the breaking strength of a certain alloy containing five metals, *A*, *B*, *C*, *D*, and *E*. Two different percentages of each metal were used in forming the $2^5 = 32$ different alloys. Since only eight alloys could be tested on a given day, the experiment was conducted over a period of 4 days in which the *ABDE* and the *AE* effects were confounded with days. The experimental data were recorded as follows:

Treatment Combination	Breaking Strength	Treatment Combination	Breaking Strength
(1)	21.4	*e*	29.5
a	32.5	*ae*	31.3
b	28.1	*be*	33.0
ab	25.7	*abe*	23.7
c	34.2	*ce*	26.1
ac	34.0	*ace*	25.9
bc	23.5	*bce*	35.2
abc	24.7	*abce*	30.4
d	32.6	*de*	28.5
ad	29.0	*ade*	36.2
bd	30.1	*bde*	24.7
abd	27.3	*abde*	29.0
cd	22.0	*cde*	31.3
acd	35.8	*acde*	34.7
bcd	26.8	*bcde*	26.8
abcd	36.4	*abcde*	23.7

(a) Set up the blocking scheme for the 4 days.

(b) What additional effect is confounded with days?

(c) Use Yates' technique to obtain the sums of squares for all main effects.

5. By confounding *ABC* in two replicates and *AB* in the third, show the block arrangement and the analysis-of-variance table for a 2^3 factorial experiment with three replicates. What is the relative information on the confounded effects?

6. The following experiment was run to study main effects and all interactions. Four factors are used at two levels each. The experiment is replicated and two blocks are necessary in each replication. The data are as follows:

Replicate I	
Block 1	Block 2
(1) = 17.1	*a* = 15.5
d = 16.8	*b* = 14.8
ab = 16.4	*c* = 16.2
ac = 17.2	*ad* = 17.2
bc = 16.8	*bd* = 18.3
abd = 18.1	*cd* = 17.3
acd = 19.1	*abc* = 17.7
bcd = 18.4	*abcd* = 19.2

Replicate II	
Block 3	Block 4
(1) = 18.7	*a* = 17.0
ab = 18.6	*b* = 17.1
ac = 18.5	*c* = 17.2
ad = 18.7	*d* = 17.6
bc = 18.9	*abc* = 17.5
bd = 17.0	*abd* = 18.3
cd = 18.7	*acd* = 18.4
abcd = 19.8	*bcd* = 18.3

(a) What effect is confounded with blocks in the first replication of the experiment? In the second replication?

(b) Conduct an appropriate analysis of variance showing tests on all main effects and interaction effects. Use a 0.05 level of significance.

7. Construct a design involving 12 runs in which two factors are varied at two levels each. You are further restricted in that blocks of size 2 must be used and you must be able to make significance tests on both main effects and the interaction effect.

8. Show the blocking scheme for a 2^7 factorial experiment in eight blocks of size 16 each, using *ABCD*, *CDEFG*, and *BDF* as defining contrasts. Indicate what interactions are completely sacrificed in the experiment.

13.6 Fractional Factorial Experiments

The 2^k factorial experiment can become quite demanding, in terms of the number of experimental units required, when k is large. One of the real advantages with this experimental plan is that it allows a degree of freedom for each interaction. However, in many experimental situations, it is known that certain interactions are negligible, and thus it would be a waste of experimental effort to use the complete factorial experiment. In fact, the experimenter may have an economic constraint that disallows taking observations at all of the 2^k treatment combinations. When k is large, we can often make use of a **fractional factorial experiment** in which perhaps one-half, one-fourth, or even one-eighth of the total factorial plan is actually carried out.

The construction of the half-replicate design is identical to the allocation of the 2^k factorial experiment into two blocks. We begin by selecting a defining contrast that is to be completely sacrificed. We then construct the two blocks accordingly and choose either of them as the experimental plan.

Consider a 2^4 factorial experiment in which we wish to use a half-replicate. The defining contrast $ABCD$ is chosen and thus an appropriate experimental plan would be to select the principal block consisting of the following treatment combinations:

$$\{(1),\ ab,\ ac,\ ad,\ bc,\ bd,\ cd,\ abcd\}.$$

With this plan, we have contrasts on all effects except $ABCD$. Clearly,

$$A \text{ contrast} = ab + ac + ad + abcd - (1) - bc - bd - cd$$
$$AB \text{ contrast} = abcd + ab + (1) + cd - ac - ad - bc - bd,$$

with similar expressions for the contrasts of the remaining main effects and interaction effects. However, with no more than 8 of the 16 observations in our fractional design, only 7 of the 14 unconfounded contrasts are orthogonal. Consider, for example, the CD contrast given by

$$CD \text{ contrast} = abcd + cd + (1) + ab - ac - ad - bc - bd.$$

Observe that this is also the single-degree-of-freedom contrast for AB. The word **aliases** is given to two factorial effects that have the same contrast. Therefore, AB and CD are aliases. In the 2^k factorial experiments, the alias of any factorial effect is its generalized interaction with the defining contrast. For example, if $ABCD$ is the defining contrast, then the alias of A is

$A(ABCD) = BCD$. It can be seen then that the complete alias structure in a half-replicate of a 2^4 factorial experiment, using $ABCD$ as the defining contrast, is (the symbol $\equiv$ implies *aliased with*)

$$A \equiv BCD$$
$$B \equiv ACD$$
$$C \equiv ABD$$
$$D \equiv ABC$$
$$AB \equiv CD$$
$$AC \equiv BD$$
$$AD \equiv BC.$$

Without supplementary statistical evidence, there is no way of explaining which of two aliased effects are actually providing the influence on the response. In a sense they *share a degree of freedom.* Herein lies the disadvantage in fractional factorial experiments. They have their greatest use when k is quite large and there is some a priori knowledge concerning the interactions. In the example presented, the main effects can be estimated if the three-factor interactions are known to be negligible. For testing purposes the only possible procedure, in the absence of either an outside measure of experimental error or a replication of the experiment, would be to pool the sums of squares associated with the two-factor interactions. This, of course, is desirable only if these interactions represent negligible effects.

The construction of the 1/4 fraction or quarter-replicate is identical to the procedure whereby one assigns 2^k treatment combinations to four blocks. This involves the sacrificing of two defining contrasts along with their generalized interaction. Any of the four resulting blocks serves as an appropriate set of experimental runs. Each effect has three aliases, which are given by the generalized interaction with the three defining contrasts. Suppose in a 1/4 fraction of a 2^6 factorial experiment, we use $ACEF$ and $BDEF$ as the defining contrast, resulting in

$$(ACEF)(BDEF) = ABCD$$

also being sacrificed. Using $L_1 = 0$, $L_2 = 0$ (modulo 2), where

$$L_1 = \gamma_1 + \gamma_3 + \gamma_5 + \gamma_6$$
$$L_2 = \gamma_2 + \gamma_4 + \gamma_5 + \gamma_6,$$

we have an appropriate set of experimental runs given by

$$\{(1), abcd, ef, abcdef, cde, cdf, abe, abf, acef, bdef, ac, bd, adf, ade, bcf, bce\}$$

and the alias structure for the main effects is written

$$A \equiv CEF \equiv ABDEF \equiv BCD$$
$$B \equiv ABCEF \equiv DEF \equiv ACD$$
$$C \equiv AEF \equiv BCDEF \equiv ABD$$
$$D \equiv ACDEF \equiv BEF \equiv ABC$$
$$E \equiv ACF \equiv BDF \equiv ABCDE$$
$$F \equiv ACE \equiv BDE \equiv ABCDF,$$

each with a single degree of freedom. For the two-factor interactions,

$$AB \equiv BCEF \equiv ADEF \equiv CD$$
$$AC \equiv EF \equiv ABCDEF \equiv BD$$
$$AD \equiv CDEF \equiv ABEF \equiv BC$$
$$AE \equiv CF \equiv ABDF \equiv BCDE$$
$$AF \equiv CE \equiv ABDE \equiv BCDF$$
$$BE \equiv ABCF \equiv DF \equiv ACDE$$
$$BF \equiv ABCE \equiv DE \equiv ACDF.$$

Here, of course, there is some aliasing among the two-factor interactions. The remaining 2 degrees of freedom are accounted for by the following groups:

$$ADF \equiv CDE \equiv ABE \equiv BCF$$
$$ABF \equiv BCE \equiv ADE \equiv CDF.$$

It becomes evident that one should always be aware of what the alias structure is for a fractional experiment before he or she finally recommends the experimental plan. Proper choice of defining contrasts is important, since it dictates the alias structure. For example, if one would like to study main effects and all two-factor interactions in an experiment involving eight factors and it is known that interactions involving three or more factors are negligible, a very practical design would be one in which the defining contrasts are *ACEGH* and *BDEFGH*, resulting in a third,

$$(ACEGH)(BDEFGH) = ABCDF.$$

All main effects and two-factor interactions are not aliased with one another and are therefore estimable. The analysis of variance would contain the following:

Main effects	8 single degrees of freedom
Two-factor interactions	28 single degrees of freedom
Error	27 pooled degrees of freedom
Total	63 degrees of freedom

For the 1/8 and higher fractional factorials, the method of constructing the design generalizes. Of course, the aliasing can become quite extensive. For example, with a 1/8 fraction, each effect has seven aliases. The design is constructed by selecting three defining contrasts as if eight blocks were being constructed. Four additional effects are sacrificed and any one of the eight blocks can be properly used as the design.

13.7 *Analysis of Fractional Factorial Experiments*

The difficulty in making formal significance tests using data from fractional factorial experiments lies in the determination of the proper error term. Unless there are data available from prior experiments, the error must come from a pooling of contrasts representing effects that are presumed to be negligible.

Sums of squares for individual effects are found using essentially the same procedures given for the complete factorial. One can form a contrast in the treatment combinations by constructing the usual table of positive and negative signs. For example, for a half-replicate of a 2^3 factorial experiment, with ABC the defining contrast, one possible set of treatment combinations and the appropriate algebraic signs for each contrast used in computing the sums of squares for the various effects are presented in Table 13.10.

Note that in Table 13.10 the A and BC contrasts are identical, illustrating the aliasing. Also, $B \equiv AC$ and $C \equiv AB$. In this situation we have three orthogonal contrasts representing the 3 degrees of freedom available. If two observations are obtained for each of the four treatment combinations, we would then have an estimate of the error variance with 4 degrees of freedom. Assuming the interaction effects to be negligible, we could test all the main effects for significance.

Table 13.10 Signs for Contrasts in a Half-Replicate of a 2^3 Factorial Experiment

Treatment Combination	Factorial Effect						
	A	B	C	AB	AC	BC	ABC
a	+	−	−	−	−	+	+
b	−	+	−	−	+	−	+
c	−	−	+	+	−	−	+
abc	+	+	+	+	+	+	+

The sum of squares for any main effect, say A, is given by

$$SSA = \frac{(a - b - c + abc)^2}{2^2 n}.$$

In general, the single-degree-of-freedom sum of squares for any effect in a 2^{-p} fraction of a 2^k factorial experiment $(k > p)$, is obtained by squaring contrasts in the treatment totals selected and dividing by $2^{k-p}n$, where n is the number of replications of these treatment combinations.

Example 13.4   Suppose that we wish to use a half-replicate to study the effects of five factors, each at two levels, on some response and it is known that whatever the effect of each factor, it will be constant for each level of the other factors. Let the defining contrast be $ABCDE$, causing main effects to be aliased with four-factor interactions. The pooling of contrasts involving interactions provides $15 - 5 = 10$ degrees of freedom for error. Perform an analysis of variance on the following data, testing all main effects for significance at the 0.05 level:

Treatment	Response	Treatment	Response
a	11.3	bcd	14.1
b	15.6	abe	14.2
c	12.7	ace	11.7
d	10.4	ade	9.4
e	9.2	bce	16.2
abc	11.0	bde	13.9
abd	8.9	cde	14.7
acd	9.6	$abcde$	13.2

Solution. The sums of squares for the main effects are

$$SSA = \frac{(11.3 - 15.6 - \cdots - 14.7 + 13.2)^2}{2^{5-1}} = \frac{(-17.5)^2}{16} = 19.14$$

$$SSB = \frac{(-11.3 + 15.6 - \cdots - 14.7 + 13.2)^2}{2^{5-1}} = \frac{(18.1)^2}{16} = 20.48$$

$$SSC = \frac{(-11.3 - 15.6 + \cdots + 14.7 + 13.2)^2}{2^{5-1}} = \frac{(10.3)^2}{16} = 6.63$$

$$SSD = \frac{(-11.3 - 15.6 - \cdots + 14.7 + 13.2)^2}{2^{5-1}} = \frac{(-7.7)^2}{16} = 3.71$$

$$SS(E) = \frac{(-11.3 - 15.6 - \cdots + 14.7 + 13.2)^2}{2^{5-1}} = \frac{(8.9)^2}{16} = 4.95,$$

where the factor E is enclosed in parentheses to avoid confusion with the error sum of squares. The total sum of squares is

$$SST = 11.3^2 + 15.6^2 + \cdots + 13.2^2 - \frac{196.1^2}{16} = 85.74.$$

All other calculations and tests of significance are summarized in Table 13.11. The tests indicate that factor A has a significant negative effect on the response, while factor B has a significant positive effect. Factors C, D, and E are not significant at the 0.05 level.

Table 13.11 Analysis of Variance for the Data of a Half-Replicate of a 2^5 Factorial Experiment

Source of Variation	Sum of Squares	Degrees of Freedom	Mean Square	Computed f
Main effect				
A	19.14	1	19.14	6.21
B	20.48	1	20.48	6.65
C	6.63	1	6.63	2.15
D	3.71	1	3.71	1.20
E	4.95	1	4.95	1.61
Error	30.83	10	3.08	
Total	85.74	15		

Exercises

1. List the aliases for the various effects in a 2^5 factorial experiment when the defining contrast is *ACDE*.

2. (a) Obtain a 1/2 fraction of a 2^4 factorial design using *BCD* as the defining contrast.
   (b) Divide the 1/2 fraction into 2 blocks of 4 units each by confounding *ABC*.
   (c) Show the analysis-of-variance table (sources of variation and degrees of freedom) for testing all unconfounded main effects, assuming all interaction effects are negligible.

3. Construct a 1/4 fraction of a 2^6 factorial design using *ABCD* and *BDEF* as the defining contrasts. Show what effects are aliased with the six main effects.

4. Using the defining contrasts *ABCE* and *ABDF*, obtain a 1/4 fraction of a 2^6 design.
   (b) Show the analysis-of-variance table (sources of variation and degrees of freedom) for all appropriate tests assuming that E and F do not interact and all three-factor and higher interactions are negligible.

5. Seven factors are varied at two levels in an experiment involving only 16 trials. A 1/8 fraction of a 2^7 factorial experiment is used with the defining contrasts being *ACD*, *BEF*, and *CEG*. The data are as follows:

Treatment Combination	Response	Treatment Combination	Response
(1)	31.6	*acg*	31.1
ad	28.7	*cdg*	32.0
abce	33.1	*beg*	32.8
cdef	33.6	*adefg*	35.3
acef	33.7	*efg*	32.4
bcde	34.2	*abdeg*	35.3
abdf	32.5	*bcdfg*	35.6
bf	27.8	*abcfg*	35.1

Perform an analysis of variance on all seven main effects, assuming that interactions are negligible. Use a 0.05 level of significance.

6. In the study "*The Use of Regression Analysis for Correcting Matrix Effects in the X-Ray Fluorescence Analysis of Pyrotechnic Compositions*," published in the *Proceedings of the Tenth Conference on the Design of Experiments in Army Research Development and Testing*, ARO-D Report 65-3 (1965), an experiment was conducted in which the concentrations of 4 components of a propellant mixture and the weights of fine and course particles in the slurry were each allowed to vary. Factors *A*, *B*, *C*, and *D*, each at two levels, represent the con-

centrations of the 4 components and factors *E* and *F*, also at two levels, represent the weights of the fine and course particles present in the slurry. The goal of the analysis was to determine if the X-ray intensity ratios associated with component 1 of the propellant was significantly influenced by varying the concentrations of the various components and the weights of the particle sizes in the mixture. A 1/8 fraction of a 2^6 factorial experiment was used with the defining contrasts being *ADE*, *BCE*, and *ACF*. The following data represent the total of a pair of intensity readings:

Batch	Treatment Combination	Intensity Ratio Total
1	*abef*	2.2480
2	*cdef*	1.8570
3	(1)	2.2428
4	*ace*	2.3270
5	*bde*	1.8830
6	*abcd*	1.8078
7	*adf*	2.1424
8	*bcf*	1.9122

The pooled error mean square with 8 degrees of freedom is given by 0.02005. Analyze the data using a 0.05 level of significance to determine if the concentrations of the components and the weights of the fine and course particles present in the slurry have a significant influence on the intensity ratios associated with component 1. Assume that no interaction exists among the 6 factors.

14

Nonparametric Statistics

14.1 Nonparametric Tests

Most of the hypothesis-testing procedures discussed in previous chapters are based on the assumption that the random samples are selected from normal populations. Fortunately, most of these tests are still reliable when we experience slight departures from normality, particularly when the sample size is large. Traditionally, these testing procedures have been referred to as **parametric methods**. In this chapter we shall consider a number of alternative test procedures, called **nonparametric** or **distribution-free methods**, that often assume no knowledge whatsoever about the distributions of the underlying populations, except perhaps that they are continuous.

Nonparametric tests have gained a certain appeal in recent years for several reasons. First, the computations involved are usually very quick and easy to carry out. Second, the data need not be quantitative measurements but could be in the form of qualitative responses such as "defective" versus "nondefective," "yes" versus "no," and so forth, or frequently are values of an ordinal scale to which we assign ranks. On an ordinal scale the subjects are ranked according to a specified order, and a nonparametric test analyzes the various ranks. For example, two judges might rank five brands of premium beer by assigning a rank of 1 to the brand believed to have the best overall quality, a rank of 2 to the second best, and so forth. A nonparametric test could then be used to determine whether there is any agreement between the two judges. A third and perhaps the most important advantage in using nonparametric tests is that they are encumbered with less restrictive assumptions than their parametric counterparts.

We should also point out that there are a number of disadvantages associated with nonparametric tests. Primarily, they do not utilize all the information provided by the sample. As a result of this wastefulness, a nonparametric test will be less efficient than the corresponding parametric procedure when both methods are applicable. Consequently, a nonparametric test will require a larger sample size than will the corresponding parametric test in order to achieve the same probability of committing a type II error.

In summary, if a parametric and a nonparametric test are both applicable to the same set of data, we should probably avoid the "quick and easy" nonparametric test and carry out the more efficient parametric technique. However, recognizing the fact that the assumptions of normality often cannot be justified, and also the fact that we do not always have quantitative measurements, it is fortunate that statisticians have provided us with a number of useful nonparametric procedures.

14.2 Sign Test

The procedures discussed in Section 8.4 for testing the null hypothesis that $\mu = \mu_0$ are valid only if the population is approximately normal or if the sample is large. However, if $n < 30$ and the population is decidedly non-normal, we must resort to a nonparametric test. Perhaps the easiest and quickest to perform is a test called the **sign test**. In testing the null hypothesis H_0 that $\mu = \mu_0$ against an appropriate alternative on the basis of a random sample of size n, we replace each sample value exceeding μ_0 with a *plus* sign and each sample value less than μ_0 with a *minus* sign. If the null hypothesis is true and the population is symmetric, the sum of the plus signs should be approximately equal to the sum of the minus signs. When one sign appears more frequently than it should, based on chance alone, we reject the hypothesis that the population mean μ is equal to μ_0.

In theory the sign test is applicable only in situations where μ_0 cannot equal the value of any of the observations. Although there is a zero probability of obtaining a sample observation exactly equal to μ_0 when the population is continuous, nevertheless in practice a sample value equal to μ_0 will often occur from a lack of precision in recording the data. When sample values equal to μ_0 are observed, they are excluded from the analysis and the sample size is correspondingly reduced.

The appropriate test statistic for the sign test is the binomial random variable X, representing the number of plus signs in our random sample. If the null hypothesis that $\mu = \mu_0$ is true, the probability that a sample value results in either a plus or a minus sign is equal to 1/2. Therefore, to test the null hypothesis that $\mu = \mu_0$, we are actually testing the null hypothesis that the number of plus signs is a value of a random variable having the binomial distribution with the parameter $p = 1/2$. P values for both one-sided and two-sided alternatives can then be calculated using this binomial distribution. For example, in testing

$$H_0: \quad \mu = \mu_0,$$
$$H_1: \quad \mu < \mu_0,$$

we shall reject H_0 in favor of H_1 only if the proportion of plus signs is sufficiently less than $1/2$, that is, when the value x of our random variable is small. Hence, if the computed P value

$$P = P(X \leq x \text{ when } p = \tfrac{1}{2})$$

is less than or equal to some preselected significance level α, we reject H_0 in favor of H_1. For example, when $n = 15$ and $x = 3$, we find from Table A.1 that

$$P = P(X \leq 3 \text{ when } p = \tfrac{1}{2})$$

$$= \sum_{x=0}^{3} b(x; 15, \tfrac{1}{2}) = 0.0176$$

so that the null hypothesis $\mu = \mu_0$ can be rejected at the 0.05 level of significance but not at the 0.01 level.

To test the hypothesis

$$H_0: \quad \mu = \mu_0,$$
$$H_1: \quad \mu > \mu_0,$$

we reject H_0 in favor of H_1 only if the proportion of plus signs is sufficiently greater than $1/2$, that is, when x is large. Hence, if the computed P value

$$P = P(X \geq x \text{ when } p = \tfrac{1}{2})$$

is less than α, we reject H_0 in favor of H_1. Finally, to test the hypothesis

$$H_0: \quad \mu = \mu_0,$$
$$H_1: \quad \mu \neq \mu_0,$$

we reject H_0 in favor of H_1 when the proportion of plus signs is significantly less than or greater than $1/2$. This, of course, is equivalent to x being sufficiently small or sufficiently large. Therefore, if $x < n/2$ and the computed P value

$$P = 2P(X \leq x \text{ when } p = \tfrac{1}{2})$$

is less than or equal to α, or if $x > n/2$ and the computed P value

$$P = 2P(X \geq x \text{ when } p = \tfrac{1}{2})$$

is less than or equal to α, we reject H_0 in favor of H_1.

Whenever $n > 10$, binomial probabilities with $p = 1/2$ can be approximated from the normal curve, since $np = nq > 5$. Suppose, for example, that we wish to test the hypothesis

$$H_0: \quad \mu = \mu_0,$$
$$H_1: \quad \mu < \mu_0,$$

at the $\alpha = 0.05$ level of significance for a random sample of size $n = 20$ that yields $x = 6$ plus signs. Using the normal-curve approximation with

$$\mu = np = (20)(0.5) = 10,$$

and

$$\sigma = \sqrt{npq} = \sqrt{(20)(0.5)(0.5)} = 2.236,$$

we find that

$$z = \frac{6.5 - 10}{2.236} = -1.57.$$

Therefore,

$$P = P(X \le 6) \simeq P(Z < -1.57) = 0.0582,$$

which leads to the acceptance of the null hypothesis.

Example 14.1 The following data represent the number of hours that a rechargeable hedge trimmer operates before a recharge is required: 1.5, 2.2, 0.9, 1.3, 2.0, 1.6, 1.8, 1.5, 2.0, 1.2, and 1.7. Use the sign test to test the hypothesis at the 0.05 level of significance that this particular trimmer operates, on the average, 1.8 hours before requiring a recharge.

Solution. Using the six-step procedure, we have

1. $H_0: \quad \mu = 1.8$.
2. $H_1: \quad \mu \ne 1.8$.
3. $\alpha = 0.05$.
4. Test statistic: Binomial variable X with $p = \frac{1}{2}$.
5. Computations: Replacing each value by the symbol " $+$ " if it exceeds 1.8, by the symbol " $-$ " if it is less than 1.8, and discarding the one measurement that equals 1.8, we obtain the sequence

$$- + - - + - - + - -$$

for which $n = 10$, $x = 3$, and $n/2 = 5$. Therefore, from Table A.1 the computed P value is

$$P = 2P(X \le 3 \text{ when } p = \tfrac{1}{2})$$

$$= 2 \sum_{x=0}^{3} b(x; 10, \tfrac{1}{2})$$

$$= 0.3438 > 0.05.$$

6. Decision: Accept the null hypothesis and conclude that the average operating time is not significantly different from 1.8 hours.

The sign test for testing $\mu = \mu_0$ on the basis of a random sample from a single population can also be used when n pairs of observations are selected from two nonnormal populations defined over a continuous sample space. In testing the null hypothesis H_0 that $\mu_1 = \mu_2$, or $\mu_D = 0$, we simply replace each difference d_i of the paired observations with a plus or minus sign depending on whether d_i is positive or negative and then proceed as before. One can also use the sign test to test the null hypothesis $\mu_1 - \mu_2 = d_0$ for paired observations. Here we replace each difference, d_i, with a plus or minus sign depending on whether the adjusted difference, $d_i - d_0$, is positive or negative. Throughout this section we have assumed that the populations are symmetric. However, even if populations are skewed, we can carry out the same test procedure, but the hypotheses refer to the population medians rather than the means.

Example 14.2 A taxi company is trying to decide whether the use of radial tires instead of regular belted tires improves fuel economy. Sixteen cars were equipped with radial tires and driven over a prescribed test course. Without changing drivers, the same cars were then equipped with the regular belted tires and driven once again over the test course. The gasoline consumption, in kilometers per liter, was recorded as follows:

Car	Radial Tires	Belted Tires
1	4.2	4.1
2	4.7	4.9
3	6.6	6.2
4	7.0	6.9
5	6.7	6.8
6	4.5	4.4
7	5.7	5.7
8	6.0	5.8
9	7.4	6.9
10	4.9	4.9
11	6.1	6.0
12	5.2	4.9
13	5.7	5.3
14	6.9	6.5
15	6.8	7.1
16	4.9	4.8

Can we conclude at the 0.05 level of significance that cars equipped with radial tires give better fuel economy than those equipped with regular belted tires?

Solution. Let μ_1 and μ_2 represent the mean kilometers per liter for cars equipped with radial and belted tires, respectively.

1. H_0: $\mu_1 - \mu_2 = 0$.
2. H_1: $\mu_1 - \mu_2 > 0$.
3. $\alpha = 0.05$.
4. Test statistic: Binomial variable X with $p = 1/2$.
5. Computations: After replacing each positive difference by a " $+$ " symbol and each negative difference by a " $-$ " symbol, and then discarding the two zero differences, we obtain the sequence

$$+ - + + - + + + + + + + - +$$

for which $n = 14$ and $x = 11$. Using the normal-curve approximation, we find

$$z = \frac{10.5 - 7}{\sqrt{14/2}} = 1.87,$$

and then

$$P = P(X \geq 11) \simeq P(Z > 1.87) = 0.0307.$$

6. Decision: Reject H_0 and conclude that, on the average, radial tires do improve fuel economy.

Not only is the sign test one of our simplest nonparametric procedures to apply, it has the additional advantage of being applicable to dichotomous data that cannot be recorded on a numerical scale but can be represented by positive and negative responses. For example, the sign test is applicable in experiments where a qualitative response such as "hit" or "miss" is recorded, and in sensory-type experiments where a plus or minus sign is recorded depending on whether the taste tester correctly or incorrectly identifies the desired ingredient.

14.3 *Signed-Rank Test*

The sign test utilizes only the plus and minus signs of the differences between the observations and μ_0 in the one-sample case, or the plus and minus signs of the differences between the pairs of observations in the paired-sample case, but it does not take into consideration the magnitudes of these differences. A test utilizing both direction and magnitude was proposed in 1945 by Frank Wilcoxon and is now commonly referred to as the **Wilcoxon signed-rank test**.

To test the null hypothesis that we are sampling a continuous symmetric population with mean $\mu = \mu_0$, we first subtract μ_0 from each sample value, discarding all differences equal to zero. The remaining differences are then ranked without regard to sign. A rank of 1 is assigned to the smallest absolute difference (i.e., without sign), a rank of 2 to the next smallest, and so on. When the absolute value of two or more differences is the same, assign to each the average of the ranks that would have been assigned if the differences were distinguishable. For example, if the fifth and sixth smallest differences are equal in absolute value, each would be assigned a rank of 5.5. If the hypothesis $\mu = \mu_0$ is true, the total of the ranks corresponding to the positive differences should be almost equal to the total of the ranks corresponding to the negative differences. Let us represent these totals by w_+ and w_-, respectively. We shall designate the smaller of the w_+ and w_- by w.

In selecting repeated samples, we would expect w_+ and w_-, and therefore w, to vary. Thus we may think of w_+, w_-, and w as values of the corresponding random variables W_+, W_-, and W. The null hypothesis $\mu = \mu_0$ can be rejected in favor of the alternative $\mu < \mu_0$ only if w_+ is small and w_- is large. Likewise, the alternative $\mu > \mu_0$ can be accepted only if w_+ is large and w_- is small. For a two-sided alternative we may reject H_0 in favor of H_1 if either w_+ or w_- and hence if w is sufficiently small. Therefore, no matter what the alternative hypothesis may be, we reject the null hypothesis when the value of the appropriate statistic W_+, W_-, or W is sufficiently small.

To test the null hypothesis that we are sampling two continuous symmetric populations with $\mu_1 = \mu_2$ for the paired-sample case, we rank the differences of the paired observations without regard to sign and proceed as with the single-sample case. The various test procedures for both the single- and paired-sample cases are summarized in Table 14.1.

It is not difficult to show that whenever $n < 5$ and the level of significance does not exceed 0.05 for a one-tailed test or 0.10 for a two-tailed test, all possible values of w_+, w_-, or w will lead to the acceptance of the null hypothesis. However, when $5 \leq n \leq 30$, Table A.16 gives approximate critical values of W_+ and W_- for levels of significance equal to 0.01, 0.025, and 0.05 for

Table 14.1 Signed-Rank Test

To Test H_0	Versus H_1	Compute
$\mu = \mu_0$	$\mu < \mu_0$ $\mu > \mu_0$ $\mu \neq \mu_0$	w_+ w_- w
$\mu_1 = \mu_2$	$\mu_1 < \mu_2$ $\mu_1 > \mu_2$ $\mu_1 \neq \mu_2$	w_+ w_- w

a one-tailed test, and critical values of W for levels of significance equal to 0.02, 0.05, and 0.10 for a two-tailed test. The null hypothesis is rejected if the computed value w_+, w_-, or w is **less than or equal** to the appropriate tabled value. For example, when $n = 12$ Table A.16 shows that a value of $w_+ \leq 17$ is required for the one-sided alternative $\mu < \mu_0$ to be significant at the 0.05 level.

Example 14.3   Rework Example 14.1 on page 532 by using the signed-rank test.

Solution. Following the six-step procedure, we have

1. H_0: $\mu = 1.8$.
2. H_1: $\mu \neq 1.8$.
3. $\alpha = 0.05$.
4. Critical region: Since $n = 10$, after discarding the one measurement that equals 1.8, Table A.16 shows the critical region to be $w \leq 8$.
5. Computations: Subtracting 1.8 from each measurement and then ranking the differences without regard to sign, we have

d_i	-0.3	0.4	-0.9	-0.5	0.2	-0.2	-0.3	0.2	-0.6	-0.1
Ranks	5.5	7	10	8	3	3	5.5	3	9	1

Now $w_+ = 13$ and $w_- = 42$ so that $w = 13$, the smaller of w_+ and w_-.

6. Decision: Accept H_0 as before and conclude that the average operating time is not significantly different from 1.8 hours.

When $n \geq 15$, the sampling distribution of W_+ (or W_-) approaches the normal distribution with mean

$$\mu_{W_+} = \frac{n(n + 1)}{4}$$

and variance

$$\sigma^2_{W_+} = \frac{n(n + 1)(2n + 1)}{24}.$$

Therefore, when n exceeds the largest value in Table A.16 the statistic

$$Z = \frac{(W_+ - \mu_{W_+})}{\sigma_{W_+}}$$

can be used to determine the critical region for our test.

The signed-rank test can also be used to test the null hypothesis that $\mu_1 - \mu_2 = d_0$. In this case the populations need not be symmetric. As with the sign test we subtract d_0 from each difference, rank the adjusted differences without regard to sign, and apply the same procedure as above.

Example 14.4 It is claimed that a college senior can increase his score in the major field area of the graduate record examination by at least 50 points if he is provided sample problems in advance. To test this claim, 20 college seniors were divided into 10 pairs such that each matched pair had almost the same overall quality point average for their first 3 years in college. Sample problems and answers were provided at random to one member of each pair 1 week prior to the examination. The following examination scores were recorded:

	Pair									
	1	2	3	4	5	6	7	8	9	10
With sample problems	531	621	663	579	451	660	591	719	543	575
Without sample problems	509	540	688	502	424	683	568	748	530	524

Test the null hypothesis at the 0.05 level of significance that sample problems increase the scores by 50 points against the alternative hypothesis that the increase is less than 50 points.

Solution. Let μ_1 and μ_2 represent the mean score of all students taking the test in question with and without sample problems, respectively. We follow the six-step procedure already outlined:

1. H_0: $\mu_1 - \mu_2 = 50$.
2. H_1: $\mu_1 - \mu_2 < 50$.
3. $\alpha = 0.05$.
4. Critical region: Since $n = 10$, Table A.16 shows the critical region to be $w_+ \leq 11$.
5. Computations:

	Pair									
	1	2	3	4	5	6	7	8	9	10
d_i	22	81	−25	77	27	−23	23	−29	13	51
$d_i - d_0$	−28	31	−75	27	−23	−73	−27	−79	−37	1
Ranks	5	6	9	3.5	2	8	3.5	10	7	1

Now we find $w_+ = 6 + 3.5 + 1 = 10.5$.

6. Decision: Reject H_0 and conclude that sample problems do not, on the average, increase one's graduate record score by as much as 50 points.

Exercises

1. The following data represent the time, in minutes, that a patient had to wait on 12 visits to a doctor's office before being seen by the doctor:

17	15	20	20
32	28	12	26
25	25	35	24

   Use the sign test at the 0.05 level of significance to test the doctor's claim that, on the average, her patients do not wait more than 20 minutes before being admitted to the examination room.

2. The following data represent the number of hours of flight training received by 18 student pilots from a certain instructor prior to their first solo flight:

9	12	18	14	12	14
12	10	16	11	9	11
13	11	13	15	13	14

   Using binomial probabilities from Table A.1, perform a sign test at the 0.02 level of significance to test the instructor's claim that, on the average, his students solo after 12 hours of flight training.

3. A food inspector examined 16 jars of a certain brand of jam to determine the percent of foreign impurities. The following data were recorded:

2.4	2.3	3.1	2.2
2.3	1.2	1.0	2.4
1.7	1.1	4.2	1.9
1.7	3.6	1.6	2.3

   Using the normal approximation to the binomial distribution, perform a sign test at the 0.05 level of significance to test the null hypothesis that the average percent of impurities in this brand of jam is 2.5% against the alternative that the average percent of impurities is not 2.5%.

4. A paint supplier claims that a new additive will reduce the drying time of its acrylic paint. To test this claim, 12 panels of wood are painted, one-half of each panel with paint containing the regular additive and the other half with paint containing the new additive. The drying times, in hours, were recorded as follows:

Panel	Drying Time (hours)	
	New Additive	Regular Additive
1	6.4	6.6
2	5.8	5.8
3	7.4	7.8
4	5.5	5.7
5	6.3	6.0
6	7.8	8.4
7	8.6	8.8
8	8.2	8.4
9	7.0	7.3
10	4.9	5.8
11	5.9	5.8
12	6.5	6.5

   Use the sign test at the 0.05 level to test the null hypothesis that the new additive is no better than the regular additive in reducing the drying time of this kind of paint.

5. It is claimed that a new diet will reduce a person's weight by 4.5 kilograms on the average in a period of 2 weeks. The weights of 10 women who followed this diet were recorded before and after a 2-week period, yielding the following data:

Woman	Weight Before	Weight After
1	58.5	60.0
2	60.3	54.9
3	61.7	58.1
4	69.0	62.1
5	64.0	58.5
6	62.6	59.9

Woman	Weight Before	Weight After
7	56.7	54.4
8	63.6	60.2
9	68.2	62.3
10	59.4	58.7

Use the sign test at the 0.05 level of significance to test the hypothesis that the diet reduces a person's weight by 4.5 kilograms on the average, against the alternative hypothesis that the mean difference in weight is less than 4.5 kilograms.

6. Two types of instruments for measuring the amount of sulfur monoxide in the atmosphere are being compared in an air-pollution experiment. The following readings were recorded daily for a period of 2 weeks:

	Sulfur Monoxide	
Day	Instrument A	Instrument B
1	0.96	0.87
2	0.82	0.74
3	0.75	0.63
4	0.61	0.55
5	0.89	0.76
6	0.64	0.70
7	0.81	0.69
8	0.68	0.57
9	0.65	0.53
10	0.84	0.88
11	0.59	0.51
12	0.94	0.79
13	0.91	0.84
14	0.77	0.63

Using the normal approximation to the binomial distribution, perform a sign test to determine whether the different instruments lead to different results. Use a 0.05 level of significance.

7. The following figures give the systolic blood pressure of 16 joggers before and after an 8-kilometer run:

Jogger	Before	After
1	158	164
2	149	158
3	160	163
4	155	160
5	164	172
6	138	147
7	163	167
8	159	169
9	165	173
10	145	147
11	150	156
12	161	164
13	132	133
14	155	161
15	146	154
16	159	170

Use the sign test at the 0.05 level of significance to test the null hypothesis that jogging 8 kilometers increases a runner's systolic blood pressure by 8 points on the average against the alternative that the average increase is less than 8 points.

8. Analyze the data of Exercise 1 using the signed-rank test.

9. Analyze the data of Exercise 2 using the signed-rank test.

10. The weights of 4 people before they stopped smoking and 5 weeks after they stopped smoking, in kilograms, are as follows:

	Individual				
	1	2	3	4	5
Before	66	80	69	52	75
After	71	82	68	56	73

Use the signed-rank test for paired observations to test the hypothesis, at the 0.05 level of significance, that giving up smoking has no effect on a person's weight against the alternative that one's weight increases if he or she quits smoking.

11. Rework Exercise 5 using the signed-rank test.

12. The following are the numbers of prescriptions filled by two pharmacies over a 20-day period:

Day	Pharmacy A	Pharmacy B
1	19	17
2	21	15
3	15	12
4	17	12
5	24	16
6	12	15
7	19	11
8	14	13
9	20	14
10	18	21
11	23	19
12	21	15
13	17	11
14	12	10
15	16	20
16	15	12
17	20	13
18	18	17
19	14	16
20	22	18

Use the signed-rank test at the 0.01 level of significance to determine whether the two pharmacies, on the average, fill the same number of prescriptions against the alternative that Pharmacy A fills more prescriptions than Pharmacy B.

13. Rework Exercise 7 using the signed-rank test.

14. The following data represent the number of hours of sleep obtained by two college students for 20 nights that precede school days:

Night	Student A	Student B
1	7.3	6.5
2	8.2	8.2
3	8.5	7.4
4	9.0	8.1
5	9.3	7.6
6	7.6	9.0
7	8.1	6.8
8	7.7	7.4
9	6.9	8.2
10	9.1	7.9
11	7.9	7.3
12	6.8	8.0
13	8.1	6.9
14	7.9	8.2
15	7.8	7.6
16	8.2	9.0
17	7.5	6.6
18	8.4	7.2
19	8.3	8.3
20	8.8	7.9

Use the signed-rank test at the 0.05 level of significance to determine whether these two students, on the average, get the same number of hours of sleep on nights preceding school days.

14.4 Rank-Sum Test

In this section we consider a very simple nonparametric procedure proposed by Wilcoxon for the comparison of the means of two continuous nonnormal populations when independent samples are selected from the populations. This nonparametric alternative to the two-sample t test discussed in Section 8.4 is called the **Wilcoxon rank-sum test** or the **Wilcoxon two-sample test**.

We shall test the null hypothesis H_0 that $\mu_1 = \mu_2$ against some suitable alternative. First we select a random sample from each of the populations. Let n_1 be the number of observations in the smaller sample, and n_2 the number of observations in the larger sample. When the samples are of equal size, n_1 and n_2 may be randomly assigned. Arrange the $n_1 + n_2$ observations of the combined samples in ascending order and substitute a rank of $1, 2, \ldots, n_1 + n_2$ for each observation. In the case of ties (identical observations), we replace the observations by the mean of the ranks that the observations would have if they were distinguishable. For example, if the seventh and eighth observations are identical, we would assign a rank of 7.5 to each of the two observations.

The sum of the ranks corresponding to the n_1 observations in the smaller sample is denoted by w_1. Similarly, the value w_2 represents the sum of the n_2 ranks corresponding to the larger sample. The total $w_1 + w_2$ depends only on the number of observations in the two samples and is in no way affected by the results of the experiment. Hence, if $n_1 = 3$ and $n_2 = 4$, then $w_1 + w_2 = 1 + 2 + \cdots + 7 = 28$, regardless of the numerical values of the observations. In general,

$$w_1 + w_2 = \frac{(n_1 + n_2)(n_1 + n_2 + 1)}{2},$$

the arithmetic sum of the integers $1, 2, \ldots, n_1 + n_2$. Once we have determined w_1 it may be easier to find w_2 by the formula

$$w_2 = \frac{(n_1 + n_2)(n_1 + n_2 + 1)}{2} - w_1.$$

In choosing repeated samples of size n_1 and n_2, we would expect w_1, and therefore w_2, to vary. Thus we may think of w_1 and w_2 as values of the random variables W_1 and W_2, respectively. The null hypothesis $\mu_1 = \mu_2$ will be rejected in favor of the alternative $\mu_1 < \mu_2$ only if w_1 is small and w_2 is large. Likewise, the alternative $\mu_1 > \mu_2$ can be accepted only if w_1 is large and w_2 is small. For a two-tailed test, we may reject H_0 in favor of H_1 if w_1 is small and w_2 is large or if w_1 is large and w_2 is small. In other words, the alternative $\mu_1 < \mu_2$ is accepted if w_1 is sufficiently small; the alternative $\mu_1 > \mu_2$ is accepted if w_2 is sufficiently small; and the alternative $\mu_1 \neq \mu_2$ is accepted if the minimum of w_1 and w_2 is sufficiently small. In actual practice we usually base our decision on the value

$$u_1 = w_1 - \frac{n_1(n_1 + 1)}{2}$$

or

$$u_2 = w_2 - \frac{n_2(n_2 + 1)}{2}$$

of the related statistic U_1 or U_2, or on the value u of the statistic U, the minimum of U_1 and U_2. These statistics simplify the construction of tables of critical values, since both U_1 and U_2 have symmetric sampling distributions and assume values in the interval from 0 to $n_1 n_2$ such that $u_1 + u_2 = n_1 n_2$.

From the formulas for u_1 and u_2 we see that u_1 will be small when w_1 is small and u_2 will be small when w_2 is small. Consequently, the null hypothesis

Table 14.2 Rank-Sum Test

To Test H_0	Versus H_1	Compute
$\mu_1 = \mu_2$	$\mu_1 < \mu_2$ $\mu_1 > \mu_2$ $\mu_1 \neq \mu_2$	u_1 u_2 u

will be rejected whenever the appropriate statistic U_1, U_2, or U assumes a value less than or equal to the desired critical value given in Table A.17. The various test procedures are summarized in Table 14.2.

Table A.17 gives critical values of U_1 and U_2 for levels of significance equal to 0.001, 0.01, 0.025, and 0.05 for a one-tailed test, and critical values of U for levels of significance equal to 0.002, 0.02, 0.05, and 0.10 for a two-tailed test. If the observed value of u_1, u_2, or u is **less than or equal** to the tabled critical value, the null hypothesis is rejected at the level of significance indicated by the table. Suppose, for example, that we wish to test the null hypothesis that $\mu_1 = \mu_2$ against the one-sided alternative that $\mu_1 < \mu_2$ at the 0.05 level of significance for random samples of size $n_1 = 3$ and $n_2 = 5$ that yield the value $w_1 = 8$. It follows that

$$u_1 = 8 - \frac{(3)(4)}{2} = 2.$$

Our one-tailed test is based on the statistic U_1. Using Table A.17, we reject the null hypothesis of equal means when $u_1 \leq 1$. Since $u_1 = 2$ falls in the acceptance region, the null hypothesis cannot be rejected.

Example 14.5   The nicotine content of two brands of cigarettes, measured in milligrams, was found to be as follows:

Brand A	2.1	4.0	6.3	5.4	4.8	3.7	6.1	3.3	
Brand B	4.1	0.6	3.1	2.5	4.0	6.2	1.6	2.2	1.9	5.4

Test the hypothesis, at the 0.05 level of significance, that the average nicotine contents of the two brands are equal against the alternative that they are unequal.

Solution. We proceed by the six-step rule with $n_1 = 8$ and $n_2 = 10$.

1. H_0: $\mu_1 = \mu_2$.
2. H_1: $\mu_1 \neq \mu_2$.
3. $\alpha = 0.05$.
4. Critical region: $u \leq 17$ (from Table A.17).

5. Computations: The observations are arranged in ascending order and ranks from 1 to 18 assigned.

Original Data	Ranks
0.6	1
1.6	2
1.9	3
2.1	4
2.2	5
2.5	6
3.1	7
3.3	8
3.7	9
4.0	10.5
4.0	10.5
4.1	12
4.8	13
5.4	14.5
5.4	14.5
6.1	16
6.2	17
6.3	18

The ranks of the observations belonging to sample A, the smaller sample, appear in color. Now

$$w_1 = 4 + 8 + 9 + 10.5 + 13 + 14.5 + 18 = 93$$

and

$$w_2 = \left[\frac{(18)(19)}{2} \right] - 93 = 78.$$

Therefore,

$$u_1 = 93 - \left[\frac{(8)(9)}{2} \right] = 57$$

$$u_2 = 78 - \left[\frac{(10)(11)}{2} \right] = 23$$

so that $u = 23$.

6. Decision: Accept H_0 and conclude that there is no difference in the average nicotine contents of the two brands of cigarettes.

When both n_1 and n_2 exceed 8, the sampling distribution of U_1 (or U_2) approaches the normal distribution with mean

$$\mu_{U_1} = \frac{n_1 n_2}{2}$$

and variance

$$\sigma_{U_1}^2 = \frac{n_1 n_2 (n_1 + n_2 + 1)}{12}.$$

Consequently, when n_2 is greater than 20, the maximum value in Table A.17, and n_1 is at least 9, one could use the statistic

$$Z = \frac{U_1 - \mu_{U_1}}{\sigma_{U_1}}$$

for our test, with the critical region falling in either or both tails of the standard normal distribution, depending on the form of H_1.

The use of the Wilcoxon rank-sum test is not restricted to nonnormal populations. It can be used in place of the two-sample t test when the populations are normal, although the probability of committing a type II error will be larger. The Wilcoxon rank-sum test is always superior to the t test for decidedly nonnormal populations.

14.5 *Kruskal–Wallis Test*

The **Kruskal–Wallis test**, also called the **Kruskal–Wallis H test**, is a generalization of the rank-sum test to the case of $k > 2$ samples. It is used to test the null hypothesis H_0 that k independent samples are from identical populations. Introduced in 1952 by W. H. Kruskal and W. A. Wallis, the test is an alternative nonparametric procedure to the F test for testing the equality of means in the one-factor analysis of variance when the experimenter wishes to avoid the assumption that the samples were selected from normal populations.

Let n_i ($i = 1, 2, \ldots, k$) be the number of observations in the ith sample. First we combine all k samples and arrange the $n = n_1 + n_2 + \cdots + n_k$ observations in ascending order, substituting the appropriate rank from $1, 2, \ldots, n$ for each observation. In the case of ties (identical observations), we follow the usual procedure of replacing the observations by the means of the ranks that the observations would have if they were distinguishable. The sum of the ranks corresponding to the n_i observations in the ith sample is denoted by the random variable R_i. Now let us consider the statistic

$$H = \frac{12}{n(n + 1)} \sum_{i=1}^{k} \frac{R_i^2}{n_i} - 3(n + 1),$$

which is approximated very well by a chi-square distribution with $k - 1$ degrees of freedom when H_0 is true and if each sample consists of at least 5 observations. Note that the statistic H assumes the value h, where

$$h = \frac{12}{n(n + 1)} \sum_{i=1}^{k} \frac{r_i^2}{n_i} - 3(n + 1),$$

when R_1 assumes the value r_1, R_2 assumes the value r_2, and so forth. The fact that h is large when the independent samples come from populations that are not identical allows us to establish the following decision criterion for testing H_0:

Kruskal–Wallis Test   *To test the null hypothesis H_0 that k independent samples are from identical populations, compute*

$$h = \frac{12}{n(n + 1)} \sum_{i=1}^{k} \frac{r_i^2}{n_i} - 3(n + 1).$$

If h falls in the critical region $H > \chi_\alpha^2$ with $v = k - 1$ degrees of freedom, reject H_0 at the α level of significance; otherwise, accept H_0.

Example 14.6   In an experiment to determine which of three different missile systems is preferable, the propellant burning rate was measured. The data, after coding, are given in Table 14.3.

Table 14.3 Propellant
Burning Rates

Missile System		
1	2	3
24.0	23.2	18.4
16.7	19.8	19.1
22.8	18.1	17.3
19.8	17.6	17.3
18.9	20.2	19.7
	17.8	18.9
		18.8
		19.3

Use the Kruskal–Wallis test and a significance level of $\alpha = 0.05$ to test the hypothesis that the propellant burning rates are the same for the three missile systems.

Solution

1. H_0: $\mu_1 = \mu_2 = \mu_3$.
2. H_1: the three means are not all equal.
3. $\alpha = 0.05$.
4. Critical region: $h > \chi^2_{0.05} = 5.991$, for $v = 2$ degrees of freedom.
5. Computations: In Table 14.4 we convert the 19 observations to ranks and sum the ranks for each missile system.

Table 14.4 Ranks for Propellant Burning Rates

Missile System		
1	2	3
19	18	7
1	14.5	11
17	6	2.5
14.5	4	2.5
9.5	16	13
$r_1 = 61.0$	5	9.5
	$r_2 = 63.5$	8
		12
		$r_3 = 65.5$

Now, substituting $n_1 = 5$, $n_2 = 6$, $n_3 = 8$, and $r_1 = 61.0$, $r_2 = 63.5$, $r_3 = 65.5$, our test statistic H assumes the value

$$h = \frac{12}{(19)(20)} \left(\frac{61.0^2}{5} + \frac{63.5^2}{6} + \frac{65.5^2}{8} \right) - (3)(20)$$

$$= 1.66.$$

6. Decision: Since $h = 1.66$ does not fall in the critical region $h > 5.991$, we have insufficient evidence to reject the hypothesis that the propellant burning rates are the same for the three missile systems.

Exercises

---

1. A cigarette manufacturer claims that the tar content of brand B cigarettes is lower than that of brand A. To test this claim, the following determi- nations of tar content, in milligrams, were recorded:

Brand A	12	9	13	11	14
Brand B	8	10	7		

Use the rank-sum test with $\alpha = 0.05$ to test whether the claim is valid.

2. To find out whether a new serum will arrest leukemia, 9 patients, who have all reached an advanced stage of the disease, are selected. Five patients receive the treatment and 4 do not. The survival times, in years, from the time the experiment commenced are

Treatment	2.1	5.3	1.4	4.6	0.9
No treatment	1.9	0.5	2.8	3.1	

Use the rank-sum test, at the 0.05 level of significance, to determine if the serum is effective.

3. The following data represent the number of hours that two different types of scientific pocket calculators operate before a recharge is required:

Calculator A	5.5 5.6 6.3 4.6 5.3 5.0 6.2 5.8 5.1
Calculator B	3.8 4.8 4.3 4.2 4.0 4.9 4.5 5.2 4.5

Use the rank-sum test with $\alpha = 0.01$ to determine if calculator A operates longer than calculator B on a full battery charge.

4. A fishing line is being manufactured by two processes. To determine if there is a difference in the mean breaking strength of the lines, 10 pieces by each process are selected and tested for breaking strength. The results are as follows:

Process 1	Process 2
10.4	8.7
9.8	11.2
11.5	9.8
10.0	10.1
9.9	10.8
9.6	9.5
10.9	11.0
11.8	9.8
9.3	10.5
10.7	9.9

Use the rank-sum test with $\alpha = 0.1$ to determine if there is a difference between the mean breaking strengths of the lines manufactured by the two processes.

5. From a mathematics class of 12 equally capable students using programmed materials, 5 are selected at random and given additional instruction by the teacher. The results on the final examination were as follows:

	Grade
Additional instruction	87 69 78 91 80
No additional instruction	75 88 64 82 93 79 67

Use the rank-sum test with $\alpha = 0.05$ to determine if the additional instruction affects the average grade.

6. The following data represent the weights, in kilograms, of personal luggage carried on various flights by a member of a baseball team and a member of a basketball team:

Luggage Weight (kilograms)				
Baseball Player			Basketball Player	
16.3	20.0	18.6	15.4	16.3
18.1	15.0	15.4	17.7	18.1
15.9	18.6	15.6	18.6	16.8
14.1	14.5	18.3	12.7	14.1
17.7	19.1	17.4	15.0	13.6
16.3	13.6	14.8	15.9	16.3
13.2	17.2	16.5		

Use the rank-sum test with $\alpha = 0.05$ to test the null hypothesis that the two athletes carry the same amount of luggage on the average against the alternative hypothesis that the average weights of luggage for the two athletes are different.

7. The following data represent the operating times in hours for three types of scientific pocket calcu-

lators before a recharge is required:

Calculator		
A	*B*	*C*
4.9	5.5	6.4
6.1	5.4	6.8
4.3	6.2	5.6
4.6	5.8	6.5
5.3	5.5	6.3
	5.2	6.6
	4.8	

Use the Kruskal–Wallis test, at the 0.01 level of significance, to test the hypothesis that the operating times for all three calculators are equal.

8. Random samples of four brands of cigarettes were tested for tar content. The following figures show the milligrams of tar found in the 16 cigarettes tested:

Brand *A*	Brand *B*	Brand *C*	Brand *D*
14	16	16	17
10	18	15	20
11	14	14	19
13	15	12	21

Use the Kruskal–Wallis test, at the 0.05 level of significance, to test whether there is a significant difference in tar content among the four brands of cigarettes.

9. In Exercise 8 on page 421 use the Kruskal–Wallis test, at the 0.05 level of significance to determine if the grade distributions given by the 3 teachers differ significantly.

10. In Exercise 10 on page 422, use the Kruskal–Wallis test, at the 0.05 level of significance, to determine if the chemical analyses performed by the four laboratories give, on the average, the same results.

14.6 *Runs Test*

In applying the many statistical concepts that were discussed throughout this book, it was always assumed that our sample data had been collected by some randomization procedure. The **runs test**, based on the order in which the sample observations are obtained, is a useful technique for testing the null hypothesis H_0 that the observations have indeed been drawn at random.

To illustrate the runs test, let us suppose that 12 people have been polled to find out if they use a certain product. One would seriously question the assumed randomness of the sample if all 12 people were of the same sex. We shall designate a male and female by the symbols *M* and *F*, respectively, and record the outcomes according to their sex in the order in which they occur. A typical sequence for the experiment might be

$$M \ \ M \quad F \ \ F \ \ F \quad M \quad F \ \ F \quad M \ \ M \ \ M \ \ M,$$

where we have grouped subsequences of similar symbols, Such groupings are called **runs**.

Definition 14.1   *A* **run** *is a subsequence of one or more identical symbols representing a common property of the data.*

Regardless of whether our sample measurements represent qualitative or quantitative data, the runs test divides the data into two mutually exclusive categories: male or female; defective or nondefective; heads or tails; above or below the median; and so forth. Consequently, a sequence will always be limited to two distinct symbols. Let n_1 be the number of symbols associated with the category that occurs the least and n_2 be the number of symbols that belong to the other category. Then the sample size $n = n_1 + n_2$.

For the $n = 12$ symbols in our poll we have five runs with the first containing two M's, the second containing three F's, and so on. If the number of runs is larger of smaller than what we would expect by chance, the hypothesis that the sample was drawn at random should be rejected. Certainly, a sample resulting in only two runs,

$$M \quad M \quad M \quad M \quad M \quad M \quad M \quad F \quad F \quad F \quad F \quad F,$$

or the reverse, is most unlikely to occur from a random selection process. Such a result indicates that the first seven people interviewed were all males followed by five females. Likewise, if the sample resulted in the maximum number of 12 runs, as in the alternating sequence

$$M \quad F \quad M \quad F \quad M \quad F \quad M \quad F \quad M \quad F \quad M \quad F,$$

we would again be suspicious of the order in which the individuals were selected for the poll.

The runs test for randomness is based on the random variable V, the total number of runs that occur in the complete sequence of our experiment. In Table A.18, values of $P(V \leq v^*$ when H_0 is true) are given for $v^* = 2, 3, \ldots, 20$ runs, and values of n_1 and n_2 less than or equal to 10. The P values for both one-tailed and two-tailed tests can be obtained using these tabled values.

In the poll taken above we exhibit a total of 5 F's and 7 M's. Hence, with $n_1 = 5, n_2 = 7$, and $v = 5$, we note from Table A.18 for a two-tailed test that the P value is

$$P = 2P(V \leq 5 \text{ when } H_0 \text{ is true})$$

$$= 0.394 > 0.05.$$

That is, the value $v = 5$ is reasonable at the 0.05 level of significance when H_0 is true, and therefore we have insufficient evidence to reject the hypothesis of randomness in our sample.

When the number of runs is large, for example if $v = 11$ and $n_1 = 5$ and $n_2 = 7$, then the P value in a two-tailed test is

$$P = 2P(V \geq 11 \text{ when } H_0 \text{ is true})$$

$$= 2[1 - P(V \leq 10) \text{ when } H_0 \text{ is true}]$$

$$= 2(1 - 0.992) = 0.016 < 0.05,$$

which leads us to reject the hypothesis that the sample values occurred at random.

The runs test can also be used to detect departures in randomness of a sequence of quantitative measurements over time, caused by trends or periodicities. Replacing each measurement in the order in which they are collected by a *plus* symbol if it falls above the median, by a *minus* symbol if it falls below the median, and omitting all measurements that are exactly equal to the median, we generate a sequence of plus and minus symbols that are tested for randomness as illustrated in the following example.

Example 14.7   A machine is adjusted to dispense acrylic paint thinner into a container. Would you say that the amount of paint thinner being dispensed by this machine varies randomly if the contents of the next 15 containers are measured and found to be 3.6, 3.9, 4.1, 3.6, 3.8, 3.7, 3.4, 4.0, 3.8, 4.1, 3.9, 4.0, 3.8, 4.2, and 4.1 liters? Use a 0.1 level of significance.

Solution.   Using the six-step procedure, we have

1. H_0:   Sequence is random.
2. H_1:   Sequence is not random.
3. $\alpha = 0.1$.
4. Test statistic: V, the total number of runs.
5. Computations: For the given sample we find $\tilde{x} = 3.9$. Replacing each measurement by the symbol " $+$ " if it falls above 3.9, by the symbol " $-$ " if it falls below 3.9, and omitting the two measurements that equal 3.9, we obtain the sequence

$$- + - - - - + + + + - + +$$

for which $n_1 = 6$, $n_2 = 7$, and $v = 6$. Therefore, from Table A.18, the computed P value is

$$P = 2P(V \le 6 \text{ when } H_0 \text{ is true})$$
$$= 0.596 > 0.1.$$

6. Decision: Accept the hypothesis that the sequence of measurements vary randomly.

The runs test, although less powerful, can also be used as an alternative to the Wilcoxon two-sample test to test the claim that two random samples come from populations having the same distributions and therefore equal means. If the populations are symmetric, rejection of the claim of equal distributions is equivalent to accepting the alternative hypothesis that the means are not equal. In performing the test, we first combine the observations from both samples and arrange them in ascending order. Now assign the letter A to each observation taken from one of the populations and the letter B to each observation from the second population, thereby generating a sequence consisting of the symbols A and B. If observations from one population are tied with observations from the other population, the sequence of A and B symbols generated will not be unique and consequently the number of runs is unlikely to be unique. Procedures for breaking ties usually result in additional tedious computations, and for this reason one might prefer to apply the Wilcoxon rank-sum test whenever these situations occur.

To illustrate the use of runs in testing for equal means, consider the survival times of the leukemia patients of Exercise 2 on page 547 for which we have

$$0.5 \quad 0.9 \quad 1.4 \quad 1.9 \quad 2.1 \quad 2.8 \quad 3.1 \quad 4.6 \quad 5.3$$
$$B \quad A \quad A \quad B \quad A \quad B \quad B \quad A \quad A$$

resulting in $v = 6$ runs. If the two symmetric populations have equal means the observations from the two samples will be intermingled resulting in many runs. However, if the population means are significantly different, we would expect most of the observations for one of the two samples to be smaller than those for the other sample. In the extreme case where the populations do not overlap, we would obtain a sequence of the form

$$A \quad A \quad A \quad A \quad A \quad B \quad B \quad B \quad B \quad \text{or} \quad B \quad B \quad B \quad B \quad A \quad A \quad A \quad A$$

and in either case there are only two runs. Consequently, the hypothesis of equal populations means will be rejected at the α level of significance only when v is small enough so that

$$P = P(V \leq v \text{ when } H_0 \text{ is true}) \leq \alpha,$$

implying a one-tailed test.

Returning to the data of Exercise 2 on page 547, for which $n_1 = 4$, $n_2 = 5$, and $v = 6$, we find from Table A.18 that

$$P = P(V \leq 6 \text{ when } H_0 \text{ is true})$$
$$= 0.786 > 0.05$$

and therefore fail to reject the null hypothesis of equal means. Hence we conclude that the new serum does not prolong life by arresting leukemia.

When n_1 and n_2 increase in size, the sampling distribution of V approaches the normal distribution with mean

$$\mu_V = \frac{2n_1 n_2}{n_1 + n_2} + 1$$

and variance

$$\sigma_V^2 = \frac{2n_1 n_2 (2n_1 n_2 - n_1 - n_2)}{(n_1 + n_2)^2 (n_1 + n_2 - 1)}.$$

Consequently, when n_1 and n_2 are both greater than 10, one could use the statistic

$$Z = \frac{V - \mu_V}{\sigma_V}$$

to establish the critical region for the runs test.

14.7 Tolerance Limits

Tolerance limits for a normal distribution of measurements were discussed in Chapter 7. In this section we shall consider a method for constructing tolerance intervals that are independent of the shape of the underlying distribution. As one might suspect, for a reasonable degree of confidence they will be substantially longer than those constructed assuming normality, and the sample size required is generally very large. Nonparametric tolerance limits are stated in terms of the smallest and largest observations in our sample.

Two-Sided Tolerance Limits   *For any distribution of measurements, two-sided tolerance limits are given by the smallest and largest observations in a sample of size n, where n is determined so that one can assert with 100γ% confidence that **at least** the proportion 1 − α of the distribution is included between the sample extremes.*

Table A.19 gives required sample sizes for selected values of γ and $1 - \alpha$. For example, when $\gamma = 0.99$ and $1 - \alpha = 0.95$, we must choose a random sample of size $n = 130$ in order to be 99% confident that at least 95% of the distribution of measurements is included between the sample extremes.

Instead of determining the sample size n such that a specified proportion of measurements are contained between the sample extremes, it is desirable in many industrial processes to determine the sample size such that a fixed proportion of the population falls below the largest (or above the smallest) observation in the sample. Such limits are called one-sided tolerance limits.

One-Sided Tolerance Limits   *For any distribution of measurements, a one-sided tolerance limit is given by the smallest (largest) observation in a sample of size n, where n is determined so that one can assert with 100γ% confidence that **at least** the proportion 1 − α of the distribution will exceed the smallest (be less than the largest) observation in the sample.*

Table A.20 gives required sample sizes corresponding to selected values of γ and $1 - \alpha$. Hence, when $\gamma = 0.95$ and $1 - \alpha = 0.70$, we must choose a sample of size $n = 9$ in order to be 95% confident that 70% of our distribution of measurements will exceed the smallest observation in the sample.

14.8 Rank Correlation Coefficient

In Chapter 9 we used the sample correlation coefficient r to measure the linear relationship between two continuous variables X and Y. If ranks $1, 2, \ldots, n$ are assigned to the x observations in order of magnitude and similarly to the y observations, and if these ranks are then substituted for the actual numerical values into the formula for r on page 347, we obtain the nonparametric counterpart of the conventional correlation coefficient. A correlation coefficient calculated in this manner is known as the **Spearman rank correlation coefficient** and is denoted by r_S. When there are no ties among either set of measurements, the formula for r_S reduces to a much simpler expression involving the differences d_i between the ranks assigned to the n pairs of x's and y's, which we now state.

Rank Correlation Coefficient   *A nonparametric measure of association between two variables X and Y is given by the* **rank correlation coefficient**

$$r_S = 1 - \frac{6\sum_{i=1}^{n} d_i^2}{n(n^2 - 1)},$$

where d_i is the difference between the ranks assigned to x_i and y_i, and n is the number of pairs of data.

In practice the preceding formula is also used when there are ties either among the x or y observations. The ranks for tied observations are assigned as in the signed-rank test by averaging the ranks that would have been assigned if the observations were distinguishable.

The value of r_S will usually be close to the value obtained by finding r based on numerical measurements and is interpreted in much the same way. As before, the values of r_S will range from -1 to $+1$. A value of $+1$ or -1 indicates perfect association between X and Y, the plus sign occurring for identical rankings and the minus sign occurring for reverse rankings. When r_S is close to zero, we would conclude that the variables are uncorrelated.

Example 14.8   The figures listed in Table 14.5, released by the Federal Trade Commission, show the milligrams of tar and nicotine found in 10 brands of cigarettes.

Table 14.5 Tar and Nicotine Contents

Cigarette Brand	Tar Content	Nicotine Content
Viceroy	14	0.9
Marlboro	17	1.1
Chesterfield	28	1.6
Kool	17	1.3
Kent	16	1.0
Raleigh	13	0.8
Old Gold	24	1.5
Philip Morris	25	1.4
Oasis	18	1.2
Players	31	2.0

Calculate the rank correlation coefficient to measure the degree of relationship between tar and nicotine content in cigarettes.

Solution. Let X and Y represent the tar and nicotine contents, respectively. First we assign ranks to each set of measurements with the rank of 1 assigned to the lowest number in each set, the rank of 2 to the second lowest number in each set, and so forth, until the rank of 10 is assigned to the largest number. Table 14.6 shows the individual rankings of the measurements and the differences in ranks for the 10 pairs of observations.

Substituting into the formula for r_S, we find that

$$r_S = 1 - \frac{(6)(5.5)}{(10)(100 - 1)} = 0.97,$$

Table 14.6 Rankings for Tar and Nicotine Contents

Cigarette Brand	x_i	y_i	d_i
Viceroy	2	2	0
Marlboro	4.5	4	0.5
Chesterfield	9	9	0
Kool	4.5	6	−1.5
Kent	3	3	0
Raleigh	1	1	0
Old Gold	7	8	−1
Philip Morris	8	7	1
Oasis	6	5	1
Players	10	10	0

indicating a high positive correlation between the amount of tar and nicotine found in cigarettes.

Some advantages in using r_S rather than r do exist. For instance, we no longer assume the underlying relationship between X and Y to be linear and therefore, when the data possess a distinct curvilinear relationship, the rank correlation coefficient will likely be more reliable than the conventional measure. A second advantage in using the rank correlation coefficient is the fact that no assumptions of normality are made concerning the distributions of X and Y. Perhaps the greatest advantage occurs when one is unable to make meaningful numerical measurements but nevertheless can establish rankings. Such is the case, for example, when different judges rank a group of individuals according to some attribute. The rank correlation coefficient can be used in this situation as a measure of the consistency of the two judges.

To test the hypothesis that $\rho = 0$ by using a rank correlation coefficient, one needs to consider the sampling distribution of the r_S values under the assumption of no correlation. Critical values for $\alpha = 0.05, 0.025, 0.01,$ and 0.05 have been calculated and are given in Table A.21. The setup of this table is similar to the table of critical values for the t distribution except for the left column, which now gives the number of pairs of observations rather than the degrees of freedom. Since the distribution of the r_S values is symmetric about zero when, $\rho = 0$, the r_S value that leaves an area of α to the left is equal to the negative of the r_S value that leaves an area of α to the right. For a two-sided alternative hypothesis, the critical region of size α falls equally in the two tails of the distribution. For a test in which the alternative hypothesis is negative, the critical region is entirely in the left tail of the distribution, and when the alternative is positive, the critical region is placed entirely in the right tail.

Example 14.9 Refer to Example 14.8 and test the hypothesis that the correlation between the amount of tar and nicotine found in cigarettes is zero against the alternative that it is greater than zero. Use a 0.01 level of significance.

Solution

1. H_0: $\rho = 0$.
2. H_1: $\rho > 0$.
3. $\alpha = 0.01$.
4. Critical region: $r_S > 0.745$, from Table A.21.
5. Computations: From Example 14.8, $r_S = 0.97$.
6. Decision: Reject H_0 and conclude that there is a significant correlation or relationship between the amount of tar and nicotine found in cigarettes.

Under the assumption of no correlation, it can be shown that the distribution of the r_S values approaches a normal distribution with a mean of 0 and a standard deviation of $1/\sqrt{n-1}$ as n increases. Consequently, when n exceeds the values given in Table A.21, one could test for a significant correlation by computing

$$z = \frac{r_s - 0}{1/\sqrt{n-1}} = r_s \sqrt{n-1}$$

and comparing with critical values of the standard normal distribution given in Table A.3 of the Appendix.

Exercises

1. A random sample of 15 adults living in a small town are selected to estimate the proportion of voters favoring a certain candidate for mayor. Each individual was also asked if he or she was a college graduate. By letting Y and N designate the responses of "yes" and "no" to the education question, the following sequence was obtained:

   $N\ N\ N\ N\ Y\ Y\ N\ Y\ Y\ N\ Y\ N\ N\ N\ N$

   Use the runs test at the 0.1 level of significance to determine if the sequence supports the contention that the sample was selected at random.

2. A silver-plating process is being used to coat a certain type of serving tray. When the process is in control, the thickness of the silver on the trays will vary randomly following a normal distribution with a mean of 0.02 millimeter and a standard deviation of 0.005 millimeter. Suppose that the next 12 trays examined show the following thicknesses of silver: 0.019, 0.021, 0.020, 0.019, 0.020, 0.018, 0.023, 0.021, 0.024, 0.022, 0.023, 0.022. Use the runs test to determine if the fluctuations in thickness from one tray to another is random. Let $\alpha = 0.05$.

3. Use the runs test to test whether there is a difference in the average operating time for the two calculators of Exercise 3 on page 547.

4. In an industrial production line, items are inspected periodically for defectives. The following is a sequence of defective items, D, and nondefective items, N, produced by this production line:

   $D\ D\ N\ N\ N\ D\ N\ N\ D\ D\ N\ N\ N$
   $N\ D\ D\ D\ N\ N\ D\ N\ N\ N\ N\ D\ N\ D$

   Use the large-sample theory for the runs test, with a significance level of 0.05, to determine whether the defectives are occurring at random or not.

5. Assuming that the measurements of Exercise 2 on page 59 were recorded in successive rows from left to right as they were collected, use the runs test, with $\alpha = 0.05$, to test the hypothesis that the data represent a random sequence.

6. How large a sample is required to be 95% confident that at least 85% of the distribution of measurements is included between the sample extremes?

7. What is the probability that the range of a random sample of size 24 includes at least 90% of the population?

8. How large a sample is required to be 99% confident that at least 80% of the population will be less than the largest observation in the sample?

9. What is the probability that at least 95% of a population will exceed the smallest value in a random sample of size $n = 135$?

10. The following table gives the recorded grades for 10 students on a midterm test and the final examination in a calculus course:

Student	Midterm Test	Final Examination
L.S.A.	84	73
W.P.B.	98	63
R.W.K.	91	87
J.R.L.	72	66
J.K.L.	86	78
D.L.P.	93	78
B.L.P.	80	91
D.W.M.	0	0
M.N.M.	92	88
R.H.S.	87	77

(a) Calculate the rank correlation coefficient.

(b) Test the null hypothesis that $\rho = 0$ against the alternative that $\rho > 0$. Use $\alpha = 0.025$.

11. With reference to the data of Exercise 1 on page 321,

(a) calculate the rank correlation coefficient;

(b) test the null hypothesis at the 0.05 level of significance that $\rho = 0$ against the alternative that $\rho \neq 0$. Compare your results with those obtained in Exercise 5 on page 352.

12. Calculate the rank correlation coefficient for the daily rainfall and amount of particulate removed in Exercise 9 on page 323.

13. The following data compare the rankings on November 4, 1981, of the top 15 major college football teams reported in the *Associated Press* poll with the rankings reported in the *United Press* poll:

Team	AP poll	UPI poll
Pittsburgh	1	1
Clemson	2	3
Southern Cal.	3	2
Georgia	4	4
Texas	5	5
Penn St.	6	6
Alabama	7	7
North Carolina	8	9
Nebraska	9	8
Michigan	10	10
Miami, Fla.	11	11
Florida St.	12	14
Mississippi St.	13	15
Washington	14	12
Oklahoma	15	13

(a) Calculate the rank correlation coefficient.

(b) Test the null hypothesis that $\rho = 0$ against the alternative that $\rho > 0$. Use a 0.01 level of significance.

14. With reference to the weights and chest sizes of infants in Exercise 4 on page 351.

(a) calculate the rank correlation coefficient;

(b) test the hypothesis at the 0.025 level of significance that $\rho = 0$ against the alternative that $\rho > 0$.

15. A consumer panel tested 9 makes of microwave ovens for overall quality. The ranks assigned by the panel and the suggested retail prices were as follows:

Manufacturer	Panel Rating	Suggested Price
A	6	$480
B	9	395
C	2	575
D	8	550
E	5	510
F	1	545
G	7	400
H	4	465
I	3	420

Is there a significant relationship between the quality and the price of a microwave oven? Use a 0.05 level of significance.

16. Two judges at a college homecoming parade ranked 8 floats in the following order:

	Float							
	1	2	3	4	5	6	7	8
Judge A	5	8	4	3	6	2	7	1
Judge B	7	5	4	2	8	1	6	3

(a) Calculate the rank correlation.

(b) Test the null hypothesis that $\rho = 0$ against the alternative that $\rho > 0$. Use $\alpha = 0.05$.

17. In the article called *Risky Assumptions* by Paul Slovic, Baruch Fischoff, and Sarah Lichtenstein, published in *Psychology Today* (June 1980), the risk of dying in the United States from 30 activities and technologies are ranked by members of the League of Women Voters and also by experts who are professionally involved in assessing risks.

The rankings are as follows:

Activity or Technology Risk	Voters	Experts
Nuclear power	1	20
Motor vehicles	2	1
Handguns	3	4
Smoking	4	2
Motorcycles	5	6
Alcoholic beverages	6	3
Private aviation	7	12
Police work	8	17
Pesticides	9	8
Surgery	10	5
Fire fighting	11	18
Large construction	12	13
Hunting	13	23
Spray cans	14	26
Mountain climbing	15	29
Bicycles	16	15
Commercial aviation	17	16
Electric power	18	9

Activity or Technology Risk	Voters	Experts
Swimming	19	10
Contraceptives	20	11
Skiing	21	30
X-rays	22	7
Football	23	27
Railroads	24	19
Food preservatives	25	14
Food coloring	26	21
Power mowers	27	28
Antibiotics	28	24
Home appliances	29	22
Vaccinations	30	25

(a) Calculate the rank correlation coefficient.

(b) Test the null hypothesis of zero correlation between the rankings of the League of Women Voters and the experts against the alternative that the correlation is not zero. Use a 0.05 level of significance.

Bibliography

BOWKER, A. H., and G. J. LIEBERMAN. *Engineering Statistics*, 2nd ed. Englewood Cliffs, N.J.: Prentice-Hall, Inc., 1972.

BOX, G. E. P., W. G. HUNTER, and J. S. HUNTER. *Statistics for Experimenters*. New York: John Wiley & Sons, Inc., 1978.

BROWNLEE, K. A. *Statistical Theory and Methodology in Science and Engineering*, 2nd ed. New York: John Wiley & Sons, Inc., 1965.

CHATTERJEE, S., and B. PRICE. *Regression Analysis by Example*. New York: John Wiley & Sons, Inc., 1977.

COOK, R. D., and S. WEISBERG. *Residuals and Influence in Regression*. New York: Chapman and Hall, 1982.

DANIEL, C., and F. WOOD. *Fitting Equations to Data*, 2nd ed. New York: John Wiley & Sons, Inc., 1980.

DANIEL, W. W. *Applied Nonparametric Statistics*. Boston: Houghton Mifflin Company, 1978.

DERMAN, C., L. GLASER, and I. OLKIN. *Probability Models and Applications*. New York: Macmillan Publishing Company, 1980.

DEVORE, J. L. *Probability and Statistics for Engineering and the Sciences*. Monterey, Calif.: Brooks/Cole Publishing Co., 1982.

DIXON, W. J., and F. J. MASSEY, JR. *Introduction to Statistical Analysis*, 3rd ed. New York: McGraw-Hill Book Company, 1969.

DRAPER, N., and H. SMITH. *Applied Regression Analysis*, 2nd ed. New York: John Wiley & Sons, Inc., 1981.

DYER, D. D., and J. P. KEATING. "On the Determination of Critical Values for Bartlett's Test." *J. Am. Stat. Assoc.*, Vol. 75, 1980.

FREUND, J. E., and R. E. WALPOLE. *Mathematical Statistics*, 3rd ed. Englewood Cliffs, N.J.: Prentice-Hall, Inc., 1980.

GUNST, R. F., and R. L. MASON. *Regression Analysis and Its Application: A Data-Oriented Approach.* New York: Marcel Dekker, Inc., 1980.

GUTTMAN, I., and S. S. WILKS. *Introductory Engineering Statistics.* New York: John Wiley & Sons, Inc., 1965.

HICKS, C. R. *Fundamental Concepts in the Design of Experiments*, 2nd ed. New York: Holt, Rinehart and Winston, 1973.

HOCKING, R. R. "The Analysis and Selection of Variables in Linear Regression." *Biometrics*, Vol. 32, 1976.

HOERL, A. E., and R. W. KENNARD. "Ridge Regression: Applications to Nonorthogonal Problems." *Technometrics*, Vol. 12, no. 1, 1970.

HOGG, R. V., and A. T. CRAIG. *Introduction to Mathematical Statistics*, 4th ed. New York: Macmillan Publishing Company, 1978.

HOLLANDER, M., and D. WOLFE. *Nonparametric Statistical Methods.* Boston: Houghton Mifflin Company, 1973.

JOHNSON, N. L., and F. C. LEONE. *Statistics and Experimental Design: In Engineering and the Physical Sciences*, Vols. I and II, 2nd ed. New York: John Wiley & Sons, Inc., 1977.

KOOPMANS, L. H. *An Introduction to Contemporary Statistics.* Boston: Duxbury Press, 1981.

LARSEN, R. J., and M. L. MORRIS. *Introduction to Mathematical Statistics*, Englewood Cliffs, N.J.: Prentice-Hall, Inc., 1981.

LEHMANN, E. *Nonparametrics: Statistical Methods Based on Ranks.* San Francisco: Holden-Day, Inc., 1975.

LI, C. C. *Introduction to Experimental Statistics.* New York: McGraw-Hill Book Company, 1964.

McLAVE, J., and F. DIETRICH. *Statistics.* San Francisco: Dellen Publishing Co., 1978.

MENDENHALL, W. *An Introduction to Linear Models and the Design and Analysis of Experiments.* Belmont, Calif.: Wadsworth Publishing Co., 1968.

MILLER, I., and J. E. FREUND. *Probability and Statistics for Engineers*, 3rd ed. Englewood Cliffs, N.J.: Prentice-Hall, Inc., 1985.

MONTGOMERY, D. C., and E. A. PECK. *Introduction to Linear Regression Analysis.* New York: John Wiley & Sons, Inc., 1982.

MOSTELLER, F., and J. TUKEY. *Data Analysis and Regression.* Reading, Mass.: Addison-Wesley Publishing Co., Inc., 1977.

MYERS, R. H. *Response Surface Methodology.* Boston: Allyn & Bacon, Inc., 1971.

NETER, J., W. WASSERMAN, and G. A. WHITMORE. *Applied Statistics.* Boston: Allyn & Bacon, Inc., 1978.

NOETHER, G. E. *Introduction to Statistics: A Nonparametric Approach*, 2nd ed. Boston: Houghton Mifflin Company, 1976.

OTT, L. *An Introduction to Statistical Methods and Data Analysis.* Boston: Duxbury Press, 1977.

ROSS, S. *Introduction to Applied Probability Models*, 2nd ed. New York: Academic Press, Inc., 1980.

SNEDECOR, G., and W. G. COCHRAN. *Statistical Methods*, 7th ed. Ames, Iowa: The Iowa State University Press, 1980.

STEEL, R. G. D., and J. H. TORRIE. *Principles and Procedures of Statistics*, 2nd ed. New York: McGraw-Hill Book Company, 1979.

THOMPSON, W. O., and F. B. CADY. *Proceedings of the University of Kentucky Conference on Regression with a Large Number of Predictor Variables*, Lexington, 1973.

TUKEY, T. W. *Exploratory Data Analysis.* Reading, Mass.: Addison-Wesley Publishing Co., Inc., 1977.

WALPOLE, R. E. *Introduction to Statistics,* 3rd ed. New York: Macmillan Publishing Company, 1982.

WINER, B. J. *Statistical Principles in Experimental Design,* 2nd ed. New York: McGraw-Hill Book Company, 1971.

YOUNGER, M. S. *A Handbook for Linear Regression.* Boston: Duxbury Press, 1979.

Appendix: Statistical Tables

Table A.1 Binomial Probability Sums $\sum_{x=0}^{r} b(x; n, p)$

							p				
n	r	.10	.20	.25	.30	.40	.50	.60	.70	.80	.90
1	0	.9000	.8000	.7500	.7000	.6000	.5000	.4000	.3000	.2000	.1000
	1	1.0000	1.0000	1.0000	1.0000	1.0000	1.0000	1.0000	1.0000	1.0000	1.0000
2	0	.8100	.6400	.5625	.4900	.3600	.2500	.1600	.0900	.0400	.0100
	1	.9900	.9600	.9375	.9100	.8400	.7500	.6400	.5100	.3600	.1900
	2	1.0000	1.0000	1.0000	1.0000	1.0000	1.0000	1.0000	1.0000	1.0000	1.0000
3	0	.7290	.5120	.4219	.3430	.2160	.1250	.0640	.0270	.0080	.0010
	1	.9720	.8960	.8438	.7840	.6480	.5000	.3520	.2160	.1040	.0280
	2	.9990	.9920	.9844	.9730	.9360	.8750	.7840	.6570	.4880	.2710
	3	1.0000	1.0000	1.0000	1.0000	1.0000	1.0000	1.0000	1.0000	1.0000	1.0000
4	0	.6561	.4096	.3164	.2401	.1296	.0625	.0256	.0081	.0016	.0001
	1	.9477	.8192	.7383	.6517	.4752	.3125	.1792	.0837	.0272	.0037
	2	.9963	.9728	.9492	.9163	.8208	.6875	.5248	.3483	.1808	.0523
	3	.9999	.9984	.9961	.9919	.9744	.9375	.8704	.7599	.5904	.3439
	4	1.0000	1.0000	1.0000	1.0000	1.0000	1.0000	1.0000	1.0000	1.0000	1.0000
5	0	.5905	.3277	.2373	.1681	.0778	.0312	.0102	.0024	.0003	.0000
	1	.9185	.7373	.6328	.5282	.3370	.1875	.0870	.0308	.0067	.0005
	2	.9914	.9421	.8965	.8369	.6826	.5000	.3174	.1631	.0579	.0086
	3	.9995	.9933	.9844	.9692	.9130	.8125	.6630	.4718	.2627	.0815
	4	1.0000	.9997	.9990	.9976	.9898	.9688	.9222	.8319	.6723	.4095
	5		1.0000	1.0000	1.0000	1.0000	1.0000	1.0000	1.0000	1.0000	1.0000
6	0	.5314	.2621	.1780	.1176	.0467	.0156	.0041	.0007	.0001	.0000
	1	.8857	.6554	.5339	.4202	.2333	.1094	.0410	.0109	.0016	.0001
	2	.9841	.9011	.8306	.7443	.5443	.3438	.1792	.0705	.0170	.0013
	3	.9987	.9830	.9624	.9295	.8208	.6563	.4557	.2557	.0989	.0158
	4	.9999	.9984	.9954	.9891	.9590	.8906	.7667	.5798	.3447	.1143
	5	1.0000	.9999	.9998	.9993	.9959	.9844	.9533	.8824	.7379	.4686
	6		1.0000	1.0000	1.0000	1.0000	1.0000	1.0000	1.0000	1.0000	1.0000
7	0	.4783	.2097	.1335	.0824	.0280	.0078	.0016	.0002	.0000	
	1	.8503	.5767	.4449	.3294	.1586	.0625	.0188	.0038	.0004	.0000
	2	.9743	.8520	.7564	.6471	.4199	.2266	.0963	.0288	.0047	.0002
	3	.9973	.9667	.9294	.8740	.7102	.5000	.2898	.1260	.0333	.0027
	4	.9998	.9953	.9871	.9712	.9037	.7734	.5801	.3529	.1480	.0257
	5	1.0000	.9996	.9987	.9962	.9812	.9375	.8414	.6706	.4233	.1497
	6		1.0000	.9999	.9998	.9984	.9922	.9720	.9176	.7903	.5217
	7			1.0000	1.0000	1.0000	1.0000	1.0000	1.0000	1.0000	1.0000

Table A.1 (*continued*) Binomial Probability Sums $\sum_{x=0}^{r} b(x; n, p)$

n	r	.10	.20	.25	.30	.40	.50	.60	.70	.80	.90
8	0	.4305	.1678	.1001	.0576	.0168	.0039	.0007	.0001	.0000	
	1	.8131	.5033	.3671	.2553	.1064	.0352	.0085	.0013	.0001	
	2	.9619	.7969	.6785	.5518	.3154	.1445	.0498	.0113	.0012	.0000
	3	.9950	.9437	.8862	.8059	.5941	.3633	.1737	.0580	.0104	.0004
	4	.9996	.9896	.9727	.9420	.8263	.6367	.4059	.1941	.0563	.0050
	5	1.0000	.9988	.9958	.9887	.9502	.8555	.6846	.4482	.2031	.0381
	6		.9991	.9996	.9987	.9915	.9648	.8936	.7447	.4967	.1869
	7		1.0000	1.0000	.9999	.9993	.9961	.9832	.9424	.8322	.5695
	8				1.0000	1.0000	1.0000	1.0000	1.0000	1.0000	1.0000
9	0	.3874	.1342	.0751	.0404	.0101	.0020	.0003	.0000		
	1	.7748	.4362	.3003	.1960	.0705	.0195	.0038	.0004	.0000	
	2	.9470	.7382	.6007	.4628	.2318	.0898	.0250	.0043	.0003	.0000
	3	.9917	.9144	.8343	.7297	.4826	.2539	.0994	.0253	.0031	.0001
	4	.9991	.9804	.9511	.9012	.7334	.5000	.2666	.0988	.0196	.0009
	5	.9999	.9969	.9900	.9747	.9006	.7461	.5174	.2703	.0856	.0083
	6	1.0000	.9997	.9987	.9957	.9750	.9102	.7682	.5372	.2618	.0530
	7		1.0000	.9999	.9996	.9962	.9805	.9295	.8040	.5638	.2252
	8			1.0000	1.0000	.9997	.9980	.9899	.9596	.8658	.6126
	9					1.0000	1.0000	1.0000	1.0000	1.0000	1.0000
10	0	.3487	.1074	.0563	.0282	.0060	.0010	.0001	.0000		
	1	.7361	.3758	.2440	.1493	.0464	.0107	.0017	.0001	.0000	
	2	.9298	.6778	.5256	.3828	.1673	.0547	.0123	.0016	.0001	
	3	.9872	.8791	.7759	.6496	.3823	.1719	.0548	.0106	.0009	.0000
	4	.9984	.9672	.9219	.8497	.6331	.3770	.1662	.0474	.0064	.0002
	5	.9999	.9936	.9803	.9527	.8338	.6230	.3669	.1503	.0328	.0016
	6	1.0000	.9991	.9965	.9894	.9452	.8281	.6177	.3504	.1209	.0128
	7		.9999	.9996	.9984	.9877	.9453	.8327	.6172	.3222	.0702
	8		1.0000	1.0000	.9999	.9983	.9893	.9536	.8507	.6242	.2639
	9				1.0000	.9999	.9990	.9940	.9718	.8926	.6513
	10					1.0000	1.0000	1.0000	1.0000	1.0000	1.0000
11	0	.3138	.0859	.0422	.0198	.0036	.0005	.0000			
	1	.6974	.3221	.1971	.1130	.0302	.0059	.0007	.0000		
	2	.9104	.6174	.4552	.3127	.1189	.0327	.0059	.0006	.0000	
	3	.9815	.8369	.7133	.5696	.2963	.1133	.0293	.0043	.0002	
	4	.9972	.9496	.8854	.7897	.5328	.2744	.0994	.0216	.0020	.0000
	5	.9997	.9883	.9657	.9218	.7535	.5000	.2465	.0782	.0117	.0003
	6	1.0000	.9980	.9924	.9784	.9006	.7256	.4672	.2103	.0504	.0028
	7		.9998	.9988	.9957	.9707	.8867	.7037	.4304	.1611	.0185
	8		1.0000	.9999	.9994	.9941	.9673	.8811	.6873	.3826	.0896
	9			1.0000	1.0000	.9993	.9941	.9698	.8870	.6779	.3026
	10					1.0000	.9995	.9964	.9802	.9141	.6862
	11						1.0000	1.0000	1.0000	1.0000	1.0000

Table A.1 (*continued*) Binomial Probability Sums $\sum\limits_{x=0}^{r} b(x; n, p)$

							p				
n	*r*	.10	.20	.25	.30	.40	.50	.60	.70	.80	.90
12	0	.2824	.0687	.0317	.0138	.0022	.0002	.0000			
	1	.6590	.2749	.1584	.0850	.0196	.0032	.0003	.0000		
	2	.8891	.5583	.3907	.2528	.0834	.0193	.0028	.0002	.0000	
	3	.9744	.7946	.6488	.4925	.2253	.0730	.0153	.0017	.0001	
	4	.9957	.9274	.8424	.7237	.4382	.1938	.0573	.0095	.0006	.0000
	5	.9995	.9806	.9456	.8821	.6652	.3872	.1582	.0386	.0039	.0001
	6	.9999	.9961	.9857	.9614	.8418	.6128	.3348	.1178	.0194	.0005
	7	1.0000	.9994	.9972	.9905	.9427	.8062	.5618	.2763	.0726	.0043
	8		.9999	.9996	.9983	.9847	.9270	.7747	.5075	.2054	.0256
	9		1.0000	1.0000	.9998	.9972	.9807	.9166	.7472	.4417	.1109
	10				1.0000	.9997	.9968	.9804	.9150	.7251	.3410
	11					1.0000	.9998	.9978	.9862	.9313	.7176
	12						1.0000	1.0000	1.0000	1.0000	1.0000
13	0	.2542	.0550	.0238	.0097	.0013	.0001	.0000			
	1	.6213	.2336	.1267	.0637	.0126	.0017	.0001	.0000		
	2	.8661	.5017	.3326	.2025	.0579	.0112	.0013	.0001		
	3	.9658	.7473	.5843	.4206	.1686	.0461	.0078	.0007	.0000	
	4	.9935	.9009	.7940	.6543	.3530	.1334	.0321	.0040	.0002	
	5	.9991	.9700	.9198	.8346	.5744	.2905	.0977	.0182	.0012	.0000
	6	.9999	.9930	.9757	.9376	.7712	.5000	.2288	.0624	.0070	.0001
	7	1.0000	.9980	.9944	.9818	.9023	.7095	.4256	.1654	.0300	.0009
	8		.9998	.9990	.9960	.9679	.8666	.6470	.3457	.0991	.0065
	9		1.0000	.9999	.9993	.9922	.9539	.8314	.5794	.2527	.0342
	10			1.0000	.9999	.9987	.9888	.9421	.7975	.4983	.1339
	11				1.0000	.9999	.9983	.9874	.9363	.7664	.3787
	12					1.0000	.9999	.9987	.9903	.9450	.7458
	13						1.0000	1.0000	1.0000	1.0000	1.0000
14	0	.2288	.0440	.0178	.0068	.0008	.0001	.0000			
	1	.5846	.1979	.1010	.0475	.0081	.0009	.0001			
	2	.8416	.4481	.2811	.1608	.0398	.0065	.0006	.0000		
	3	.9559	.6982	.5213	.3552	.1243	.0287	.0039	.0002		
	4	.9908	.8702	.7415	.5842	.2793	.0898	.0175	.0017	.0000	
	5	.9985	.9561	.8883	.7805	.4859	.2120	.0583	.0083	.0004	
	6	.9998	.9884	.9617	.9067	.6925	.3953	.1501	.0315	.0024	.0000
	7	1.0000	.9976	.9897	.9685	.8499	.6047	.3075	.0933	.0116	.0002
	8		.9996	.9978	.9917	.9417	.7880	.5141	.2195	.0439	.0015
	9		1.0000	.9997	.9983	.9825	.9102	.7207	.4158	.1298	.0092
	10			1.0000	.9998	.9961	.9713	.8757	.6448	.3018	.0441
	11				1.0000	.9994	.9935	.9602	.8392	.5519	.1584
	12					.9999	.9991	.9919	.9525	.8021	.4154
	13					1.0000	.9999	.9992	.9932	.9560	.7712
	14						1.0000	1.0000	1.0000	1.0000	1.0000

Table A.1 (*continued*) Binomial Probability Sums $\sum_{x=0}^{r} b(x; n, p)$

n	r	.10	.20	.25	.30	.40	.50	.60	.70	.80	.90
15	0	.2059	.0352	.0134	.0047	.0005	.0000				
	1	.5490	.1671	.0802	.0353	.0052	.0005	.0000			
	2	.8159	.3980	.2361	.1268	.0271	.0037	.0003	.0000		
	3	.9444	.6482	.4613	.2969	.0905	.0176	.0019	.0001		
	4	.9873	.8358	.6865	.5155	.2173	.0592	.0094	.0007	.0000	
	5	.9978	.9389	.8516	.7216	.4032	.1509	.0338	.0037	.0001	
	6	.9997	.9819	.9434	.8689	.6098	.3036	.0951	.0152	.0008	
	7	1.0000	.9958	.9827	.9500	.7869	.5000	.2131	.0500	.0042	.0000
	8		.9992	.9958	.9848	.9050	.6964	.3902	.1311	.0181	.0003
	9		.9999	.9992	.9963	.9662	.8491	.5968	.2784	.0611	.0023
	10		1.0000	.9999	.9993	.9907	.9408	.7827	.4845	.1642	.0127
	11			1.0000	.9999	.9981	.9824	.9095	.7031	.3518	.0556
	12				1.0000	.9997	.9963	.9729	.8732	.6020	.1841
	13					1.0000	.9995	.9948	.9647	.8329	.4510
	14						1.0000	.9995	.9953	.9648	.7941
	15							1.0000	1.0000	1.0000	1.0000
16	0	.1853	.0281	.0100	.0033	.0003	.0000				
	1	.5147	.1407	.0635	.0261	.0033	.0003	.0000			
	2	.7892	.3518	.1971	.0994	.0183	.0021	.0001			
	3	.9316	.5981	.4050	.2459	.0651	.0106	.0009	.0000		
	4	.9830	.7982	.6302	.4499	.1666	.0384	.0049	.0003		
	5	.9967	.9183	.8103	.6598	.3288	.1051	.0191	.0016	.0000	
	6	.9995	.9733	.9204	.8247	.5272	.2272	.0583	.0071	.0002	
	7	.9999	.9930	.9729	.9256	.7161	.4018	.1423	.0257	.0015	.0000
	8	1.0000	.9985	.9925	.9743	.8577	.5982	.2839	.0744	.0070	.0001
	9		.9998	.9984	.9929	.9417	.7728	.4728	.1753	.0267	.0005
	10		1.0000	.9997	.9984	.9809	.8949	.6712	.3402	.0817	.0033
	11			1.0000	.9997	.9951	.9616	.8334	.5501	.2018	.0170
	12				1.0000	.9991	.9894	.9349	.7541	.4019	.0684
	13					.9999	.9979	.9817	.9006	.6482	.2108
	14					1.0000	.9997	.9967	.9739	.8593	.4853
	15						1.0000	.9997	.9967	.9719	.8147
	16							1.0000	1.0000	1.0000	1.0000

Table A.1 (*continued*) Binomial Probability Sums $\sum_{x=0}^{r} b(x; n, p)$

							p					
n	r	.10	.20	.25	.30	.40	.50	.60	.70	.80	.90
17	0	.1668	.0225	.0075	.0023	.0002	.0000				
	1	.4818	.1182	.0501	.0193	.0021	.0001	.0000			
	2	.7618	.3096	.1637	.0774	.0123	.0012	.0001			
	3	.9174	.5489	.3530	.2019	.0464	.0064	.0005	.0000		
	4	.9779	.7582	.5739	.3887	.1260	.0245	.0025	.0001		
	5	.9953	.8943	.7653	.5968	.2639	.0717	.0106	.0007	.0000	
	6	.9992	.9623	.8929	.7752	.4478	.1662	.0348	.0032	.0001	
	7	.9999	.9891	.9598	.8954	.6405	.3145	.0919	.0127	.0005	
	8	1.0000	.9974	.9876	.9597	.8011	.5000	.1989	.0403	.0026	.0000
	9		.9995	.9969	.9873	.9081	.6855	.3595	.1046	.0109	.0001
	10		.9999	.9994	.9968	.9652	.8338	.5522	.2248	.0377	.0008
	11		1.0000	.9999	.9993	.9894	.9283	.7361	.4032	.1057	.0047
	12			1.0000	.9999	.9975	.9755	.8740	.6113	.2418	.0221
	13				1.0000	.9995	.9936	.9536	.7981	.4511	.0826
	14					.9999	.9988	.9877	.9226	.6904	.2382
	15					1.0000	.9999	.9979	.9807	.8818	.5182
	16						1.0000	.9998	.9977	.9775	.8332
	17							1.0000	1.0000	1.0000	1,0000
18	0	.1501	.0180	.0056	.0016	.0001	.0000				
	1	.4503	.0991	.0395	.0142	.0013	.0001				
	2	.7338	.2713	.1353	.0600	.0082	.0007	.0000			
	3	.9018	.5010	.3057	.1646	.0328	.0038	.0002			
	4	.9718	.7164	.5787	.3327	.0942	.0154	.0013	.0000		
	5	.9936	.8671	.7175	.5344	.2088	.0481	.0058	.0003		
	6	.9988	.9487	.8610	.7217	.3743	.1189	.0203	.0014	.0000	
	7	.9998	.9837	.9431	.8593	.5634	.2403	.0576	.0061	.0002	
	8	1.0000	.9957	.9807	.9404	.7368	.4073	.1347	.0210	.0009	
	9		.9991	.9946	.9790	.8653	.5927	.2632	.0596	.0043	.0000
	10		.9998	.9988	.9939	.9424	.7597	.4366	.1407	.0163	.0002
	11		1.0000	.9998	.9986	.9797	.8811	.6257	.2783	.0513	.0012
	12			1.0000	.9997	.9942	.9519	.7912	.4656	.1329	.0064
	13				1.0000	.9987	.9846	.9058	.6673	.2836	.0282
	14					.9998	.9962	.9672	.8354	.4990	.0982
	15					1.0000	.9993	.9918	.9400	.7287	.2662
	16						.9999	.9987	.9858	.9009	.5497
	17						1.0000	.9999	.9984	.9820	.8499
	18							1.0000	1.0000	1.0000	1.0000

Table A.1 (*continued*) Binomial Probability Sums $\sum_{x=0}^{r} b(x; n, p)$

n	r						p				
		.10	.20	.25	.30	.40	.50	.60	.70	.80	.90
19	0	.1351	.0144	.0042	.0011	.0001					
	1	.4203	.0829	.0310	.0104	.0008	.0000				
	2	.7054	.2369	.1113	.0462	.0055	.0004	.0000			
	3	.8850	.4551	.2631	.1332	.0230	.0022	.0001			
	4	.9648	.6733	.4654	.2822	.0696	.0096	.0006	.0000		
	5	.9914	.8369	.6678	.4739	.1629	.0318	.0031	.0001		
	6	.9983	.9324	.8251	.6655	.3081	.0835	.0116	.0006		
	7	.9997	.9767	.9225	.8180	.4878	.1796	.0352	.0028	.0000	
	8	1.0000	.9933	.9713	.9161	.6675	.3238	.0885	.0105	.0003	
	9		.9984	.9911	.9674	.8139	.5000	.1861	.0326	.0016	
	10		.9997	.9977	.9895	.9115	.6762	.3325	.0839	.0067	.0000
	11		.9999	.9995	.9972	.9648	.8204	.5122	.1820	.0233	.0003
	12		1.0000	.9999	.9994	.9884	.9165	.6919	.3345	.0676	.0017
	13			1.0000	.9999	.9969	.9682	.8371	.5261	.1631	.0086
	14				1.0000	.9994	.9904	.9304	.7178	.3267	.0352
	15					.9999	.9978	.9770	.8668	.5449	.1150
	16					1.0000	.9996	.9945	.9538	.7631	.2946
	17						1.0000	.9992	.9896	.9171	.5797
	18							.9999	.9989	.9856	.8649
	19							1.0000	1.0000	1.0000	1.0000
20	0	.1216	.0115	.0032	.0008	.0000					
	1	.3917	.0692	.0243	.0076	.0005	.0000				
	2	.6769	.2061	.0913	.0355	.0036	.0002	.0000			
	3	.8670	.4114	.2252	.1071	.0160	.0013	.0001			
	4	.9568	.6296	.4148	.2375	.0510	.0059	.0003			
	5	.9887	.8042	.6172	.4164	.1256	.0207	.0016	.0000		
	6	.9976	.9133	.7858	.6080	.2500	.0577	.0065	.0003		
	7	.9996	.9679	.8982	.7723	.4159	.1316	.0210	.0013	.0000	
	8	.9999	.9900	.9591	.8867	.5956	.2517	.0565	.0051	.0001	
	9	1.0000	.9974	.9861	.9520	.7553	.4119	.1275	.0171	.0006	
	10		.9994	.9961	.9829	.8725	.5881	.2447	.0480	.0026	.0000
	11		.9999	.9991	.9949	.9435	.7483	.4044	.1133	.0100	.0001
	12		1.0000	.9998	.9987	.9790	.8684	.5841	.2277	.0321	.0004
	13			1.0000	.9997	.9935	.9423	.7500	.3920	.0867	.0024
	14				1.0000	.9984	.9793	.8744	.5836	.1958	.0113
	15					.9997	.9941	.9490	.7625	.3704	.0432
	16					1.0000	.9987	.9840	.8929	.5886	.1330
	17						.9998	.9964	.9645	.7939	.3231
	18						1.0000	.9995	.9924	.9308	.6083
	19							1.0000	.9992	.9885	.8784
	20								1.0000	1.0000	1.0000

Table A.2 Poisson Probability Sums $\sum\limits_{x=0}^{r} p(x; \mu)$

					μ				
r	0.1	0.2	0.3	0.4	0.5	0.6	0.7	0.8	0.9
0	0.9048	0.8187	0.7408	0.6730	0.6065	0.5488	0.4966	0.4493	0.4066
1	0.9953	0.9825	0.9631	0.9384	0.9098	0.8781	0.8442	0.8088	0.7725
2	0.9998	0.9989	0.9964	0.9921	0.9856	0.9769	0.9659	0.9526	0.9371
3	1.0000	0.9999	0.9997	0.9992	0.9982	0.9966	0.9942	0.9909	0.9865
4		1.0000	1.0000	0.9999	0.9998	0.9996	0.9992	0.9986	0.9977
5				1.0000	1.0000	1.0000	0.9999	0.9998	0.9997
6							1.0000	1.0000	1.0000

					μ				
r	1.0	1.5	2.0	2.5	3.0	3.5	4.0	4.5	5.0
0	0.3679	0.2231	0.1353	0.0821	0.0498	0.0302	0.0183	0.0111	0.0067
1	0.7358	0.5578	0.4060	0.2873	0.1991	0.1359	0.0916	0.0611	0.0404
2	0.9197	0.8088	0.6767	0.5438	0.4232	0.3208	0.2381	0.1736	0.1247
3	0.9810	0.9344	0.8571	0.7576	0.6472	0.5366	0.4335	0.3423	0.2650
4	0.9963	0.9814	0.9473	0.8912	0.8153	0.7254	0.6288	0.5321	0.4405
5	0.9994	0.9955	0.9834	0.9580	0.9161	0.8576	0.7851	0.7029	0.6160
6	0.9999	0.9991	0.9955	0.9858	0.9665	0.9347	0.8893	0.8311	0.7622
7	1.0000	0.9998	0.9989	0.9958	0.9881	0.9733	0.9489	0.9134	0.8666
8		1.0000	0.9998	0.9989	0.9962	0.9901	0.9786	0.9597	0.9319
9			1.0000	0.9997	0.9989	0.9967	0.9919	0.9829	0.9682
10				0.9999	0.9997	0.9990	0.9972	0.9933	0.9863
11				1.0000	0.9999	0.9997	0.9991	0.9976	0.9945
12					1.0000	0.9999	0.9997	0.9992	0.9980
13						1.0000	0.9999	0.9997	0.9993
14							1.0000	0.9999	0.9998
15								1.0000	0.9999
16									1.0000

Table A.2 (*continued*) Poisson Probability Sums $\sum\limits_{x=0}^{r} p(x;\mu)$

	μ								
r	5.5	6.0	6.5	7.0	7.5	8.0	8.5	9.0	9.5
0	0.0041	0.0025	0.0015	0.0009	0.0006	0.0003	0.0002	0.0001	0.0001
1	0.0266	0.0174	0.0113	0.0073	0.0047	0.0030	0.0019	0.0012	0.0008
2	0.0884	0.0620	0.0430	0.0296	0.0203	0.0138	0.0093	0.0062	0.0042
3	0.2017	0.1512	0.1118	0.0818	0.0591	0.0424	0.0301	0.0212	0.0149
4	0.3575	0.2851	0.2237	0.1730	0.1321	0.0996	0.0744	0.0550	0.0403
5	0.5289	0.4457	0.3690	0.3007	0.2414	0.1912	0.1496	0.1157	0.0885
6	0.6860	0.6063	0.5265	0.4497	0.3782	0.3134	0.2562	0.2068	0.1649
7	0.8095	0.7440	0.6728	0.5987	0.5246	0.4530	0.3856	0.3239	0.2687
8	0.8944	0.8472	0.7916	0.7291	0.6620	0.5925	0.5231	0.4557	0.3918
9	0.9462	0.9161	0.8774	0.8305	0.7764	0.7166	0.6530	0.5874	0.5218
10	0.9747	0.9574	0.9332	0.9015	0.8622	0.8159	0.7634	0.7060	0.6453
11	0.9890	0.9799	0.9661	0.9466	0.9208	0.8881	0.8487	0.8030	0.7520
12	0.9955	0.9912	0.9840	0.9730	0.9573	0.9362	0.9091	0.8758	0.8364
13	0.9983	0.9964	0.9929	0.9872	0.9784	0.9658	0.9486	0.9261	0.8981
14	0.9994	0.9986	0.9970	0.9943	0.9897	0.9827	0.9726	0.9585	0.9400
15	0.9998	0.9995	0.9988	0.9976	0.9954	0.9918	0.9862	0.9780	0.9665
16	0.9999	0.9998	0.9996	0.9990	0.9980	0.9963	0.9934	0.9889	0.9823
17	1.0000	0.9999	0.9998	0.9996	0.9992	0.9984	0.9970	0.9947	0.9911
18		1.0000	0.9999	0.9999	0.9997	0.9994	0.9987	0.9976	0.9957
19			1.0000	1.0000	0.9999	0.9997	0.9995	0.9989	0.9980
20					1.0000	0.9999	0.9998	0.9996	0.9991
21						1.0000	0.9999	0.9998	0.9996
22							1.0000	0.9999	0.9999
23								1.0000	0.9999
24									1.0000

Table A.2 (*continued*) Poisson Probability Sums $\sum\limits_{x=0}^{r} p(x;\mu)$

	μ								
r	10.0	11.0	12.0	13.0	14.0	15.0	16.0	17.0	18.0
0	0.0000	0.0000	0.0000						
1	0.0005	0.0002	0.0001	0.0000	0.0000				
2	0.0028	0.0012	0.0005	0.0002	0.0001	0.0000	0.0000		
3	0.0103	0.0049	0.0023	0.0010	0.0005	0.0002	0.0001	0.0000	0.0000
4	0.0293	0.0151	0.0076	0.0037	0.0018	0.0009	0.0004	0.0002	0.0001
5	0.0671	0.0375	0.0203	0.0107	0.0055	0.0028	0.0014	0.0007	0.0003
6	0.1301	0.0786	0.0458	0.0259	0.0142	0.0076	0.0040	0.0021	0.0010
7	0.2202	0.1432	0.0895	0.0540	0.0316	0.0180	0.0100	0.0054	0.0029
8	0.3328	0.2320	0.1550	0.0998	0.0621	0.0374	0.0220	0.0126	0.0071
9	0.4579	0.3405	0.2424	0.1658	0.1094	0.0699	0.0433	0.0261	0.0154
10	0.5830	0.4599	0.3472	0.2517	0.1757	0.1185	0.0774	0.0491	0.0304
11	0.6968	0.5793	0.4616	0.3532	0.2600	0.1848	0.1270	0.0847	0.0549
12	0.7916	0.6887	0.5760	0.4631	0.3585	0.2676	0.1931	0.1350	0.0917
13	0.8645	0.7813	0.6815	0.5730	0.4644	0.3632	0.2745	0.2009	0.1426
14	0.9165	0.8540	0.7720	0.6751	0.5704	0.4657	0.3675	0.2808	0.2081
15	0.9513	0.9074	0.8444	0.7636	0.6694	0.5681	0.4667	0.3715	0.2867
16	0.9730	0.9441	0.8987	0.8355	0.7559	0.6641	0.5660	0.4677	0.3750
17	0.9857	0.9678	0.9370	0.8905	0.8272	0.7489	0.6593	0.5640	0.4686
18	0.9928	0.9823	0.9626	0.9302	0.8826	0.8195	0.7423	0.6550	0.5622
19	0.9965	0.9907	0.9787	0.9573	0.9235	0.8752	0.8122	0.7363	0.6509
20	0.9984	0.9953	0.9884	0.9750	0.9521	0.9170	0.8682	0.8055	0.7307
21	0.9993	0.9977	0.9939	0.9859	0.9712	0.9469	0.9108	0.8615	0.7991
22	0.9997	0.9990	0.9970	0.9924	0.9833	0.9673	0.9418	0.9047	0.8551
23	0.9999	0.9995	0.9985	0.9960	0.9907	0.9805	0.9633	0.9367	0.8989
24	1.0000	0.9998	0.9993	0.9980	0.9950	0.9888	0.9777	0.9594	0.9317
25		0.9999	0.9997	0.9990	0.9974	0.9938	0.9869	0.9748	0.9554
26		1.0000	0.9999	0.9995	0.9987	0.9967	0.9925	0.9848	0.9718
27			0.9999	0.9998	0.9994	0.9983	0.9959	0.9912	0.9827
28			1.0000	0.9999	0.9997	0.9991	0.9978	0.9950	0.9897
29				1.0000	0.9999	0.9996	0.9989	0.9973	0.9941
30					0.9999	0.9998	0.9994	0.9986	0.9967
31					1.0000	0.9999	0.9997	0.9993	0.9982
32						1.0000	0.9999	0.9996	0.9990
33							0.9999	0.9998	0.9995
34							1.0000	0.9999	0.9998
35								1.0000	0.9999
36									0.9999
37									1.0000

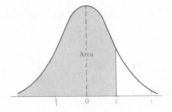

Table A.3 Areas Under the Normal Curve

z	.00	.01	.02	.03	.04	.05	.06	.07	.08	.09
−3.4	.0003	.0003	.0003	.0003	.0003	.0003	.0003	.0003	.0003	.0002
−3.3	.0005	.0005	.0005	.0004	.0004	.0004	.0004	.0004	.0004	.0003
−3.2	.0007	.0007	.0006	.0006	.0006	.0006	.0006	.0005	.0005	.0005
−3.1	.0010	.0009	.0009	.0009	.0008	.0008	.0008	.0008	.0007	.0007
−3.0	.0013	.0013	.0013	.0012	.0012	.0011	.0011	.0011	.0010	.0010
−2.9	.0019	.0018	.0017	.0017	.0016	.0016	.0015	.0015	.0014	.0014
−2.8	.0026	.0025	.0024	.0023	.0023	.0022	.0021	.0021	.0020	.0019
−2.7	.0035	.0034	.0033	.0032	.0031	.0030	.0029	.0028	.0027	.0026
−2.6	.0047	.0045	.0044	.0043	.0041	.0040	.0039	.0038	.0037	.0036
−2.5	.0062	.0060	.0059	.0057	.0055	.0054	.0052	.0051	.0049	.0048
−2.4	.0082	.0080	.0078	.0075	.0073	.0071	.0069	.0068	.0066	.0064
−2.3	.0107	.0104	.0102	.0099	.0096	.0094	.0091	.0089	.0087	.0084
−2.2	.0139	.0136	.0132	.0129	.0125	.0122	.0119	.0116	.0113	.0110
−2.1	.0179	.0174	.0170	.0166	.0162	.0158	.0154	.0150	.0146	.0143
−2.0	.0228	.0222	.0217	.0212	.0207	.0202	.0197	.0192	.0188	.0183
−1.9	.0287	.0281	.0274	.0268	.0262	.0256	.0250	.0244	.0239	.0233
−1.8	.0359	.0352	.0344	.0336	.0329	.0322	.0314	.0307	.0301	.0294
−1.7	.0446	.0436	.0427	.0418	.0409	.0401	.0392	.0384	.0375	.0367
−1.6	.0548	.0537	.0526	.0516	.0505	.0495	.0485	.0475	.0465	.0455
−1.5	.0668	.0655	.0643	.0630	.0618	.0606	.0594	.0582	.0571	.0559
−1.4	.0808	.0793	.0778	.0764	.0749	.0735	.0722	.0708	.0694	.0681
−1.3	.0968	.0951	.0934	.0918	.0901	.0885	.0869	.0853	.0838	.0823
−1.2	.1151	.1131	.1112	.1093	.1075	.1056	.1038	.1020	.1003	.0985
−1.1	.1357	.1335	.1314	.1292	.1271	.1251	.1230	.1210	.1190	.1170
−1.0	.1587	.1562	.1539	.1515	.1492	.1469	.1446	.1423	.1401	.1379
−0.9	.1841	.1814	.1788	.1762	.1736	.1711	.1685	.1660	.1635	.1611
−0.8	.2119	.2090	.2061	.2033	.2005	.1977	.1949	.1922	.1894	.1867
−0.7	.2420	.2389	.2358	.2327	.2296	.2266	.2236	.2206	.2177	.2148
−0.6	.2743	.2709	.2676	.2643	.2611	.2578	.2546	.2514	.2483	.2451
−0.5	.3085	.3050	.3015	.2981	.2946	.2912	.2877	.2843	.2810	.2776
−0.4	.3446	.3409	.3372	.3336	.3300	.3264	.3228	.3192	.3156	.3121
−0.3	.3821	.3783	.3745	.3707	.3669	.3632	.3594	.3557	.3520	.3483
−0.2	.4207	.4168	.4129	.4090	.4052	.4013	.3974	.3936	.3897	.3859
−0.1	.4602	.4562	.4522	.4483	.4443	.4404	.4364	.4325	.4286	.4247
−0.0	.5000	.4960	.4920	.4880	.4840	.4801	.4761	.4721	.4681	.4641

Table A.3 (*continued*) Areas Under the Normal Curve

z	.00	.01	.02	.03	.04	.05	.06	.07	.08	.09
0.0	.5000	.5040	.5080	.5120	.5160	.5199	.5239	.5279	.5319	.5359
0.1	.5398	.5438	.5478	.5517	.5557	.5596	.5636	.5675	.5714	.5753
0.2	.5793	.5832	.5871	.5910	.5948	.5987	.6026	.6064	.6103	.6141
0.3	.6179	.6217	.6255	.6293	.6331	.6368	.6406	.6443	.6480	.6517
0.4	.6554	.6591	.6628	.6664	.6700	.6736	.6772	.6808	.6844	.6879
0.5	.6915	.6950	.6985	.7019	.7054	.7088	.7123	.7157	.7190	.7224
0.6	.7257	.7291	.7324	.7357	.7389	.7422	.7454	.7486	.7517	.7549
0.7	.7580	.7611	.7642	.7673	.7704	.7734	.7764	.7794	.7823	.7852
0.8	.7881	.7910	.7939	.7967	.7995	.8023	.8051	.8078	.8106	.8133
0.9	.8159	.8186	.8212	.8238	.8264	.8289	.8315	.8340	.8365	.8389
1.0	.8413	.8438	.8461	.8485	.8508	.8531	.8554	.8577	.8599	.8621
1.1	.8643	.8665	.8686	.8708	.8729	.8749	.8770	.8790	.8810	.8830
1.2	.8849	.8869	.8888	.8907	.8925	.8944	.8962	.8980	.8997	.9015
1.3	.9032	.9049	.9066	.9082	.9099	.9115	.9131	.9147	.9162	.9177
1.4	.9192	.9207	.9222	.9236	.9251	.9265	.9278	.9292	.9306	.9319
1.5	.9332	.9345	.9357	.9370	.9382	.9394	.9406	.9418	.9429	.9441
1.6	.9452	.9463	.9474	.9484	.9495	.9505	.9515	.9525	.9535	.9545
1.7	.9554	.9564	.9573	.9582	.9591	.9599	.9608	.9616	.9625	.9633
1.8	.9641	.9649	.9656	.9664	.9671	.9678	.9686	.9693	.9699	.9706
1.9	.9713	.9719	.9726	.9732	.9738	.9744	.9750	.9756	.9761	.9767
2.0	.9772	.9778	.9783	.9788	.9793	.9798	.9803	.9808	.9812	.9817
2.1	.9821	.9826	.9830	.9834	.9838	.9842	.9846	.9850	.9854	.9857
2.2	.9861	.9864	.9868	.9871	.9875	.9878	.9881	.9884	.9887	.9890
2.3	.9893	.9896	.9898	.9901	.9904	.9906	.9909	.9911	.9913	.9916
2.4	.9918	.9920	.9922	.9925	.9927	.9929	.9931	.9932	.9934	.9936
2.5	.9938	.9940	.9941	.9943	.9945	.9946	.9948	.9949	.9951	.9952
2.6	.9953	.9955	.9956	.9957	.9959	.9960	.9961	.9962	.9963	.9964
2.7	.9965	.9966	.9967	.9968	.9969	.9970	.9971	.9972	.9973	.9974
2.8	.9974	.9975	.9976	.9977	.9977	.9978	.9979	.9979	.9980	.9981
2.9	.9981	.9982	.9982	.9983	.9984	.9984	.9985	.9985	.9986	.9986
3.0	.9987	.9987	.9987	.9988	.9988	.9989	.9989	.9989	.9990	.9990
3.1	.9990	.9991	.9991	.9991	.9992	.9992	.9992	.9992	.9993	.9993
3.2	.9993	.9993	.9994	.9994	.9994	.9994	.9994	.9995	.9995	.9995
3.3	.9995	.9995	.9995	.9996	.9996	.9996	.9996	.9996	.9996	.9997
3.4	.9997	.9997	.9997	.9997	.9997	.9997	.9997	.9997	.9997	.9998

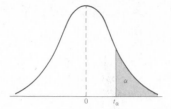

Table A.4* Critical Values of the *t* Distribution

	α				
ν	0.10	0.05	0.025	0.01	0.005
1	3.078	6.314	12.706	31.821	63.657
2	1.886	2.920	4.303	6.965	9.925
3	1.638	2.353	3.182	4.541	5.841
4	1.533	2.132	2.776	3.747	4.604
5	1.476	2.015	2.571	3.365	4.032
6	1.440	1.943	2.447	3.143	3.707
7	1.415	1.895	2.365	2.998	3.499
8	1.397	1.860	2.306	2.896	3.355
9	1.383	1.833	2.262	2.821	3.250
10	1.372	1.812	2.228	2.764	3.169
11	1.363	1.796	2.201	2.718	3.106
12	1.356	1.782	2.179	2.681	3.055
13	1.350	1.771	2.160	2.650	3.012
14	1.345	1.761	2.145	2.624	2.977
15	1.341	1.753	2.131	2.602	2.947
16	1.337	1.746	2.120	2.583	2.921
17	1.333	1.740	2.110	2.567	2.898
18	1.330	1.734	2.101	2.552	2.878
19	1.328	1.729	2.093	2.539	2.861
20	1.325	1.725	2.086	2.528	2.845
21	1.323	1.721	2.080	2.518	2.831
22	1.321	1.717	2.074	2.508	2.819
23	1.319	1.714	2.069	2.500	2.807
24	1.318	1.711	2.064	2.492	2.797
25	1.316	1.708	2.060	2.485	2.787
26	1.315	1.706	2.056	2.479	2.779
27	1.314	1.703	2.052	2.473	2.771
28	1.313	1.701	2.048	2.467	2.763
29	1.311	1.699	2.045	2.462	2.756
inf.	1.282	1.645	1.960	2.326	2.576

* From Table IV of R. A. Fisher, *Statistical Methods for Research Workers*, published by Oliver & Boyd, Edinburgh, by permission of the author and publishers.

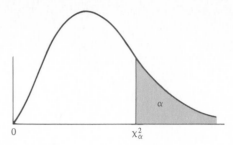

Table A.5 Critical Values of the Chi-Square Distribution

v	\multicolumn{10}{c}{α}
	.995	.99	.98	.975	.95	.90	.80	.75	.70	.50
1	$.0^4393$	$.0^3157$	$.0^3628$	$.0^3982$	.00393	.0158	.0642	.102	.148	.455
2	.0100	.0201	.0404	.0506	.103	.211	.446	.575	.713	1.386
3	.0717	.115	.185	.216	.352	.584	1.005	1.213	1.424	2.366
4	.207	.297	.429	.484	.711	1.064	1.649	1.923	2.195	3.357
5	.412	.554	.752	.831	1.145	1.610	2.343	2.675	3.000	4.351
6	.676	.872	1.134	1.237	1.635	2.204	3.070	3.455	3.828	5.348
7	.989	1.239	1.564	1.690	2.167	2.833	3.822	4.255	4.671	6.346
8	1.344	1.646	2.032	2.180	2.733	3.490	4.594	5.071	5.527	7.344
9	1.735	2.088	2.532	2.700	3.325	4.168	5.380	5.899	6.393	8.343
10	2.156	2.558	3.059	3.247	3.940	4.865	6.179	6.737	7.267	9.342
11	2.603	3.053	3.609	3.816	4.575	5.578	6.989	7.584	8.148	10.341
12	3.074	3.571	4.178	4.404	5.226	6.304	7.807	8.438	9.034	11.340
13	3.565	4.107	4.765	5.009	5.892	7.042	8.634	9.299	9.926	12.340
14	4.075	4.660	5.368	5.629	6.571	7.790	9.467	10.165	10.821	13.339
15	4.601	5.229	5.985	6.262	7.261	8.547	10.307	11.036	11.721	14.339
16	5.142	5.812	6.614	6.908	7.962	9.312	11.152	11.912	12.624	15.338
17	5.697	6.408	7.255	7.564	8.672	10.085	12.002	12.792	13.531	16.338
18	6.265	7.015	7.906	8.231	9.390	10.865	12.857	13.675	14.440	17.338
19	6.844	7.633	8.567	8.907	10.117	11.651	13.716	14.562	15.352	18.338
20	7.434	8.260	9.237	9.591	10.851	12.443	14.578	15.452	16.266	19.337
21	8.034	8.897	9.915	10.283	11.591	13.240	15.445	16.344	17.182	20.337
22	8.643	9.542	10.600	10.982	12.338	14.041	16.314	17.240	18.101	21.337
23	9.260	10.196	11.293	11.688	13.091	14.848	17.187	18.137	19.021	22.337
24	9.886	10.856	11.992	12.401	13.848	15.659	18.062	19.037	19.943	23.337
25	10.520	11.524	12.697	13.120	14.611	16.473	18.940	19.939	20.867	24.337
26	11.160	12.198	13.409	13.844	15.379	17.292	19.820	20.843	21.792	25.336
27	11.808	12.879	14.125	14.573	16.151	18.114	20.703	21.749	22.719	26.336
28	12.461	13.565	14.847	15.308	16.928	18.939	21.588	22.657	23.647	27.336
29	13.121	14.256	15.574	16.047	17.708	19.768	22.475	23.567	24.577	28.336
30	13.787	14.953	16.306	16.791	18.493	20.599	23.364	24.478	25.508	29.336

Table A.5 (*continued*) Critical Values of the Chi-Square Distribution

ν	α									
	.30	.25	.20	.10	.05	.025	.02	.01	.005	.001
1	1.074	1.323	1.642	2.706	3.841	5.024	5.412	6.635	7.879	10.827
2	2.408	2.773	3.219	4.605	5.991	7.378	7.824	9.210	10.597	13.815
3	3.665	4.108	4.642	6.251	7.815	9.348	9.837	11.345	12.838	16.268
4	4.878	5.385	5.989	7.779	9.488	11.143	11.668	13.277	14.860	18.465
5	6.064	6.626	7.289	9.236	11.070	12.832	13.388	15.086	16.750	20.517
6	7.231	7.841	8.558	10.645	12.592	14.449	15.033	16.812	18.548	22.457
7	8.383	9.037	9.803	12.017	14.067	16.013	16.622	18.475	20.278	24.322
8	9.524	10.219	11.030	13.362	15.507	17.535	18.168	20.090	21.955	26.125
9	10.656	11.389	12.242	14.684	16.919	19.023	19.679	21.666	23.589	27.877
10	11.781	12.549	13.442	15.987	18.307	20.483	21.161	23.209	25.188	29.588
11	12.899	13.701	14.631	17.275	19.675	21.920	22.618	24.725	26.757	31.264
12	14.011	14.845	15.812	18.549	21.026	23.337	24.054	26.217	28.300	32.909
13	15.119	15.984	16.985	19.812	22.362	24.736	25.472	27.688	29.819	34.528
14	16.222	17.117	18.151	21.064	23.685	26.119	26.873	29.141	31.319	36.123
15	17.322	18.245	19.311	22.307	24.996	27.488	28.259	30.578	32.801	37.697
16	18.418	19.369	20.465	23.542	26.296	28.845	29.633	32.000	34.267	39.252
17	19.511	20.489	21.615	24.769	27.587	30.191	30.995	33.409	35.718	40.790
18	20.601	21.605	22.760	25.989	28.869	31.526	32.346	34.805	37.156	42.312
19	21.689	22.718	23.900	27.204	30.144	32.852	33.687	36.191	38.582	43.820
20	22.775	23.828	25.038	28.412	31.410	34.170	35.020	37.566	39.997	45.315
21	23.858	24.935	26.171	29.615	32.671	35.479	36.343	38.932	41.401	46.797
22	24.939	26.039	27.301	30.813	33.924	36.781	37.659	40.289	42.796	48.268
23	26.018	27.141	28.429	32.007	35.172	38.076	38.968	41.638	44.181	49.728
24	27.096	28.241	29.553	33.196	36.415	39.364	40.270	42.980	45.558	51.179
25	28.172	29.339	30.675	34.382	37.652	40.646	41.566	44.314	46.928	52.620
26	29.246	30.434	31.795	35.563	38.885	41.923	42.856	45.642	48.290	54.052
27	30.319	31.528	32.912	36.741	40.113	43.194	44.140	46.963	49.645	55.476
28	31.391	32.620	34.027	37.916	41.337	44.461	45.419	48.278	50.993	56.893
29	32.461	33.711	35.139	39.087	42.557	45.722	46.693	49.588	52.336	58.302
30	33.530	34.800	36.250	40.256	43.773	46.979	47.962	50.892	53.672	59.703

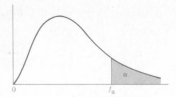

Table A.6* Critical Values of the F Distribution

$$f_{0.05}(v_1, v_2)$$

v_2	v_1								
	1	2	3	4	5	6	7	8	9
1	161.4	199.5	215.7	224.6	230.2	234.0	236.8	238.9	240.5
2	18.51	19.00	19.16	19.25	19.30	19.33	19.35	19.37	19.38
3	10.13	9.55	9.28	9.12	9.01	8.94	8.89	8.85	8.81
4	7.71	6.94	6.59	6.39	6.26	6.16	6.09	6.04	6.00
5	6.61	5.79	5.41	5.19	5.05	4.95	4.88	4.82	4.77
6	5.99	5.14	4.76	4.53	4.39	4.28	4.21	4.15	4.10
7	5.59	4.74	4.35	4.12	3.97	3.87	3.79	3.73	3.68
8	5.32	4.46	4.07	3.84	3.69	3.58	3.50	3.44	3.39
9	5.12	4.26	3.86	3.63	3.48	3.37	3.29	3.23	3.18
10	4.96	4.10	3.71	3.48	3.33	3.22	3.14	3.07	3.02
11	4.84	3.98	3.59	3.36	3.20	3.09	3.01	2.95	2.90
12	4.75	3.89	3.49	3.26	3.11	3.00	2.91	2.85	2.80
13	4.67	3.81	3.41	3.18	3.03	2.92	2.83	2.77	2.71
14	4.60	3.74	3.34	3.11	2.96	2.85	2.76	2.70	2.65
15	4.54	3.68	3.29	3.06	2.90	2.79	2.71	2.64	2.59
16	4.49	3.63	3.24	3.01	2.85	2.74	2.66	2.59	2.54
17	4.45	3.59	3.20	2.96	2.81	2.70	2.61	2.55	2.49
18	4.41	3.55	3.16	2.93	2.77	2.66	2.58	2.51	2.46
19	4.38	3.52	3.13	2.90	2.74	2.63	2.54	2.48	2.42
20	4.35	3.49	3.10	2.87	2.71	2.60	2.51	2.45	2.39
21	4.32	3.47	3.07	2.84	2.68	2.57	2.49	2.42	2.37
22	4.30	3.44	3.05	2.82	2.66	2.55	2.46	2.40	2.34
23	4.28	3.42	3.03	2.80	2.64	2.53	2.44	2.37	2.32
24	4.26	3.40	3.01	2.78	2.62	2.51	2.42	2.36	2.30
25	4.24	3.39	2.99	2.76	2.60	2.49	2.40	2.34	2.28
26	4.23	3.37	2.98	2.74	2.59	2.47	2.39	2.32	2.27
27	4.21	3.35	2.96	2.73	2.57	2.46	2.37	2.31	2.25
28	4.20	3.34	2.95	2.71	2.56	2.45	2.36	2.29	2.24
29	4.18	3.33	2.93	2.70	2.55	2.43	2.35	2.28	2.22
30	4.17	3.32	2.92	2.69	2.53	2.42	2.33	2.27	2.21
40	4.08	3.23	2.84	2.61	2.45	2.34	2.25	2.18	2.12
60	4.00	3.15	2.76	2.53	2.37	2.25	2.17	2.10	2.04
120	3.92	3.07	2.68	2.45	2.29	2.17	2.09	2.02	1.96
∞	3.84	3.00	2.60	2.37	2.21	2.10	2.01	1.94	1.88

Table A.6 (*continued*) Critical Values of the *F* Distribution

$$f_{0.05}(v_1, v_2)$$

v_2	\multicolumn{10}{c}{v_1}
	10	12	15	20	24	30	40	60	120	∞
1	241.9	243.9	245.9	248.0	249.1	250.1	251.1	252.2	253.3	254.3
2	19.40	19.41	19.43	19.45	19.45	19.46	19.47	19.48	19.49	19.50
3	8.79	8.74	8.70	8.66	8.64	8.62	8.59	8.57	8.55	8.53
4	5.96	5.91	5.86	5.80	5.77	5.75	5.72	5.69	5.66	5.63
5	4.74	4.68	4.62	4.56	4.53	4.50	4.46	4.43	4.40	4.36
6	4.06	4.00	3.94	3.87	3.84	3.81	3.77	3.74	3.70	3.67
7	3.64	3.57	3.51	3.44	3.41	3.38	3.34	3.30	3.27	3.23
8	3.35	3.28	3.22	3.15	3.12	3.08	3.04	3.01	2.97	2.93
9	3.14	3.07	3.01	2.94	2.90	2.86	2.83	2.79	2.75	2.71
10	2.98	2.91	2.85	2.77	2.74	2.70	2.66	2.62	2.58	2.54
11	2.85	2.79	2.72	2.65	2.61	2.57	2.53	2.49	2.45	2.40
12	2.75	2.69	2.62	2.54	2.51	2.47	2.43	2.38	2.34	2.30
13	2.67	2.60	2.53	2.46	2.42	2.38	2.34	2.30	2.25	2.21
14	2.60	2.53	2.46	2.39	2.35	2.31	2.27	2.22	2.18	2.13
15	2.54	2.48	2.40	2.33	2.29	2.25	2.20	2.16	2.11	2.07
16	2.49	2.42	2.35	2.28	2.24	2.19	2.15	2.11	2.06	2.01
17	2.45	2.38	2.31	2.23	2.19	2.15	2.10	2.06	2.01	1.96
18	2.41	2.34	2.27	2.19	2.15	2.11	2.06	2.02	1.97	1.92
19	2.38	2.31	2.23	2.16	2.11	2.07	2.03	1.98	1.93	1.88
20	2.35	2.28	2.20	2.12	2.08	2.04	1.99	1.95	1.90	1.84
21	2.32	2.25	2.18	2.10	2.05	2.01	1.96	1.92	1.87	1.81
22	2.30	2.23	2.15	2.07	2.03	1.98	1.94	1.89	1.84	1.78
23	2.27	2.20	2.13	2.05	2.01	1.96	1.91	1.86	1.81	1.76
24	2.25	2.18	2.11	2.03	1.98	1.94	1.89	1.84	1.79	1.73
25	2.24	2.16	2.09	2.01	1.96	1.92	1.87	1.82	1.77	1.71
26	2.22	2.15	2.07	1.99	1.95	1.90	1.85	1.80	1.75	1.69
27	2.20	2.13	2.06	1.97	1.93	1.88	1.84	1.79	1.73	1.67
28	2.19	2.12	2.04	1.96	1.91	1.87	1.82	1.77	1.71	1.65
29	2.18	2.10	2.03	1.94	1.90	1.85	1.81	1.75	1.70	1.64
30	2.16	2.09	2.01	1.93	1.89	1.84	1.79	1.74	1.68	1.62
40	2.08	2.00	1.92	1.84	1.79	1.74	1.69	1.64	1.58	1.51
60	1.99	1.92	1.84	1.75	1.70	1.65	1.59	1.53	1.47	1.39
120	1.91	1.83	1.75	1.66	1.61	1.55	1.50	1.43	1.35	1.25
∞	1.83	1.75	1.67	1.57	1.52	1.46	1.39	1.32	1.22	1.00

Table A.6 (*continued*) Critical Values of the *F* Distribution

$$f_{0.01}(\nu_1, \nu_2)$$

ν_2					ν_1				
	1	2	3	4	5	6	7	8	9
1	4052	4999.5	5403	5625	5764	5859	5928	5981	6022
2	98.50	99.00	99.17	99.25	99.30	99.33	99.36	99.37	99.39
3	34.12	30.82	29.46	28.71	28.24	27.91	27.67	27.49	27.35
4	21.20	18.00	16.69	15.98	15.52	15.21	14.98	14.80	14.66
5	16.26	13.27	12.06	11.39	10.97	10.67	10.46	10.29	10.16
6	13.75	10.92	9.78	9.15	8.75	8.47	8.26	8.10	7.98
7	12.25	9.55	8.45	7.85	7.46	7.19	6.99	6.84	6.72
8	11.26	8.65	7.59	7.01	6.63	6.37	6.18	6.03	5.91
9	10.56	8.02	6.99	6.42	6.06	5.80	5.61	5.47	5.35
10	10.04	7.56	6.55	5.99	5.64	5.39	5.20	5.06	4.94
11	9.65	7.21	6.22	5.67	5.32	5.07	4.89	4.74	4.63
12	9.33	6.93	5.95	5.41	5.06	4.82	4.64	4.50	4.39
13	9.07	6.70	5.74	5.21	4.86	4.62	4.44	4.30	4.19
14	8.86	6.51	5.56	5.04	4.69	4.46	4.28	4.14	4.03
15	8.68	6.36	5.42	4.89	4.56	4.32	4.14	4.00	3.89
16	8.53	6.23	5.29	4.77	4.44	4.20	4.03	3.89	3.78
17	8.40	6.11	5.18	4.67	4.34	4.10	3.93	3.79	3.68
18	8.29	6.01	5.09	4.58	4.25	4.01	3.84	3.71	3.60
19	8.18	5.93	5.01	4.50	4.17	3.94	3.77	3.63	3.52
20	8.10	5.85	4.94	4.43	4.10	3.87	3.70	3.56	3.46
21	8.02	5.78	4.87	4.37	4.04	3.81	3.64	3.51	3.40
22	7.95	5.72	4.82	4.31	3.99	3.76	3.59	3.45	3.35
23	7.88	5.66	4.76	4.26	3.94	3.71	3.54	3.41	3.30
24	7.82	5.61	4.72	4.22	3.90	3.67	3.50	3.36	3.26
25	7.77	5.57	4.68	4.18	3.85	3.63	3.46	3.32	3.22
26	7.72	5.53	4.64	4.14	3.82	3.59	3.42	3.29	3.18
27	7.68	5.49	4.60	4.11	3.78	3.56	3.39	3.26	3.15
28	7.64	5.45	4.57	4.07	3.75	3.53	3.36	3.23	3.12
29	7.60	5.42	4.54	4.04	3.73	3.50	3.33	3.20	3.09
30	7.56	5.39	4.51	4.02	3.70	3.47	3.30	3.17	3.07
40	7.31	5.18	4.31	3.83	3.51	3.29	3.12	2.99	2.89
60	7.08	4.98	4.13	3.65	3.34	3.12	2.95	2.82	2.72
120	6.85	4.79	3.95	3.48	3.17	2.96	2.79	2.66	2.56
∞	6.63	4.61	3.78	3.32	3.02	2.80	2.64	2.51	2.41

Table A.6 (*continued*) Critical Values of the *F* Distribution

$$f_{0.01}(v_1, v_2)$$

| v_2 | \multicolumn{10}{c}{v_1} |
	10	12	15	20	24	30	40	60	120	∞
1	6056	6106	6157	6209	6235	6261	6287	6313	6339	6366
2	99.40	99.42	99.43	99.45	99.46	99.47	99.47	99.48	99.49	99.50
3	27.23	27.05	26.87	26.69	26.60	26.50	26.41	26.32	26.22	26.13
4	14.55	14.37	14.20	14.02	13.93	13.84	13.75	13.65	13.56	13.46
5	10.05	9.89	9.72	9.55	9.47	9.38	9.29	9.20	9.11	9.02
6	7.87	7.72	7.56	7.40	7.31	7.23	7.14	7.06	6.97	6.88
7	6.62	6.47	6.31	6.16	6.07	5.99	5.91	5.82	5.74	5.65
8	5.81	5.67	5.52	5.36	5.28	5.20	5.12	5.03	4.95	4.86
9	5.26	5.11	4.96	4.81	4.73	4.65	4.57	4.48	4.40	4.31
10	4.85	4.71	4.56	4.41	4.33	4.25	4.17	4.08	4.00	3.91
11	4.54	4.40	4.25	4.10	4.02	3.94	3.86	3.78	3.69	3.60
12	4.30	4.16	4.01	3.86	3.78	3.70	3.62	3.54	3.45	3.36
13	4.10	3.96	3.82	3.66	3.59	3.51	3.43	3.34	3.25	3.17
14	3.94	3.80	3.66	3.51	3.43	3.35	3.27	3.18	3.09	3.00
15	3.80	3.67	3.52	3.37	3.29	3.21	3.13	3.05	2.96	2.87
16	3.69	3.55	3.41	3.26	3.18	3.10	3.02	2.93	2.84	2.75
17	3.59	3.46	3.31	3.16	3.08	3.00	2.92	2.83	2.75	2.65
18	3.51	3.37	3.23	3.08	3.00	2.92	2.84	2.75	2.66	2.57
19	3.43	3.30	3.15	3.00	2.92	2.84	2.76	2.67	2.58	2.49
20	3.37	3.23	3.09	2.94	2.86	2.78	2.69	2.61	2.52	2.42
21	3.31	3.17	3.03	2.88	2.80	2.72	2.64	2.55	2.46	2.36
22	3.26	3.12	2.98	2.83	2.75	2.67	2.58	2.50	2.40	2.31
23	3.21	3.07	2.93	2.78	2.70	2.62	2.54	2.45	2.35	2.26
24	3.17	3.03	2.89	2.74	2.66	2.58	2.49	2.40	2.31	2.21
25	3.13	2.99	2.85	2.70	2.62	2.54	2.45	2.36	2.27	2.17
26	3.09	2.96	2.81	2.66	2.58	2.50	2.42	2.33	2.23	2.13
27	3.06	2.93	2.78	2.63	2.55	2.47	2.38	2.29	2.20	2.10
28	3.03	2.90	2.75	2.60	2.52	2.44	2.35	2.26	2.17	2.06
29	3.00	2.87	2.73	2.57	2.49	2.41	2.33	2.23	2.14	2.03
30	2.98	2.84	2.70	2.55	2.47	2.39	2.30	2.21	2.11	2.01
40	2.80	2.66	2.52	2.37	2.29	2.20	2.11	2.02	1.92	1.80
60	2.63	2.50	2.35	2.20	2.12	2.03	1.94	1.84	1.73	1.60
120	2.47	2.34	2.19	2.03	1.95	1.86	1.76	1.66	1.53	1.38
∞	2.32	2.18	2.04	1.88	1.79	1.70	1.59	1.47	1.32	1.00

Table A.7* Tolerance Factors for Normal Distributions

| | | $v = 0.95$ | | | | $v = 0.99$ | |
| | 1 − α | | | | 1 − α | | |
n	0.90	0.95	0.99	n	0.90	0.95	0.99
2	32.019	37.674	48.430	2	160.193	188.491	242.300
3	8.380	9.916	12.861	3	18.930	22.401	29.055
4	5.369	6.370	8.299	4	9.398	11.150	14.527
5	4.275	5.079	6.634	5	6.612	7.855	10.260
6	3.712	4.414	5.775	6	5.337	6.345	8.301
7	3.369	4.007	5.248	7	4.613	5.488	7.187
8	3.136	3.732	4.891	8	4.147	4.936	6.468
9	2.967	3.532	4.631	9	3.822	4.550	5.966
10	2.839	3.379	4.433	10	3.582	4.265	5.594
11	2.737	3.259	4.277	11	3.397	4.045	5.308
12	2.655	3.162	4.150	12	3.250	3.870	5.079
13	2.587	3.081	4.044	13	3.130	3.727	4.893
14	2.529	3.012	3.955	14	3.029	3.608	4.737
15	2.480	2.954	3.878	15	2.945	3.507	4.605
16	2.437	2.903	3.812	16	2.872	3.421	4.492
17	2.400	2.858	3.754	17	2.808	3.345	4.393
18	2.366	2.819	3.702	18	2.753	3.279	4.307
19	2.337	2.784	3.656	19	2.703	3.221	4.230
20	2.310	2.752	3.615	20	2.659	3.168	4.161
25	2.208	2.631	3.457	25	2.494	2.972	3.904
30	2.140	2.549	3.350	30	2.385	2.841	3.733
35	2.090	2.490	3.272	35	2.306	2.748	3.611
40	2.052	2.445	3.213	40	2.247	2.677	3.518

* Adapted from C. Eisenhart, M. W. Hastay, and W. A. Wallis, *Techniques of Statistical Analysis*, Chapter 2, McGraw-Hill Book Company, New York, 1947. Used with permission of McGraw-Hill Book Company.

Table A.7 (*continued*) Tolerance Factors for Normal Distributions

	$v = 0.95$				$v = 0.99$		
		$1 - \alpha$				$1 - \alpha$	
n	0.90	0.95	0.99	n	0.90	0.95	0.99
45	2.021	2.408	3.165	45	2.200	2.621	3.444
50	1.996	2.379	3.126	50	2.162	2.576	3.385
55	1.976	2.354	3.094	55	2.130	2.538	3.335
60	1.958	2.333	3.066	60	2.103	2.506	3.293
65	1.943	2.315	3.042	65	2.080	2.478	3.257
70	1.929	2.299	3.021	70	2.060	2.454	3.225
75	1.917	2.285	3.002	75	2.042	2.433	3.197
80	1.907	2.272	2.986	80	2.026	2.414	3.173
85	1.897	2.261	2.971	85	2.012	2.397	3.150
90	1.889	2.251	2.958	90	1.999	2.382	3.130
95	1.881	2.241	2.945	95	1.987	2.368	3.112
100	1.874	2.233	2.934	100	1.977	2.355	3.096
150	1.825	2.175	2.859	150	1.905	2.270	2.983
200	1.798	2.143	2.816	200	1.865	2.222	2.921
250	1.780	2.121	2.788	250	1.839	2.191	2.880
300	1.767	2.106	2.767	300	1.820	2.169	2.850
400	1.749	2.084	2.739	400	1.794	2.138	2.809
500	1.737	2.070	2.721	500	1.777	2.117	2.783
600	1.729	2.060	2.707	600	1.764	2.102	2.763
700	1.722	2.052	2.697	700	1.755	2.091	2.748
800	1.717	2.046	2.688	800	1.747	2.082	2.736
900	1.712	2.040	2.682	900	1.741	2.075	2.726
1000	1.709	2.036	2.676	1000	1.736	2.068	2.718
∞	1.645	1.960	2.576	∞	1.645	1.960	2.576

Table A.8* Sample Size for the t Test of the Mean

	Level of t-test																				
Single-sided test	$\alpha = 0.005$					$\alpha = 0.01$					$\alpha = 0.025$					$\alpha = 0.05$					
Double-sided test	$\alpha = 0.01$					$\alpha = 0.02$					$\alpha = 0.05$					$\alpha = 0.1$					
$\beta =$	0.01	0.05	0.1	0.2	0.5	0.01	0.05	0.1	0.2	0.5	0.01	0.05	0.1	0.2	0.5	0.01	0.05	0.1	0.2	0.5	
0.05																					0.05
0.10																					0.10
0.15																				122	0.15
0.20										139					99					70	0.20
0.25					110					90				128	64			139	101	45	0.25
0.30				134	78				115	63			119	90	45		122	97	71	32	0.30
0.35			125	99	58			109	85	47		109	88	67	34		90	72	52	24	0.35
0.40		115	97	77	45		101	85	66	37	117	84	68	51	26	101	70	55	40	19	0.40
0.45		92	77	62	37	110	81	68	53	30	93	67	54	41	21	80	55	44	33	15	0.45
0.50	100	75	63	51	30	90	66	55	43	25	76	54	44	34	18	65	45	36	27	13	0.50
0.55	83	63	53	42	26	75	55	46	36	21	63	45	37	28	15	54	38	30	22	11	0.55
0.60	71	53	45	36	22	63	47	39	31	18	53	38	32	24	13	46	32	26	19	9	0.60
0.65	61	46	39	31	20	55	41	34	27	16	46	33	27	21	12	39	28	22	17	8	0.65
0.70	53	40	34	28	17	47	35	30	24	14	40	29	24	19	10	34	24	19	15	8	0.70
0.75	47	36	30	25	16	42	31	27	21	13	35	26	21	16	9	30	21	17	13	7	0.75
0.80	41	32	27	22	14	37	28	24	19	12	31	22	19	15	9	27	19	15	12	6	0.80
0.85	37	29	24	20	13	33	25	21	17	11	28	21	17	13	8	24	17	14	11	6	0.85
0.90	34	26	22	18	12	29	23	19	16	10	25	19	16	12	7	21	15	13	10	5	0.90
Value of 0.05	31	24	20	17	11	27	21	18	14	9	23	17	14	11	7	19	14	11	9	5	0.95
$\Delta = \frac{\lvert\delta\rvert}{\sigma}$ 1.00	28	22	19	16	10	25	19	16	13	9	21	16	13	10	6	18	13	11	8	5	1.00
1.1	24	19	16	14	9	21	16	14	12	8	18	13	11	9	6		15	11	9	7	1.1
1.2	21	16	14	12	8	18	14	12	10	7	15	12	10	8	5		13	10	8	6	1.2
1.3	18	15	13	11	8	16	13	11	9	6		14	10	9	7		11	8	7	6	1.3
1.4	16	13	12	10	7	14	11	10	9	6		12	9	8	7		10	8	7	5	1.4
1.5	15	12	11	9	7	13	10	9	8	6		11	8	7	6			9	7	6	1.5
1.6	13	11	10	8	6	12	10	9	7	5		10	8	7	6			8	6	6	1.6
1.7	12	10	9	8	6		11	9	8	7		9	7	6	5			8	6	5	1.7
1.8	12	10	9	8	6		10	8	7	7			8	7	6				7	6	1.8
1.9	11	9	8	7	6		10	8	7	6			8	6	6				7	5	1.9
2.0	10	8	8	7	5		9	7	7	6			7	6	5					6	2.0
2.1		10	8	7	7		8	7	6	6				7	6					6	2.1
2.2		9	8	7	6		8	7	6	5				7	6					6	2.2
2.3		9	7	7	6			8	6	6				6	5					5	2.3
2.4		8	7	7	6			7	6	6					6						2.4
2.5		8	7	6	6			7	6	6					6						2.5
3.0		7	6	6	5			6	5	5					5						3.0
3.5			6	5	5					5											3.5
4.0					6																4.0

* Reproduced with permission from O. L. Davies, ed., *Design and Analysis of Industrial Experiments*, Oliver & Boyd, Edinburgh, 1956.

Table A.9* Sample Size for the *t* Test of the Difference Between Two Means

Level of *t*-test

$\Delta = \frac{\|\delta\|}{\sigma}$	Single-sided test $\alpha = 0.005$ / Double-sided test $\alpha = 0.01$					$\alpha = 0.01$ / $\alpha = 0.02$					$\alpha = 0.025$ / $\alpha = 0.05$					$\alpha = 0.05$ / $\alpha = 0.1$					
$\beta =$	0.01	0.05	0.1	0.2	0.5	0.01	0.05	0.1	0.2	0.5	0.01	0.05	0.1	0.2	0.5	0.01	0.05	0.1	0.2	0.5	
0.05																					0.05
0.10																					0.10
0.15																					0.15
0.20																			137		0.20
0.25															124					88	0.25
0.30										123					87					61	0.30
0.35					110					90					64				102	45	0.35
0.40					85					70				100	50			108	78	35	0.40
0.45				118	68				101	55			105	79	39		108	86	62	28	0.45
0.50				96	55			106	82	45		106	86	64	32		88	70	51	23	0.50
0.55			101	79	46		106	88	68	38		87	71	53	27	112	73	58	42	19	0.55
0.60		101	85	67	39		90	74	58	32	104	74	60	45	23	89	61	49	36	16	0.60
0.65		87	73	57	34	104	77	64	49	27	88	63	51	39	20	76	52	42	30	14	0.65
0.70	100	75	63	50	29	90	66	55	43	24	76	55	44	34	17	66	45	36	26	12	0.70
0.75	88	66	55	44	26	79	58	48	38	21	67	48	39	29	15	57	40	32	23	11	0.75
0.80	77	58	49	39	23	70	51	43	33	19	59	42	34	26	14	50	35	28	21	10	0.80
0.85	69	51	43	35	21	62	46	38	30	17	52	37	31	23	12	45	31	25	18	9	0.85
0.90	62	46	39	31	19	55	41	34	27	15	47	34	27	21	11	40	28	22	16	8	0.90
0.95	55	42	35	28	17	50	37	31	24	14	42	30	25	19	10	36	25	20	15	7	0.95
1.00	50	38	32	26	15	45	33	28	22	13	38	27	23	17	9	33	23	18	14	7	1.00
1.1	42	32	27	22	13	38	28	23	19	11	32	23	19	14	8	27	19	15	12	6	1.1
1.2	36	27	23	18	11	32	24	20	16	9	27	20	16	12	7	23	16	13	10	5	1.2
1.3	31	23	20	16	10	28	21	17	14	8	23	17	14	11	6	20	14	11	9	5	1.3
1.4	27	20	17	14	9	24	18	15	12	8	20	15	12	10	6	17	12	10	8	4	1.4
1.5	24	18	15	13	8	21	16	14	11	7	18	13	11	9	5	15	11	9	7	4	1.5
1.6	21	16	14	11	7	19	14	12	10	6	16	12	10	8	5	14	10	8	6	4	1.6
1.7	19	15	13	10	7	17	13	11	9	6	14	11	9	7	4	12	9	7	6	3	1.7
1.8	17	13	11	10	6	15	12	10	8	5	13	10	8	6	4	11	8	7	5		1.8
1.9	16	12	11	9	6	14	11	9	8	5	12	9	7	6	4	10	7	6	5		1.9
2.0	14	11	10	8	6	13	10	9	7	5	11	8	7	6	4	9	7	6	4		2.0
2.1	13	10	9	8	5	12	9	8	7	5	10	8	6	5	3	8	6	5	4		2.1
2.2	12	10	8	7	5	11	9	7	6	4	9	7	6	5		8	6	5	4		2.2
2.3	11	9	8	7	5	10	8	7	6	4	9	7	6	5		7	5	5	4		2.3
2.4	11	9	8	6	5	10	8	7	6	4	8	6	5	4		7	5	4	4		2.4
2.5	10	8	7	6	4	9	7	6	5	4	8	6	5	4		6	5	4	3		2.5
3.0	8	6	6	5	4	7	6	5	4	3	6	5	4	4		5	4	3			3.0
3.5	6	5	5	4	3	6	5	4	4		5	4	4	3		4	3				3.5
4.0	6	5	4	4		5	4	4	3		4	4	3			4					4.0

Value of $\Delta = \frac{|\delta|}{\sigma}$

* Reproduced with permission from O. L. Davies, ed., *Design and Analysis of Industrial Experiments*, Oliver & Boyd, Edinburgh, 1956.

Table A.10* Critical Values for Bartlett's Test

$$b_k(0.01; n)$$

	Number of Populations, k								
n	2	3	4	5	6	7	8	9	10
3	.1411	.1672	*	*	*	*	*	*	*
4	.2843	.3165	.3475	.3729	.3937	.4110	*	*	*
5	.3984	.4304	.4607	.4850	.5046	.5207	.5343	.5458	.5558
6	.4850	.5149	.5430	.5653	.5832	.5978	.6100	.6204	.6293
7	.5512	.5787	.6045	.6248	.6410	.6542	.6652	.6744	.6824
8	.6031	.6282	.6518	.6704	.6851	.6970	.7069	.7153	.7225
9	.6445	.6676	.6892	.7062	.7197	.7305	.7395	.7471	.7536
10	.6783	.6996	.7195	.7352	.7475	.7575	.7657	.7726	.7786
11	.7063	.7260	.7445	.7590	.7703	.7795	.7871	.7935	.7990
12	.7299	.7483	.7654	.7789	.7894	.7980	.8050	.8109	.8160
13	.7501	.7672	.7832	.7958	.8056	.8135	.8201	.8256	.8303
14	.7674	.7835	.7985	.8103	.8195	.8269	.8330	.8382	.8426
15	.7825	.7977	.8118	.8229	.8315	.8385	.8443	.8491	.8532
16	.7958	.8101	.8235	.8339	.8421	.8486	.8541	.8586	.8625
17	.8076	.8211	.8338	.8436	.8514	.8576	.8627	.8670	.8707
18	.8181	.8309	.8429	.8523	.8596	.8655	.8704	.8745	.8780
19	.8275	.8397	.8512	.8601	.8670	.8727	.8773	.8811	.8845
20	.8360	.8476	.8586	.8671	.8737	.8791	.8835	.8871	.8903
21	.8437	.8548	.8653	.8734	.8797	.8848	.8890	.8926	.8956
22	.8507	.8614	.8714	.8791	.8852	.8901	.8941	.8975	.9004
23	.8571	.8673	.8769	.8844	.8902	.8949	.8988	.9020	.9047
24	.8630	.8728	.8820	.8892	.8948	.8993	.9030	.9061	.9087
25	.8684	.8779	.8867	.8936	.8990	.9034	.9069	.9099	.9124
26	.8734	.8825	.8911	.8977	.9029	.9071	.9105	.9134	.9158
27	.8781	.8869	.8951	.9015	.9065	.9105	.9138	.9166	.9190
28	.8824	.8909	.8988	.9050	.9099	.9138	.9169	.9196	.9219
29	.8864	.8946	.9023	.9083	.9130	.9167	.9198	.9224	.9246
30	.8902	.8981	.9056	.9114	.9159	.9195	.9225	.9250	.9271
40	.9175	.9235	.9291	.9335	.9370	.9397	.9420	.9439	.9455
50	.9339	.9387	.9433	.9468	.9496	.9518	.9536	.9551	.9564
60	.9449	.9489	.9527	.9557	.9580	.9599	.9614	.9626	.9637
80	.9586	.9617	.9646	.9668	.9685	.9699	.9711	.9720	.9728
100	.9669	.9693	.9716	.9734	.9748	.9759	.9769	.9776	.9783

* Reproduced from D. D. Dyer and J. P. Keating, "On the determination of critical values for Bartlett's test," *J. Am. Stat. Assoc.*, Vol. 75, 1980, by permission of the Board of Directors.

Table A.10 (*continued*) Critical Values for Bartlett's Test

$$b_k(0.05; n)$$

n	| Number of Populations, k								
	2	3	4	5	6	7	8	9	10
3	.3123	.3058	.3173	.3299	*	*	*	*	*
4	.4780	.4699	.4803	.4921	.5028	.5122	.5204	.5277	.5341
5	.5845	.5762	.5850	.5952	.6045	.6126	.6197	.6260	.6315
6	.6563	.6483	.6559	.6646	.6727	.6798	.6860	.6914	.6961
7	.7075	.7000	.7065	.7142	.7213	.7275	.7329	.7376	.7418
8	.7456	.7387	.7444	.7512	.7574	.7629	.7677	.7719	.7757
9	.7751	.7686	.7737	.7798	.7854	.7903	.7946	.7984	.8017
10	.7984	.7924	.7970	.8025	.8076	.8121	.8160	.8194	.8224
11	.8175	.8118	.8160	.8210	.8257	.8298	.8333	.8365	.8392
12	.8332	.8280	.8317	.8364	.8407	.8444	.8477	.8506	.8531
13	.8465	.8415	.8450	.8493	.8533	.8568	.8598	.8625	.8648
14	.8578	.8532	.8564	.8604	.8641	.8673	.8701	.8726	.8748
15	.8676	.8632	.8662	.8699	.8734	.8764	.8790	.8814	.8834
16	.8761	.8719	.8747	.8782	.8815	.8843	.8868	.8890	.8909
17	.8836	.8796	.8823	.8856	.8886	.8913	.8936	.8957	.8975
18	.8902	.8865	.8890	.8921	.8949	.8975	.8997	.9016	.9033
19	.8961	.8926	.8949	.8979	.9006	.9030	.9051	.9069	.9086
20	.9015	.8980	.9003	.9031	.9057	.9080	.9100	.9117	.9132
21	.9063	.9030	.9051	.9078	.9103	.9124	.9143	.9160	.9175
22	.9106	.9075	.9095	.9120	.9144	.9165	.9183	.9199	.9213
23	.9146	.9116	.9135	.9159	.9182	.9202	.9219	.9235	.9248
24	.9182	.9153	.9172	.9195	.9217	.9236	.9253	.9267	.9280
25	.9216	.9187	.9205	.9228	.9249	.9267	.9283	.9297	.9309
26	.9246	.9219	.9236	.9258	.9278	.9296	.9311	.9325	.9336
27	.9275	.9249	.9265	.9286	.9305	.9322	.9337	.9350	.9361
28	.9301	.9276	.9292	.9312	.9330	.9347	.9361	.9374	.9385
29	.9326	.9301	.9316	.9336	.9354	.9370	.9383	.9396	.9406
30	.9348	.9325	.9340	.9358	.9376	.9391	.9404	.9416	.9426
40	.9513	.9495	.9506	.9520	.9533	.9545	.9555	.9564	.9572
50	.9612	.9597	.9606	.9617	.9628	.9637	.9645	.9652	.9658
60	.9677	.9665	.9672	.9681	.9690	.9698	.9705	.9710	.9716
80	.9758	.9749	.9754	.9761	.9768	.9774	.9779	.9783	.9787
100	.9807	.9799	.9804	.9809	.9815	.9819	.9823	.9827	.9830

Table A.11* Critical Values for Cochran's Test

$$\alpha = 0.01$$

k \ n	2	3	4	5	6	7	8	9	10	11	17	37	145	∞
2	0.9999	0.9950	0.9794	0.9586	0.9373	0.9172	0.8988	0.8823	0.8674	0.8539	0.7949	0.7067	0.6062	0.5000
3	0.9933	0.9423	0.8831	0.8335	0.7933	0.7606	0.7335	0.7107	0.6912	0.6743	0.6059	0.5153	0.4230	0.3333
4	0.9676	0.8643	0.7814	0.7212	0.6761	0.6410	0.6129	0.5897	0.5702	0.5536	0.4884	0.4057	0.3251	0.2500
5	0.9279	0.7885	0.6957	0.6329	0.5875	0.5531	0.5259	0.5037	0.4854	0.4697	0.4094	0.3351	0.2644	0.2000
6	0.8828	0.7218	0.6258	0.5635	0.5195	0.4866	0.4608	0.4401	0.4229	0.4084	0.3529	0.2858	0.2229	0.1667
7	0.8376	0.6644	0.5685	0.5080	0.4659	0.4347	0.4105	0.3911	0.3751	0.3616	0.3105	0.2494	0.1929	0.1429
8	0.7945	0.6152	0.5209	0.4627	0.4226	0.3932	0.3704	0.3522	0.3373	0.3248	0.2779	0.2214	0.1700	0.1250
9	0.7544	0.5727	0.4810	0.4251	0.3870	0.3592	0.3378	0.3207	0.3067	0.2950	0.2514	0.1992	0.1521	0.1111
10	0.7175	0.5358	0.4469	0.3934	0.3572	0.3308	0.3106	0.2945	0.2813	0.2704	0.2297	0.1811	0.1376	0.1000
12	0.6528	0.4751	0.3919	0.3428	0.3099	0.2861	0.2680	0.2535	0.2419	0.2320	0.1961	0.1535	0.1157	0.0833
15	0.5747	0.4069	0.3317	0.2882	0.2593	0.2386	0.2228	0.2104	0.2002	0.1918	0.1612	0.1251	0.0934	0.0667
20	0.4799	0.3297	0.2654	0.2288	0.2048	0.1877	0.1748	0.1646	0.1567	0.1501	0.1248	0.0960	0.0709	0.0500
24	0.4247	0.2871	0.2295	0.1970	0.1759	0.1608	0.1495	0.1406	0.1338	0.1283	0.1060	0.0810	0.0595	0.0417
30	0.3632	0.2412	0.1913	0.1635	0.1454	0.1327	0.1232	0.1157	0.1100	0.1054	0.0867	0.0658	0.0480	0.0333
40	0.2940	0.1915	0.1508	0.1281	0.1135	0·1033	0.0957	0.0898	0.0853	0.0816	0.0668	0.0503	0.0363	0.0250
60	0.2151	0.1371	0.1069	0.0902	0.0796	0.0722	0.0668	0.0625	0.0594	0.0567	0.0461	0.0344	0.0245	0.0167
120	0.1225	0.0759	0.0585	0.0489	0.0429	0.0387	0.0357	0.0334	0.0316	0.0302	0.0242	0.0178	0.0125	0.0083
∞	0	0	0	0	0	0	0	0	0	0	0	0	0	0

* Reproduced from C. Eisenhart, M. W. Hastay, and W. A. Wallis, *Techniques of Statistical Analysis*, Chapter 15, McGraw-Hill Book Company, New York, 1947. Used with permission of McGraw-Hill Book Company.

Table A.11 (*continued*) Critical Values for Cochran's Test

$$\alpha = 0.05$$

k \ n	2	3	4	5	6	7	8	9	10	11	17	37	145	∞
2	0.9985	0.9750	0.9392	0.9057	0.8772	0.8534	0.8332	0.8159	0.8010	0.7880	0.7341	0.6602	0.5813	0.5000
3	0.9669	0.8709	0.7977	0.7457	0.7071	0.6771	0.6530	0.6333	0.6167	0.6025	0.5466	0.4748	0.4031	0.3333
4	0.9065	0.7679	0.6841	0.6287	0.5895	0.5598	0.5365	0.5175	0.5017	0.4884	0.4366	0.3720	0.3093	0.2500
5	0.8412	0.6838	0.5981	0.5441	0.5065	0.4783	0.4564	0.4387	0.4241	0.4118	0.3645	0.3066	0.2513	0.2000
6	0.7808	0.6161	0.5321	0.4803	0.4447	0.4184	0.3980	0.3817	0.3682	0.3568	0.3135	0.2612	0.2119	0.1667
7	0.7271	0.5612	0.4800	0.4307	0.3974	0.3726	0.3535	0.3384	0.3259	0.3154	0.2756	0.2278	0.1833	0.1429
8	0.6798	0.5157	0.4377	0.3910	0.3595	0.3362	0.3185	0.3043	0.2926	0.2829	0.2462	0.2022	0.1616	0.1250
9	0.6385	0.4775	0.4027	0.3584	0.3286	0.3067	0.2901	0.2768	0.2659	0.2568	0.2226	0.1820	0.1446	0.1111
10	0.6020	0.4450	0.3733	0.3311	0.3029	0.2823	0.2666	0.2541	0.2439	0.2353	0.2032	0.1655	0.1308	0.1000
12	0.5410	0.3924	0.3264	0.2880	0.2624	0.2439	0.2299	0.2187	0.2098	0.2020	0.1737	0.1403	0.1100	0.0833
15	0.4709	0.3346	0.2758	0.2419	0.2195	0.2034	0.1911	0.1815	0.1736	0.1671	0.1429	0.1144	0.0889	0.0667
20	0.3894	0.2705	0.2205	0.1921	0.1735	0.1602	0.1501	0.1422	0.1357	0.1303	0.1108	0.0879	0.0675	0.0500
24	0.3434	0.2354	0.1907	0.1656	0.1493	0.1374	0.1286	0.1216	0.1160	0.1113	0.0942	0.0743	0.0567	0.0417
30	0.2929	0.1980	0.1593	0.1377	0.1237	0.1137	0.1061	0.1002	0.0958	0.0921	0.0771	0.0604	0.0457	0.0333
40	0.2370	0.1576	0.1259	0.1082	0.0968	0.0887	0.0827	0.0780	0.0745	0.0713	0.0595	0.0462	0.0347	0.0250
60	0.1737	0.1131	0.0895	0.0765	0.0682	0.0623	0.0583	0.0552	0.0520	0.0497	0.0411	0.0316	0.0234	0.0167
120	0.0998	0.0632	0.0495	0.0419	0.0371	0.0337	0.0312	0.0292	0.0279	0.0266	0.0218	0.0165	0.0120	0.0083
∞	0	0	0	0	0	0	0	0	0	0	0	0	0	0

Table A.12* Least Significant Studentized Ranges r_p

$$\alpha = 0.05$$

ν	p								
	2	3	4	5	6	7	8	9	10
1	17.97	17.97	17.97	17.97	17.97	17.97	17.97	17.97	17.97
2	6.085	6.085	6.085	6.085	6.085	6.085	6.085	6.085	6.085
3	4.501	4.516	4.516	4.516	4.516	4.516	4.516	4.516	4.516
4	3.927	4.013	4.033	4.033	4.033	4.033	4.033	4.033	4.033
5	3.635	3.749	3.797	3.814	3.814	3.814	3.814	3.814	3.814
6	3.461	3.587	3.649	3.680	3.694	3.697	3.697	3.697	3.697
7	3.344	3.477	3.548	3.588	3.611	3.622	3.626	3.626	3.626
8	3.261	3.399	3.475	3.521	3.549	3.566	3.575	3.579	3.579
9	3.199	3.339	3.420	3.470	3.502	3.523	3.536	3.544	3.547
10	3.151	3.293	3.376	3.430	3.465	3.489	3.505	3.516	3.522
11	3.113	3.256	3.342	3.397	3.435	3.462	3.480	3.493	3.501
12	3.082	3.225	3.313	3.370	3.410	3.439	3.459	3.474	3.484
13	3.055	3.200	3.289	3.348	3.389	3.419	3.442	3.458	3.470
14	3.033	3.178	3.268	3.329	3.372	3.403	3.426	3.444	3.457
15	3.014	3.160	3.250	3.312	3.356	3.389	3.413	3.432	3.446
16	2.998	3.144	3.235	3.298	3.343	3.376	3.402	3.422	3.437
17	2.984	3.130	3.222	3.285	3.331	3.366	3.392	3.412	3.429
18	2.971	3.118	3.210	3.274	3.321	3.356	3.383	3.405	3.421
19	2.960	3.107	3.199	3.264	3.311	3.347	3.375	3.397	3.415
20	2.950	3.097	3.190	3.255	3.303	3.339	3.368	3.391	3.409
24	2.919	3.066	3.160	3.226	3.276	3.315	3.345	3.370	3.390
30	2.888	3.035	3.131	3.199	3.250	3.290	3.322	3.349	3.371
40	2.858	3.006	3.102	3.171	3.224	3.266	3.300	3.328	3.352
60	2.829	2.976	3.073	3.143	3.198	3.241	3.277	3.307	3.333
120	2.800	2.947	3.045	3.116	3.172	3.217	3.254	3.287	3.314
∞	2.772	2.918	3.017	3.089	3.146	3.193	3.232	3.265	3.294

* Abridged from H. Leon Harter, "Critical Values for Duncan's New Multiple Range Test," *Biometrics*, Vol. 16, no. 4, 1960, by permission of the author and the edtior.

Table A.12 (*continued*) Least Significant Studentized Ranges r_p

$\alpha = 0.01$

ν	p								
	2	3	4	5	6	7	8	9	10
1	90.03	90.03	90.03	90.03	90.03	90.03	90.03	90.03	90.03
2	14.04	14.04	14.04	14.04	14.04	14.04	14.04	14.04	14.04
3	8.261	8.321	8.321	8.321	8.321	8.321	8.321	8.321	8.321
4	6.512	6.677	6.740	6.756	6.756	6.756	6.756	6.756	6.756
5	5.702	5.893	5.989	6.040	6.065	6.074	6.074	6.074	6.074
6	5.243	5.439	5.549	5.614	5.655	5.680	5.694	5.701	5.703
7	4.949	5.145	5.260	5.334	5.383	5.416	5.439	5.454	5.464
8	4.746	4.939	5.057	5.135	5.189	5.227	5.256	5.276	5.291
9	4.596	4.787	4.906	4.986	5.043	5.086	5.118	5.142	5.160
10	4.482	4.671	4.790	4.871	4.931	4.975	5.010	5.037	5.058
11	4.392	4.579	4.697	4.780	4.841	4.887	4.924	4.952	4.975
12	4.320	4.504	4.622	4.706	4.767	4.815	4.852	4.883	4.907
13	4.260	4.442	4.560	4.644	4.706	4.755	4.793	4.824	4.850
14	4.210	4.391	4.508	4.591	4.654	4.704	4.743	4.775	4.802
15	4.168	4.347	4.463	4.547	4.610	4.660	4.700	4.733	4.760
16	4.131	4.309	4.425	4.509	4.572	4.622	4.663	4.696	4.724
17	4.099	4.275	4.391	4.475	4.539	4.589	4.630	4.664	4.693
18	4.071	4.246	4.362	4.445	4.509	4.560	4.601	4.635	4.664
19	4.046	4.220	4.335	4.419	4.483	4.534	4.575	4.610	4.639
20	4.024	4.197	4.312	4.395	4.459	4.510	4.552	4.587	4.617
24	3.956	4.126	4.239	4.322	4.386	4.437	4.480	4.516	4.546
30	3.889	4.056	4.168	4.250	4.314	4.366	4.409	4.445	4.477
40	3.825	3.988	4.098	4.180	4.244	4.296	4.339	4.376	4.408
60	3.762	3.922	4.031	4.111	4.174	4.226	4.270	4.307	4.340
120	3.702	3.858	3.965	4.044	4.107	4.158	4.202	4.239	4.272
∞	3.643	3.796	3.900	3.978	4.040	4.091	4.135	4.172	4.205

Table A.13* Values of $d_{\alpha/2}(k, v)$ for Two-Sided Comparisons Between k Treatments and a Control

$$\alpha = 0.05$$

v	k = number of treatment means (excluding control)								
	1	2	3	4	5	6	7	8	9
5	2.57	3.03	3.29	3.48	3.62	3.73	3.82	3.90	3.97
6	2.45	2.86	3.10	3.26	3.39	3.49	3.57	3.64	3.71
7	2.36	2.75	2.97	3.12	3.24	3.33	3.41	3.47	3.53
8	2.31	2.67	2.88	3.02	3.13	3.22	3.29	3.35	3.41
9	2.26	2.61	2.81	2.95	3.05	3.14	3.20	3.26	3.32
10	2.23	2.57	2.76	2.89	2.99	3.07	3.14	3.19	3.24
11	2.20	2.53	2.72	2.84	2.94	3.02	3.08	3.14	3.19
12	2.18	2.50	2.68	2.81	2.90	2.98	3.04	3.09	3.14
13	2.16	2.48	2.65	2.78	2.87	2.94	3.00	3.06	3.10
14	2.14	2.46	2.63	2.75	2.84	2.91	2.97	3.02	3.07
15	2.13	2.44	2.61	2.73	2.82	2.89	2.95	3.00	3.04
16	2.12	2.42	2.59	2.71	2.80	2.87	2.92	2.97	3.02
17	2.11	2.41	2.58	2.69	2.78	2.85	2.90	2.95	3.00
18	2.10	2.40	2.56	2.68	2.76	2.83	2.89	2.94	2.98
19	2.09	2.39	2.55	2.66	2.75	2.81	2.87	2.92	2.96
20	2.09	2.38	2.54	2.65	2.73	2.80	2.86	2.90	2.95
24	2.06	2.35	2.51	2.61	2.70	2.76	2.81	2.86	2.90
30	2.04	2.32	2.47	2.58	2.66	2.72	2.77	2.82	2.86
40	2.02	2.29	2.44	2.54	2.62	2.68	2.73	2.77	2.81
60	2.00	2.27	2.41	2.51	2.58	2.64	2.69	2.73	2.77
120	1.98	2.24	2.38	2.47	2.55	2.60	2.65	2.69	2.73
∞	1.96	2.21	2.35	2.44	2.51	2.57	2.61	2.65	2.69

* Reproduced from Charles W. Dunnett, "New Tables for Multiple Comparison with a Control," *Biometrics*, Vol. 20, no. 3, 1964, by permission of the author and the editor.

Table A.13 (*continued*) Values of $d_{\alpha/2}(k, v)$ for Two-Sided Comparisons Between k Treatments and a Control

$$\alpha = 0.01$$

v	\multicolumn{9}{c}{k = number of treatment means (excluding control)}
	1	2	3	4	5	6	7	8	9
5	4.03	4.63	4.98	5.22	5.41	5.56	5.69	5.80	5.89
6	3.71	4.21	4.51	4.71	4.87	5.00	5.10	5.20	5.28
7	3.50	3.95	4.21	4.39	4.53	4.64	4.74	4.82	4.89
8	3.36	3.77	4.00	4.17	4.29	4.40	4.48	4.56	4.62
9	3.25	3.63	3.85	4.01	4.12	4.22	4.30	4.37	4.43
10	3.17	3.53	3.74	3.88	3.99	4.08	4.16	4.22	4.28
11	3.11	3.45	3.65	3.79	3.89	3.98	4.05	4.11	4.16
12	3.05	3.39	3.58	3.71	3.81	3.89	3.96	4.02	4.07
13	3.01	3.33	3.52	3.65	3.74	3.82	3.89	3.94	3.99
14	2.98	3.29	3.47	3.59	3.69	3.76	3.83	3.88	3.93
15	2.95	3.25	3.43	3.55	3.64	3.71	3.78	3.83	3.88
16	2.92	3.22	3.39	3.51	3.60	3.67	3.73	3.78	3.83
17	2.90	3.19	3.36	3.47	3.56	3.63	3.69	3.74	3.79
18	2.88	3.17	3.33	3.44	3.53	3.60	3.66	3.71	3.75
19	2.86	3.15	3.31	3.42	3.50	3.57	3.63	3.68	3.72
20	2.85	3.13	3.29	3.40	3.48	3.55	3.60	3.65	3.69
24	2.80	3.07	3.22	3.32	3.40	3.47	3.52	3.57	3.61
30	2.75	3.01	3.15	3.25	3.33	3.39	3.44	3.49	3.52
40	2.70	2.95	3.09	3.19	3.26	3.32	3.37	3.41	3.44
60	2.66	2.90	3.03	3.12	3.19	3.25	3.29	3.33	3.37
120	2.62	2.85	2.97	3.06	3.12	3.18	3.22	3.26	3.29
∞	2.58	2.79	2.92	3.00	3.06	3.11	3.15	3.19	3.22

Table A.14* Values of $d_\alpha(k, v)$ for One-Sided Comparisons Between k Treatments and a Control

$$\alpha = 0.05$$

v	\multicolumn{9}{c}{k = number of treatment means (excluding control)}								
	1	2	3	4	5	6	7	8	9
5	2.02	2.44	2.68	2.85	2.98	3.08	3.16	3.24	3.30
6	1.94	2.34	2.56	2.71	2.83	2.92	3.00	3.07	3.12
7	1.89	2.27	2.48	2.62	2.73	2.82	2.89	2.95	3.01
8	1.86	2.22	2.42	2.55	2.66	2.74	2.81	2.87	2.92
9	1.83	2.18	2.37	2.50	2.60	2.68	2.75	2.81	2.86
10	1.81	2.15	2.34	2.47	2.56	2.64	2.70	2.76	2.81
11	1.80	2.13	2.31	2.44	2.53	2.60	2.67	2.72	2.77
12	1.78	2.11	2.29	2.41	2.50	2.58	2.64	2.69	2.74
13	1.77	2.09	2.27	2.39	2.48	2.55	2.61	2.66	2.71
14	1.76	2.08	2.25	2.37	2.46	2.53	2.59	2.64	2.69
15	1.75	2.07	2.24	2.36	2.44	2.51	2.57	2.62	2.67
16	1.75	2.06	2.23	2.34	2.43	2.50	2.56	2.61	2.65
17	1.74	2.05	2.22	2.33	2.42	2.49	2.54	2.59	2.64
18	1.73	2.04	2.21	2.32	2.41	2.48	2.53	2.58	2.62
19	1.73	2.03	2.20	2.31	2.40	2.47	2.52	2.57	2.61
20	1.72	2.03	2.19	2.30	2.39	2.46	2.51	2.56	2.60
24	1.71	2.01	2.17	2.28	2.36	2.43	2.48	2.53	2.57
30	1.70	1.99	2.15	2.25	2.33	2.40	2.45	2.50	2.54
40	1.68	1.97	2.13	2.23	2.31	2.37	2.42	2.47	2.51
60	1.67	1.95	2.10	2.21	2.28	2.35	2.39	2.44	2.48
120	1.66	1.93	2.08	2.18	2.26	2.32	2.37	2.41	2.45
∞	1.64	1.92	2.06	2.16	2.23	2.29	2.34	2.38	2.42

* Reproduced from Charles W. Dunnett, "A Multiple Comparison Procedure for Comparing Several Treatments with a Control," *J. Am. Stat. Assoc.*, Vol. 50, 1955, 1096–1121, by permission of the author and the editor.

Table A.14 (*continued*) Values of $d_\alpha(k, v)$ for One-Sided Comparisons Between k Treatments and a Control

$$\alpha = 0.01$$

v	k = number of treatment means (excluding control)								
	1	2	3	4	5	6	7	8	9
5	3.37	3.90	4.21	4.43	4.60	4.73	4.85	4.94	5.03
6	3.14	3.61	3.88	4.07	4.21	4.33	4.43	4.51	4.59
7	3.00	3.42	3.66	3.83	3.96	4.07	4.15	4.23	4.30
8	2.90	3.29	3.51	3.67	3.79	3.88	3.96	4.03	4.09
9	2.82	3.19	3.40	3.55	3.66	3.75	3.82	3.89	3.94
10	2.76	3.11	3.31	3.45	3.56	3.64	3.71	3.78	3.83
11	2.72	3.06	3.25	3.38	3.48	3.56	3.63	3.69	3.74
12	2.68	3.01	3.19	3.32	3.42	3.50	3.56	3.62	3.67
13	2.65	2.97	3.15	3.27	3.37	3.44	3.51	3.56	3.61
14	2.62	2.94	3.11	3.23	3.32	3.40	3.46	3.51	3.56
15	2.60	2.91	3.08	3.20	3.29	3.36	3.42	3.47	3.52
16	2.58	2.88	3.05	3.17	3.26	3.33	3.39	3.44	3.48
17	2.57	2.86	3.03	3.14	3.23	3.30	3.36	3.41	3.45
18	2.55	2.84	3.01	3.12	3.21	3.27	3.33	3.38	3.42
19	2.54	2.83	2.99	3.10	3.18	3.25	3.31	3.36	3.40
20	2.53	2.81	2.97	3.08	3.17	3.23	3.29	3.34	3.38
24	2.49	2.77	2.92	3.03	3.11	3.17	3.22	3.27	3.31
30	2.46	2.72	2.87	2.97	3.05	3.11	3.16	3.21	3.24
40	2.42	2.68	2.82	2.92	2.99	3.05	3.10	3.14	3.18
60	2.39	2.64	2.78	2.87	2.94	3.00	3.04	3.08	3.12
120	2.36	2.60	2.73	2.82	2.89	2.94	2.99	3.03	3.06
∞	2.33	2.56	2.68	2.77	2.84	2.89	2.93	2.97	3.00

Table A.15* Power of the Analysis-of-Variance Test

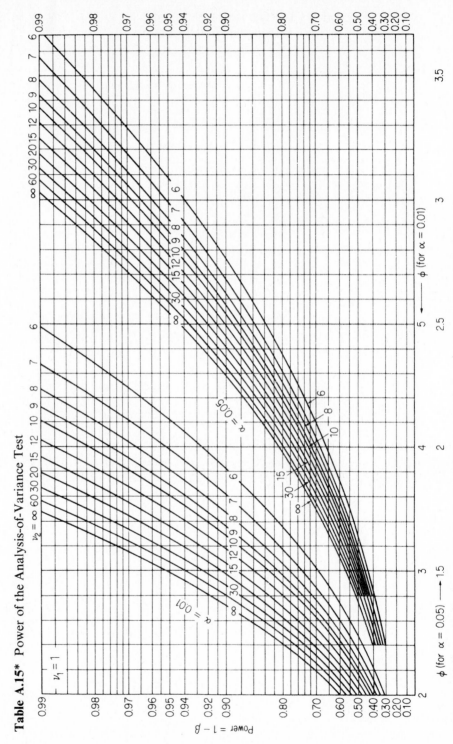

* Reproduced from E. S. Pearson and H. O. Hartley, "Charts of the Power Function for Analysis-of-Variance Tests, Derived from the Non-central F Distribution," *Biometrika*, Vol. 38, 1951, 112, by permission of the editor.

Table A.15 (*continued*) Power of the Analysis-of-Variance Test

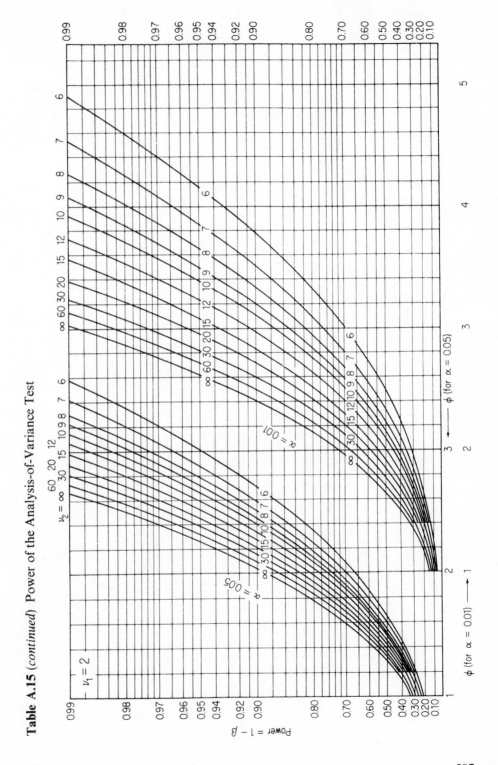

597

Table A.15 (*continued*) Power of the Analysis-of-Variance Test

$\nu_1 = 3$

Power $= 1 - \beta$

ϕ (for $\alpha = 0.01$) —

ϕ (for $\alpha = 0.05$)

$\nu_2 = \infty$

$\alpha = 0.05$

$\alpha = 0.01$

Table A.15 (*continued*) Power of the Analysis-of-Variance Test

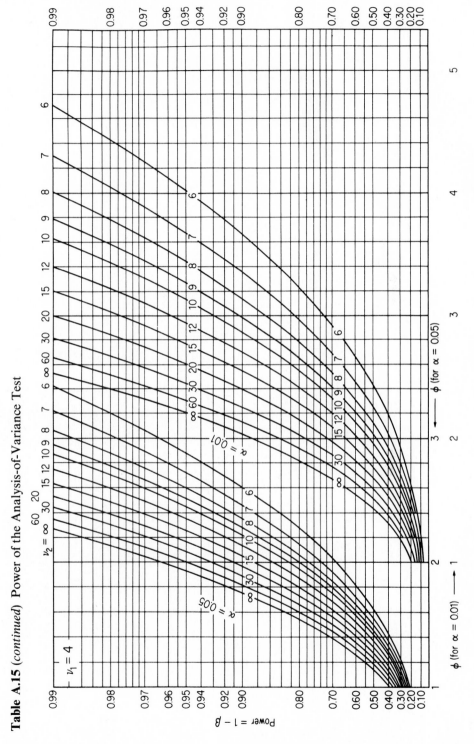

Table A.15 (continued) Power of the Analysis-of-Variance Test

Table A.15 (*continued*) Power of the Analysis-of-Variance Test

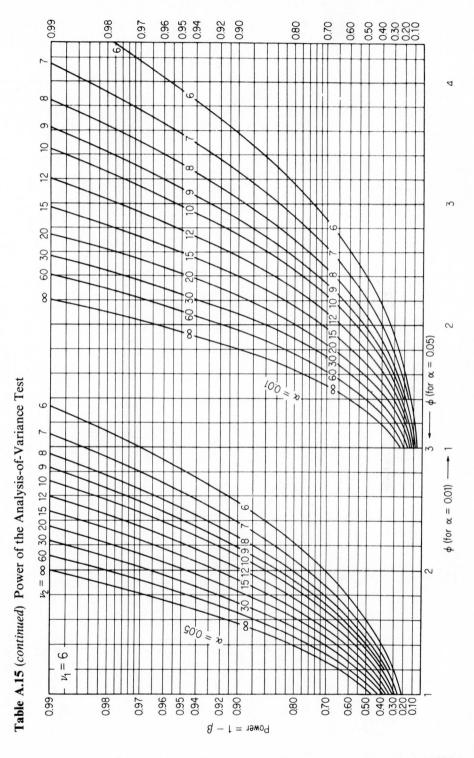

Table A.15 (*continued*) Power of the Analysis-of-Variance Test

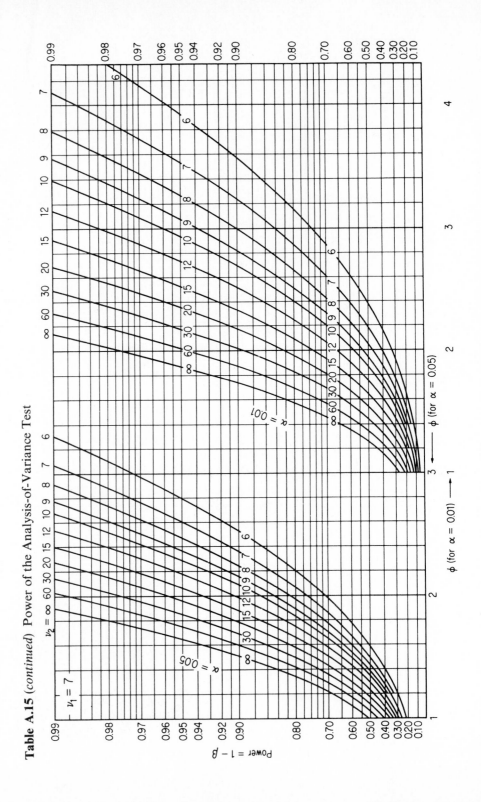

Table A.15 (*continued*) Power of the Analysis-of-Variance Test

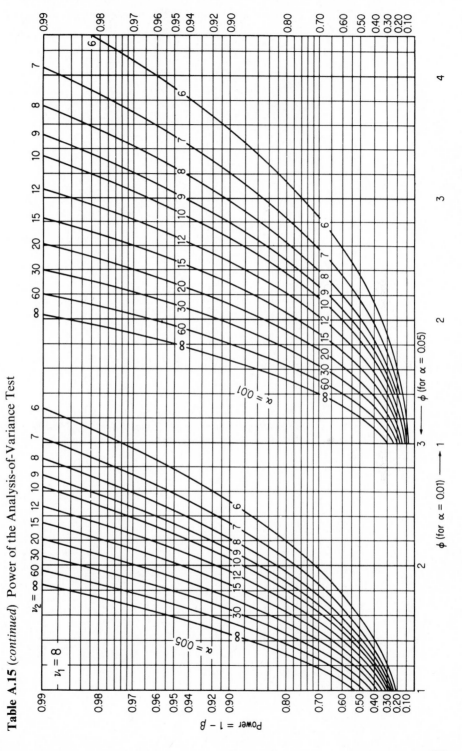

Table A.16* Critical Values for the Signed-Rank Test

n	One-sided $\alpha = 0.01$ Two-sided $\alpha = 0.02$	One-sided $\alpha = 0.025$ Two-sided $\alpha = 0.05$	One-sided $\alpha = 0.05$ Two-sided $\alpha = 0.10$
5			1
6		1	2
7	0	2	4
8	2	4	6
9	3	6	8
10	5	8	11
11	7	11	14
12	10	14	17
13	13	17	21
14	16	21	26
15	20	25	30
16	24	30	36
17	28	35	41
18	33	40	47
19	38	46	54
20	43	52	60
21	49	59	68
22	56	66	75
23	62	73	83
24	69	81	92
25	77	90	101
26	85	98	110
27	93	107	120
28	102	117	130
29	111	127	141
30	120	137	152

* Reproduced from F. Wilcoxon and R. A. Wilcox, *Some Rapid Approximate Statistical Procedures*, American Cyanamid Company, Pearl River, N.Y., 1964, by permission of the American Cyanamid Company.

Table A.17* Critical Values for the Rank-Sum Test

One-Tailed Test at $\alpha = 0.001$ or Two-Tailed Test at $\alpha = 0.002$

n_1	\	\	\	\	\	\	n_2	\	\	\	\	\	\	\	\
	6	7	8	9	10	11	12	13	14	15	16	17	18	19	20
1															
2															
3												0	0	0	0
4					0	0	0	1	1	1	2	2	3	3	3
5		0	0	1	1	2	2	3	3	4	5	5	6	7	7
6	0	1	2	2	3	4	4	5	6	7	8	9	10	11	12
7		2	3	3	5	6	7	8	9	10	11	13	14	15	16
8			5	5	6	8	9	11	12	14	15	17	18	20	21
9				7	8	10	12	14	15	17	19	21	23	25	26
10					10	12	14	17	19	21	23	25	27	29	32
11						15	17	20	22	24	27	29	32	34	37
12							20	23	25	28	31	34	37	40	42
13								26	29	32	35	38	42	45	48
14									32	36	39	43	46	50	54
15										40	43	47	51	55	59
16											48	52	56	60	65
17												57	61	66	70
18													66	71	76
19														77	82
20															88

One-Tailed Test at $\alpha = 0.01$ or Two-Tailed Test at $\alpha = 0.02$

n_1	\	\	\	\	\	\	\	n_2	\	\	\	\	\	\	\	\
	5	6	7	8	9	10	11	12	13	14	15	16	17	18	19	20
1																
2									0	0	0	0	0	0	1	1
3			0	0	1	1	1	2	2	2	3	3	4	4	4	5
4	0	1	1	2	3	3	4	5	5	6	7	7	8	9	9	10
5	1	2	3	4	5	6	7	8	9	10	11	12	13	14	15	16
6		3	4	6	7	8	9	11	12	13	15	16	18	19	20	22
7			6	8	9	11	12	14	16	17	19	21	23	24	26	28
8				10	11	13	15	17	20	22	24	26	28	30	32	34
9					14	16	18	21	23	26	28	31	33	36	38	40
10						19	22	24	27	30	33	36	38	41	44	47
11							25	28	31	34	37	41	44	47	50	53
12								31	35	38	42	46	49	53	56	60
13									39	43	47	51	55	59	63	67
14										47	51	56	60	65	69	73
15											56	61	66	70	75	80
16												66	71	76	82	87
17													77	82	88	93
18														88	94	100
19															101	107
20																114

* Based in part on Tables 1, 3, 5, and 7 of D. Auble, "Extended tables for the Mann–Whitney statistic," *Bulletin of the Institute of Educational Research at Indiana University*, Vol. 1, no. 2, 1953, by permission of the director.

Table A.17 (*continued*) Critical Values for the Rank-Sum Test

One-Tailed Test at $\alpha = 0.025$ or Two-Tailed Test at $\alpha = 0.05$

n_1								n_2									
	4	5	6	7	8	9	10	11	12	13	14	15	16	17	18	19	20
1																	
2					0	0	0	0	1	1	1	1	1	2	2	2	2
3		0	1	1	2	2	3	3	4	4	5	5	6	6	7	7	8
4	0	1	2	3	4	4	5	6	7	8	9	10	11	11	12	13	13
5		2	3	5	6	7	8	9	11	12	13	14	15	17	18	19	20
6			5	6	8	10	11	13	14	16	17	19	21	22	24	25	27
7				8	10	12	14	16	18	20	22	24	26	28	30	32	34
8					13	15	17	19	22	24	26	29	31	34	36	38	41
9						17	20	23	26	28	31	34	37	39	42	45	48
10							23	26	29	33	36	39	42	45	48	52	55
11								30	33	37	40	44	47	51	55	58	62
12									37	41	45	49	53	57	61	65	69
13										45	50	54	59	63	67	72	76
14											55	59	64	67	74	78	83
15												64	70	75	80	85	90
16													75	81	86	92	98
17														87	93	99	105
18															99	106	112
19																113	119
20																	127

One-Tailed Test at $\alpha = 0.05$ or Two-Tailed Test at $\alpha = 0.10$

n_1									n_2									
	3	4	5	6	7	8	9	10	11	12	13	14	15	16	17	18	19	20
1																	0	0
2			0	0	0	1	1	1	1	2	2	3	3	3	3	4	4	4
3	0	0	1	2	2	3	4	4	5	5	6	7	7	8	9	9	10	11
4		1	2	3	4	5	6	7	8	9	10	11	12	14	15	16	17	18
5			4	5	6	8	9	11	12	13	15	16	18	19	20	22	23	25
6				7	8	10	12	14	16	17	19	21	23	25	26	28	30	32
7					11	13	15	17	19	21	24	26	28	30	33	35	37	39
8						15	18	20	23	26	28	31	33	36	39	41	44	47
9							21	24	27	30	33	36	39	42	45	48	51	54
10								27	31	34	37	41	44	48	51	55	58	62
11									34	38	42	46	50	54	57	61	65	69
12										42	47	51	55	60	64	68	72	77
13											51	56	61	65	70	75	80	84
14												61	66	71	77	82	87	92
15													72	77	83	88	94	100
16														83	89	95	101	107
17															96	102	109	115
18																109	116	123
19																	123	130
20																		138

Table A.18* $P(V \leq v^*$ when H_0 is true) in the Runs Test

(n_1, n_2)	2	3	4	5	6	7	8	9	10
					v^*				
(2, 3)	0.200	0.500	0.900	1.000					
(2, 4)	0.133	0.400	0.800	1.000					
(2, 5)	0.095	0.333	0.714	1.000					
(2, 6)	0.071	0.286	0.643	1.000					
(2, 7)	0.056	0.250	0.583	1.000					
(2, 8)	0.044	0.222	0.533	1.000					
(2, 9)	0.036	0.200	0.491	1.000					
(2, 10)	0.030	0.182	0.455	1.000					
(3, 3)	0.100	0.300	0.700	0.900	1.000				
(3, 4)	0.057	0.200	0.543	0.800	0.971	1.000			
(3, 5)	0.036	0.143	0.429	0.714	0.929	1.000			
(3, 6)	0.024	0.107	0.345	0.643	0.881	1.000			
(3, 7)	0.017	0.083	0.283	0.583	0.833	1.000			
(3, 8)	0.012	0.067	0.236	0.533	0.788	1.000			
(3, 9)	0.009	0.055	0.200	0.491	0.745	1.000			
(3, 10)	0.007	0.045	0.171	0.455	0.706	1.000			
(4, 4)	0.029	0.114	0.371	0.629	0.886	0.971	1.000		
(4, 5)	0.016	0.071	0.262	0.500	0.786	0.929	0.992	1.000	
(4, 6)	0.010	0.048	0.190	0.405	0.690	0.881	0.976	1.000	
(4, 7)	0.006	0.033	0.142	0.333	0.606	0.833	0.954	1.000	
(4, 8)	0.004	0.024	0.109	0.279	0.533	0.788	0.929	1.000	
(4, 9)	0.003	0.018	0.085	0.236	0.471	0.745	0.902	1.000	
(4, 10)	0.002	0.014	0.068	0.203	0.419	0.706	0.874	1.000	
(5, 5)	0.008	0.040	0.167	0.357	0.643	0.833	0.960	0.992	1.000
(5, 6)	0.004	0.024	0.110	0.262	0.522	0.738	0.911	0.976	0.998
(5, 7)	0.003	0.015	0.076	0.197	0.424	0.652	0.854	0.955	0.992
(5, 8)	0.002	0.010	0.054	0.152	0.347	0.576	0.793	0.929	0.984
(5, 9)	0.001	0.007	0.039	0.119	0.287	0.510	0.734	0.902	0.972
(5, 10)	0.001	0.005	0.029	0.095	0.239	0.455	0.678	0.874	0.958
(6, 6)	0.002	0.013	0.067	0.175	0.392	0.608	0.825	0.933	0.987
(6, 7)	0.001	0.008	0.043	0.121	0.296	0.500	0.733	0.879	0.966
(6, 8)	0.001	0.005	0.028	0.086	0.226	0.413	0.646	0.821	0.937
(6, 9)	0.000	0.003	0.019	0.063	0.175	0.343	0.566	0.762	0.902
(6, 10)	0.000	0.002	0.013	0.047	0.137	0.288	0.497	0.706	0.864
(7, 7)	0.001	0.004	0.025	0.078	0.209	0.383	0.617	0.791	0.922
(7, 8)	0.000	0.002	0.015	0.051	0.149	0.296	0.514	0.704	0.867
(7, 9)	0.000	0.001	0.010	0.035	0.108	0.231	0.427	0.622	0.806
(7, 10)	0.000	0.001	0.006	0.024	0.080	0.182	0.355	0.549	0.743
(8, 8)	0.000	0.001	0.009	0.032	0.100	0.214	0.405	0.595	0.786
(8, 9)	0.000	0.001	0.005	0.020	0.069	0.157	0.319	0.500	0.702
(8, 10)	0.000	0.000	0.003	0.013	0.048	0.117	0.251	0.419	0.621
(9, 9)	0.000	0.000	0.003	0.012	0.044	0.109	0.238	0.399	0.601
(9, 10)	0.000	0.000	0.002	0.008	0.029	0.077	0.179	0.319	0.510
(10, 10)	0.000	0.000	0.001	0.004	0.019	0.051	0.128	0.242	0.414

* Reproduced from C. Eisenhart and F. Swed, "Tables for Testing Randomness of Grouping in a Sequence of Alternatives," *Ann. Math. Stat.*, Vol. 14, 1943, by permission of the editor.

Table A.18 (*continued*) $P(V \leq v^*$ when H_0 is true) in the Runs Test

(n_1, n_2)	v^*									
	11	12	13	14	15	16	17	18	19	20
(2, 3)										
(2, 4)										
(2, 5)										
(2, 6)										
(2, 7)										
(2, 8)										
(2, 9)										
(2, 10)										
(3, 3)										
(3, 4)										
(3, 5)										
(3, 6)										
(3, 7)										
(3, 8)										
(3, 9)										
(3, 10)										
(4, 4)										
(4, 5)										
(4, 6)										
(4, 7)										
(4, 8)										
(4, 9)										
(4, 10)										
(5, 5)										
(5, 6)	1.000									
(5, 7)	1.000									
(5, 8)	1.000									
(5, 9)	1.000									
(5, 10)	1.000									
(6, 6)	0.998	1.000								
(6, 7)	0.992	0.999	1.000							
(6, 8)	0.984	0.998	1.000							
(6, 9)	0.972	0.994	1.000							
(6, 10)	0.958	0.990	1.000							
(7, 7)	0.975	0.996	0.999	1.000						
(7, 8)	0.949	0.988	0.998	1.000	1.000					
(7, 9)	0.916	0.975	0.994	0.999	1.000					
(7, 10)	0.879	0.957	0.990	0.998	1.000					
(8, 8)	0.900	0.968	0.991	0.999	1.000	1.000				
(8, 9)	0.843	0.939	0.980	0.996	0.999	1.000	1.000			
(8, 10)	0.782	0.903	0.964	0.990	0.998	1.000	1.000			
(9, 9)	0.762	0.891	0.956	0.988	0.997	1.000	1.000	1.000		
(9, 10)	0.681	0.834	0.923	0.974	0.992	0.999	1.000	1.000	1.000	
(10, 10)	0.586	0.758	0.872	0.949	0.981	0.996	0.999	1.000	1.000	1.000

Table A.19* Sample Size for Two-Sided Nonparametric Tolerance Limits

	γ					
$1 - \alpha$	0.50	0.70	0.90	0.95	0.99	0.995
0.995	336	488	777	947	1,325	1,483
0.99	168	244	388	473	662	740
0.95	34	49	77	93	130	146
0.90	17	24	38	46	64	72
0.85	11	16	25	30	42	47
0.80	9	12	18	22	31	34
0.75	7	10	15	18	24	27
0.70	6	8	12	14	20	22
0.60	4	6	9	10	14	16
0.50	3	5	7	8	11	12

* Reproduced from Tables A-25d of Wilfrid J. Dixon and Frank J. Massey, Jr., *Introduction to Statistical Analysis*, 3rd ed., McGraw-Hill Book Company, New York, 1969. Used with permission of McGraw-Hill Book Company.

Table A.20* Sample Size for One-Sided Nonparametric Tolerance Limits

	γ				
$1 - \alpha$	0.50	0.70	0.95	0.99	0.995
0.995	139	241	598	919	1,379
0.99	69	120	299	459	688
0.95	14	24	59	90	135
0.90	7	12	29	44	66
0.85	5	8	19	29	43
0.80	4	6	14	21	31
0.75	3	5	11	17	25
0.70	2	4	9	13	20
0.60	2	3	6	10	14
0.50	1	2	5	7	10

* Reproduced from Tables A-25e of Wilfrid J. Dixon and Frank J. Massey, Jr., *Introduction to Statistical Analysis*, 3rd ed., McGraw-Hill Book Company, New York, 1969. Used with permission of McGraw-Hill Book Company.

Table A.21* Critical Values of Spearman's Rank Correlation Coefficient

n	$\alpha = 0.05$	$\alpha = 0.025$	$\alpha = 0.01$	$\alpha = 0.005$
5	0.900	—	—	—
6	0.829	0.886	0.943	—
7	0.714	0.786	0.893	—
8	0.643	0.738	0.833	0.881
9	0.600	0.683	0.783	0.833
10	0.564	0.648	0.745	0.794
11	0.523	0.623	0.736	0.818
12	0.497	0.591	0.703	0.780
13	0.475	0.566	0.673	0.745
14	0.457	0.545	0.646	0.716
15	0.441	0.525	0.623	0.689
16	0.425	0.507	0.601	0.666
17	0.412	0.490	0.582	0.645
18	0.399	0.476	0.564	0.625
19	0.388	0.462	0.549	0.608
20	0.377	0.450	0.534	0.591
21	0.368	0.438	0.521	0.576
22	0.359	0.428	0.508	0.562
23	0.351	0.418	0.496	0.549
24	0.343	0.409	0.485	0.537
25	0.336	0.400	0.475	0.526
26	0.329	0.392	0.465	0.515
27	0.323	0.385	0.456	0.505
28	0.317	0.377	0.448	0.496
29	0.311	0.370	0.440	0.487
30	0.305	0.364	0.432	0.478

* Reproduced from E. G. Olds, "Distribution of Sums of Squares of Rank Differences for Small Samples," *Ann. Math. Stat.*, Vol. 9, 1938, by permission of the editor.

Answers to Exercises

Chapter 1

PAGE 8

1. (a) $\{8, 16, 24, 32, 40, 48\}$.
   (b) $\{-5, 1\}$.
   (c) $\{T, HT, HHT, HHH\}$.
   (d) {North America, South America, Europe, Asia, Africa, Australia, Antarctica}.
   (e) $\varnothing$.

2. $\{(x, y) \mid x^2 + y^2 < 9; x > 0, y > 0\}$.

3. $A = C$.

4. (a)

	Red					
Green	1	2	3	4	5	6
1	(1, 1)	(1, 2)	(1, 3)	(1, 4)	(1, 5)	(1, 6)
2	(2, 1)	(2, 2)	(2, 3)	(2, 4)	(2, 5)	(2, 6)
3	(3, 1)	(3, 2)	(3, 3)	(3, 4)	(3, 5)	(3, 6)
4	(4, 1)	(4, 2)	(4, 3)	(4, 4)	(4, 5)	(4, 6)
5	(5, 1)	(5, 2)	(5, 3)	(5, 4)	(5, 5)	(5, 6)
6	(6, 1)	(6, 2)	(6, 3)	(6, 4)	(6, 5)	(6, 6)

   (b) $S = \{(x, y) \mid x = 1, 2, \ldots, 6; y = 1, 2, \ldots, 6\}$.

5. $S = \{1HH, 1HT, 1TH, 1TT, 2H, 2T, 3HH, 3HT, 3TH, 3TT, 4H, 4T, 5HH, 5HT, 5TH, 5TT, 6H, 6T\}$.

6. $S = \{A_1A_2, A_1A_3, A_1A_4, A_2A_3, A_2A_4, A_3A_4\}$.

7. $S_1 = \{MMMM, MMMF, MMFM, MFMM, FMMM, MMFF, MFMF, MFFM, FMFM, FFMM, FMMF, MFFF, FMFF, FFMF, FFFM, FFFF\}$; $S_2 = \{0, 1, 2, 3, 4\}$.

8. (a) $A = \{(3, 6), (4, 5), (4, 6), (5, 4), (5, 5), (5, 6), (6, 3), (6, 4), (6, 5), (6, 6)\}$.
   (b) $B = \{(1, 2), (2, 2), (3, 2), (4, 2), (5, 2), (6, 2), (2, 1), (2, 3), (2, 4), (2, 5), (2, 6)\}$.
   (c) $C = \{(5, 1), (5, 2), (5, 3), (5, 4), (5, 5), (5, 6), (6, 1), (6, 2), (6, 3), (6, 4), (6, 5), (6, 6)\}$.
   (d) $A \cap C = \{(5, 4), (5, 5), (5, 6), (6, 3), (6, 4), (6, 5), (6, 6)\}$.
   (e) $A \cap B = \varnothing$.
   (f) $B \cap C = \{(5, 2), (6, 2)\}$.

(g)

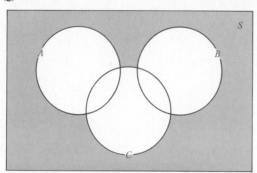

9. (a) $A = \{1HH, 1HT, 1TH, 1TT, 2H, 2T\}$.
   (b) $B = \{1TT, 3TT, 5TT\}$.
   (c) $A' = \{3HH, 3HT, 3TH, 3TT, 4H, 4T, 5HH,$ $5HT, 5TH, 5TT, 6H, 6T\}$.
   (d) $A' \cap B = \{3TT, 5TT\}$.
   (e) $A \cup B = \{1HH, 1HT, 1TH, 1TT, 2H, 2T,$ $3TT, 5TT\}$.

10. (a) $S = \{YYY, YYN, YNY, NYY, YNN, NYN,$ $NNY, NNN\}$.
    (b) $E = \{YYY, YYN, YNY, NYY\}$.
    (c) One possible event: "The second woman interviewed uses brand X."

11. (a) $S = \{M_1M_2, M_1F_1, M_1F_2, M_2M_1, M_2F_1,$ $M_2F_2, F_1M_1, F_1M_2, F_1F_2, F_2M_1, F_2M_2,$ $F_2F_1\}$.
    (b) $A = \{M_1M_2, M_1F_1, M_1F_2, M_2M_1, M_2F_1,$ $M_2F_2\}$.
    (c) $B = \{M_1F_1, M_1F_2, M_2F_1, M_2F_2, F_1M_1,$ $F_1M_2, F_2M_1, F_2M_2\}$.
    (d) $C = \{F_1F_2, F_2F_1\}$.
    (e) $A \cap B = \{M_1F_1, M_1F_2, M_2F_1, M_2F_2\}$.
    (f) $A \cup C = \{M_1M_2, M_1F_1, M_1F_2, M_2M_1,$ $M_2F_1, M_2F_2, F_1F_2, F_2F_1\}$.
    (g)

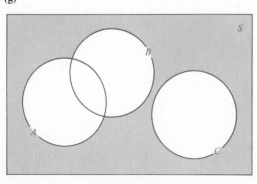

12. $S = \{Vhr, Vhb, Vmr, Vmb, Vcr, Vcb, Nhr, Nhb,$ $Nmr, Nmb, Ncr, Ncb, Chr, Chb, Cmr, Cmb, Ccr,$ $Ccb, Mhr, Mhb, Mmr, Mmb, Mcr, Mcb\}$.

13.

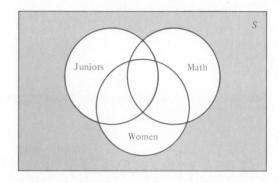

14. (a) $\{0, 2, 3, 4, 5, 6, 8\}$.
    (b) $\varnothing$.   (c) $\{0, 1, 6, 7, 8, 9\}$.
    (d) $\{1, 3, 5, 6, 7, 9\}$.
    (e) $\{0, 1, 6, 7, 8, 9\}$.   (f) $\{2, 4\}$.

15. (a) {nitrogen, potassium, uranium, oxygen}.
    (b) {copper, sodium, zinc, oxygen}.
    (c) {copper, sodium, nitrogen, potassium, uranium, zinc}.
    (d) copper, uranium, zinc}.
    (e) $\varnothing$.   (f) {oxygen}.

16. (a) $M \cup N = \{x \mid 0 < x < 9\}$.
    (b) $M \cap N = \{x \mid 1 < x < 5\}$.
    (c) $M' \cap N' = \{x \mid 9 < x < 12\}$.

17. (a)

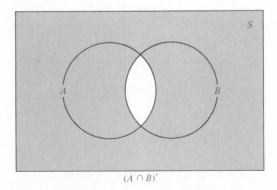

$(A \cap B)'$

(b)

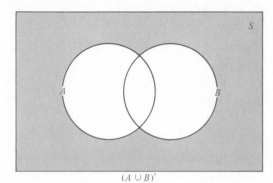

$(A \cup B)'$

(c)

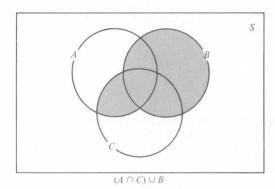

$(A \cap C) \cup B$

18. (a) Not mutually exclusive.
    (b) Mutually exclusive.
    (c) Not mutually exclusive.
    (d) Mutually exclusive.

19. (a) The family will experience mechanical problems but will receive no ticket for a traffic violation and will not arrive at a campsite that has no vacancies.
    (b) The family will receive a traffic ticket and arrive at a campsite that has no vacancies but will not experience mechanical problems.
    (c) The family will experience mechanical problems and will arrive at a campsite that has no vacancies.
    (d) The family will receive a traffic ticket but will not arrive at a campsite that has no vacancies.
    (e) The family will not experience mechanical problems.

20. (a) 6.   (b) 2.   (c) 2, 5, 6.   (d) 4, 5, 7, 8.

PAGE 17

1. 18.
2. 24.
3. 156.
4. 8.
5. 20.
6. 96.
7. 48.
8. 30.
9. 210.
10. 512.
11. (a) 1024.   (b) 243.
12. (a) 5040.   (b) 720.
13. 72.
14. (a) 720.   (b) 144.   (c) 480.
15. 362,880.
16. (a) 180.   (b) 75.   (c) 105.
17. 2880.
18. (a) 40,320.   (b) 384.   (c) 576.
19. (a) 40,320.   (b) 336.
20. 6720.
21. 360.
22. 59,280.
23. 24.
24. 5040.
25. 3360.
26. 1260.
27. 7920.
28. 4400.
29. 56.
30. (a) 21.   (b) 10.
31. (a) 84.   (b) 40.   (c) 15.
32. 1,244,117,160.
33. 800.
34. 288.

PAGE 24

1. (a) Sum of the probabilities exceeds 1.
    (b) Sum of the probabilities is less than 1.
    (c) A negative probability.
    (d) Probability of both a heart and a black card is zero.

2. (a) 5/18.   (b) 1/3.   (c) 7/36.

3. S = {$10, $25, $100}; 17/20.

4. (a) 4/9.   (b) 7/18.   (c) 2/9.

5. (a) 0.8.   (b) 0.7.   (c) 0.5.

6. (a) 0.7   (b) 0.5.   (c) 0.2.

7. (a) 5/26.   (b) 9/26.   (c) 19/26.

8. (a) 3/5.   (b) 2/5.   (c) 1/10.

9. 10/117.

10. (a) 5/36.   (b) 5/18.

11. 95/663.

12. (a) 1/3.   (b) 5/14.

13. (a) 94/54,145.   (b) 143/39,984.

14. (a) 25/108.   (b) 25/1296.

15. (a) 22/25.   (b) 3/25.   (c) 17/50.

16. (a) 22/125.   (b) 31/500.   (c) 171/500.

17. (a) 0.3.   (b) 0.2.

18. (a) 0.75.   (b) 0.25.

19. 13/120.

20. (a) 0.55.   (b) 0.83.   (c) 0.27.

PAGE 33

1. (a) The probability that a convict who pushed dope also committed armed robbery.

    (b) The probability that a convict who committed armed robbery, did not push dope.

    (c) The probability that a convict who did not push dope also did not commit armed robbery.

2. 5/9.

3. (a) 14/39.   (b) 95/112.

4. (a) 30/49.   (b) 16/31.

5. (a) 5/34.   (b) 3/8.

6. (a) 2/11.   (b) 5/11.

7. 6/13.

8. (a) 0.56.   (b) 0.35.

9. (a) 0.35.   (b) 0.875.   (c) 0.55.

10. (a) 0.34.   (b) 5/7.   (c) 1/12.

11. (a) 9/28.   (b) 3/4.   (c) 0.91.

12. 0.24.

13. 0.27.

14. 35/64.

15. 5/8.

16. (a) 25/102.   (b) 20/221.

17. (a) 0.0016.   (b) 0.9984.

18. 0.03.

19. (a) 1/5.   (b) 4/15.   (c) 3/5.

20. 0.018.

21. (a) 91/323.   (b) 91/323.

22. (a) 7/25.   (b) 18/25.

23. (a) 0.384.   (b) 0.512.

24. (a) 1/4.   (b) 1/16.

25. (a) 0.0001.   (b) 0.0081.

PAGE 40

1. 0.0744.

2. 0.27.

3. 0.2097.

4. 1/9.

5. 3/10.

6. 15/23.

7. 0.174.

8. (a) 0.054.   (b) 4/9.

Chapter 2

PAGE 52

1. Discrete; continuous; continuous; discrete; discrete; continuous.

2.

Sample Space	x
NNN	0
NNB	1
NBN	1
BNN	1
NBB	2
BNB	2
BBN	2
BBB	3

3.

Sample Space	w
HHH	3
HHT	1
HTH	1
THH	1
HTT	−1
THT	−1
TTH	−1
TTT	−3

4. $S = \{HHH, \quad THHH, \quad HTHHH, \quad TTHHH,$
$TTTHHH, HTTHHH, THTHHH, HHTHHH,$
$\ldots\}$; discrete.

5. (a) 1/30.   (b) 1/10.

6.

t	20	25	30
$P(T=t)$	$\frac{1}{5}$	$\frac{3}{5}$	$\frac{1}{5}$

7.

x	0	1	2	3
$f(x)$	$\frac{8}{27}$	$\frac{4}{9}$	$\frac{2}{9}$	$\frac{1}{27}$

8.

w	−3	−1	1	3
$P(W=w)$	$\frac{1}{27}$	$\frac{2}{9}$	$\frac{4}{9}$	$\frac{8}{27}$

9. $f(x) = \dfrac{\binom{5}{x}\binom{5}{4-x}}{\binom{10}{4}}$, for $x = 0, 1, 2, 3, 4$.

10. $f(x) = 1/6, x = 1, 2, \ldots, 6$.

11.

x	0	1	2
$f(x)$	$\frac{2}{7}$	$\frac{4}{7}$	$\frac{1}{7}$

12.

x	0	1	2	3
$f(x)$	$\frac{703}{1700}$	$\frac{741}{1700}$	$\frac{117}{850}$	$\frac{11}{850}$

13.
$$F(w) = \begin{cases} 0 & \text{for } w < -3 \\ \frac{1}{27} & \text{for } -3 \le w < -1 \\ \frac{7}{27} & \text{for } -1 \le w < 1 \\ \frac{19}{27} & \text{for } 1 \le w < 3 \\ 1 & \text{for } w \ge 3 \end{cases}$$
(a) 20/27.   (b) 2/3.

15.
$$F(x) = \begin{cases} 0 & \text{for } x < 0 \\ \frac{2}{7} & \text{for } 0 \le x < 1 \\ \frac{6}{7} & \text{for } 1 \le x < 2 \\ 1 & \text{for } x \ge 2 \end{cases}$$
(a) 4/7.   (b) 5/7.

17.
$$F(x) = \begin{cases} 0 & \text{for } x < 0 \\ 0.41 & \text{for } 0 \le x < 1 \\ 0.78 & \text{for } 1 \le x < 2 \\ 0.94 & \text{for } 2 \le x < 3 \\ 0.99 & \text{for } 3 \le x < 4 \\ 1 & \text{for } x \ge 4 \end{cases}$$

18. (a) 1/4.   (b) 1/2.   (c) 1/2.

19. (b) 1/4.   (c) 0.3.

20. (a) 16/27.   (b) 1/3.

21. (b) 19/80.

22. (a) 0.68.   (b) 0.375.

23. $F(x) = (x-1)/2$; 1/4.

24. $F(x) = (x+4)(x-2)/27$; 1/3.

25. (a) 3/2.   (b) $F(x) = x^{3/2}$; 0.3004.

26. (a) 1/9.   (b) 0.1020.

27. (a) 0.7981.   (b) 0.7981.

PAGE 59

1. (a) Stems are 1, 2, 3, 4, 5, 6, 7, 8, 9, with frequencies 3, 2, 3, 4, 5, 11, 14, 14, 4.
  (b) Class intervals are 10–19, 20–29, 30–39, 40–49, 50–59, 60–69, 70–79, 80–89, 90–99 with relative frequencies 0.05, 0.03, 0.05, 0.07, 0.08, 0.18, 0.23, 0.23, 0.07.
  (c) Skewed to the left.
  (d) Relative cumulative frequencies are 0, 0.05, 0.08, 0.13, 0.20, 0.28, 0.46, 0.69, 0.92, 0.99.
  (f) 1st quartile $\simeq$ 55.5; 7th decile $\simeq$ 80.

2. (a) Stems are 0, 1, 2, 3, 4, 5, 6 with frequencies 8, 6, 3, 2, 3, 4, 4.
  (b) Relative frequencies are 0.267, 0.200, 0.100, 0.067, 0.100, 0.133, 0.133.
  (c) Relative cumulative frequencies are 0, 0.267, 0.467, 0.567, 0.634, 0.734, 0.867, 1.000.
  (e) Approximately 4.28 years.

3. (a) Stems are 0∗, 0·, 1∗, 1·, 2∗, 2·, 3∗ with frequencies 2, 17, 16, 10, 3, 1, 1.
(b) Relative frequencies are 0.04, 0.34, 0.32, 0.20, 0.06, 0.02, 0.02.
(c) Skewed to the right.
(d) Relative cumulative frequencies are 0, 0.04, 0.38, 0.70, 0.90, 0.96, 0.98, 1.00.
(f) 75th percentile $\simeq$ 15.9.

4. Stems are 1*d*, 1*e*, 2*a*, 2*b*, 2*c*, 2*d*, 2*e*, 3*a*, 3*b*, 3*c*, 3*d*, 3*e*, 4*a*, 4*b*, 4*c*, 4*d* with frequencies 1, 1, 0, 1, 1, 2, 1, 6, 6, 5, 4, 4, 2, 2, 2, 2.

5. (a) Stems are 0*d*, 0*e*, 1*a*, 1*b*, 1*c*, 1*d*, 1*e*, 2*a*, 2*b*, 2*c* with frequencies 1, 1, 1, 2, 4, 13, 8, 5, 3, 2.
(b) Relative frequencies are 0.025, 0.025, 0.025, 0.050, 0.100, 0.325, 0.200, 0.125, 0.075, 0.050.

PAGE 71

1. (a) 1/36.   (b) 1/15.

2. (a) 1/5.   (b) 7/30.   (c) 3/5.   (d) 4/15.

3. (a)

$f(x, y)$		x		
	0	1	2	3
0		$\frac{3}{70}$	$\frac{9}{70}$	$\frac{3}{70}$
y 1	$\frac{2}{70}$	$\frac{18}{70}$	$\frac{18}{70}$	$\frac{2}{70}$
2	$\frac{3}{70}$	$\frac{9}{70}$	$\frac{3}{70}$	

(b) 1/2.

4. (a)

$f(x, y)$		x	
	0	1	2
0	$\frac{16}{36}$	$\frac{8}{36}$	$\frac{1}{36}$
y 1	$\frac{8}{36}$	$\frac{2}{36}$	
2	$\frac{1}{36}$		

(b) 11/12.

5.

$f(x, y)$		x		
	0	1	2	3
-3	$\frac{1}{8}$			
y -1		$\frac{3}{8}$		
1			$\frac{3}{8}$	
3				$\frac{1}{8}$

6. (a)

$f(x, y)$		x		
	0	1	2	3
0	$\frac{1}{55}$	$\frac{6}{55}$	$\frac{6}{55}$	$\frac{1}{55}$
y 1	$\frac{6}{55}$	$\frac{16}{55}$	$\frac{6}{55}$	
2	$\frac{6}{55}$	$\frac{6}{55}$		
3	$\frac{1}{55}$			

(b) 42/55.

7. (a) 135/1024.   (b) 1/2.

8. (a) 1/50.   (b) 13/75.   (c) 14/25.
(d) 8/15.

9. 0.6534.

10. (a)

x	0	1	2	3
$g(x)$	$\frac{1}{10}$	$\frac{1}{5}$	$\frac{3}{10}$	$\frac{2}{5}$

(b)

y	0	1	2
$h(y)$	$\frac{1}{5}$	$\frac{1}{3}$	$\frac{7}{15}$

11. (a)

$f(w, z)$		w	
	0	1	2
z 0	0.36	0.24	
1		0.24	0.16

(b)

w	0	1	2
$g(w)$	0.36	0.48	0.16

(c)

z	0	1
$h(z)$	0.60	0.40

(d) 0.64.

12. (a)

y	0	1	2
$f(y \mid 2)$	$\frac{3}{10}$	$\frac{3}{5}$	$\frac{1}{10}$

(b) 3/10.

13. (a)

x	1	2	3
$g(x)$	$\frac{1}{3}$	$\frac{19}{36}$	$\frac{5}{36}$

(b)

y	1	2	3
$h(y)$	$\frac{1}{4}$	$\frac{14}{45}$	$\frac{79}{180}$

(c) 9/19.

14. (a)

x	2	4
$g(x)$	0.40	0.60

(b)

y	1	3	5
$h(y)$	0.25	0.50	0.25

15.

(a) $g(x) = \begin{cases} \dfrac{2(x+1)}{3}, & 0 \le x \le 1 \\ 0, & \text{elsewhere.} \end{cases}$

(b) $h(y) = \begin{cases} \dfrac{(1+4y)}{3}, & 0 \le y \le 1 \\ 0, & \text{elsewhere.} \end{cases}$

(c) 5/12.

16. (a) 1/16. (b) $g(x) = 12x(1-x)^2$, $0 \le x \le 1$.
(c) 1/4.

17. 3/4.

18. 0.6321.

19. Dependent.

20. Independent.

21. (a) Dependent. (b) 1/3.

22. (b) 0.64.

23. (a) 3. (b) 21/512.

24. Independent.

25. Dependent.

26. (a) $g(y, z) = 2yz^2/9$,
$0 < y < 1, 0 < z < 3$.
(b) $h(y) = 2y, 0 < y < 1$.
(c) 7/162.
(d) 1/4.

Chapter 3

PAGE 81

1. 6/7.

2. 3/4.

3. 25¢.

4. 1/2.

5. 0.88.

6. $12.67.

7. $500.

8. $88.

9. $1.23.

10. Fair game.

11. $2100.

12. $333.

13. $(\ln 4)/\pi$.

14. 8/15.

15. 100 hours.

16. 49.5.

17. 209.

18. 9/8.

19. $1855.

20. 3.

21. $167.

22. 8.

23. (a) 35.2.   (b) $\mu_X = 3.20; \mu_Y = 3.00$.

24. (a) $-3/7$.   (b) $\mu_X = 3/2; \mu_Y = 1$.

25. 2.

26. 0.9752.

27. 8/3.

28. (a) $\mu_X = 5/9; \mu_Y = 11/18$.   (b) 7/12.

29. 4/5 kilogram.

30. 333 liters.

PAGE 90

1. 20/49.

2. 3.041.

3. 0.74.

4. $\mu = 1; \sigma^2 = 1$.

5. 1/18.

6. 5/30.

7. 1/6.

8. 1342.25.

9. 118.9.

10. 7/180.

11. $\mu_Y = 10$; $\sigma_Y^2 = 144$.

12. $-3/14$.

13. -0.1244.

14. -0.036.

15. $-1/162$.

16. $-2/75$.

17. $-1/144$.

PAGE 101

1. 10.33.

2. 8.

3. 80 cents.

4. $\mu_Y = 10$; $\sigma_Y^2 = 144$.

5. 209.

6. 109 kilowatt hours.

7. $\mu = 7/2$; $\sigma^2 = 15/4$.

8. (a) -2.60.　　(b) 9.60.

9. 3/14.

10. (a) 7.　　(b) 0.　　(c) 12.25.

11. 8/3.

12. 1.

13. (a) 175/12.　　(b) 175/6.

14. 68.

15. 52.

16. (a) At least 3/4.　　(b) At least 8/9.

Chapter 4

PAGE 111

1. 3/10.

2. $f(x) = 1/25$, $x = 1, 2, \ldots, 25$.

3. $\mu = 5.5$; $\sigma^2 = 8.25$.

4. (a) 0.0879.　　(b) 0.3672.

5. (a) 16/81.　　(b) 64/81.

6. (a) 0.647.　　(b) 0.680.

7. $f(x) = 1/2$, $x = 0, 1$. Uniform and binomial.

8. (a) 0.1239.　　(b) 0.5941.

9. (a) 0.7073.　　(b) 0.4613.　　(c) 0.1484.

10. (a) 0.6294.　　(b) 0.0386.　　(c) 0.7237.

11. 0.1240.

12. 0.8343.

13. 0.8369.

14. 0.0006.

15. (a) 0.0778.　　(b) 0.3370.　　(c) 0.0870.

16. 0.8208 and 0.8400; 2-engine plane.

17. 0.9728 and 0.9600; 4-engine plane.

18. $\mu = 6.3$; $\sigma^2 = 0.63$.

19. $\mu = 4$; $\sigma^2 = 3.2$.

20. (a) $\mu = 8.4$; $\sigma^2 = 2.52$.　　(b) At least 3/4.

21. $\mu \pm 2\sigma = 3.5 \pm 2.05$.

22. (a) 3.75.　　(b) From 0.396 to 7.104.

23. (a) 32.　　(b) At least 8/9.

24. 15/128.

25. 0.0095.

26. 21/256.

27. 0.0077.

PAGE 120

1. (a) 0.3246.　　(b) 0.4773.

2. 53/65.

3. 5/14.

4. (a) 1/6.　　(b) 29/30.

5. $h(x; 6, 3, 4) = \dfrac{\binom{4}{x}\binom{2}{2-x}}{\binom{6}{3}}$, $x = 1, 2, 3$;

　　$P(2 \leq X \leq 3) = 4/5$.

6. 10/21.

7. 0.9517.

8. (a) 71/115.　　(b) 3/25.

9. (a) 0.6815.　　(b) 0.1153.

10. $\mu = 1.2$.

11. 3.25; From 0.52 to 5.98.

12. 0.2131.

13. 0.9453.

14. 0.3222.

15. 0.7758.

16. 0.0129.

17. (a) 4/33.   (b) 8/165.

18. 17/63.

19. 0.0308.

PAGE 127

1. 0.0515.

2. 0.0651.

3. (a) 0.1638.   (b) 0.032.

4. (a) 0.1172.   (b) 1/16.

5. 63/64.

6. (a) 2/243.   (b) 16/81.

7. (a) 0.0630.   (b) 0.9730.

8. (a) 0.1008.   (b) 0.4232.   (c) 0.8009.

9. (a) 0.1429.   (b) 0.1353.

10. (a) 0.1512.   (b) 0.4015.

11. (a) 0.3840.   (b) 0.0067.

12. (a) 0.0458.   (b) 0.0060.

13. (a) 0.3840.   (b) 0.1395.   (c) 0.0553.

14. 0.6288.

15. 0.2657.

16. (a) 0.1321.   (b) 0.3376.

17. (a) $\mu = 4; \sigma^2 = 4$.   (b) From 0 to 8.

18. (a) $\mu = 10; \sigma^2 = 10$.
     (b) From 0.51 to 19.49.

Chapter 5

PAGE 144

1. (a) 0.9236.   (b) 0.8133.
     (c) 0.2424.   (d) 0.0823.
     (e) 0.0250.   (f) 0.6435.

2. (a) 0.35.   (b) -1.21.
     (c) 2.14.   (d) 1.96.

3. (a) -1.72.   (b) 0.54.   (c) 1.28.

4. (a) 0.9850.   (b) 0.0918.
     (c) 0.3371.   (d) 35.04.
     (e) 23.1 and 36.9.

5. (a) 0.1151.   (b) 16.1.
     (c) 20.275.   (d) 0.5403.

6. 0.9974.

7. (a) 0.8980.   (b) 0.0287.   (c) 0.6080.

8. (a) 19.77%.   (b) 59.67%.   (c) 1.22%.

9. (a) 0.0548.   (b) 0.4514.
     (c) 23.   (d) 189.95 milliliters.

10. (a) 0.0062.   (b) 0.6826.
     (c) 9.969 centimeters.

11. (a) 0.0571.   (b) 99.11%.
     (c) 0.3974.   (d) 27.952 minutes.
     (e) 0.0092.

12. (a) 64.   (b) 86.   (c) 78.

13. 62.

14. (a) 16.   (b) 549.   (c) 28.   (d) 27.

15. (a) 56.99%.   (b) $10.23.

16. (a) 0.0427.   (b) 0.7642.   (c) 0.6964.

17. (a) 0.0401.   (b) 0.0244.

18. (a) 19.36%.   (b) 39.70%.

19. 26.

20. (a) 0.0045.   (b) 0.1496.   (c) 0.0526.

21. 6.24 years.

PAGE 151

1. (a) 0.8006.   (b) 0.7803.

2. (a) 0.7925.   (b) 0.0352.   (c) 0.0101.

3. (a) 0.8643.   (b) 0.2978.   (c) 0.0796.

4. (a) 0.1210.   (b) 0.2033.

5. (a) 0.9514.   (b) 0.0668.

6. (a) 0.9966.   (b) 0.1841.

7. (a) 0.1171.   (b) 0.2049.

8. (a) 0.0838.   (b) 0.1635.

9. 0.1357.

10. 0.4364.

11. (a) 0.0778.   (b) 0.0571.   (c) 0.6811.

12. 0.6086.

13. 0.1737.

PAGE 160

1. $2.8e^{-1.8} - 3.4e^{-2.4} = 0.1545$.

2. $4e^{-3} = 0.1992$.

4. (a) $1 - 3e^{-2} = 0.5940$.
     (b) $5e^{-4} = 0.0916$.

5. (a) $\mu = 6; \sigma^2 = 18$.
     (b) From 0 to 14.485 million liters.

6. (a) $\alpha = 3; \beta = 2$.
   (b) $25e^{-6} = 0.0620$.

7. $\sum_{x=4}^{6} \binom{6}{x}(1 - e^{-3/4})^x (e^{-3/4})^{6-x} = 0.3968$.

8. 0.0352.

9. (a) 3/5.   (b) 1/2.

10. (a) 0.6.   (b) 0.7.   (c) 0.5.

12. (a) $\sqrt{\pi/2} = 1.2533$.   (b) $e^{-2} = 0.1353$.

14. $e^{-4} = 0.0183$.

15. 0.1808.

16. $e^{-1.5} = 0.223$.

17. 0.577.

Chapter 6

PAGE 178

1. $g(y) = 1/3$, $y = 1, 3, 5$.

2. $g(y) = \binom{3}{\sqrt{y}}\left(\frac{2}{5}\right)^{\sqrt{y}}\left(\frac{3}{5}\right)^{3-\sqrt{y}}$, $y = 0, 1, 4, 9$.

3. $g(y_1, y_2) = \left(\dfrac{\overset{2}{\overbrace{y_1 + y_2}}}{2}, \dfrac{y_1 - y_2}{2}, 2 - y_1\right)$
$\times \left(\dfrac{1}{4}\right)^{(y_1+y_2)/2}\left(\dfrac{1}{3}\right)^{(y_1-y_2)/2}\left(\dfrac{5}{12}\right)^{2-y_1}$;

$y_1 = 0, 1, 2$; $y_2 = -2, -1, 0, 1, 2$;
$y_2 \le y_1$; $y_1 + y_2 = 0, 2, 4$.

4.

y	1	2	3	4	6
$h(y)$	$\frac{1}{18}$	$\frac{2}{9}$	$\frac{1}{6}$	$\frac{2}{9}$	$\frac{1}{3}$

6. $g(y) = 1/6y^{1/3}$, $0 < y < 8$.

7. Gamma distribution with $\alpha = 3/2$ and $\beta = m/2b$.

8. (a) $g(y) = y^{-1/2} - 1$, $0 < y < 1$.
   (b) 3/4.

9. (a) $g(y) = 32/y^3$, $y > 4$.
   (b) 1/4.

10. (a) $g(z) = 4z^3$, $0 < z < 1$.
   (b) 65/256.

11. $h(z) = 2(1 - z)$, $0 < z < 1$.

13. $h(w) = 6 + 6w - 12w^{1/2}$, $0 < w < 1$.

14. $g(y) = 1/2\sqrt{y}$, $0 < y < 1$.

15. $g(y) = \begin{cases} \dfrac{2}{9\sqrt{y}}, & 0 < y < 1 \\ \dfrac{(\sqrt{y}+1)}{9\sqrt{y}}, & 1 < y < 4. \end{cases}$

18. $\mu = 1/p; \sigma^2 = q/p^2$.

19. Both equal μ.

20. 0.9306.

24. (a) 0.0724.   (b) 0.2424.

25. (a) 0.4751.   (b) 0.5697.

PAGE 189

1. (a) Responses of all people in Richmond who have a telephone.
   (b) Outcomes for a large or infinite number of tosses of a coin.
   (c) Length of life of such tennis shoes when worn on the professional tour.
   (d) All possible time intervals for this lawyer to drive from her home to her office.

2. (a) Number of tickets issued by all state troopers in Montgomery County during the Memorial Day weekend.
   (b) Number of tickets issued by all state troopers in South Carolina during the Memorial Day weekend.

3. (a) $\bar{x} = 2.4$.   (b) $\tilde{x} = 2$.   (c) $m = 3$.

4. (a) $\bar{x} = 8.6$ minutes.   (b) $\tilde{x} = 9.5$ minutes.
   (c) Modes are 5 and 10 minutes.

5. (a) $\bar{x} = 3.2$ seconds.   (b) $\tilde{x} = 3.1$ seconds.

6. (a) $\bar{x} = 35.7$ grams.   (b) $\tilde{x} = 32.5$ grams.
   (c) $m = 29$ grams.

7. (a) $\bar{x} = \$22.50$.   (b) Modes are \$10 and \$25.

8. $\bar{x} = 22.2$ days; $\tilde{x} = 14$ days; $m = 8$ days.

9. (a) Range is 10.   (b) $s = 3.307$.

10. (a) Range is 2.0.   (b) $s^2 = 0.498$.

11. (a) 2.971.   (b) 2.971.

12. (a) 11.69 milligrams.   (b) $s^2 = 10.776$.

13. $s = 0.585$.

15. (a) 45.9.   (b) 5.1.

16. $s = 2.653$.

17. $s^2 = \$1.147 \times 10^8$.

18. $s = 5801$ kilometers.

PAGE 198

1. 0.3159.

2. 0.1912.

3. (a) Reduced from 0.7 to 0.4.
   (b) Increased from 0.2 to 0.8.

4. 100.

5. Yes.

6. (a) $\mu_{\bar{X}} = 174.5; \sigma_{\bar{X}} = 1.38$.
   (b) Approximately 154.
   (c) Approximately 6.

7. (a) $\mu = 5.3; \sigma^2 = 0.81$.
   (b) $\mu_{\bar{X}} = 5.3; \sigma_{\bar{X}}^2 = 0.0225$.
   (c) 0.9082.

8. 0.0668.

9. (a) 0.6898.   (b) 5.35 years.

10. (a) 0.0062.   (b) 0.0668.   (c) 0.3413.

12. 0.7070.

13. 0.5596.

14. (a) 0.0768.   (b) 0.2812.

15. 0.9052.

PAGE 210

1. (a) 27.488.   (b) 18.475.   (c) 36.415.

2. (a) 16.750.   (b) 30.144.   (c) 26.217.

3. (a) 13.277.   (b) 32.852.   (c) 46.928.

4. (a) 38.932.   (b) 12.592.   (c) 20.483.

5. (a) 0.05.   (b) 0.94.

6. Not valid.

8. (a) 2.145.   (b) -1.372.   (c) -3.499.

9. (a) 0.975.   (b) 0.10.
   (c) 0.875.   (d) 0.99.

10. (a) 0.985.   (b) 0.975.

11. (a) 2.500.   (b) 1.319.   (c) 1.714.

12. $t = -2.000$; valid claim.

13. No; $\mu > 20$.

14. $t = 1.64$; yes.

15. (a) 2.71.   (b) 3.51.   (c) 2.92.
    (d) 0.47.   (e) 0.34.

16. 0.05.

17. 0.99.

Chapter 7

PAGE 224

4. $765 < \mu < 795$.

5. $2.20 < \mu < 2.30$.

6. (a) $172.23 < \mu < 176.77$.   (b) Error ≤ 2.27.

7. (a) $22{,}496 < \mu < 24{,}504$.   (b) Error ≤ 1004.

8. 68.

9. 11.

10. 28.

11. 56.

12. $10.15 < \mu < 12.45$.

13. $0.978 < \mu < 1.033$.

14. $1.49 < \mu < 3.71$.

15. $47.722 < \mu < 49.278$.

16. $74.34 < \mu < 84.26$.

17. 0.925 to 1.675.

18. 0.382 to 7.192.

19. 11,426 to 35,574.

20. 44.52 to 52.48.

PAGE 233

1. $2.9 < \mu_1 - \mu_2 < 7.1$.

2. $6.56 < \mu_B - \mu_A < 11.24$.

3. $2.80 < \mu_1 - \mu_2 < 3.40$.

4. $0.69 < \mu_1 - \mu_2 < 7.31$.

5. $1.5 < \mu_1 - \mu_2 < 12.5$.

6. $0.033 < \mu_2 - \mu_1 < 0.299$.

7. $0.70 < \mu_2 - \mu_1 < 3.30$; valid claim.

8. $4.3 < \mu_1 - \mu_2 < 5.7$.

9. $-6522 < \mu_1 - \mu_2 < 2922$.

10. $-11.9 < \mu_{II} - \mu_I < 36.5$.

11. $-1.3 < \mu_D < 6.8$.

12. $-2912 < \mu_D < 687$.

13. $0.990 < \mu_D < 6.124$; valid claim.

14. $-0.12 < \mu_D < 3.12$; valid claim.

15. $30.55 < \mu_D < 50.61$.

16. $-197.7 < \mu_D < 1032.7$.

PAGE 243

1. (a) $0.498 < p < 0.642$.   (b) Error ≤ 0.072.

2. (a) $0.1442 < p < 0.1998$.

   (b) Error ≤ 0.0278.

3. $0.194 < p < 0.262$.

4. $0.017 < p < 0.143$.

5. (a) $0.739 < p < 0.961$.   (b) No.

6. (a) $0.130 < p < 0.350$.   (b) Error ≤ 0.110.

7. (a) $0.644 < p < 0.690$.   (b) Error ≤ 0.023.

8. 2090.

9. 2576.

10. 467.

11. 160.

12. 9604.

13. 16,577.

14. 601.

15. $-0.0136 < p_F - p_M < 0.0636$.

16. $0.016 < p_A - p_B < 0.164$; valid claim.

17. $0.0011 < p_1 - p_2 < 0.0869$.

18. $-0.057 < p_B - p_A < 0.177$.

19. $0.067 < p_1 - p_2 < 0.193$.

PAGE 248

1. $0.293 < \sigma^2 < 6.736$; valid claim.

2. $8.400 < \sigma^2 < 39.827$.

3. $1.863 < \sigma < 3.578$.

4. $0.00022 < \sigma^2 < 0.00357$.

5. $1.410 < \sigma < 6.385$.

6. $1.258 < \sigma^2 < 5.410$.

7. $0.549 < \sigma_1/\sigma_2 < 2.690$.

8. $0.238 < \sigma_1^2/\sigma_2^2 < 1.895$; yes.

9. $0.016 < \sigma_1^2/\sigma_2^2 < 0.454$; no.

PAGE 257

1. $p^* = 0.173$.

2. (a)

p	0.05	0.10	0.15
$f(p\,|\,x=2)$	0.12	0.55	0.33

   (b) $p^* = 0.111$.

3. (a) $f(p\,|\,x=1) = 40p(1-p)^3/0.2844$.

   (b) $p^* = 0.106$.

4. (a)

p	0.6	0.7
$f(p\,|\,x=12)$	0.228	0.772

   (b) $p^* = 0.677$.

5. $8.077 < \mu < 8.692$.

6. (a) \$7.04.   (b) $\$6.65 < \mu < \$7.43$.

   (c) 0.6532.

7. (a) 0.2509.   (b) $68.71 < \mu < 71.69$.

   (c) 0.0174.

8. $f(\mu\,|\,x_1, x_2, \ldots, x_{25}) = \dfrac{1}{\sqrt{2\pi}\,13.706}$

   $\times\, e^{-1/2[(\mu-780)/20]^2}$, $770 < \mu < 830$.

10. $R(\hat{P}; p) = pq/n$.

11.
$$R(\Theta; \theta) = \begin{cases} 0 & \text{for } \theta = 0 \\ \frac{2}{3} & \text{for } \theta = 1 \\ \frac{2}{3} & \text{for } \theta = 2 \\ 0 & \text{for } \theta = 3. \end{cases}$$

12.
$$R(\Theta_2; \theta) = \begin{cases} 0 & \text{for } \theta = 0 \\ \frac{1}{3} & \text{for } \theta = 1 \\ 1 & \text{for } \theta = 2 \\ 0 & \text{for } \theta = 3. \end{cases}$$

13. $\hat{\Theta}_1$.

14. $\hat{\Theta}_2$.

Chapter 8

PAGE 272

1. (a) Conclude that fewer than 30% of the public are allergic to some cheese products when in fact 30% or more are allergic.

   (b) Conclude that at least 30% of the public are allergic to some cheese products when in fact fewer than 30% are allergic.

2. (a) The training course is effective.

   (b) The training course is effective.

3. (a) The firm is not guilty.

   (b) The firm is guilty.

4. (a) $\alpha = 0.0853$.

   (b) $\beta = 0.8287$; $\beta = 0.7817$.

   (c) No.

5. (a) $\alpha = 0.0536$.

   (b) $\beta = 0.0918$; $\beta = 0.1401$.

   (c) Fair.

6. (a) $\alpha = 0.0548$.

   (b) $\beta = 0.3504$; $\beta = 0.6177$; $\beta = 0.8281$.

7. (a) $\alpha = 0.0559$.
   (b) $\beta = 0.0017$; $\beta = 0.0968$; $\beta = 0.5557$.

8. (a) $\alpha = 0.0850$. (b) $\beta = 0.3409$.

9. (a) $\alpha = 0.0032$. (b) $\beta = 0.0062$.

10. (a) $\alpha = 0.4199$. (b) $\beta = 0.3529$.

11. (a) $\alpha = 0.1357$. (b) $\beta = 0.2578$.

12. (a) $\alpha = 0.0466$. (b) $\beta = 0.0022$.

13. $\alpha = 0.0094$; $\beta = 0.0122$.

14. (a) $\alpha = 0.0793$. (b) $\beta = 0.0793$; $\beta = 0.5$.

15. (a) $\alpha = 0.0718$. (b) $\beta = 0.1151$.

16. (a) $\alpha = 0.0026$. (b) $\beta = 0.0228$.

17. (a) $\alpha = 0.0384$. (b) $\beta = 0.5$; $\beta = 0.2776$.

18.

Value of μ	Probability of Accepting H_0
184	0.08
188	0.27
192	0.58
196	0.84
200	0.93
204	0.84
208	0.58
212	0.27
216	0.08

19. (a) H_0: $\mu = 21.8$, H_1: $\mu \neq 21.8$; critical region in both tails.
    (b) H_0: $p = 0.2$, H_1: $p > 0.2$; critical region in right tail.
    (c) H_0: $\mu = 6.2$, H_1: $\mu > 6.2$; critical region in right tail.
    (d) H_0: $p = 0.7$, H_1: $p < 0.7$; critical region in left tail.
    (e) H_0: $p = 0.58$, H_1: $p \neq 0.58$; critical region in both tails.
    (f) H_0: $\mu = 340$, H_1: $\mu < 340$; critical region in left tail.

20. (a) H_0: $p = 0.2$, H_1: $p > 0.2$; critical region in right tail.
    (b) H_0: $\mu = 3$, H_1: $\mu \neq 3$; critical region in both tails.
    (c) H_0: $p = 0.15$, H_1: $p < 0.15$; critical region in left tail.

(d) H_0: $\mu = \$10$, H_1: $\mu > \$10$; critical region in right tail.
(e) H_0: $\mu = 9$, H_1: $\mu \neq 9$; critical region in both tails.

PAGE 285

1. $z = -1.64$; accept $\mu = 800$ hours.

2. $z = -1.27$; accept $\mu = 22.2$ deciliters.

3. $z = -2.76$; yes, $\mu < 40$ months.

4. $z = 2.72$; yes, $\mu > 162.5$ centimeters.

5. $z = 8.97$; yes, $\mu > 20{,}000$ kilometers.

6. $z = 1.58$; no, accept $\mu = 125$ hours.

7. $t = 0.77$; accept $\mu = 10$ liters.

8. $t = 4.38$; yes, $\mu > 220$ milligrams.

9. $t = 1.41$; accept $\mu = 3.5$ milligrams.

10. $t = 1.78$; accept $\mu = \$10.00$.

11. $t = -1.98$; accept $\mu = 35$ minutes.

12. $z = 4.22$; reject H_0, $\mu_1 > \mu_2$.

13. $z = -2.60$; reject H_0, $\mu_A - \mu_B < 12$ kilograms.

14. $z = 2.45$; reject H_0, $\mu_1 - \mu_2 > \$500$.

15. $t = 1.50$; no, $\mu_1 - \mu_2 = 0.5$ micromoles per 30 minutes.

16. $t = -0.92$; yes, $\mu_1 - \mu_2 = 8$.

17. $t = +0.70$; not effective.

18. $t = -0.84$; accept $\mu_1 = \mu_2$.

19. $t = 2.55$; reject H_0, $\mu_1 - \mu_2 > 4$ kilometers.

20. $t = -2.07$; reject H_0, $\mu_1 - \mu_2 < 8$ months.

21. $t' = 0.22$; accept $\mu_{II} - \mu_I = 10$ minutes.

22. $t' = -0.42$; accept $\mu_S = \mu_N$.

23. $t' = 2.76$; no, $\mu_1 > \mu_2$.

24. $t = -1.58$; yes, $\mu_1 = \mu_2$.

25. $t = 2.45$; yes.

26. $t = -0.90$; accept $\mu_1 - \mu_2 = 4.5$.

27. $t = -2.53$; valid claim.

28. $t = 2.64$; yes.

29. 22.

30. 21.

31. 79.

32. 48.

33. 10.

34. 68.

PAGE 294

1. $P = 0.3916$; accept $p = 0.2$.
2. $P = 0.2131$; no, $p = 0.4$.
3. $P = 0.0207$; yes, the coin is not balanced.
4. $z = -1.44$; $p = 0.6$.
5. $z = 2.85$; reject H_0, $p > 1/5$.
6. $z = 1.34$; valid estimate.
7. $z = 1.44$; valid claim.
8. $z = 1.33$; no increase.
9. $z = 2.40$; yes.
10. $z = 2.42$; yes.
11. $z = 1.88$; yes.
12. $z = 1.11$; no.
13. $z = 2.12$; yes.
14. (a) $z = 2.18$; reject H_0, $p_1 - p_2 > 0.3$.
    (b) $z = -0.29$; accept $p_A - p_B = 0.10$.

PAGE 298

1. $\chi^2 = 18.12$; accept $\sigma^2 = 0.03$.
2. $\chi^2 = 10.74$; accept $\sigma = 6$.
3. $\chi^2 = 17.45$; reject H_0, $\sigma^2 > 1.3$.
4. $\chi^2 = 17.19$; accept $\sigma = 1.40$.
5. $\chi^2 = 42.37$; machine is out of control.
6. (a) $z = 2.64$; reject H_0, $\sigma > 7.5$.
    (b) $z = -1.92$; no.
7. $f = 1.33$; accept $\sigma_1^2 = \sigma_2^2$.
8. $f = 10.09$; reject H_0, $\sigma_1^2 > \sigma_2^2$.
9. $f = 0.75$; accept $\sigma_1 = \sigma_2$.
10. $f = 0.086$; reject H_0, $\sigma_1^2 < \sigma_2^2$.
11. $f = 1.18$; accept $\sigma_A = \sigma_B$.

PAGE 310

1. $\chi^2 = 4.47$; yes.
2. $\chi^2 = 6.76$; no.
3. $\chi^2 = 10.14$; reject H_0, ratio is not $5 : 2 : 2 : 1$.
4. $\chi^2 = 10.00$; reject H_0, distribution is not uniform.
5. $\chi^2 = 2.33$; accept H_0, binomial distribution.
6. $\chi^2 = 1.67$; accept H_0, hypergeometric distribution.
7. $\chi^2 = 2.57$; accept H_0, geometric distribution.
10. $\chi^2 = 12.78$; reject H_0, not normal.

11. $\chi^2 = 5.19$; accept H_0, normal distribution.
12. $\chi^2 = 14.60$; not independent.
13. $\chi^2 = 5.40$; independent.
14. $\chi^2 = 7.54$; independent.
15. $\chi^2 = 124.59$; yes.
16. $\chi^2 = 3.81$; equally effective.
17. $\chi^2 = 31.17$; attitudes are not homogeneous.
18. $\chi^2 = 5.78$; no.
19. $\chi^2 = 5.92$; proportions are equal.
20. $\chi^2 = 12.56$; proportions are different.
21. $\chi^2 = 1.84$; proportions are the same.
22. $\chi^2 = 1.39$; no difference.
23. $\chi^2 = 10.19$; proportions are different.
24. $\chi^2 = 8.84$; yes.

Chapter 9

PAGE 321

1. (a) $a = 64.52916$, $b = 0.56090$.
    (b) $\hat{y} = 81.4$.
2. (a) $\hat{y} = 12.0623 + 0.7771x$.
    (b) $\hat{y} = 78$.
3. (a) $\hat{y} = 6.4136 + 1.8091x$.
    (b) $\hat{y} = 9.580$.
4. (a) $\hat{y} = 42.5818 - 0.6861x$.
    (b) $\hat{y} = 25.7724$.
5. (a) $\hat{y} = 5.8254 + 0.5676x$.
    (c) $\hat{y} = 34.205$.
6. (b) $\hat{y} = 32.5059 + 0.4711x$.
    (d) $x = 59$.
7. (b) $\hat{y} = 343.706 + 3.221x$.
    (c) $\hat{y} = \$456$.
8. (a) $\hat{y} = 2.776 - 0.180x$.
    (b) $\hat{y} = 2.24$.
9. (a) $\hat{y} = 153.175 - 6.324x$.
    (b) $\hat{y} = 123$.
10. (a) $\hat{z} = 6461.392 \times 0.947^w$.
    (b) $\hat{z} = \$5197$.
11. (a) $\hat{y} = 2.654$; $\hat{C} = 2.57 \times 10^6$.
    (b) $\hat{P} = 22.9$ kg/cm^2.
12. (a) $\hat{v} = 4.604 \times s^{-0.059}$.
    (b) $\hat{v} = 4.0$ meters per second.

PAGE 334

3. (a) $s^2 = 176.362$.
  (b) $t = 2.04$, accept $\beta = 0$.

4. (a) $s^2 = 379.150$.
  (b) $-69.913 < \alpha < 94.038$.
  (c) $-0.248 < \beta < 1.802$.

5. (a) $s^2 = 0.40$.
  (b) $4.324 < \alpha < 8.503$.
  (c) $0.446 < \beta < 3.172$.

6. (a) $s^2 = 2.69$.
  (b) $21.958 < \alpha < 63.205$.
  (c) $-1.478 < \beta < 0.106$.

7. (a) $s^2 = 6.626$.
  (b) $2.684 < \alpha < 8.968$.
  (c) $0.498 < \beta < 0.637$.

8. $t = 1.78$; reject H_0, $\alpha > 10$.

9. $t = -2.24$; reject H_0, $\beta < 6$.

10. $58.808 < \mu_{Y|80} < 89.652$.

11. (a) $24.444 < \mu_{Y|24.5} < 27.112$.
  (b) $21.889 < y_0 < 29.668$.

13. $7.815 < y_0 < 10.801$.

14. (a) $32.233 < \mu_{Y|50} < 36.177$.
  (b) $26.432 < y_0 < 41.978$.

15. (a) $36.908 < \mu_{Y|35} < 61.080$.
  (b) $12.925 < y_0 < 85.063$.

16. (a) $\$452.84 < \mu_{Y|45} < \$524.45$.
  (b) $\$390.85 < y_0 < \$586.44$.

PAGE 344

1. (a) $b = \sum_{i=1}^{n} x_i y_i \Big/ \sum_{i=1}^{n} x_i^2$. (b) $\hat{y} = 2.003x$.

2. $\hat{y} = 0.349 + 1.929x$; $t = 1.40$; accept H_0.

3. $E(B) = \beta + \gamma \sum_{i=1}^{n} (x_{1i} - \bar{x}_1)x_{2i} \Big/ \sum_{i=1}^{n} (x_{1i} - \bar{x}_1)^2$.

4. $f = 9.00$; reject H_0.

5. (a) $a = 10.81153$, $b = -0.34370$.
  (b) $f = 0.43$; regression is linear.

6. $f = 1.58$; regression is linear.

7. $f = 1.12$; regression is linear.

8. (a) $\hat{y} = 3.1266 + 1.8429x$.
  (b) $f = 2.60$; regression is linear.

PAGE 351

1. $r = 0.240$.

2. $t = 0.51$; accept $\rho = 0$.

4. (a) $r = 0.784$.
  (b) $t = 3.34$; reject H_0, $\rho > 0$.
  (c) 61.5%.

5. (a) $r = 0.392$.
  (b) $t = 2.04$; accept $\rho = 0$.

6. (a) $r = -0.979$.
  (b) $z = -4.22$; reject H_0, $\rho < -0.5$.
  (c) 95.8%.

7. (a) $r = 0.862$.
  (b) $z = 2.26$; accept $\rho = 0.5$.

Chapter 10

PAGE 363

1. (a) $\hat{y} = 27.547 + 0.922x_1 + 0.284x_2$.
  (b) $\hat{y} = 84$.

2. $\hat{y} = 55.2266 - 0.0378x_1 + 1.6816x_2$.

3. $\hat{y} = 0.5800 + 2.7122x_1 + 2.0497x_2$.

4. (a) $\hat{y} = -22.9932 + 1.3957x_1 + 0.2176x_2$.
  (b) $\hat{y} = 80$ kg.

5. (a) $\hat{y} = 56.4633 + 0.1525x - 0.00008x^2$.
  (b) $\hat{y} = 86.7\%$.

6. (a) $\hat{d} = 13.3587 - 0.3394v + 0.011825v^2$.
  (b) $\hat{d} = 47.54$.

7. $\hat{y} = 141.6118 - 0.2819x + 0.0003x^2$.

8. (a) $\hat{y} = 19.033333 + 1.0085714x - 0.020380952x^2$.
  (b) $f = 0.018$; model is adequate.

9. $\hat{y} = 19.98519 + 0.30363x_1 + 0.59635x_2 - 0.49706x_3 - 0.70378x_4$.

10. (a) $\hat{y} = 1.0714 + 4.6032x - 1.8452x^2 + 0.1944x^3$.
  (b) $\hat{y} = 4.45$.

11. $\hat{y} = 3.3205 + 0.4210x_1 - 0.2958x_2 + 0.0164x_3 + 0.1247x_4$.

12. $\hat{y} = 14.3 - 1.4167x_1 + 14.7x_2 - 6.7x_2^2 - 0.2x_1x_2$.

PAGE 372

1. 34.3699.

2. 0.4316.

3. 0.000995.

4. $\hat{\sigma}_{B_1}^2 = 0.000071$; $\hat{\sigma}_{B_2}^2 = 0.063523$;
  $\hat{\sigma}_{B_1 B_2} = -0.001134$.

5. (a) $\hat{\sigma}_{B_2}^2 = 0.00002.$ (b) $\hat{\sigma}_{B_1 B_4} = -0.000003.$

6. $26.2352 < y_0 < 57.1516;$
$34.8580 < \mu_{Y|2500, 48.0} < 48.5288.$

7. $29.93 < \mu_{Y|19.5} < 31.97.$

8. $16.78 < y_0 < 16.93;$ $16.83 < \mu_{Y|8.2, 6.0, 10.3, 5.8} < 16.88.$

9. $t = 2.86;$ reject H_0, $\beta_2 > 0.$

10. $t = -4.48;$ reject H_0, $\beta_1 < 0.$

11. $t = 3.35;$ reject H_0, $\beta_1 > 2.$

PAGE 382

1. $R^2 = 0.9997.$

2. $f = 12{,}689;$ regression is significant.

3. $f = 13{,}409;$ regression is significant.

4. $f = 11{,}759;$ reject H_0.

5. $f = 20.07;$ reject H_0, $\beta_1 < 0.$

6. (a) $\hat{y} = 9.9 + 0.575x_1 + 0.550x_2 + 1.150x_3.$
(b) $f = 7.69$ for β_1; not significant.
$f = 7.04$ for β_2; not significant.
$f = 30.77$ for β_3; significant at the 0.01 level.

PAGE 396

1. (b) $\hat{y} = 4.690$ seconds.
(c) $4.450 < \mu_{Y|180, 260} < 4.930.$

2. (a) $\hat{y} = -6.33592 + 0.33738x_1.$
(b) Same as (a).
(c) Same as (a).

3. $\hat{y} = 2.1833 + 0.9576x_2 + 3.3253x_3.$

4. (a) $y = -29.5805 + 0.27877x_1 + 0.06971x_2 + 1.24146x_3 - 0.39535x_4 + 0.22369x_5.$
(b) $\hat{y} = -56.93515 + 1.63432x_3 + 0.24862x_5.$
(c) (x_2, x_5), (x_1, x_5), and $(x_1, x_3, x_5).$

5. (a) $\hat{y} = -587.211 + 428.433x.$
(b) $\hat{y} = 1180 - 191.691x + 35.20945x^2.$
(c) Quadratic model.

6. $t = -0.53;$ accept $\beta_4 = 0.$

7. $\hat{\sigma}_{B_1}^2 = 20{,}588.04;$ $\hat{\sigma}_{B_{11}}^2 = 62.6502.$

PAGE 405

1. $\hat{y} = -4.2468 + 0.3561x_1 - 0.1415x_2 + 0.1966x_3 - 0.0185x_4.$

2. The computations were carried out on a computer and recorded as follows:

d	b_1^*	b_2^*	b_3^*	b_4^*
0.01	13.3876	−4.5420	0.7203	2.8689
0.02	11.7907	−3.1648	1.0330	2.7538
0.03	10.6232	−2.2076	1.2560	2.7160
0.06	8.4567	−0.5425	1.6669	2.7229
0.08	6.5865	0.0812	1.8328	2.7450
0.10	6.9526	0.5159	1.9537	2.7643
0.15	5.9229	1.1761	2.1463	2.7933
0.17	5.6401	1.3444	2.1967	2.7985

Chapter 11

PAGE 420

3. $f = 0.31;$ no significant difference.

4. $f = 6.90;$ mean number of hours of relief differ.

5. $f = 14.52;$ yes, significant.

6. $f = 5.46;$ average times are the same.

7. $f = 2.25;$ average heights are the same.

8. $f = 0.46;$ no significant difference.

9. $f = 8.38;$ average specific gravities differ.

10. (a) $b = 0.77;$ variances are equal.
(b) $f = 13.33;$ significant differences among the laboratories.

11. $b = 0.99;$ variances are equal.

12. $b = 0.79;$ variances are equal.

13. $g = 0.4237;$ variances are equal.

14. $g = 0.6771;$ variances are equal.

PAGE 430

1. (a) $f = 14.27;$ reject H_0.
(b) $f = 23.23;$ reject H_0.
(c) $f = 2.48;$ accept H_0.

2. (a) $f = 5.15;$ significant.
(b) $f = 9.86;$ significant.

3. (a) $f = 13.50;$ treatment means differ.
(b) $f(1 \text{ versus } 2) = 29.35;$ significant.
$f(3 \text{ versus } 4) = 3.59;$ not significant.

4.

$\bar{x}_4$	$\bar{x}_3$	$\bar{x}_1$	$\bar{x}_5$	$\bar{x}_2$
2.8	4.0	5.2	6.6	7.8

5.

$\bar{x}_3$	$\bar{x}_1$	$\bar{x}_4$	$\bar{x}_2$
56.52	59.66	61.12	61.96

6. (a) $f = 7.10$; reject H_0.
   (b) Blend 4 differs significantly from all others.

7. (a) $f = 9.01$; yes, significant.
   (b) Depletion and Modified Hess are significantly different from the other three procedures.

8. (a) $f = 5.34$; significant.
   (b) $d_1 = 0.5553$; not significant.
   $d_2 = 3.1158$; significant.
   $d_3 = 3.0464$; significant.

9. $d_1 = 9.0878$; significant.
   $d_2 = 6.8498$; significant.
   $d_3 = 2.3059$; significant.
   $d_4 = 2.5093$; significant.

10. $d_1 = 4.2439$; significant.
   $d_2 = 4.5301$; significant.
   $d_3 = 4.0586$; significant.

PAGE 450

3. (a) $f(\text{blocks}) = 8.30$; significant.
   $f(\text{fertilizers}) = 6.11$; significant.
   (b) $f = 17.37$; significant.
   $f = 0.96$; not significant.

4. $f(\text{varieties}) = 1.74$; no difference in the yielding capabilities of the different varieties.
   $f(\text{locations}) = 8.13$; yes, it was necessary to plant each variety at each location.

5. (a) $f = 1.78$; no difference in the percent of foreign additives due to different analysts.
   (b) $f = 5.99$; percent of foreign additives is not the same for all three brands of jam.

6. $f(\text{subjects}) = 0.15$; not significant.
   $f(\text{students}) = 4.37$; significant.

7. $f(\text{stations}) = 26.14$; significant.
   $f(\text{dates}) = 0.37$; not significant.

8. (a) $f(\text{stations}) = 0.15$; not significant.
   (b) $f(\text{months}) = 1.44$; not significant.

9. $f(\text{diet}) = 11.86$; significant.

10. (a) $f = 3.00$; not significant.
   (b) $f = 30.32$; significant.

11. $f = 0.58$; no differences among treatment means.

12. (a) $f(\text{treatments}) = 8.60$; mean weight losses are different.
   (b) $f(\text{subjects}) = 1.77$; mean weight losses are not different.

16. (a) $f = 3.30$; no difference in the grades due to different time periods.
   (b) $f = 1.76$; courses are of equal difficulty.
   (c) $f = 5.03$; grades are affected by different professors.

17. $f = 1.29$; color additives have no effect on setting time.

18. $f(\text{rations}) = 12.55$; yes, significant.

PAGE 469

1. (a) $f = 14.9$; operators differ significantly.
   (b) $\hat{\sigma}_\alpha^2 = 28.91$; $s^2 = 8.32$.

3. (a) $f = 3.33$; no significant difference.
   (b) $\hat{\sigma}_\alpha^2 = 1.08$; $\hat{\sigma}_\beta^2 = 2.25$.

6. No; 16.

7. 9.

Chapter 12

PAGE 482

1. (a) $f = 8.13$; significant.
   (b) $f = 5.18$; significant.
   (c) $f = 1.63$; not significant.

2. (a) $f = 2.36$; not significant.
   (b) $f = 16.59$; significant.
   (c) $f = 0.41$; not significant.

3. (a) $f = 14.81$; significant.
   (b) $f = 9.04$; significant.
   (c) $f = 0.61$; not significant.

4. (a) $f(\text{humidity}) = 4.57$; significant.
   $f(\text{coating}) = 6.87$; significant.
   $f(\text{humidity} \times \text{coating}) = 2.44$; not significant.
   (b) Corrosion damage is different for medium humidity than for low or high humidity.

5. (a) $f = 34.40$; significant.
   (b) $f = 26.95$; significant.
   (c) $f = 20.30$; significant.

6. (a) $f = 5.29$; significant.
   (b) $f = 1.97$; not significant.
   (c) $f = 33.94$; significant.
   (d) $f = 0.50$; not significant.

PAGE 494

1. (a) AB: $f = 3.83$; significant.
   AC: $f = 3.79$; significant.
   BC: $f = 1.31$; not significant.
   ABC: $f = 1.63$; not significant.

(b) A: $f = 0.54$; not significant.
  B: $f = 6.85$; significant.
  C: $f = 2.15$; not significant.

2. Significant effects:
  A: $f = 8.28$ C: $f = 507.57$
  Insignificant effects:
  B: $f = 0.29$ BC: $f = 5.92$
  AB: $f = 3.85$ ABC: $f = 1.84$
  AC: $f = 2.33$

3. Significant effects:
  Stress: $f = 45.96$
  Insignificant effects:
  Coating: $f = 0.05$
  Humidity: $f = 2.13$
  Coating × Humidity: $f = 3.41$
  Coating × Stress: $f = 0.08$
  Humidity × Stress: $f = 3.15$
  Coating × Humidity × Stress: $f = 1.93$

4. (a) $f = 2.69$; no significant interaction.
  (b) A: $f = 3.37$; significant.
  B: $f = 5.63$; significant.
  C: $f = 4.82$; significant.
  The pooled error includes BC.

PAGE 501

1. (a) $f = 1.49$; no significant interaction.
  (b) f(operators) $= 12.45$; significant.
  f(filters) $= 8.39$; significant.
  (c) $\hat{\sigma}_\alpha^2 = 0.1701$ (filters);
  $\hat{\sigma}_\beta^2 = 0.3514$ (operators);
  $s^2 = 0.1867$.

2. $\hat{\sigma}_\alpha^2 = 1.4678$ (brand); $\hat{\sigma}_\beta^2 = 12.1642$ (time);
  $s^2 = 9.0237$.

3. (a) $\hat{\sigma}_\beta^2, \hat{\sigma}_\gamma^2, \hat{\sigma}_{\alpha\gamma}^2$ are significant.
  (b) $\hat{\sigma}_\gamma^2, \hat{\sigma}_{\alpha\gamma}^2$ are significant.

4. Yes.

5. 0.59.

Chapter 13

PAGE 511

1. $SSA = 2.6667$, $SSB = 170.6667$, $SSC = 104.1667$,
  $SS(AB) = 1.5000$, $SS(AC) = 42.6667$,
  $SS(BC) = 0.0000$, $SS(ABC) = 1.5000$.

2. Significant effects:
  A: $f = 1294.65$ AB: $f = 20.88$
  B: $f = 43.56$ AC: $f = 16.21$
  C: $f = 116.49$ ABC: $f = 289.23$
  Insignificant effect:
  BC: $f = 0.00$

3. Significant effects:
  A: $f = 1940.64$
  B: $f = 4411.62$
  C: $f = 1098.38$
  D: $f = 458.50$

4. Significant effects:
  A: $f = 57.85$ AC: $f = 7.08$
  B: $f = 7.52$ AD: $f = 4.85$
  C: $f = 127.87$ BC: $f = 10.96$
  D: $f = 44.72$ BD: $f = 4.85$
  AB: $f = 6.94$ CD: $f = 6.52$
  Insignificant effects:
  ABC: $f = 1.26$ BCD: $f = 1.20$
  ABD: $f = 1.14$ $ABCD$: $f = 0.87$.
  ACD: $f = 1.72$

5. Significant effects:
  A: $f = 9.98$ BC: $f = 19.03$
  Insignificant effects:
  B: $f = 0.20$ AC: $f = 0.20$
  C: $f = 6.54$ AD: $f = 0.57$
  D: $f = 0.02$ BD: $f = 1.83$
  AB: $f = 1.83$ CD: $f = 0.02$

PAGE 520

1. A, B, C, AC, BC, and ABC each with 1 degree of freedom can be tested using an error mean square with 12 degrees of freedom.

2. (a) A: $f = 1.55$; not significant.
  B: $f = 1.27$; not significant.
  C: $f = 3.49$; not significant.
  D: $f = 0.79$; not significant.
  (b) ABC

3.

Block 1	Block 2	Block 3	Block 4
(1)	c	d	a
ab	abc	ac	b
acd	ad	bc	cd
bcd	bd	abd	$abcd$

CD is also confounded.

4. (a)

Block 1	Block 2	Block 3	Block 4
(1)	*a*	*b*	*ab*
c	*ac*	*bc*	*abc*
ae	*e*	*abe*	*be*
bd	*abd*	*d*	*ad*
ace	*ce*	*abce*	*bce*
bcd	*abcd*	*cd*	*acd*
abde	*bde*	*ade*	*de*
abcde	*bcde*	*acde*	*cde*

(b) *BD*.

(c) $SSA = 21.9453$; $SSC = 2.4753$;
$SS(E) = 1.0878$; $SSB = 40.2753$;
$SSD = 7.7028$

5.

Block		Block		Block	
1	2	1	2	1	2
abc	*ab*	*abc*	*ab*	(1)	*a*
a	*ac*	*a*	*ac*	*c*	*b*
b	*bc*	*b*	*bc*	*ab*	*ac*
c	(1)	*c*	(1)	*abc*	*bc*

Replicate 1	Replicate 2	Replicate 3
ABC	*ABC*	*AB*
confounded	confounded	confounded

6. (a) *ABD*; *ABCD*.

(b) Significant effects:
   C: $f = 7.53$ D: $f = 13.38$
   Insignificant effects:

A:	$f = 3.35$	CD:	$f = 0.30$
B:	$f = 0.84$	ABC:	$f = 1.94$
AB:	$f = 0.84$	ABD:	$f = 0.54$
AC:	$f = 0.54$	ACD:	$f = 0.03$
AD:	$f = 1.20$	BCD:	$f = 0.30$
BC:	$f = 0.84$	$ABCD$:	$f = 0.89$
BD:	$f = 0.54$		

7.

Block		Block		Block	
1	2	1	2	1	2
(1)	*a*	*a*	(1)	(1)	*ab*
ab	*b*	*ab*	*b*	*a*	*b*

Replicate 1	Replicate 2	Replicate 3

8. Block 1 = {(1), *eg*, *abcd*, *bdg*, *adf*, *bcf*, *cdef*, *abcdeg*, *bde*, *adefg*, *bcefg*, *cdfg*, *acg*, *abef*, *ace*, *abfg*}

Block 2 = {*a*, *aeg*, *bcd*, *abdg*, *df*, *abcf*, *acdef*, *bcdeg*, *abde*, *defg*, *abcefg*, *acdfg*, *cg*, *bef*, *ce*, *bfg*}

Block 3 = {*b*, *beg*, *acd*, *dg*, *abdf*, *cf*, *bcdef*, *acdeg*, *de*, *abdefg*, *cefg*, *bcdfg*, *abcg*, *aef*, *abce*, *afg*}

Block 4 = {*c*, *ceg*, *abd*, *bcdg*, *acdf*, *bf*, *def*, *abdeg*, *bcde*, *acdefg*, *befg*, *dfg*, *ag*, *abcef*, *ae*, *abcfg*}

Block 5 = {*d*, *deg*, *abc*, *bg*, *af*, *bcdf*, *cef*, *abceg*, *be*, *aefg*, *bcdefg*, *cfg*, *acdg*, *abdef*, *acde*, *abdfg*}

Block 6 = {*e*, *g*, *abcde*, *bdeg*, *adef*, *bcef*, *cdf*, *abcdg*, *bd*, *adfg*, *bcfg*, *cdefg*, *aceg*, *abf*, *ac*, *abefg*}

Block 7 = {*f*, *efg*, *abcdf*, *bdfg*, *ad*, *bc*, *cde*, *abcdefg*, *bdef*, *adeg*, *bceg*, *cdg*, *acfg*, *abe*, *acef*, *abg*}

Block 8 = {*ab*, *abeg*, *cd*, *adg*, *bdf*, *acf*, *abcdef*, *cdeg*, *ade*, *bdefg*, *acefg*, *abcdfg*, *bcg*, *ef*, *bce*, *fg*}

Interactions sacrificed: *ABCD*, *CDEFG*, *BDF*, *ABEFG*, *ACF*, *BCEG*, *ADEG*.

PAGE 527

1.

$A \equiv CDE$		$AE \equiv CD$	
$B \equiv ABCDE$		$BC \equiv ABDE$	
$C \equiv ADE$		$BD \equiv ABCE$	
$D \equiv ACE$		$BE \equiv ABCD$	
$E \equiv ACD$		$ABC \equiv BDE$	
$AB \equiv BCDE$		$ABD \equiv BCE$	
$AC \equiv DE$		$ABE \equiv BCD$	
$AD \equiv CE$			

2. (a) Principal block = {(1), *a*, *bc*, *abc*, *bd*, *abd*, *cd*, *acd*}.

(b)

Block 1	Block 2
(1)	*a*
bc	*abc*
abd	*bd*
acd	*cd*

(c)

Source of Variation	Degrees of Freedom
A	1
B	1
C	1
D	1
Blocks	1
Error	2
Total	7

3. Principal block $= \{(1),\ ac,\ bd,\ abcd,\ abe,\ bce,\ ade,$
   $cde,\ abf,\ bcf,\ adf,\ cdf,\ ef,\ acef,$
   $bdef,\ abcdef\}.$

   $A \equiv BCD \equiv ABDEF \equiv CEF$
   $B \equiv ACD \equiv DEF \equiv ABCEF$
   $C \equiv ABD \equiv BCDEF \equiv AEF$
   $D \equiv ABC \equiv BEF \equiv ACDEF$
   $E \equiv ABCDE \equiv BDF \equiv ACF$
   $F \equiv ABCDF \equiv BDE \equiv ACE.$

4. (a) Principal block $= \{(1),\ ab,\ acd,\ bcd,\ ce,\ abce,$
   $ade,\ bde,\ acf,\ bcf,\ df,\ abdf,$
   $aef,\ bef,\ cdef,\ abcdef\}.$

(b)

Source of Variation	Degrees of Freedom
A	1
B	1
C	1
D	1
E	1
F	1
AB	1
AC	1
AD	1
BC	1
BD	1
CD	1
Error	3
Total	15

5. Significant effect:
   $E:\ f = 5.39$
   Insignificant effects:
   $A:\ f = 0.48$ $D:\ f = 1.94$
   $B:\ f = 1.35$ $F:\ f = 1.09$
   $C:\ f = 3.03$ $G:\ f = 4.36$

6. No significant effects:
   $A:\ f = 1.24$ $D:\ f = 3.37$
   $B:\ f = 1.61$ $E:\ f = 0.14$
   $C:\ f = 1.17$ $F:\ f = 0.03$

Chapter 14

PAGE 538

1. $x = 7$; $P = 0.1719$, accept H_0.
2. $x = 10$; $P = 0.4544$; accept H_0.
3. $x = 3$; $P = 0.0244$, reject H_0.
4. $x = 2$; $P = 0.0547$, accept H_0.
5. $x = 4$; $P = 0.3770$, accept H_0.
6. $x = 12$; $P = 0.0160$, reject H_0.
7. $x = 4$; $P = 0.1335$, accept H_0.
8. $w_- = 12.5$; accept H_0.
9. $w = 43$; accept H_0.
10. $w_+ = 3.5$; accept H_0.
11. $w_+ = 17.5$; accept H_0.
12. $z = 2.80$; reject H_0.
13. $z = -2.13$; reject H_0, $\mu_1 - \mu_2 < 8$.
14. $z = 1.42$; accept H_0.

PAGE 546

1. $u_1 = 1$; claim is valid.
2. $u_1 = 8$; serum is not effective.
3. $u_2 = 5$: A operates longer.
4. $u = 43.5$; $\mu_1 = \mu_2$.
5. $u = 15$; no, $\mu_1 = \mu_2$.
6. $z = -0.84$; accept H_0.
7. $h = 10.47$; operating times are different.
8. $h = 11.27$; tar contents are different.
9. $h = 1.07$; no significant difference.
10. $h = 12.83$; laboratories give different analyses.

PAGE 556

1. $v = 7$; $P = 0.910$, random sample.

2. $v = 2$; $P = 0.016$, reject randomness.

3. $v = 6$; $P = 0.044$, $\mu_A = \mu_B$.

4. $z = -0.55$; defectives occur at random.

5. $z = 1.11$; random sample.

6. 30.

7. 0.70.

8. 21.

9. 0.995.

10. (a) 0.24. (b) Accept H_0.

11. (a) $r_S = 0.39$. (b) Accept H_0.

12. $r_S = 0.99$.

13. (a) $r_S = 0.96$. (b) Reject H_0, $\rho > 0$.

14. (a) $r_S = 0.72$. (b) Reject H_0, $\rho > 0$.

15. $r = -0.47$; no significant relationship.

16. (a) $r_S = 0.71$. (b) Reject H_0, $\rho > 0$.

17. (a) $r_S = 0.59$. (b) Reject H_0, $\rho > 0$.

Index